Special Bonus Software Offer

As A Buyer Of A Building News 1993 Costbook You Are Eligible To Receive

— AT NO COST! —

AF712694

CONSTRUCTION ESTIMATOR

The Complete, Stand Alone, Easy-To-Use, Construction Estimating System

- This program provides the basic tools to develop accurate, comprehensive and organized cost estimates.
- The reliable cost data can be used to supplement your own records and provides a basis for accurate estimating.
- IBM Compatible — 5¼″ diskette.
- Immediate on-screen access to data through "Look-up" windows. "Pull-down" menus.
- Easy to learn and master — very user-friendly.
- Different markups can be applied to each estimate.

Act NOW And Receive – FREE – An Electronic Database With Over 1,200 Lines Of Building News Costbook Data

Order Form

NAME ______________________

COMPANY ______________________

ADDRESS ______________________

CITY, STATE, ZIP ______________________

TELEPHONE ______________________

Simply fill out this form, read and sign the statement below and send this complete page to our East Coast office:

Building News Bookstore
77 Wexford Street
Needham Heights, MA 02194

I purchased my 1993 Costbook at: ______________________
STORE CITY, STATE

If you have received the Bonus Software through our mail offering, there is no need to order another. You may copy and share the software and data with co-workers and associates.

Bonus Software Qualifications

The CONSTRUCTION™ ESTIMATOR is a software product developed, produced and distributed by CPR International, Inc., 3195 Adeline St., Suite A, Berkeley, CA 94703, (510) 654-8338. The electronic cost data provided with the CONSTRUCTION ESTIMATOR is derived from the Building News 1993 Construction Costbooks.

CPR International has no affiliation with BNI Publications Inc. (Building News). All questions regarding the software should be directed to CPR International, Inc.

BNI Publications Inc., its authors and editors, do not warrant or guarantee the correctness of the data or information contained in the costbooks. BNI Publications Inc., and its authors and editors, do hereby disclaim any responsibility or liability in connection with the use of data in electronic or printed form published by BNI Publications Inc., or other information in the database or costbooks.

I have read and agree to the above information.

NAME - PLEASE SIGN ______________________

NAME - PLEASE PRINT ______________________

BNi Building News™

BNi Building News™

MECHANICAL/ELECTRICAL 1993 COSTBOOK

THIRD EDITION

Edited by

A.M. FOGARTY & ASSOCIATES

Estimating Consultants
To The Construction Industry
For Over 35 Years

Co-Sponsored By

Consulting-Specifying Engineer®

BNi Building News™
Los Angeles • Boston

BNI Publications, Inc.
3055 OVERLAND AVENUE
LOS ANGELES, CA 90034
(310) 202-7775

77 WEXFORD STREET
NEEDHAM HEIGHTS, MA 02194
(617) 455-1466

A. M. Fogarty & Assoc., Inc.
175 DERBY STREET, SUITE 5
HINGHAM, MA 02043
(617) 749-7272

ISBN 1-55701-073-0

Distributor to the trade, Macmillan Publishing Company, 866 Third Avenue, New York, New York 10022.

PREFACE

For the past 47 years, Building News has been dedicated to providing construction professionals with timely and reliable information. Based on this experience, our staff has researched and compiled thousands of up-to-the-minute costs for the **Building News 1993 Costbooks.** This book is an essential reference for contractors, engineers, architects, facilities managers — any construction professional who must provide an estimate on any type of building project.

Whether working up a preliminary estimate or submitting a formal bid, the costs listed here can quickly and easily be tailored to your needs. All costs are based on national averages, while a table of modifiers is provided for regional adjustments. Overhead and profit are included in all costs.

Complete man-hour tables follow the unit costs to provide data on typical durations of specific tasks. This information can be used to schedule projects as well as to determine specific labor costs based on local labor rates.

All data is categorized according to the MASTERFORMAT of the Construction Specifications Institute (CSI). This industry standard provides an all-inclusive checklist to ensure that no element of a project is overlooked. In addition, to make specific items even easier to locate, there is a complete alphabetical index.

This costbook contains an appendix with reference charts and tables taken from an array of sources. Text explains the costs in certain categories and provides helpful pointers that should be taken into account with every estimate.

A section on square foot costs provides an overview of project costs for different building types — commercial, residential, etc. — with summaries of the actual projects. Square foot costs are invaluable for making budget estimates and checking prices when time is a factor. For a complete source of square foot costs refer to the **Building News Square Foot 1993 Costbook.**

The "Features in this Book" section presents a clear overview of the many features of this book. Included is an explanation of the data, sample page layout and discussion of how to best use the information in the book.

Of course, all buildings and construction projects are unique. The costs provided in this book are based on averages from well-managed projects with good labor productivity under normal working conditions (eight hours a day). Other circumstances affecting costs such as overtime, unusual working conditions, savings from buying bulk quantities for large projects, and unusual or hidden costs must be factored in as they arise.

The data provided in this book is for estimating purposes only. Check all applicable federal, state and local codes and regulations for specific requirements.

About the Editors

As a leader in construction cost estimating since 1972, A.M. Fogarty & Associates has estimated literally billions of dollars worth of construction projects. Over 250 years of combined experience provides constantly and accuracy, assuming you the most reliable information that is so valuable to the success of your business. In addition to construction cost estimating, A.M. Fogarty & Associates provides consulting services in the areas of construction management, scheduling and planning, specifications and building evaluation.

BNi Building News™

TABLE OF CONTENTS

CSI MASTERFORMAT

All data in the Costbook pages and Man-Hour tables is organized according to the CSI MASTERFORMAT — the industry standard numbering/classification system. The data is divided into the 16 divisions as shown below. The five digit numbers within each division correspond to the MASTERFORMAT Broadscope designations. Each section is further broken down into MASTERFORMAT Mediumscope designations. These numbers, in most cases, are the same as those used in architectural and engineering specifications.

The construction estimating information in this book is divided into two main sections: Costbook Pages and Man-Hour Tables. Each is organized to the 16 divisions of the CSI MASTERFORMAT as shown below. In addition, there are estensive Construction Reference Tables, Geographic Cost Modifiers, Square Foot Tables and a detailed Index. Sample pages with graphic explanations are included before the Costbook pages, Man-Hour Tables and Square Foot Tables. These explanations along with the discussions below will provide a good understanding of what is included in this book and how it can best be used for construction estimating.

GENERAL REQUIREMENTS	*Division 1*
Summary of Work	01010
Allowances	01020
Measurement and Payment	01025
Alternates/Alternatives	01030
Modification Procedures	01035
Coordination	01040
Field Engineering	01050
Regulatory Requirements	01060
Identification Systems	01070
References	01090
Special Project Procedures	01100
Project Meetings	01200
Submittals	01300
Quality Control	01400
Construction Facilities & Temporary Controls	01500
Material and Equipment	01600
Facility Startup/Commissioning	01650
Contract Closeout	01700
Maintenance	01800

SITEWORK	*Division 2*
Subsurface Investigation	02010
Demolition	02050
Site Preparation	02100
Dewatering	02140
Shoring and Underpinning	02150
Excavation Support Systems	02160
Cofferdams	02170
Earthwork	02200
Tunneling	02300
Piles and Caissons	02350
Railroad Work	02450
Marine Work	02480
Paving and Surfacing	02500
Utility Piping Materials	02600
Water Distribution	02660
Fuel and Steam Distribution	02680
Sewerage and Drainage	02700
Restoration of Underground Pipe	02760
Ponds and Reservoirs	02770
Power and Communications	02780
Site Improvements	02800
Landscaping	02900

CONCRETE	*Division 3*
Concrete Formwork	03100
Concrete Reinforcement	03200
Concrete Accessories	03250
Cast-in-Place Concrete	03300
Concrete Curing	03370
Precast Concrete	03400
Cementitious Decks and Toppings	03500
Grout	03600
Concrete Restoration and Cleaning	03700
Mass Concrete	03800

MASONRY	*Division 4*
Mortar and Masonry Grout	04100
Masonry Accessories	04150
Unit Masonry	04200
Stone	04400
Masonry Restoration and Cleaning	04500
Refractories	04550
Corrosion Resistant Masonry	04600
Simulated Masonry	04700

FEATURES IN THIS BOOK

The construction estimating information in this book is divided into two main sections: Costbook Pages and Man-Hour Tables. Each section is organized according to the 16 divisions of the CSI MASTERFORMAT as shown on the previous pages. In addition, there are extensive Supporting Construction Reference tables, Geographic Costs Modifiers, Square Foot tables and a detailed Index.

Sample pages with graphic explanations are included before the Costbook pages and Man-Hour tables. These explanations along with the discussions below, will provide a good understanding of what is included in this book and how it can best be used in construction estimating.

Material Costs

The material costs used in this book represent national averages for prices that a contractor would expect to pay plus an allowance for freight (if applicable), handling and storage. These costs reflect neither the lowest or highest prices, but rather a typical average cost over time. Periodic fluctuations in availability and in certain commodities (eg. copper, lumber) can significantly affect local material pricing. In the final estimating and bidding stages of a project when the highest degree of accuracy is required, it is best to check local, current prices.

Labor Costs

Labor costs include the basic wage, plus commonly applicable taxes, insurance and markups for overhead and profit. The labor rates used here to develop the costs are typical average prevailing wage rates. Rates for different trades are used where appropriate for each type of work.

Taxes and insurance which are most often applied to labor rates include employer-paid Social Security/Medicare taxes (FICA), Worker's Compensation insurance, state and federal unemployment taxes, and business insurance. Fixed government rates as well as average allowances are included in the labor costs. However, most of these items vary significantly from state to state and within states. For more specific data, local agencies and sources should be consulted.

Equipment Costs

Costs for various types and pieces of equipment are included in Division 1 - General Requirements and can be included in an estimate when required either as a total "Equipment" category or with specific appropriate trades. Costs for equipment are included when appropriate in the installation of costs in the Costbook pages.

Overhead And Profit

Included in the labor costs are allowances for overhead and profit for the contractor/employer whose workers are performing the specific tasks. No cost allowances or fees are included for management of subcontractors by the general contractor or construction manager. These costs, where appropriate, must be added to the costs as listed in the book.

The allowance for overhead is included to account for office overhead, the contractors' typical costs of doing business. These costs normally include in-house office staff salaries and benefits, office rent and operating expenses, professional fees, vehicle costs and other operating costs which are not directly applicable to specific jobs. It should be noted for this book that office overhead as included should be distinguished from project overhead, the General Requirements (CSI Division 1) which are specific to particular projects. Project overhead should be included on an item by item basis for each job.

Depending on the trade, an allowance of 10-15 percent is incorporated into the labor/installation costs to account for typical profit of the installing contractor. See Division 1, General Requirements, for a more detailed review of typical profit allowances.

Adjustments to Costs

The costs as presented in this book attempt to represent national averages. Costs, however, vary among regions, states and even between adjacent localities.

In order to more closely approximate the probable costs for specific locations throughout the U.S., a table of Geographic Cost Modifiers is provided. These adjustment factors are used to modify costs obtained from this book to help account for regional variations of construction costs. Whenever local current costs are known, whether material or equipment prices or labor rates, they should be used if more accuracy is required.

Man-Hour Tables

The man-hour data used to develop the labor costs are listed in the second main section of this book, the Man-Hour Tables. These productivities represent typical installation labor for thousands of construction items. The data takes into account all activities involved in normal construction under commonly experienced working conditions such as site movement, material handling, start-up, etc. As with the Costbook pages, these items are listed according to the CSI MASTERFORMAT.

Square Foot Tables

Included as an additional reference are Square Foot Tables which list hundreds of actual projects for dozens of building types each with associated building size and total square foot building cost. This data provides an overview of construction costs by building type. These costs are for actual projects. The variations within similar building types may be due, among other factors, to size, location, quality and specified components, materials and processes. Depending upon all such factors, specific building costs can vary significantly and may not necessarily fall within the range of costs as presented.

The data has been updated to reflect current construction costs and is a partial summary of the projects and costs presented in the **Building News Square Foot 1993 Costbook.** The source of this data is *Design Cost & Data* magazine which has over 25 years of history in tracking, collecting and publishing construction cost data.

Editors' Note: The **Building News 1993 Costbooks** are intended to provide accurate, reliable, average costs and typical productivities for thousands of common construction components. The data is developed and compiled from various industry sources, including government, manufacturers, suppliers and working professionals. The intent of the information is to provide assistance and guidelines to construction professionals in estimating. The user should be aware that local conditions, material and labor availability and cost variations, economic considerations, weather, local codes and regulations, etc., all affect the actual cost of construction. These and other such factors must be considered and incorporated into any and all construction estimates.

Sample Costbook Page

In order to best use the information in this book, please review this sample page and read the "Features In This Book" section.

CSI MASTERFORMAT Division

CSI Broadscope Category

CSI Mediumscope Category (First 5 Digits)

Detailed Descriptions
Complete descriptions of items may include information listed above a particular line. Review of the whole category is recommended for a complete description.

03 CONCRETE

PLACING CONCRETE	UNIT	MAT.	INST.	TOTAL
03380.05 BEAM CONCRETE				
By crane	C.Y.	65.00	38.50	103.50
By pump	"	65.00	35.00	100.00
By hand buggy	"	65.00	20.40	85.40
Bond beam, 3000# concrete				
By pump				
8" high				
4" wide	L.F.	0.15	0.77	0.92
6" wide	"	0.36	0.87	1.23
8" wide	"	0.47	0.96	1.43
10" wide	"	0.63	1.05	1.68
12" wide	"	0.84	1.20	2.04
16" high				
8" wide	L.F.	1.15	1.20	2.35
10" wide	"	1.55	1.35	2.90
12" wide	"	2.05	1.60	3.65
By crane				
8" high				
4" wide	L.F.	0.18	0.84	1.02
6" wide	"	0.36	0.92	1.28
8" wide	"	0.47	0.96	1.43
10" wide	"	0.63	1.05	1.68
12" wide	"	0.84	1.20	2.04
16" high				
8" wide	L.F.	1.15	1.20	2.35
10" wide	"	1.55	1.30	2.85
12" wide	"	2.05	1.50	3.55
03380.15 COLUMN CONCRETE				
Columns				
2500# or 3000# concrete				
By crane	C.Y.	56.50	35.00	91.50
By pump	"	56.50	32.10	88.60
3500# or 4000# concrete				
By crane	C.Y.	60.50	35.00	95.50
By pump	"	60.50	32.10	92.60
5000# concrete				
By crane	C.Y.	65.00	35.00	100.00
By pump	"	65.00	32.10	97.10
03380.20 ELEVATED SLAB CONCRETE				
Elevated slab				
2500# or 3000# concrete				
By crane	C.Y.	56.50	19.25	75.75
By pump	"	56.50	14.80	71.30
By hand buggy	"	56.50	20.40	76.90
3500# or 4000# concrete				
By crane	C.Y.	60.50	19.25	79.75
By pump	"	60.50	14.80	75.30
By hand buggy	"	60.50	20.40	80.90
5000# concrete				
By crane	C.Y.	65.00	19.25	84.25
By pump	"	65.00	14.80	79.80

Unit of Measurement
Each item (and cost) is defined in terms of the common estimating unit. All costs are listed in dollars per unit.

Material Cost
Material costs represent average contractor prices plus an allowance for freight, handling and storage.

Installation Cost
Installation costs include basic wage rates, markups for taxes, insurance overhead and profit and also include equipment costs where appropriate.

Total Cost
The total cost is the sum of material and installation costs. This total represents typical contractors' costs including overhead and profit, but does not include markups for the general contractor or construction management fees.

BNi Building News™

01 GENERAL

REQUIREMENTS	UNIT	MAT.	INST.	TOTAL
01020.10 ALLOWANCES				
Overhead				
$20,000 project				
Minimum	PCT.			15.00
Average	"			20.00
Maximum	"			40.00
$100,000 project				
Minimum	PCT.			12.00
Average	"			15.00
Maximum	"			25.00
$500,000 project				
Minimum	PCT.			10.00
Average	"			12.00
Maximum	"			20.00
$1,000,000 project				
Minimum	PCT.			6.00
Average	"			10.00
Maximum	"			12.00
Profit				
$20,000 project				
Minimum	PCT.			10.00
Average	"			15.00
Maximum	"			25.00
$100,000 project				
Minimum	PCT.			10.00
Average	"			12.00
Maximum	"			20.00
$500,000 project				
Minimum	PCT.			5.00
Average	"			10.00
Maximum	"			15.00
$1,000,000 project				
Minimum	PCT.			3.00
Average	"			8.00
Maximum	"			15.00
Professional fees				
Architectural				
$100,000 project				
Minimum	PCT.			5.00
Average	"			10.00
Maximum	"			20.00
$500,000 project				
Minimum	PCT.			5.00
Average	"			8.00
Maximum	"			12.00
$1,000,000 project				
Minimum	PCT.			3.50
Average	"			7.00
Maximum	"			10.00
Taxes				
Sales tax				
Minimum	PCT.			4.00
Average	"			5.00

01 GENERAL

REQUIREMENTS	UNIT	MAT.	INST.	TOTAL
01020.10 ALLOWANCES				
Maximum	PCT.			8.50
Unemployment				
Minimum	PCT.			2.00
Average	"			5.00
Maximum	"			7.00
Social security (FICA)	"			
Average	"			
01050.10 FIELD STAFF				
Superintendent				
Minimum	YEAR			44,000
Average	"			68,200
Maximum	"			99,000
Field engineer				
Minimum	YEAR			38,500
Average	"			57,200
Maximum	"			82,500
Foreman				
Minimum	YEAR			27,500
Average	"			40,700
Maximum	"			55,000
Bookkeeper/timekeeper				
Minimum	YEAR			13,200
Average	"			20,900
Maximum	"			27,500
Watchman				
Minimum	YEAR			11,000
Average	"			13,750
Maximum	"			16,500
01310.10 SCHEDULING				
Scheduling for				
$100,000 project				
Minimum	PCT.			1.00
Average	"			2.00
Maximum	"			4.00
$500,000 project				
Minimum	PCT.			0.40
Average	"			1.00
Maximum	"			2.00
$1,000,000 project				
Minimum	PCT.			0.30
Average	"			0.80
Maximum	"			1.50
Scheduling software, not including computer				
Minimum	EA.			500.00
Average	"			2,500
Maximum	"			40,000

01 GENERAL

REQUIREMENTS	UNIT	MAT.	INST.	TOTAL
01330.10 SURVEYING				
Surveying				
Small crew	DAY	0.00	424.00	424.00
Average crew	"	0.00	642.00	642.00
Large crew	"	0.00	848.00	848.00
01380.10 JOB REQUIREMENTS				
Job photographs, small jobs				
Minimum	EA.			55.00
Average	"			110.00
Large projects				
Minimum	EA.			385.00
Average	"			550.00
Maximum	"			1,100
01410.10 TESTING				
Testing concrete, per test				
Minimum	EA.			12.10
Average	"			22.00
Maximum	"			33.00
Welding, per test				
Minimum	EA.			11.00
Average	"			22.00
Maximum	"			66.00
01500.10 TEMPORARY FACILITIES				
Barricades, temporary				
Highway				
Concrete	L.F.	6.60	2.10	8.70
Wood	"	1.75	0.85	2.60
Steel	"	2.10	0.71	2.81
Pedestrian barricades				
Plywood	L.F.	1.20	0.71	1.91
Chain link fence	"	1.75	0.71	2.46
Trailers, general office type, per month				
Minimum	EA.			165.00
Average	"			242.00
Maximum	"			550.00
Crew change trailers, per month				
Minimum	EA.			66.00
Average	"			77.00
Maximum	"			110.00
01505.10 MOBILIZATION				
Equipment mobilization				
Bulldozer				
Minimum	EA.			110.00
Average	"			220.00
Maximum	"			330.00
Backhoe/front-end loader				
Minimum	EA.			55.00

01 GENERAL

REQUIREMENTS	UNIT	MAT.	INST.	TOTAL
01505.10 MOBILIZATION				
Average	EA.			110.00
Maximum	"			220.00
Crane, crawler type				
Minimum	EA.			1,100
Average	"			2,750
Maximum	"			5,500
Truck crane				
Minimum	EA.			275.00
Average	"			412.00
Maximum	"			660.00
Pile driving rig				
Minimum	EA.			5,500
Average	"			11,000
Maximum	"			16,500
01525.10 CONSTRUCTION AIDS				
Scaffolding/staging, rent per month				
Measured by lineal feet of base				
10' high	L.F.			6.60
20' high	"			11.00
30' high	"			16.50
40' high	"			19.80
50' high	"			23.10
Measured by square foot of surface				
Minimum	S.F.			0.44
Average	"			0.55
Maximum	"			0.88
Safety nets, heavy duty, per job				
Minimum	S.F.			0.22
Average	"			0.28
Maximum	"			0.55
Tarpaulins, fabric, per job				
Minimum	S.F.			0.11
Average	"			0.22
Maximum	"			0.55
01570.10 SIGNS				
Construction signs, temporary				
Signs, 2' x 4'				16.50
Minimum	EA.			19.80
Average	"			44.00
Maximum	"			165.00
Signs, 4' x 8'				27.50
Minimum	EA.			38.50
Average	"			110.00
Maximum	"			440.00
Signs, 8' x 8'				
Minimum	EA.			44.00
Average	"			165.00
Maximum	"			1,650

01 GENERAL

REQUIREMENTS

01600.10 EQUIPMENT	UNIT	MAT.	INST.	TOTAL
Air compressor				
60 cfm				
By day	EA.			55.00
By week	"			165.00
By month	"			495.00
300 cfm				
By day	EA.			99.00
By week	"			297.00
By month	"			891.00
600 cfm				
By day	EA.			330.00
By week	"			990.00
By month	"			2,970
Air tools, per compressor, per day				
Minimum	EA.			22.00
Average	"			27.50
Maximum	"			38.50
Generators, 5 kw				
By day	EA.			44.00
By week	"			132.00
By month	"			396.00
Heaters, salamander type, per week				
Minimum	EA.			55.00
Average	"			82.50
Maximum	"			110.00
Pumps, submersible				
50 gpm				
By day	EA.			44.00
By week	"			132.00
By month	"			396.00
100 gpm				
By day	EA.			55.00
By week	"			165.00
By month	"			495.00
500 gpm				
By day	EA.			88.00
By week	"			264.00
By month	"			792.00
Diaphragm pump, by week				
Minimum	EA.			88.00
Average	"			110.00
Maximum	"			132.00
Pickup truck				
By day	EA.			77.00
By week	"			231.00
By month	"			693.00
Dump truck				
6 cy truck				
By day	EA.			220.00
By week	"			660.00
By month	"			1,980
Backhoe, track mounted				

01 GENERAL

REQUIREMENTS	UNIT	MAT.	INST.	TOTAL
01600.10 EQUIPMENT				
1/2 cy capacity				
By day	EA.			440.00
By week	"			1,320
By month	"			3,960
1 cy capacity				
By day	EA.			715.00
By week	"			2,150
By month	"			6,430
Backhoe/loader, rubber tired				
1/2 cy capacity				
By day	EA.			275.00
By week	"			825.00
By month	"			2,480
3/4 cy capacity				
By day	EA.			330.00
By week	"			990.00
By month	"			2,970
Bulldozer				
75 hp				
By day	EA.			440.00
By week	"			1,320
By month	"			3,960
Cranes, crawler type				
15 ton capacity				
By day	EA.			440.00
By week	"			1,320
By month	"			3,960
25 ton capacity				
By day	EA.			550.00
By week	"			1,650
By month	"			4,950
50 ton capacity				
By day	EA.			990.00
By week	"			2,970
By month	"			8,910
100 ton capacity				
By day	EA.			1,650
By week	"			4,950
By month	"			14,850
Truck mounted, hydraulic				
15 ton capacity				
By day	EA.			440.00
By week	"			1,320
By month	"			3,960
Loader, rubber tired				
1 cy capacity				
By day	EA.			385.00
By week	"			1,160
By month	"			3,960

01 GENERAL

REQUIREMENTS	UNIT	MAT.	INST.	TOTAL
01740.10 BONDS				
Performance bonds				
Minimum	PCT			1.00
Average	"			2.00
Maximum	"			3.00

02 SITEWORK

DEMOLITION

	UNIT	MAT.	INST.	TOTAL
02060.10 BUILDING DEMOLITION				
Cut-outs				
Concrete, elevated slabs, mesh reinforcing				
Under 5 cf	C.F.	0.00	21.20	21.20
Over 5 cf	"	0.00	17.65	17.65
Bar reinforcing				
Under 5 cf	C.F.	0.00	35.30	35.30
Over 5 cf	"	0.00	26.50	26.50
Rubbish handling				
Load in dumpster or truck				
Minimum	C.F.	0.00	0.47	0.47
Maximum	"	0.00	0.71	0.71
For use of elevators, add				
Minimum	C.F.	0.00	0.11	0.11
Maximum	"	0.00	0.21	0.21
Rubbish hauling				
Hand loaded on trucks, 2 mile trip	C.Y.	0.00	16.80	16.80
Machine loaded on trucks, 2 mile trip	"	0.00	11.50	11.50
02075.80 CORE DRILLING				
Concrete				
6" thick				
3" dia.	EA.	0.00	19.05	19.05
4" dia.	"	0.00	22.25	22.25
6" dia.	"	0.00	26.70	26.70
8" dia.	"	0.00	44.50	44.50
8" thick				
3" dia.	EA.	0.00	26.70	26.70
4" dia.	"	0.00	33.40	33.40
6" dia.	"	0.00	38.10	38.10
8" dia.	"	0.00	53.50	53.50
10" thick				
3" dia.	EA.	0.00	33.40	33.40
4" dia.	"	0.00	38.10	38.10
6" dia.	"	0.00	44.50	44.50
8" dia.	"	0.00	67.00	67.00
12" thick				
3" dia.	EA.	0.00	44.50	44.50
4" dia.	"	0.00	53.50	53.50
6" dia.	"	0.00	67.00	67.00
8" dia.	"	0.00	89.00	89.00

02 SITEWORK

HAZARDOUS WASTE	UNIT	MAT.	INST.	TOTAL
02080.10 ASBESTOS REMOVAL				
Enclosure using wood studs & poly, install & remove	S.F.	0.50	0.42	0.92
Trailer (change room)	DAY			55.00
Disposal suits (4 suits per man day)	"			22.00
Type C respirator mask, includes hose & filters, per man	"			11.00
Respirator mask & filter, light contamination	"			4.40
Air monitoring test, 12 tests per day				
Off job testing	DAY			660.00
On the job testing	"			880.00
Asbestos vacuum with attachments	EA.			303.00
Hydraspray piston pump	"			385.00
Negative air pressure system	"			440.00
Grade D breathing air equipment	"			1,100
Glove bag, 44" x 60" x 6 mil plastic	"			3.30
40 CY asbestos dumpster				
Weekly rental	EA.			385.00
Pick up/delivery	"			165.00
Asbestos dump fee	"			82.50
02080.12 DUCT INSULATION REMOVAL				
Remove duct insulation, duct size				
6" x 12"	L.F.	0.00	1.05	1.05
x 18"	"	0.00	1.40	1.40
x 24"	"	0.00	1.75	1.75
8" x 12"	"	0.00	1.50	1.50
x 18"	"	0.00	1.65	1.65
x 24"	"	0.00	1.95	1.95
12" x 12"	"	0.00	1.50	1.50
x 18"	"	0.00	1.75	1.75
x 24"	"	0.00	2.10	2.10
02080.15 PIPE INSULATION REMOVAL				
Removal, asbestos insulation				
2" thick, pipe				
1" to 3" dia.	L.F.	0.00	1.75	1.75
4" to 6" dia.	"	0.00	2.00	2.00
3" thick				
7" to 8" dia.	L.F.	0.00	2.10	2.10
9" to 10" dia.	"	0.00	2.25	2.25
11" to 12" dia.	"	0.00	2.35	2.35
13" to 14" dia.	"	0.00	2.50	2.50
15" to 18" dia.	"	0.00	2.65	2.65

02 SITEWORK

SITE DEMOLITION

SITE DEMOLITION	UNIT	MAT.	INST.	TOTAL
02105.43 GAS PIPING				
Remove welded steel pipe, not including excavation				
4" dia.	L.F.	0.00	7.20	7.20
5" dia.	"	0.00	11.50	11.50
6" dia.	"	0.00	14.35	14.35
8" dia.	"	0.00	23.00	23.00
10" dia.	EA.	0.00	28.70	28.70
02105.45 SANITARY PIPING				
Remove sewer pipe, not including excavation				
4" dia.	L.F.	0.00	4.60	4.60
6" dia.	"	0.00	5.20	5.20
8" dia.	"	0.00	5.75	5.75
10" dia.	"	0.00	6.05	6.05
12" dia.	"	0.00	6.40	6.40
15" dia.	"	0.00	6.75	6.75
18" dia.	"	0.00	7.65	7.65
24" dia.	"	0.00	9.55	9.55
30" dia.	"	0.00	11.50	11.50
36" dia.	"	0.00	14.35	14.35
02105.48 WATER PIPING				
Remove water pipe, not including excavation				
4" dia.	L.F.	0.00	5.20	5.20
6" dia.	"	0.00	5.45	5.45
8" dia.	"	0.00	6.05	6.05
10" dia.	"	0.00	6.40	6.40
12" dia.	"	0.00	6.75	6.75
14" dia.	"	0.00	7.20	7.20
16" dia.	"	0.00	7.65	7.65
18" dia.	"	0.00	8.20	8.20
20" dia.	"	0.00	8.85	8.85
Remove valves				
6"	EA.	0.00	57.50	57.50
10"	"	0.00	64.00	64.00
14"	"	0.00	72.00	72.00
18"	"	0.00	95.50	95.50
02105.60 UNDERGROUND TANKS				
Remove underground storage tank, and backfill				
50 to 250 gals	EA.	0.00	383.00	383.00
600 gals	"	0.00	383.00	383.00
1000 gals	"	0.00	574.00	574.00
4000 gals	"	0.00	919.00	919.00
5000 gals	"	0.00	919.00	919.00
10,000 gals	"	0.00	1,530	1,530
12,000 gals	"	0.00	1,910	1,910
15,000 gals	"	0.00	2,300	2,300
20,000 gals	"	0.00	2,870	2,870

02 SITEWORK

SITE DEMOLITION

SITE DEMOLITION	UNIT	MAT.	INST.	TOTAL
02105.66 SEPTIC TANKS				
Remove septic tank				
1000 gals	EA.	0.00	95.50	95.50
2000 gals	"	0.00	115.00	115.00
5000 gals	"	0.00	144.00	144.00
15,000 gals	"	0.00	1,150	1,150
25,000 gals	"	0.00	1,530	1,530
40,000 gals	"	0.00	2,300	2,300
02162.10 TRENCH SHEETING				
Closed timber, including pull and salvage, excavation				
8' deep	S.F.	1.60	3.80	5.40
20' deep	"	1.60	5.40	7.00

EARTHWORK

EARTHWORK	UNIT	MAT.	INST.	TOTAL
02210.10 HAULING MATERIAL				
Haul material by 10 cy dump truck, round trip distance				
1 mile	C.Y.	0.00	2.35	2.35
2 mile	"	0.00	2.80	2.80
5 mile	"	0.00	3.80	3.80
10 mile	"	0.00	4.20	4.20
20 mile	"	0.00	4.65	4.65
30 mile	"	0.00	5.60	5.60
Site grading, cut & fill, sandy clay, 200' haul, 75 hp dozer	"	0.00	1.70	1.70
Spread topsoil by equipment on site	"	0.00	1.85	1.85
Site grading (cut and fill to 6") less than 1 acre				
75 hp dozer	C.Y.	0.00	2.80	2.80
1.5 cy backhoe/loader	"	0.00	4.20	4.20
02210.30 BULK EXCAVATION				
Hydraulic excavator				
1 cy capacity				
Light material	C.Y.	0.00	1.90	1.90
Medium material	"	0.00	2.30	2.30
Wet material	"	0.00	2.85	2.85
Blasted rock	"	0.00	3.30	3.30
Wheel mounted front-end loader				
7/8 cy capacity				
Light material	C.Y.	0.00	1.45	1.45
Medium material	"	0.00	1.70	1.70
Wet material	"	0.00	1.95	1.95
Blasted rock	"	0.00	2.35	2.35

02 SITEWORK

EARTHWORK	UNIT	MAT.	INST.	TOTAL
02210.30 BULK EXCAVATION				
Track mounted front-end loader				
1-1/2 cy capacity				
Light material	C.Y.	0.00	0.98	0.98
Medium material	"	0.00	1.05	1.05
Wet material	"	0.00	1.15	1.15
Blasted rock	"	0.00	1.30	1.30
02220.40 BUILDING EXCAVATION				
Structural excavation, unclassified earth				
3/8 cy backhoe	C.Y.	0.00	7.85	7.85
3/4 cy backhoe	"	0.00	5.85	5.85
1 cy backhoe	"	0.00	4.90	4.90
Foundation backfill and compaction by machine	"	0.00	11.75	11.75
02220.50 UTILITY EXCAVATION				
Trencher, sandy clay, 8" wide trench				
18" deep	L.F.	0.00	0.93	0.93
24" deep	"	0.00	1.05	1.05
36" deep	"	0.00	1.20	1.20
Trench backfill, 95% compaction				
Tamp by hand	C.Y.	0.00	13.25	13.25
Vibratory compaction	"	0.00	10.60	10.60
Trench backfilling, with borrow sand, place & compact	"	4.10	10.60	14.70
02220.60 TRENCHING				
Trenching and continuous footing excavation				
By gradall				
1 cy capacity				
Light soil	C.Y.	0.00	1.70	1.70
Medium soil	"	0.00	1.80	1.80
Heavy/wet soil	"	0.00	1.95	1.95
Loose rock	"	0.00	2.15	2.15
Blasted rock	"	0.00	2.25	2.25
By hydraulic excavator				
1/2 cy capacity				
Light soil	C.Y.	0.00	1.95	1.95
Medium soil	"	0.00	2.15	2.15
Heavy/wet soil	"	0.00	2.35	2.35
Loose rock	"	0.00	2.60	2.60
Blasted rock	"	0.00	2.95	2.95
Hand excavation				
Bulk, wheeled 100'				
Normal soil	C.Y.	0.00	23.55	23.55
Sand or gravel	"	0.00	21.20	21.20
Medium clay	"	0.00	30.30	30.30
Heavy clay	"	0.00	42.40	42.40
Loose rock	"	0.00	53.00	53.00
Trenches, up to 2' deep				
Normal soil	C.Y.	0.00	26.50	26.50
Sand or gravel	"	0.00	23.55	23.55
Medium clay	"	0.00	35.30	35.30

02 SITEWORK

EARTHWORK

EARTHWORK	UNIT	MAT.	INST.	TOTAL
02220.60 TRENCHING				
Heavy clay	C.Y.	0.00	53.00	53.00
Loose rock	"	0.00	70.50	70.50
Trenches, to 6' deep				
Normal soil	C.Y.	0.00	30.30	30.30
Sand or gravel	"	0.00	26.50	26.50
Medium clay	"	0.00	42.40	42.40
Heavy clay	"	0.00	70.50	70.50
Loose rock	"	0.00	106.00	106.00
Backfill trenches				
With compaction				
By hand	C.Y.	0.00	17.65	17.65
By 60 hp tracked dozer	"	0.00	1.05	1.05
By 200 hp tracked dozer	"	0.00	0.65	0.65
By small front-end loader	"	0.00	1.20	1.20
Backfill trenches, sand bedding, no compaction				
By hand	C.Y.	4.95	17.65	22.60
By small front-end loader	"	4.95	1.70	6.65
02220.90 HAND EXCAVATION				
Excavation				
To 2' deep				
Normal soil	C.Y.	0.00	23.55	23.55
Sand and gravel	"	0.00	21.20	21.20
Medium clay	"	0.00	26.50	26.50
Heavy clay	"	0.00	30.30	30.30
Loose rock	"	0.00	35.30	35.30
To 6' deep				
Normal soil	C.Y.	0.00	30.30	30.30
Sand and gravel	"	0.00	26.50	26.50
Medium clay	"	0.00	35.30	35.30
Heavy clay	"	0.00	42.40	42.40
Loose rock	"	0.00	53.00	53.00
Compaction of backfill around structures or in trench				
By hand with air tamper	C.Y.	0.00	15.15	15.15
By hand with vibrating plate tamper	"	0.00	14.15	14.15
1 ton roller	"	0.00	21.00	21.00
Miscellaneous hand labor				
Trim slopes, sides of excavation	S.F.	0.00	0.04	0.04
Trim bottom of excavation	"	0.00	0.04	0.04
Excavation around obstructions and services	C.Y.	0.00	70.50	70.50
02240.05 SOIL STABILIZATION				
Straw bale secured with rebar	L.F.	1.10	0.71	1.81
Filter barrier, 18" high filter fabric	"	1.10	2.10	3.20
Sediment fence, 36" fabric with 6" mesh	"	2.50	2.65	5.15

02 SITEWORK

EARTHWORK

	UNIT	MAT.	INST.	TOTAL
02605.30 MANHOLES				
Precast sections, 48" dia.		88.00	91.50	179.50
Base section	EA.	98.00	95.50	193.50
1'0" riser	"	24.20	76.50	100.70
1'4" riser	"	30.80	82.00	112.80
2'8" riser	"	48.40	88.50	136.90
4'0" riser	"	98.00	95.50	193.50
2'8" cone top	"	61.50	115.00	176.50
Precast manholes, 48" dia.				
4' deep	EA.	187.00	230.00	417.00
6' deep	"	264.00	287.00	551.00
7' deep	"	314.00	328.00	642.00
8' deep	"	363.00	383.00	746.00
10' deep	"	390.00	460.00	850.00
Cast-in-place, 48" dia., with frame and cover				
5' deep	EA.	363.00	574.00	937.00
6' deep	"	478.00	656.00	1,134
8' deep	"	698.00	766.00	1,464
10' deep	"	814.00	919.00	1,733
Brick manholes, 48" dia. with cover, 8" thick				
4' deep	EA.	478.00	263.00	741.00
6' deep	"	600.00	292.00	892.00
8' deep	"	770.00	329.00	1,099
10' deep	"	957.00	376.00	1,333
12' deep	"	1,200	439.00	1,639
14' deep	"	1,460	526.00	1,986
Inverts for manholes				
Single channel	EA.	71.50	105.00	176.50
Triple channel	"	82.50	132.00	214.50
Frames and covers, 24" diameter				
300 lb	EA.	236.00	21.20	257.20
400 lb	"	248.00	23.55	271.55
500 lb	"	286.00	30.30	316.30
Watertight, 350 lb	"	297.00	70.50	367.50
For heavy equipment, 1200 lb	"	649.00	106.00	755.00
Steps for manholes				
7" x 9"	EA.	8.35	4.25	12.60
8" x 9"	"	10.75	4.70	15.45
Curb inlet, 4' throat, cast in place				
12"-30" pipe	EA.	231.00	574.00	805.00
36"-48" pipe	"	253.00	656.00	909.00
Raise exist frame and cover, repaving	"	0.00	230.00	230.00
02610.10 CAST IRON FLANGED PIPE				
Cast iron flanged sections				
4" pipe, with one bolt set				
3' section	EA.	104.00	10.45	114.45
4' section	"	110.00	11.50	121.50
5' section	"	121.00	12.75	133.75
6' section	"	132.00	14.35	146.35
8' section	"	154.00	16.40	170.40
10' section	"	170.00	23.00	193.00

EARTHWORK

EARTHWORK	UNIT	MAT.	INST.	TOTAL
02610.10 CAST IRON FLANGED PIPE				
12' section	EA.	187.00	38.30	225.30
15' section	"	214.00	57.50	271.50
18' section	"	242.00	76.50	318.50
6" pipe, with one bolt set				
3' section	EA.	132.00	11.50	143.50
4' section	"	143.00	13.50	156.50
5' section	"	154.00	15.30	169.30
6' section	"	165.00	17.65	182.65
8' section	"	187.00	25.50	212.50
10' section	"	209.00	28.70	237.70
12' section	"	231.00	38.30	269.30
15' section	"	264.00	57.50	321.50
18' section	"	319.00	82.00	401.00
8" pipe, with one bolt set				
3' section	EA.	176.00	14.35	190.35
4' section	"	198.00	16.40	214.40
5' section	"	209.00	19.15	228.15
6' section	"	220.00	23.00	243.00
8' section	"	253.00	32.80	285.80
10' section	"	286.00	38.30	324.30
12' section	"	308.00	57.50	365.50
15' section	"	363.00	76.50	439.50
18' section	"	429.00	95.50	524.50
10" pipe, with one bolt set				
3' section	EA.	231.00	14.75	245.75
4' section	"	264.00	16.90	280.90
5' section	"	286.00	19.80	305.80
6' section	"	308.00	23.95	331.95
8' section	"	341.00	34.80	375.80
10' section	"	374.00	41.00	415.00
12' section	"	429.00	64.00	493.00
15' section	"	473.00	82.00	555.00
18' section	"	550.00	115.00	665.00
12" pipe, with one bolt set				
3' section	EA.	308.00	15.95	323.95
4' section	"	330.00	18.55	348.55
5' section	"	363.00	22.10	385.10
6' section	"	385.00	26.10	411.10
8' section	"	434.00	38.30	472.30
10' section	"	484.00	44.20	528.20
12' section	"	539.00	72.00	611.00
15' section	"	616.00	95.50	711.50
18' section	"	693.00	128.00	821.00
02610.11 CAST IRON FITTINGS				
Mechanical joint, with 2 bolt kits				
90 deg bend				
4"	EA.	60.50	14.15	74.65
6"	"	77.00	16.30	93.30
8"	"	104.00	21.20	125.20
10"	"	160.00	30.30	190.30

EARTHWORK

02610.11 CAST IRON FITTINGS

EARTHWORK	UNIT	MAT.	INST.	TOTAL
12"	EA.	231.00	42.40	273.40
14"	"	319.00	53.00	372.00
16"	"	352.00	70.50	422.50
45 deg bend				
4"	EA.	49.50	14.15	63.65
6"	"	71.50	16.30	87.80
8"	"	104.00	21.20	125.20
10"	"	132.00	30.30	162.30
12"	"	187.00	42.40	229.40
14"	"	231.00	53.00	284.00
16"	"	297.00	70.50	367.50
Tee, with 3 bolt kits				
4" x 4"	EA.	93.50	21.20	114.70
6" x 6"	"	121.00	26.50	147.50
8" x 8"	"	176.00	35.30	211.30
10" x 10"	"	242.00	53.00	295.00
12" x 12"	"	319.00	70.50	389.50
Wye, with 3 bolt kits				
6" x 6"	EA.	160.00	26.50	186.50
8" x 8"	"	236.00	35.30	271.30
10" x 10"	"	346.00	53.00	399.00
12" x 12"	"	484.00	70.50	554.50
Reducer, with 2 bolt kits				
6" x 4"	EA.	60.50	26.50	87.00
8" x 6"	"	82.50	35.30	117.80
10" x 8"	"	138.00	53.00	191.00
12" x 10"	"	148.00	70.50	218.50
Flanged, 90 deg bend, 125 lb.				
4"	EA.	38.50	17.65	56.15
6"	"	60.50	21.20	81.70
8"	"	88.00	26.50	114.50
10"	"	154.00	35.30	189.30
12"	"	220.00	53.00	273.00
14"	"	440.00	70.50	510.50
16"	"	660.00	70.50	730.50
Tee				
4"	EA.	77.00	26.50	103.50
6"	"	110.00	30.30	140.30
8"	"	160.00	35.30	195.30
10"	"	314.00	42.40	356.40
12"	"	418.00	53.00	471.00
14"	"	957.00	70.50	1,028
16"	"	1,380	106.00	1,486

02610.13 GATE VALVES

	UNIT	MAT.	INST.	TOTAL
Gate valve, (AWWA) mechanical joint, with adjustable box				
4" valve	EA.	336.00	38.30	374.30
6" valve	"	390.00	46.00	436.00
8" valve	"	468.00	57.50	525.50
10" valve	"	682.00	67.50	749.50
12" valve	"	946.00	82.00	1,028

EARTHWORK	UNIT	MAT.	INST.	TOTAL
02610.13 GATE VALVES				
14" valve	EA.	2,310	95.50	2,406
16" valve	"	3,080	104.00	3,184
18" valve	"	4,180	115.00	4,295
Flanged, with box, post indicator (AWWA)				
4" valve	EA.	286.00	46.00	332.00
6" valve	"	352.00	52.00	404.00
8" valve	"	440.00	64.00	504.00
10" valve	"	682.00	76.50	758.50
12" valve	"	946.00	95.50	1,042
14" valve	"	2,310	115.00	2,425
16" valve	"	3,080	144.00	3,224
02610.15 WATER METERS				
Water meter, displacement type				
1"	EA.	121.00	30.00	151.00
1-1/2"	"	440.00	33.30	473.30
2"	"	660.00	37.50	697.50
02610.17 CORPORATION STOPS				
Stop for flared copper service pipe				
3/4"	EA.	14.30	15.00	29.30
1"	"	20.90	16.65	37.55
1-1/4"	"	60.50	20.00	80.50
1-1/2"	"	71.50	25.00	96.50
2"	"	104.00	30.00	134.00
02610.40 DUCTILE IRON PIPE				
Ductile iron pipe, cement lined, slip-on joints				
4"	L.F.	6.35	3.20	9.55
6"	"	7.85	3.40	11.25
8"	"	10.30	3.60	13.90
10"	"	14.15	3.85	18.00
12"	"	17.35	4.60	21.95
14"	"	21.95	5.75	27.70
16"	"	25.40	6.40	31.80
18"	"	28.90	7.20	36.10
20"	"	32.30	8.20	40.50
Mechanical joint pipe				
4"	L.F.	6.95	4.40	11.35
6"	"	8.65	4.80	13.45
8"	"	11.55	5.20	16.75
10"	"	15.00	5.75	20.75
12"	"	19.25	7.65	26.90
14"	"	24.20	8.85	33.05
16"	"	26.60	10.45	37.05
18"	"	30.00	11.50	41.50
20"	"	34.70	12.75	47.45
Fittings, mechanical joint				
90 degree elbow				
4"	EA.	86.00	14.15	100.15
6"	"	116.00	16.30	132.30

EARTHWORK	UNIT	MAT.	INST.	TOTAL
02610.40 DUCTILE IRON PIPE				
8"	EA.	165.00	21.20	186.20
10"	"	242.00	30.30	272.30
12"	"	330.00	42.40	372.40
14"	"	506.00	53.00	559.00
16"	"	572.00	70.50	642.50
18"	"	880.00	85.00	965.00
20"	"	1,010	106.00	1,116
45 degree elbow				
4"	EA.	87.00	14.15	101.15
6"	"	104.00	16.30	120.30
8"	"	143.00	21.20	164.20
10"	"	198.00	30.30	228.30
12"	"	275.00	42.40	317.40
14"	"	418.00	53.00	471.00
16"	"	495.00	70.50	565.50
18"	"	715.00	106.00	821.00
20"	"	803.00	106.00	909.00
Tee				
4"x3"	EA.	121.00	26.50	147.50
4"x4"	"	132.00	26.50	158.50
6"x3"	"	154.00	30.30	184.30
6"x4"	"	160.00	30.30	190.30
6"x6"	"	170.00	30.30	200.30
8"x4"	"	214.00	35.30	249.30
8"x6"	"	248.00	35.30	283.30
8"x8"	"	231.00	35.30	266.30
10"x4"	"	297.00	42.40	339.40
10"x6"	"	319.00	42.40	361.40
10"x8"	"	328.00	42.40	370.40
10"x10"	"	363.00	42.40	405.40
12"x4"	"	440.00	53.00	493.00
12"x6"	"	341.00	53.00	394.00
12"x8"	"	380.00	53.00	433.00
12"x10"	"	434.00	53.00	487.00
12"x12"	"	448.00	56.50	504.50
14"x4"	"	550.00	60.50	610.50
14"x6"	"	588.00	60.50	648.50
14"x8"	"	600.00	60.50	660.50
14"x10"	"	622.00	60.50	682.50
14"x12"	"	638.00	65.00	703.00
14"x14"	"	654.00	65.00	719.00
16"x4"	"	715.00	70.50	785.50
16"x6"	"	732.00	70.50	802.50
16"x8"	"	644.00	70.50	714.50
16"x10"	"	658.00	70.50	728.50
16"x12"	"	632.00	70.50	702.50
16"x14"	"	676.00	70.50	746.50
16"x16"	"	693.00	70.50	763.50
18"x6"	"	825.00	77.00	902.00
18"x8"	"	842.00	77.00	919.00
18"x10"	"	858.00	77.00	935.00
18"x12"	"	874.00	77.00	951.00

EARTHWORK

02610.40 DUCTILE IRON PIPE

	UNIT	MAT.	INST.	TOTAL
18"x14"	EA.	984.00	77.00	1,061
18"x16"	"	979.00	77.00	1,056
18"x18"	"	1,090	77.00	1,167
20"x6"	"	1,040	85.00	1,125
20"x8"	"	1,050	85.00	1,135
20"x10"	"	1,080	85.00	1,165
20"x12"	"	1,100	85.00	1,185
20"x14"	"	1,120	85.00	1,205
20"x16"	"	1,320	85.00	1,405
20"x18"	"	1,380	85.00	1,465
20"x20"	"	1,410	85.00	1,495
Cross				
4"x3"	EA.	147.00	35.30	182.30
4"x4"	"	165.00	35.30	200.30
6"x3"	"	170.00	42.40	212.40
6"x4"	"	183.00	42.40	225.40
6"x6"	"	206.00	42.40	248.40
8"x4"	"	233.00	47.10	280.10
8"x6"	"	253.00	47.10	300.10
8"x8"	"	284.00	47.10	331.10
10"x4"	"	319.00	53.00	372.00
10"x6"	"	341.00	53.00	394.00
10"x8"	"	375.00	53.00	428.00
10"x10"	"	442.00	53.00	495.00
12"x4"	"	407.00	60.50	467.50
12"x6"	"	462.00	60.50	522.50
12"x8"	"	451.00	60.50	511.50
12"x10"	"	522.00	65.00	587.00
12"x12"	"	561.00	65.00	626.00
14"x4"	"	534.00	70.50	604.50
14"x6"	"	606.00	70.50	676.50
14"x8"	"	632.00	70.50	702.50
14"x10"	"	679.00	70.50	749.50
14"x12"	"	726.00	77.00	803.00
14"x14"	"	788.00	77.00	865.00
16"x4"	"	690.00	85.00	775.00
16"x6"	"	723.00	85.00	808.00
16"x8"	"	759.00	85.00	844.00
16"x10"	"	803.00	85.00	888.00
16"x12"	"	842.00	85.00	927.00
16"x14"	"	908.00	85.00	993.00
16"x16"	"	968.00	85.00	1,053
18"x6"	"	891.00	94.00	985.00
18"x8"	"	924.00	94.00	1,018
18"x10"	"	966.00	94.00	1,060
18"x12"	"	1,020	94.00	1,114
18"x14"	"	1,200	94.00	1,294
18"x16"	"	1,280	94.00	1,374
18"x18"	"	1,350	94.00	1,444
20"x6"	"	1,070	101.00	1,171
20"x8"	"	1,110	101.00	1,211
20"x10"	"	1,140	101.00	1,241

02 SITEWORK

EARTHWORK	UNIT	MAT.	INST.	TOTAL
02610.40 DUCTILE IRON PIPE				
20"x12"	EA.	1,200	101.00	1,301
20"x14"	"	1,260	101.00	1,361
20"x16"	"	1,480	101.00	1,581
20"x18"	"	1,560	106.00	1,666
20"x20"	"	1,650	106.00	1,756
02610.60 PLASTIC PIPE				
PVC, class 150 pipe				
4" dia.	L.F.	2.75	2.85	5.60
6" dia.	"	5.40	3.10	8.50
8" dia.	"	8.70	3.30	12.00
10" dia.	"	12.65	3.60	16.25
12" dia.	"	18.70	3.85	22.55
Schedule 40 pipe				
1-1/2" dia.	L.F.	1.65	1.25	2.90
2" dia.	"	1.95	1.35	3.30
2-1/2" dia.	"	2.15	1.40	3.55
3" dia.	"	3.10	1.50	4.60
4" dia.	"	4.40	1.75	6.15
6" dia.	"	7.45	2.10	9.55
90 degree elbows				
1"	EA.	0.55	3.55	4.10
1-1/2"	"	0.99	3.55	4.54
2"	"	1.20	3.85	5.05
2-1/2"	"	3.40	4.25	7.65
3"	"	4.55	4.70	9.25
4"	"	6.85	5.30	12.15
6"	"	35.20	7.05	42.25
45 degree elbows				
1"	EA.	0.77	3.55	4.32
1-1/2"	"	1.20	3.55	4.75
2"	"	1.65	3.85	5.50
2-1/2"	"	3.20	4.25	7.45
3"	"	4.70	4.70	9.40
4"	"	7.60	5.30	12.90
6"	"	36.30	7.05	43.35
Tees				
1"	EA.	0.85	4.25	5.10
1-1/2"	"	1.05	4.25	5.30
2"	"	2.10	4.70	6.80
2-1/2"	"	4.30	5.30	9.60
3"	"	6.35	6.05	12.40
4"	"	9.35	7.05	16.40
6"	"	35.20	8.50	43.70
Couplings				
1"	EA.	0.50	3.55	4.05
1-1/2"	"	0.61	3.55	4.16
2"	"	1.05	3.85	4.90
2-1/2"	"	2.10	4.25	6.35
3"	"	3.60	4.70	8.30
4"	"	5.15	5.30	10.45

EARTHWORK

02610.60 PLASTIC PIPE

EARTHWORK	UNIT	MAT.	INST.	TOTAL
6"	EA.	12.65	7.05	19.70
Drainage pipe				
PVC schedule 80				
1" dia.	L.F.	0.88	1.25	2.13
1-1/2" dia.	"	1.10	1.25	2.35
ABS, 2" dia.	"	1.45	1.35	2.80
2-1/2" dia.	"	2.10	1.40	3.50
3" dia.	"	2.50	1.50	4.00
4" dia.	"	3.35	1.75	5.10
6" dia.	"	5.60	2.10	7.70
8" dia.	"	7.60	3.00	10.60
10" dia.	"	9.35	3.60	12.95
12" dia.	"	16.50	3.85	20.35
90 degree elbows				
1"	EA.	1.65	3.55	5.20
1-1/2"	"	2.10	3.55	5.65
2"	"	2.55	3.85	6.40
2-1/2"	"	6.15	4.25	10.40
3"	"	6.40	4.70	11.10
4"	"	10.55	5.30	15.85
6"	"	24.20	7.05	31.25
45 degree elbows				
1"	EA.	2.65	3.55	6.20
1-1/2"	"	3.40	3.55	6.95
2"	"	4.20	3.85	8.05
2-1/2"	"	8.25	4.25	12.50
3"	"	8.80	4.70	13.50
4"	"	15.95	5.30	21.25
6"	"	38.50	7.05	45.55
Tees				
1"	EA.	1.75	4.25	6.00
1-1/2"	"	5.70	4.25	9.95
2"	"	7.05	4.70	11.75
2-1/2"	"	8.25	5.30	13.55
3"	"	9.00	6.05	15.05
4"	"	15.95	7.05	23.00
6"	"	33.00	8.50	41.50
Couplings				33.00
1"	EA.	1.30	3.55	4.85
1-1/2"	"	2.40	3.55	5.95
2"	"	3.40	3.85	7.25
2-1/2"	"	7.70	4.25	11.95
3"	"	7.70	4.70	12.40
4"	"	8.25	5.30	13.55
6"	"	13.75	7.05	20.80
Pressure pipe				
PVC, class 200 pipe				
3/4"	L.F.	0.88	1.05	1.93
1"	"	0.99	1.10	2.09
1-1/4"	"	1.10	1.20	2.30
1-1/2"	"	1.30	1.25	2.55
2"	"	1.45	1.35	2.80

EARTHWORK

02610.60 PLASTIC PIPE	UNIT	MAT.	INST.	TOTAL
2-1/2"	L.F.	2.10	1.40	3.50
3"	"	2.70	1.50	4.20
4"	"	4.60	1.75	6.35
6"	"	8.80	2.10	10.90
8"	"	13.20	3.30	16.50
90 degree elbows				
3/4"	EA.	0.44	3.55	3.99
1"	"	0.55	3.55	4.10
1-1/4"	"	0.88	3.55	4.43
1-1/2"	"	1.05	3.55	4.60
2"	"	2.10	3.85	5.95
2-1/2"	"	3.05	4.25	7.30
3"	"	6.85	4.70	11.55
4"	"	17.60	5.30	22.90
6"	"	35.20	7.05	42.25
8"	"	66.00	10.60	76.60
45 degree elbows				
3/4"	EA.	0.66	3.55	4.21
1"	"	0.83	3.55	4.38
1-1/4"	"	1.20	3.55	4.75
1-1/2"	"	1.45	3.55	5.00
2"	"	2.05	3.85	5.90
2-1/2"	"	3.30	4.25	7.55
3"	"	7.70	4.70	12.40
4"	"	14.30	5.30	19.60
6"	"	35.20	7.05	42.25
8"	"	71.50	10.60	82.10
Tees				
3/4"	EA.	0.50	4.25	4.75
1"	"	0.71	4.25	4.96
1-1/4"	"	1.05	4.25	5.30
1-1/2"	"	1.45	4.25	5.70
2"	"	2.10	4.70	6.80
2-1/2"	"	3.30	5.30	8.60
3"	"	10.20	6.05	16.25
4"	"	14.30	7.05	21.35
6"	"	50.50	8.50	59.00
8"	"	104.00	11.80	115.80
Couplings				
3/4"	EA.	0.33	3.55	3.88
1"	"	0.50	3.55	4.05
1-1/4"	"	0.66	3.55	4.21
1-1/2"	"	0.71	3.55	4.26
2"	"	0.99	3.85	4.84
2-1/2"	"	2.10	4.25	6.35
3"	"	3.30	4.70	8.00
4"	"	4.70	4.70	9.40
6"	"	14.30	5.30	19.60
8"	"	26.40	7.05	33.45

02 SITEWORK

EARTHWORK

EARTHWORK	UNIT	MAT.	INST.	TOTAL
02610.90 VITRIFIED CLAY PIPE				
Vitrified clay pipe, extra strength				
6" dia.	L.F.	2.75	5.20	7.95
8" dia.	"	3.30	5.45	8.75
10" dia.	"	5.05	5.75	10.80
12" dia.	"	7.25	7.65	14.90
15" dia.	"	13.20	11.50	24.70
18" dia.	"	19.80	12.75	32.55
24" dia.	"	36.30	16.40	52.70
30" dia.	"	61.50	23.00	84.50
36" dia.	"	88.00	32.80	120.80
02630.10 TAPPING SADDLES & SLEEVES				
Tapping saddle, tap size to 2"				
4" saddle	EA.	35.20	10.60	45.80
6" saddle	"	41.80	13.25	55.05
8" saddle	"	48.40	17.65	66.05
10" saddle	"	57.00	21.20	78.20
12" saddle	"	68.00	30.30	98.30
14" saddle	"	77.00	42.40	119.40
Tapping sleeve				
4x4	EA.	385.00	14.15	399.15
6x4	"	495.00	16.30	511.30
6x6	"	506.00	16.30	522.30
8x4	"	517.00	21.20	538.20
8x6	"	539.00	21.20	560.20
10x4	"	825.00	46.00	871.00
10x6	"	1,100	46.00	1,146
10x8	"	1,120	46.00	1,166
10x10	"	1,160	47.90	1,208
12x4	"	1,190	47.90	1,238
12x6	"	1,230	52.00	1,282
12x8	"	1,260	57.50	1,318
12x10	"	1,320	64.00	1,384
12x12	"	1,380	72.00	1,452
Tapping valve, mechanical joint				
4" valve	EA.	352.00	144.00	496.00
6" valve	"	418.00	191.00	609.00
8" valve	"	638.00	287.00	925.00
10" valve	"	1,010	383.00	1,393
12" valve	"	1,650	574.00	2,224
Tap hole in pipe				
4" hole	EA.	0.00	26.50	26.50
6" hole	"	0.00	42.40	42.40
8" hole	"	0.00	70.50	70.50
10" hole	"	0.00	85.00	85.00
12" hole	"	0.00	106.00	106.00
02640.15 VALVE BOXES				
Valve box, adjustable, for valves up to 20"				
3' deep	EA.	88.00	7.05	95.05
4' deep	"	88.00	8.50	96.50
5' deep	"	88.00	10.60	98.60

EARTHWORK

EARTHWORK	UNIT	MAT.	INST.	TOTAL
02640.19 THRUST BLOCKS				
Thrust block, 3000# concrete				
1/4 c.y.	EA.	58.50	45.20	103.70
1/2 c.y.	"	82.50	54.00	136.50
3/4 c.y.	"	93.50	90.50	184.00
1 c.y.	"	116.00	181.00	297.00
02645.10 FIRE HYDRANTS				
Standard, 3 way post, 6" mechanical joint				
2' deep	EA.	858.00	383.00	1,241
4' deep	"	924.00	460.00	1,384
6' deep	"	979.00	574.00	1,553
8' deep	"	1,060	656.00	1,716
02665.10 CHILLED WATER SYSTEMS				
Chilled water pipe, 2" thick insulation, w/casing				
Align and tack weld on sleepers				
1-1/2" dia.	L.F.	9.05	1.05	10.10
3" dia.	"	14.70	1.65	16.35
4" dia.	"	17.00	2.30	19.30
6" dia.	"	19.25	2.85	22.10
8" dia.	"	26.10	3.30	29.40
10" dia.	"	34.00	3.85	37.85
12" dia.	"	39.70	4.60	44.30
14" dia.	"	56.50	5.00	61.50
16" dia.	"	71.50	5.75	77.25
Align and tack weld on trench bottom				
18" dia.	L.F.	73.50	6.40	79.90
20" dia.	"	99.50	7.20	106.70
Preinsulated fittings				
Align and tack weld on sleepers				
Elbows				
1-1/2"	EA.	181.00	18.75	199.75
3"	"	238.00	30.00	268.00
4"	"	306.00	37.50	343.50
6"	"	442.00	50.00	492.00
8"	"	612.00	60.00	672.00
Tees				
1-1/2"	EA.	283.00	20.00	303.00
3"	"	385.00	33.30	418.30
4"	"	487.00	42.90	529.90
6"	"	680.00	60.00	740.00
8"	"	906.00	75.00	981.00
Reducers				
3"	EA.	295.00	25.00	320.00
4"	"	499.00	30.00	529.00
6"	"	668.00	37.50	705.50
8"	"	702.00	50.00	752.00
Anchors, not including concrete				
4"	EA.	125.00	37.50	162.50
6"	"	181.00	37.50	218.50

EARTHWORK	UNIT	MAT.	INST.	TOTAL
02665.10 CHILLED WATER SYSTEMS				
Align and tack weld on trench bottom				
Elbows				
10"	EA.	793.00	72.00	865.00
12"	"	963.00	82.00	1,045
14"	"	1,130	88.50	1,219
16"	"	1,250	95.50	1,346
18"	"	1,420	104.00	1,524
20"	"	1,700	115.00	1,815
Tees				
10"	EA.	1,250	72.00	1,322
12"	"	1,640	82.00	1,722
14"	"	1,700	88.50	1,789
16"	"	1,810	95.50	1,906
18"	"	2,040	104.00	2,144
20"	"	2,380	115.00	2,495
Reducers				
10"	EA.	963.00	47.90	1,011
12"	"	1,080	52.00	1,132
14"	"	1,250	57.50	1,308
16"	"	1,470	64.00	1,534
18"	"	1,700	72.00	1,772
20"	"	1,810	82.00	1,892
Anchors, not including concrete				
10"	EA.	283.00	47.90	330.90
12"	"	312.00	52.00	364.00
14"	"	351.00	57.50	408.50
16"	"	510.00	64.00	574.00
18"	"	651.00	72.00	723.00
20"	"	963.00	82.00	1,045
02670.10 WELLS				
Domestic water, drilled and cased				
4" dia.	L.F.	7.70	108.00	115.70
6" dia.	"	8.25	124.00	132.25
8" dia.	"	10.45	162.00	172.45
02685.10 GAS DISTRIBUTION				
Gas distribution lines				
Polyethylene, 60 psi coils				
1-1/4" dia.	L.F.	0.71	2.00	2.71
1-1/2" dia.	"	0.99	2.15	3.14
2" dia.	"	1.20	2.50	3.70
3" dia.	"	2.30	3.00	5.30
30' pipe lengths				
3" dia.	L.F.	2.50	3.35	5.85
4" dia.	"	4.10	3.75	7.85
6" dia.	"	5.50	5.00	10.50
8" dia.	"	10.45	6.00	16.45
Steel, schedule 40, plain end				
1" dia.	L.F.	2.30	2.50	4.80
2" dia.	"	3.85	2.75	6.60

02 SITEWORK

EARTHWORK	UNIT	MAT.	INST.	TOTAL
02685.10 GAS DISTRIBUTION				
3" dia.	L.F.	7.20	3.00	10.20
4" dia.	"	9.90	7.65	17.55
5" dia.	"	19.25	8.20	27.45
6" dia.	"	21.45	9.55	31.00
8" dia.	"	31.90	10.45	42.35
Natural gas meters, direct digital reading, threaded				
250 cfh @ 5 lbs	EA.	93.50	60.00	153.50
425 cfh @ 10 lbs	"	220.00	60.00	280.00
800 cfh @ 20 lbs	"	319.00	75.00	394.00
1000 cfh @ 25 lbs	"	946.00	75.00	1,021
1,400 cfh @ 100 lbs	"	2,150	100.00	2,250
2,300 cfh @ 100 lbs	"	3,080	150.00	3,230
5,000 cfh @ 100 lbs	"	4,620	300.00	4,920
Gas pressure regulators				
Threaded				
3/4"	EA.	40.00	37.50	77.50
1"	"	41.80	50.00	91.80
1-1/4"	"	44.00	50.00	94.00
1-1/2"	"	275.00	50.00	325.00
2"	"	286.00	60.00	346.00
Flanged				
3"	EA.	825.00	75.00	900.00
4"	"	1,160	100.00	1,260
02690.10 STORAGE TANKS				
Oil storage tank, underground				
Steel				
500 gals	EA.	440.00	144.00	584.00
1,000 gals	"	803.00	191.00	994.00
4,000 gals	"	2,190	383.00	2,573
5,000 gals	"	2,910	574.00	3,484
10,000 gals	"	4,120	1,150	5,270
Fiberglass, double wall				
550 gals	EA.	1,650	191.00	1,841
1,000 gals	"	2,090	191.00	2,281
2,000 gals	"	2,480	287.00	2,767
4,000 gals	"	3,300	574.00	3,874
6,000 gals	"	3,850	766.00	4,616
8,000 gals	"	5,280	1,150	6,430
10,000 gals	"	6,880	1,440	8,320
12,000 gals	"	8,250	1,910	10,160
15,000 gals	"	9,620	2,550	12,170
20,000 gals	"	11,550	2,870	14,420
Above ground				
Steel				
275 gals	EA.	170.00	115.00	285.00
500 gals	"	605.00	191.00	796.00
1,000 gals	"	902.00	230.00	1,132
1,500 gals	"	1,350	287.00	1,637
2,000 gals	"	1,760	383.00	2,143
5,000 gals	"	4,840	574.00	5,414

EARTHWORK

02690.10 STORAGE TANKS	UNIT	MAT.	INST.	TOTAL
Fill cap	EA.	22.00	30.00	52.00
Vent cap	"	22.00	30.00	52.00
Level indicator	"	55.00	30.00	85.00
02695.80 STEAM METERS				
In-line turbine, direct reading, 300 lb, flanged				
2"	EA.	2,640	37.50	2,678
3"	"	2,860	50.00	2,910
4"	"	3,190	60.00	3,250
Threaded, 2"				
5" line	EA.	5,170	300.00	5,470
6" line	"	5,280	300.00	5,580
8" line	"	5,390	300.00	5,690
10" line	"	5,610	300.00	5,910
12" line	"	5,780	300.00	6,080
14" line	"	6,050	300.00	6,350
16" line	"	6,320	300.00	6,620

SEWERAGE AND DRAINAGE

02720.10 CATCH BASINS	UNIT	MAT.	INST.	TOTAL
Standard concrete catch basin				
Cast in place, 3'8" x 3'8", 6" thick wall				
2' deep	EA.	187.00	287.00	474.00
3' deep	"	253.00	287.00	540.00
4' deep	"	330.00	383.00	713.00
5' deep	"	385.00	383.00	768.00
6' deep	"	440.00	460.00	900.00
4'x4', 8" thick wall, cast in place				
2' deep	EA.	198.00	287.00	485.00
3' deep	"	275.00	287.00	562.00
4' deep	"	368.00	383.00	751.00
5' deep	"	429.00	383.00	812.00
6' deep	"	473.00	460.00	933.00
Frames and covers, cast iron				
Round				
24" dia.	EA.	165.00	53.00	218.00
26" dia.	"	182.00	53.00	235.00
28" dia.	"	220.00	53.00	273.00
Rectangular				
23"x23"	EA.	143.00	53.00	196.00
27"x20"	"	176.00	53.00	229.00
24"x24"	"	170.00	53.00	223.00

02 SITEWORK

SEWERAGE AND DRAINAGE

SEWERAGE AND DRAINAGE	UNIT	MAT.	INST.	TOTAL
02720.10 CATCH BASINS				
26"x26"	EA.	187.00	53.00	240.00
Curb inlet frames and covers				
27"x27"	EA.	352.00	53.00	405.00
24"x36"	"	209.00	53.00	262.00
24"x25"	"	192.00	53.00	245.00
24"x22"	"	165.00	53.00	218.00
20"x22"	"	220.00	53.00	273.00
Airfield catch basin frame and grating, galvanized				
2'x4'	EA.	418.00	53.00	471.00
2'x2'	"	286.00	53.00	339.00
02720.40 STORM DRAINAGE				
Concrete pipe				
Plain, bell and spigot joint, class II				
6" pipe	L.F.	2.95	5.20	8.15
8" pipe	"	3.40	5.75	9.15
10" pipe	"	3.50	6.05	9.55
12" pipe	"	4.30	6.40	10.70
15" pipe	"	5.35	6.75	12.10
18" pipe	"	7.25	7.20	14.45
21" pipe	"	9.05	7.65	16.70
24" pipe	"	11.35	8.20	19.55
Reinforced, class III, tongue and groove joint				
12" pipe	L.F.	6.60	6.40	13.00
15" pipe	"	7.90	6.75	14.65
18" pipe	"	10.00	7.20	17.20
21" pipe	"	13.20	7.65	20.85
24" pipe	"	16.05	8.20	24.25
27" pipe	"	18.70	8.85	27.55
30" pipe	"	21.00	9.55	30.55
36" pipe	"	31.90	10.45	42.35
42" pipe	"	40.70	11.50	52.20
48" pipe	"	57.00	12.75	69.75
54" pipe	"	75.00	14.35	89.35
60" pipe	"	88.00	16.40	104.40
66" pipe	"	99.00	19.15	118.15
72" pipe	"	132.00	23.00	155.00
Flared end-section, concrete				
12" pipe	L.F.	39.30	6.40	45.70
15" pipe	"	44.90	6.75	51.65
18" pipe	"	50.50	7.20	57.70
24" pipe	"	61.50	8.20	69.70
30" pipe	"	75.00	9.55	84.55
36" pipe	"	103.00	10.45	113.45
42" pipe	"	112.00	11.50	123.50
48" pipe	"	123.00	12.75	135.75
54" pipe	"	135.00	14.35	149.35
Corrugated metal pipe, coated, paved invert				
16 ga.				
8" pipe	L.F.	5.45	3.85	9.30
10" pipe	"	7.60	3.95	11.55
12" pipe	"	8.70	4.10	12.80

02 SITEWORK

SEWERAGE AND DRAINAGE

02720.40 STORM DRAINAGE

SEWERAGE AND DRAINAGE	UNIT	MAT.	INST.	TOTAL
15" pipe	L.F.	10.75	4.40	15.15
18" pipe	"	13.05	4.80	17.85
21" pipe	"	15.40	5.20	20.60
24" pipe	"	18.15	5.75	23.90
30" pipe	"	25.90	6.40	32.30
36" pipe	"	33.00	7.20	40.20
12 ga., 48" pipe	"	53.50	8.20	61.70
10 ga.				
60" pipe	L.F.	77.00	9.55	86.55
72" pipe	"	102.00	11.50	113.50
Galvanized or aluminum, plain				
16 ga.				
8" pipe	L.F.	4.95	3.85	8.80
10" pipe	"	6.60	3.95	10.55
12" pipe	"	7.70	4.10	11.80
15" pipe	"	10.10	4.40	14.50
18" pipe	"	12.10	4.80	16.90
24" pipe	"	17.05	5.75	22.80
30" pipe	"	24.20	6.40	30.60
36" pipe	"	30.80	7.20	38.00
12 ga., 48" pipe	"	50.50	8.20	58.70
10 ga., 60" pipe	"	71.50	9.55	81.05
Galvanized or aluminum, coated oval arch				
16 ga.				
17" x 13"	L.F.	14.95	5.20	20.15
21" x 15"	"	18.15	5.75	23.90
14 ga.				
28" x 20"	L.F.	28.30	6.40	34.70
35" x 24"	"	36.30	8.20	44.50
12 ga.				
42" x 29"	L.F.	49.50	9.55	59.05
57" x 38"	"	66.00	11.50	77.50
64" x 43"	"	77.00	12.10	89.10
Oval arch culverts, plain				
16 ga.				60.50
17" x 13"	L.F.	88.00	5.20	93.20
21" x 15"	"	14.45	5.75	20.20
14 ga.				
28" x 20"	L.F.	23.55	6.40	29.95
35" x 24"	"	29.50	8.20	37.70
12 ga.				
57" x 38"	L.F.	55.00	9.55	64.55
64" x 43"	"	60.50	11.50	72.00
71" x 47"	"	88.00	12.10	100.10
Nestable corrugated metal pipe				
16 ga.				
10" pipe	L.F.	7.50	3.95	11.45
12" pipe	"	9.95	4.10	14.05
15" pipe	"	12.45	4.40	16.85
18" pipe	"	14.75	4.80	19.55
24" pipe	"	20.40	5.75	26.15
30" pipe	"	23.80	6.40	30.20

02 SITEWORK

SEWERAGE AND DRAINAGE

SEWERAGE AND DRAINAGE	UNIT	MAT.	INST.	TOTAL
02720.40 STORM DRAINAGE				
14 ga., 36" pipe	L.F.	28.30	7.20	35.50
Headwalls, cast in place, 30 deg wingwall				
12" pipe	EA.	231.00	68.00	299.00
15" pipe	"	275.00	68.00	343.00
18" pipe	"	346.00	77.50	423.50
24" pipe	"	517.00	77.50	594.50
30" pipe	"	627.00	90.50	717.50
36" pipe	"	688.00	136.00	824.00
42" pipe	"	858.00	136.00	994.00
48" pipe	"	913.00	181.00	1,094
54" pipe	"	1,030	226.00	1,256
60" pipe	"	1,170	271.00	1,441
4" cleanout for storm drain				
4" pipe	EA.	302.00	26.50	328.50
6" pipe	"	368.00	26.50	394.50
8" pipe	"	517.00	26.50	543.50
Connect new drain line				
To existing manhole	EA.	58.50	70.50	129.00
To new manhole	"	47.30	42.40	89.70
02730.10 SANITARY SEWERS				
Clay				
6" pipe	L.F.	3.65	3.85	7.50
8" pipe	"	4.55	4.10	8.65
10" pipe	"	6.50	4.40	10.90
12" pipe	"	8.60	4.80	13.40
PVC				
4" pipe	L.F.	0.83	2.85	3.68
6" pipe	"	1.70	3.00	4.70
8" pipe	"	2.30	3.20	5.50
10" pipe	"	3.75	3.40	7.15
12" pipe	"	5.60	3.60	9.20
Cleanout				
4" pipe	EA.	4.20	26.50	30.70
6" pipe	"	8.90	26.50	35.40
8" pipe	"	27.20	26.50	53.70
Connect new sewer line				
To existing manhole	EA.	49.50	70.50	120.00
To new manhole	"	33.00	42.40	75.40
02740.10 DRAINAGE FIELDS				
Perforated PVC pipe, for drain field				
4" pipe	L.F.	0.83	2.55	3.38
6" pipe	"	1.55	2.75	4.30
02740.50 SEPTIC TANKS				
Septic tank, precast concrete				
1000 gals	EA.	456.00	191.00	647.00
2000 gals	"	936.00	287.00	1,223
5000 gals	"	3,850	574.00	4,424

SEWERAGE AND DRAINAGE

02740.50 SEPTIC TANKS	UNIT	MAT.	INST.	TOTAL
25,000 gals	EA.	15,400	2,300	17,700
40,000 gals	"	22,000	3,830	25,830
Leaching pit, precast concrete, 72" diameter				
3' deep	EA.	275.00	144.00	419.00
6' deep	"	330.00	164.00	494.00
8' deep	"	440.00	191.00	631.00

02760.10 PIPELINE RESTORATION	UNIT	MAT.	INST.	TOTAL
Relining existing water main				
6" dia.	L.F.	3.75	16.15	19.90
8" dia.	"	4.30	17.00	21.30
10" dia.	"	4.85	17.95	22.80
12" dia.	"	5.35	19.00	24.35
14" dia.	"	5.65	20.20	25.85
16" dia.	"	5.95	21.55	27.50
18" dia.	"	6.25	23.10	29.35
20" dia.	"	7.15	24.85	32.00
24" dia.	"	7.70	26.90	34.60
36" dia.	"	8.25	32.30	40.55
48" dia.	"	9.35	35.90	45.25
72" dia.	"	11.90	40.40	52.30
Replacing in line gate valves				
6" valve	EA.	402.00	215.00	617.00
8" valve	"	660.00	269.00	929.00
10" valve	"	991.00	323.00	1,314
12" valve	"	1,750	404.00	2,154
16" valve	"	3,960	462.00	4,422
18" valve	"	6,050	539.00	6,589
20" valve	"	8,250	646.00	8,896
24" valve	"	11,550	808.00	12,358
36" valve	"	31,900	1,080	32,980

POWER & COMMUNICATIONS

02780.20 HIGH VOLTAGE CABLE	UNIT	MAT.	INST.	TOTAL
High voltage XLP copper cable, shielded, 5000v				
#6 awg	L.F.	0.94	0.45	1.39
#4 awg	"	1.10	0.56	1.66
#2 awg	"	1.40	0.66	2.06
#1 awg	"	1.50	0.73	2.23
#1/0 awg	"	1.75	0.83	2.58
#2/0 awg	"	2.95	1.00	3.95
#3/0 awg	"	3.00	1.20	4.20

POWER & COMMUNICATIONS	UNIT	MAT.	INST.	TOTAL
02780.20 HIGH VOLTAGE CABLE				
#4/0 awg	L.F.	3.15	1.25	4.40
#250 awg	"	3.65	1.50	5.15
#300 awg	"	4.20	1.70	5.90
#350 awg	"	4.70	1.85	6.55
#500 awg	"	6.75	2.50	9.25
#750 awg	"	9.95	2.80	12.75
Ungrounded, 15,000v				
#1 awg	L.F.	2.25	1.05	3.30
#1/0 awg	"	2.65	1.20	3.85
#2/0 awg	"	3.05	1.25	4.30
#3/0 awg	"	3.50	1.40	4.90
#4/0 awg	"	3.90	1.60	5.50
#250 awg	"	4.35	1.70	6.05
#300 awg	"	4.95	1.85	6.80
#350 awg	"	5.50	2.15	7.65
#500 awg	"	7.20	2.80	10.00
#750 awg	"	10.70	3.40	14.10
#1000 awg	"	15.75	4.25	20.00
Aluminum cable, shielded, 5000v				
#6 awg	L.F.	1.15	0.38	1.53
#4 awg	"	1.30	0.45	1.75
#2 awg	"	1.40	0.52	1.92
#1 awg	"	1.55	0.59	2.14
#1/0 awg	"	1.70	0.66	2.36
#2/0 awg	"	1.85	0.69	2.54
#3/0 awg	"	2.15	0.73	2.88
#4/0 awg	"	2.35	0.83	3.18
#250 awg	"	2.55	0.90	3.45
#300 awg	"	2.75	1.05	3.80
#350 awg	"	2.95	1.20	4.15
#500 awg	"	3.60	1.25	4.85
#750 awg	"	4.65	1.55	6.20
#1000 awg	"	5.20	1.75	6.95
Ungrounded, 15,000v				
#1 awg	L.F.	2.10	0.73	2.83
#1/0 awg	"	2.25	0.87	3.12
#2/0 awg	"	2.40	0.94	3.34
#3/0 awg	"	2.60	0.97	3.57
#4/0 awg	"	2.85	1.00	3.85
#250 awg	"	3.05	1.05	4.10
#300 awg	"	3.30	1.10	4.40
#350 awg	"	3.55	1.25	4.80
#500 awg	"	4.15	1.50	5.65
#750 awg	"	5.45	1.80	7.25
#1000 awg	"	7.20	2.20	9.40
Indoor terminations, 5000v				
#6 - #4	EA.	33.00	5.45	38.45
#2 - #2/0	"	38.50	5.45	43.95
#3/0 - #250	"	49.50	5.45	54.95
#300 - #750	"	55.00	95.50	150.50
#1000	"	66.00	132.00	198.00
In-line splice, 5000v				

POWER & COMMUNICATIONS

02780.20 HIGH VOLTAGE CABLE	UNIT	MAT.	INST.	TOTAL
#6 - #4/0	EA.	71.50	132.00	203.50
#250 - #500	"	77.00	347.00	424.00
#750 - #1000	"	108.00	451.00	559.00
T-splice, 5000v				
#2 - #4/0	EA.	71.50	416.00	487.50
#250 - #500	"	77.00	694.00	771.00
#750 - #1000	"	108.00	868.00	976.00
Indoor terminations, 15,000v				
#2 - #2/0	EA.	44.00	121.00	165.00
#3/0 - #500	"	55.00	185.00	240.00
#750 - #1000	"	66.00	214.00	280.00
In-line splice, 15,000v				
#2 - #4/0	EA.	77.00	312.00	389.00
#250 - #500	"	108.00	416.00	524.00
#750 - #1000	"	132.00	625.00	757.00
T-splice, 15,000v				
#4	EA.	77.00	625.00	702.00
#250 - #500	"	104.00	1,040	1,144
#750 - #1000	"	132.00	1,560	1,692
Compression lugs, 15,000v				
#4	EA.	5.20	13.90	19.10
#2	"	6.00	18.50	24.50
#1	"	6.55	18.50	25.05
#1/0	"	10.00	23.15	33.15
#2/0	"	10.50	23.15	33.65
#3/0	"	11.80	29.50	41.30
#4/0	"	13.00	29.50	42.50
#250	"	15.20	33.00	48.20
#300	"	17.60	33.00	50.60
#350	"	18.60	40.20	58.80
#500	"	27.50	43.40	70.90
#750	"	43.70	52.50	96.20
#1000	"	64.00	66.00	130.00
Compression splices, 15,000v				
#4	EA.	5.85	23.15	29.00
#2	"	6.40	25.20	31.60
#1	"	7.30	31.20	38.50
#1/0	"	7.75	34.70	42.45
#2/0	"	8.35	40.20	48.55
#3/0	"	9.05	43.40	52.45
#4/0	"	9.90	48.70	58.60
#250	"	10.85	52.50	63.35
#350	"	12.35	60.50	72.85
#500	"	18.25	69.50	87.75
#750	"	30.00	87.00	117.00

02780.40 SUPPORTS & CONNECTORS	UNIT	MAT.	INST.	TOTAL
Cable supports for conduit				
1-1/2"	EA.	36.80	12.05	48.85
2"	"	51.00	12.05	63.05
2-1/2"	"	57.00	13.90	70.90

02 SITEWORK

POWER & COMMUNICATIONS

02780.40 SUPPORTS & CONNECTORS	UNIT	MAT.	INST.	TOTAL
3"	EA.	73.50	13.90	87.40
3-1/2"	"	96.50	17.35	113.85
4"	"	119.00	17.35	136.35
5"	"	213.00	23.15	236.15
6"	"	445.00	25.20	470.20
Split bolt connectors				
#10	EA.	1.15	6.95	8.10
#8	"	1.30	6.95	8.25
#6	"	1.45	6.95	8.40
#4	"	1.70	13.90	15.60
#3	"	2.45	13.90	16.35
#2	"	2.75	13.90	16.65
#1/0	"	3.60	23.15	26.75
#2/0	"	5.70	23.15	28.85
#3/0	"	8.70	23.15	31.85
#4/0	"	9.90	23.15	33.05
#250	"	10.20	34.70	44.90
#350	"	18.05	34.70	52.75
#500	"	23.65	34.70	58.35
#750	"	40.00	52.50	92.50
#1000	"	55.00	52.50	107.50
Single barrel lugs				
#6	EA.	0.41	8.70	9.11
#1/0	"	0.78	17.35	18.13
#250	"	1.90	23.15	25.05
#350	"	2.45	23.15	25.60
#500	"	4.75	23.15	27.90
#600	"	5.05	31.20	36.25
#800	"	5.70	31.20	36.90
#1000	"	6.85	31.20	38.05
Double barrel lugs				
#1/0	EA.	1.55	31.20	32.75
#250	"	4.50	44.80	49.30
#350	"	6.40	44.80	51.20
#600	"	9.70	66.00	75.70
#800	"	11.10	66.00	77.10
#1000	"	11.45	66.00	77.45
Three barrel lugs				
#2/0	EA.	12.40	44.80	57.20
#250	"	23.75	66.00	89.75
#350	"	38.90	66.00	104.90
#600	"	42.60	92.50	135.10
#800	"	68.50	92.50	161.00
#1000	"	97.00	92.50	189.50
Four barrel lugs				
#250	EA.	26.40	95.50	121.90
#350	"	44.00	95.50	139.50
#600	"	48.00	121.00	169.00
#800	"	75.50	121.00	196.50
Compression conductor adapters				
#6	EA.	3.45	10.30	13.75
#4	"	3.60	12.05	15.65

02 SITEWORK

POWER & COMMUNICATIONS

02780.40 SUPPORTS & CONNECTORS	UNIT	MAT.	INST.	TOTAL
#2	EA.	3.75	15.40	19.15
#1	"	4.35	15.40	19.75
#1/0	"	4.50	18.50	23.00
#250	"	8.60	27.80	36.40
#350	"	10.25	29.50	39.75
#500	"	13.25	38.00	51.25
#750	"	18.15	39.70	57.85
Terminal blocks, 2 screw				
3 circuit	EA.	9.85	6.95	16.80
6 circuit	"	13.55	6.95	20.50
8 circuit	"	15.85	6.95	22.80
10 circuit	"	18.40	10.30	28.70
12 circuit	"	20.70	10.30	31.00
18 circuit	"	28.00	10.30	38.30
24 circuit	"	35.10	12.05	47.15
36 circuit	"	49.30	12.05	61.35
Compression splice				
#8 awg	EA.	1.70	13.20	14.90
#6 awg	"	2.00	9.55	11.55
#4 awg	"	2.10	9.55	11.65
#2 awg	"	3.25	18.50	21.75
#1 awg	"	4.60	18.50	23.10
#1/0 awg	"	5.60	18.50	24.10
#2/0 awg	"	5.95	29.50	35.45
#3/0 awg	"	6.95	29.50	36.45
#4/0 awg	"	7.40	29.50	36.90
#250 awg	"	7.85	47.10	54.95
#300 awg	"	8.55	47.10	55.65
#350 awg	"	8.70	48.70	57.40
#400 awg	"	11.80	48.70	60.50
#500 awg	"	13.80	52.50	66.30
#600 awg	"	20.90	52.50	73.40
#750 awg	"	22.15	60.50	82.65
#1000 awg	"	29.10	60.50	89.60

SITE IMPROVEMENTS

02810.40 LAWN IRRIGATION	UNIT	MAT.	INST.	TOTAL
Residential system, complete				
Minimum	ACRE			11,000
Maximum	"			16,500
Commercial system, complete				
Minimum	ACRE			13,200
Maximum	"			19,800

02 SITEWORK

SITE IMPROVEMENTS

02810.40 LAWN IRRIGATION	UNIT	MAT.	INST.	TOTAL
Components				
Pop-up head	EA.	8.80	13.25	22.05
Impact head	"	8.80	13.25	22.05
Rotary impact head	"	24.20	26.50	50.70
Shrub head	"	2.30	13.25	15.55
Hose bibb	"	4.30	26.50	30.80

03 CONCRETE

FORMWORK

03110.20 ELEVATED SLAB FORMWORK	UNIT	MAT.	INST.	TOTAL
Elevated slab formwork				
Slab, with drop panels				
1 use	S.F.	2.05	2.15	4.20
5 uses	"	0.74	1.85	2.59
Floor slab, hung from steel beams				
1 use	S.F.	1.90	2.10	4.00
5 uses	"	0.68	1.80	2.48
Floor slab, with pans or domes				
1 use	S.F.	2.75	2.45	5.20
5 uses	"	1.20	2.10	3.30
Equipment curbs, 12" high				
1 use	L.F.	1.70	2.70	4.40
5 uses	"	0.71	2.25	2.96

03110.25 EQUIPMENT PAD FORMWORK	UNIT	MAT.	INST.	TOTAL
Equipment pad, job built				
1 use	S.F.	2.10	3.40	5.50
2 uses	"	1.25	3.20	4.45
3 uses	"	0.99	3.00	3.99
4 uses	"	0.77	2.85	3.62
5 uses	"	0.66	2.70	3.36

03110.35 FOOTING FORMWORK	UNIT	MAT.	INST.	TOTAL
Wall footings, job built, continuous				
1 use	S.F.	0.88	2.70	3.58
5 uses	"	0.38	2.25	2.63
Column footings, spread				
1 use	S.F.	1.05	3.40	4.45
5 uses	"	0.42	2.70	3.12

03110.55 SLAB/MAT FORMWORK	UNIT	MAT.	INST.	TOTAL
Edge forms				
6" high				
1 use	L.F.	1.55	2.45	4.00
5 uses	"	0.44	2.10	2.54
12" high				
1 use	L.F.	1.45	2.70	4.15
5 uses	"	0.41	2.25	2.66
Formwork for openings				
1 use	S.F.	2.10	5.40	7.50
5 uses	"	0.66	3.85	4.51

03110.65 WALL FORMWORK	UNIT	MAT.	INST.	TOTAL
Wall forms, exterior, job built				
Up to 8' high wall				
1 use	S.F.	1.75	2.70	4.45
5 uses	"	0.63	2.25	2.88
Over 8' high wall				
1 use	S.F.	1.90	3.40	5.30
5 uses	"	0.78	2.70	3.48

03 CONCRETE

FORMWORK	UNIT	MAT.	INST.	TOTAL
03110.65 WALL FORMWORK				
Column pier and pilaster				
1 use	S.F.	2.00	5.40	7.40
5 uses	"	0.89	3.85	4.74
Interior wall forms				
Up to 8' high				
1 use	S.F.	1.75	2.45	4.20
5 uses	"	0.63	2.10	2.73
Over 8' high				
1 use	S.F.	1.90	3.00	4.90
5 uses	"	0.78	2.45	3.23

REINFORCEMENT	UNIT	MAT.	INST.	TOTAL
03210.20 ELEVATED SLAB REINFORCING				
Elevated slab				
#3 - #4	TON	638.00	384.00	1,022
#5 - #6	"	578.00	341.00	919.00
Galvanized				
#3 - #4	TON	1,130	384.00	1,514
#5 - #6	"	1,070	341.00	1,411
03210.25 EQUIP. PAD REINFORCING				
Equipment pad				
#3 - #4	TON	638.00	614.00	1,252
#5 - #6	"	578.00	559.00	1,137
#7 - #8	"	550.00	512.00	1,062
#9 - #10	"	550.00	473.00	1,023
#11 - #12	"	550.00	439.00	989.00
03210.35 FOOTING REINFORCING				
Footings				
Grade 50				
#3 - #4	TON	638.00	512.00	1,150
#5 - #6	"	578.00	439.00	1,017
Grade 60				
#3 - #4	TON	660.00	512.00	1,172
#5 - #6	"	600.00	439.00	1,039
03210.55 SLAB/MAT REINFORCING				
Bars, slabs				
#3 - #4	TON	638.00	512.00	1,150
#5 - #6	"	578.00	439.00	1,017
#7 - #8	"	550.00	384.00	934.00

03 CONCRETE

REINFORCEMENT	UNIT	MAT.	INST.	TOTAL
03210.55 SLAB/MAT REINFORCING				
Galvanized				
#3 - #4	TON	1,130	512.00	1,642
#5 - #6	"	1,070	439.00	1,509
03210.65 WALL REINFORCING				
Walls				
#3 - #4	TON	638.00	439.00	1,077
#5 - #6	"	578.00	384.00	962.00
Galvanized				
#3 - #4	TON	1,130	439.00	1,569
#5 - #6	"	1,070	384.00	1,454
Masonry wall (horizontal)				
#3 - #4	TON	638.00	1,230	1,868
#5 - #6	"	578.00	1,020	1,598
Galvanized				
#3 - #4	TON	1,130	1,230	2,360
#5 - #6	"	1,070	1,020	2,090
Masonry wall (vertical)				
#3 - #4	TON	638.00	1,540	2,178
#5 - #6	"	578.00	1,230	1,808
Galvanized				
#3 - #4	TON	1,130	1,540	2,670
#5 - #6	"	1,070	1,230	2,300

PLACING CONCRETE	UNIT	MAT.	INST.	TOTAL
03380.20 ELEVATED SLAB CONCRETE				
Elevated slab				
2500# or 3000# concrete				
By crane	C.Y.	56.50	20.00	76.50
By pump	"	56.50	15.40	71.90
By hand buggy	"	56.50	21.20	77.70
5000# concrete				
By crane	C.Y.	65.00	20.00	85.00
By pump	"	65.00	15.40	80.40
By hand buggy	"	65.00	21.20	86.20
03380.25 EQUIPMENT PAD CONCRETE				
Equipment pad				
2500# or 3000# concrete				
By chute	C.Y.	56.50	7.05	63.55
By pump	"	56.50	28.60	85.10

03 CONCRETE

PLACING CONCRETE

	UNIT	MAT.	INST.	TOTAL
03380.25 EQUIPMENT PAD CONCRETE				
By crane	C.Y.	56.50	33.40	89.90
3500# or 4000# concrete				
By chute	C.Y.	60.50	7.05	67.55
By pump	"	60.50	28.60	89.10
By crane	"	60.50	33.40	93.90
5000# concrete				
By chute	C.Y.	65.00	7.05	72.05
By pump	"	65.00	28.60	93.60
By crane	"	65.00	33.40	98.40
03380.35 FOOTING CONCRETE				
Continuous footing				
2500# or 3000# concrete				
By chute	C.Y.	56.50	7.05	63.55
By pump	"	56.50	25.00	81.50
By crane	"	56.50	28.60	85.10
5000# concrete				
By chute	C.Y.	65.00	7.05	72.05
By pump	"	65.00	25.00	90.00
By crane	"	65.00	28.60	93.60
Spread footing				
2500# or 3000# concrete				
Under 5 cy				
By chute	C.Y.	56.50	7.05	63.55
By pump	"	56.50	26.70	83.20
By crane	"	56.50	30.80	87.30
5000# concrete				
Under 5 c.y.				
By chute	C.Y.	65.00	7.05	72.05
By pump	"	65.00	26.70	91.70
By crane	"	65.00	30.80	95.80
03380.55 SLAB/MAT CONCRETE				
Slab on grade				
2500# or 3000# concrete				
By chute	C.Y.	56.50	5.30	61.80
By crane	"	56.50	16.70	73.20
By pump	"	56.50	14.30	70.80
By hand buggy	"	56.50	14.15	70.65
5000# concrete				
By chute	C.Y.	65.00	5.30	70.30
By crane	"	65.00	16.70	81.70
By pump	"	65.00	14.30	79.30
By hand buggy	"	65.00	14.15	79.15
03380.65 WALL CONCRETE				
Walls				
2500# or 3000# concrete				
To 4'				
By chute	C.Y.	56.50	6.05	62.55

03 CONCRETE

PLACING CONCRETE	UNIT	MAT.	INST.	TOTAL
03380.65 WALL CONCRETE				
By crane	C.Y.	56.50	33.40	89.90
By pump	"	56.50	30.80	87.30
To 8'				
By crane	C.Y.	56.50	36.40	92.90
By pump	"	56.50	33.40	89.90
Filled block (CMU)				
3000# concrete, by pump				
4" wide	S.F.	0.21	1.45	1.66
6" wide	"	0.47	1.65	2.12
8" wide	"	0.74	2.00	2.74
10" wide	"	0.99	2.35	3.34
12" wide	"	1.25	2.85	4.10
Pilasters, 3000# concrete	C.F.	2.85	40.00	42.85
Wall cavity, 2" thick, 3000# concrete	S.F.	0.53	1.35	1.88

04 MASONRY

MORTAR AND GROUT

04100.10 MASONRY GROUT	UNIT	MAT.	INST.	TOTAL
Grout, non shrink, non-metallic, trowelable	C.F.	6.75	0.77	7.52
Grout-filled individual cmu cells				
4" wide	L.F.	0.13	0.57	0.70
8" wide	"	0.40	0.57	0.97

04150.50 MASONRY FLASHING	UNIT	MAT.	INST.	TOTAL
Through-wall flashing				
5 oz. coated copper	S.F.	2.30	2.20	4.50
0.030" elastomeric	"	0.50	1.75	2.25

UNIT MASONRY

04210.10 BRICK MASONRY	UNIT	MAT.	INST.	TOTAL
Standard size brick, running bond				
Face brick, red (6.4/sf)				
Veneer	S.F.	1.85	4.40	6.25
Cavity wall	"	1.85	3.75	5.60
9" solid wall	"	4.05	7.50	11.55
Chimney, standard brick, including flue				
16" x 16"	L.F.	13.40	26.30	39.70
16" x 20"	"	16.35	26.30	42.65
16" x 24"	"	19.00	26.30	45.30
20" x 20"	"	19.15	32.90	52.05
20" x 24"	"	27.50	32.90	60.40
20" x 32"	"	34.70	37.60	72.30

04220.10 CONCRETE MASONRY UNITS	UNIT	MAT.	INST.	TOTAL
Hollow, load bearing				
4"	S.F.	1.05	1.95	3.00
8"	"	1.85	2.20	4.05
Solid, load bearing				
4"	S.F.	1.55	1.95	3.50
8"	"	2.10	2.20	4.30
Back-up block, 8" x 16"				
4"	S.F.	1.10	1.55	2.65
8"	"	1.60	1.75	3.35
Steel angles and plates				
Minimum	Lb.	0.28	0.38	0.66
Maximum	"	0.31	0.66	0.97
Various size angle lintels				
1/4" stock				
3" x 3"	L.F.	1.50	1.65	3.15
3" x 3-1/2"	"	1.60	1.65	3.25

04 MASONRY

UNIT MASONRY

UNIT MASONRY	UNIT	MAT.	INST.	TOTAL
04220.10 CONCRETE MASONRY UNITS				
3/8" stock				
3" x 4"	L.F.	2.65	1.65	4.30
4" x 4"	"	2.90	1.65	4.55
1/2" stock				
6" x 4"	L.F.	4.85	1.65	6.50
04295.10 PARGING/MASONRY PLASTER				
Parging				
1/2" thick	S.F.	0.33	1.75	2.08
3/4" thick	"	0.44	2.20	2.64

MASONRY RESTORATION

MASONRY RESTORATION	UNIT	MAT.	INST.	TOTAL
04520.10 RESTORATION AND CLEANING				
Masonry cleaning				
Washing brick				
Smooth surface	S.F.	0.24	0.44	0.68
Rough surface	"	0.33	0.58	0.91
Pointing masonry				
Brick	S.F.	0.77	1.05	1.82
Concrete block	"	0.33	0.75	1.08
Brick removal and replacement				
Minimum	EA.	0.33	3.30	3.63
Average	"	0.44	4.40	4.84
Maximum	"	0.88	13.15	14.03
04550.10 REFRACTORIES				
Flue liners				
Rectangular				
8" x 12"	L.F.	4.20	4.40	8.60
12" x 12"	"	6.05	4.80	10.85
12" x 18"	"	9.35	5.25	14.60
16" x 16"	"	10.75	5.85	16.60
18" x 18"	"	13.40	6.25	19.65
20" x 20"	"	21.45	6.60	28.05
24" x 24"	"	28.60	7.50	36.10
Round				
18" dia.	L.F.	26.40	6.25	32.65
24" dia.	"	45.10	7.50	52.60

05 METALS

METAL FASTENING

05050.10 STRUCTURAL WELDING	UNIT	MAT.	INST.	TOTAL
Welding				
Single pass				
1/8"	L.F.	0.34	1.50	1.84
3/16"	"	0.68	2.00	2.68
1/4"	"	0.91	2.45	3.36
Miscellaneous steel shapes				
Plain	Lb.	0.68	0.06	0.74
Galvanized	"	1.05	0.10	1.15
Plates				
Plain	Lb.	0.63	0.07	0.70
Galvanized	"	0.94	0.12	1.06

05050.90 METAL ANCHORS	UNIT	MAT.	INST.	TOTAL
Anchor bolts				
3/8" x				
8" long	EA.			2.75
10" long	"			2.95
12" long	"			3.20
1/2" x				
8" long	EA.			3.85
10" long	"			3.95
12" long	"			4.30
18" long	"			4.60
Add 25% for galvanized anchor bolts				

05050.95 METAL LINTELS	UNIT	MAT.	INST.	TOTAL
Lintels, steel				
Plain	Lb.	0.68	0.74	1.42
Galvanized	"	0.99	0.74	1.73

COLD FORMED FRAMING

05410.10 METAL FRAMING	UNIT	MAT.	INST.	TOTAL
Furring channel, galvanized				
Beams and columns, 3/4"				
12" o.c.	S.F.	0.33	2.95	3.28
16" o.c.	"	0.28	2.70	2.98
Walls, 3/4"				
12" o.c.	S.F.	0.33	1.50	1.83
16" o.c.	"	0.28	1.25	1.53
Stud, load bearing				

COLD FORMED FRAMING

05410.10 METAL FRAMING

COLD FORMED FRAMING	UNIT	MAT.	INST.	TOTAL
16" o.c.				
16 ga.				
2-1/2"	S.F.	1.05	1.30	2.35
3-5/8"	"	1.25	1.30	2.55
24" o.c.				
16 ga.				
2-1/2"	S.F.	0.77	1.15	1.92
3-5/8"	"	0.99	1.15	2.14

06 WOOD AND PLASTICS

FASTENERS AND ADHESIVES

06050.10 ACCESSORIES

FASTENERS AND ADHESIVES	UNIT	MAT.	INST.	TOTAL
Anchors				
Bolts, threaded two ends, with nuts and washers				
1/2" dia.				
4" long	EA.	0.83	1.70	2.53
7-1/2" long	"	1.05	1.70	2.75
3/4" dia.				
7-1/2" long	EA.	1.65	1.70	3.35
15" long	"	2.15	1.70	3.85
Bolts, carriage				
1/4 x 4	EA.	0.87	2.70	3.57
5/16 x 6	"	0.92	2.85	3.77
3/8 x 6	"	1.05	2.85	3.90
1/2 x 6	"	1.20	2.85	4.05
Joist and beam hangers				
18 ga.				
2 x 4	EA.	0.41	2.70	3.11
2 x 6	"	0.46	2.70	3.16
2 x 8	"	0.55	2.70	3.25
2 x 10	"	0.66	3.00	3.66
2 x 12	"	0.71	3.40	4.11
Rafter anchors, 18 ga., 1-1/2" wide				
Strap ties, 14 ga., 1-3/8" wide				
12" long	EA.	0.88	2.25	3.13
18" long	"	1.05	2.45	3.50
24" long	"	1.25	2.70	3.95
36" long	"	1.80	3.00	4.80

ROUGH CARPENTRY

06110.10 BLOCKING

ROUGH CARPENTRY	UNIT	MAT.	INST.	TOTAL
Steel construction				
Walls				
2x4	L.F.	0.29	1.80	2.09
2x6	"	0.42	2.10	2.52
2x8	"	0.58	2.25	2.83
2x10	"	0.94	2.45	3.39
2x12	"	1.20	2.70	3.90
Ceilings				
2x4	L.F.	0.29	2.10	2.39
2x6	"	0.42	2.45	2.87
2x8	"	0.58	2.70	3.28
2x10	"	0.94	3.00	3.94
2x12	"	1.20	3.40	4.60
Wood construction				

ROUGH CARPENTRY

ROUGH CARPENTRY	UNIT	MAT.	INST.	TOTAL
06110.10 BLOCKING				
Walls				
2x4	L.F.	0.29	1.50	1.79
2x6	"	0.42	1.70	2.12
2x8	"	0.58	1.80	2.38
2x10	"	0.94	1.95	2.89
2x12	"	1.20	2.10	3.30
Ceilings				
2x4	L.F.	0.29	1.70	1.99
2x6	"	0.42	1.95	2.37
2x8	"	0.58	2.10	2.68
2x10	"	0.94	2.25	3.19
2x12	"	1.20	2.45	3.65
06110.20 CEILING FRAMING				
Ceiling joists				
12" o.c.				
2x4	S.F.	0.34	0.65	0.99
2x8	"	0.69	0.71	1.40
16" o.c.				
2x4	S.F.	0.29	0.52	0.81
2x8	"	0.58	0.56	1.14
Sister joists for ceilings				
2x4	L.F.	0.29	1.95	2.24
2x6	"	0.42	2.25	2.67
2x8	"	0.58	2.70	3.28
2x10	"	0.94	3.40	4.34
2x12	"	1.20	4.50	5.70
06110.30 FLOOR FRAMING				
Floor joists				
12" o.c.				
2x6	S.F.	0.50	0.54	1.04
2x8	"	0.69	0.55	1.24
2x10	"	1.10	0.56	1.66
2x12	"	1.45	0.59	2.04
16" o.c.				
2x6	S.F.	0.42	0.45	0.87
2x8	"	0.58	0.46	1.04
2x10	"	0.94	0.47	1.41
2x12	"	1.20	0.48	1.68
Sister joists for floors				
2x4	L.F.	0.29	1.70	1.99
2x6	"	0.38	1.95	2.33
2x8	"	0.58	2.25	2.83
2x10	"	0.94	2.70	3.64
2x12	"	1.20	3.40	4.60
3x6	"	0.99	2.70	3.69
3x8	"	1.30	3.00	4.30
3x10	"	1.65	3.40	5.05
3x12	"	2.00	3.85	5.85
4x6	"	1.60	2.70	4.30

06 WOOD AND PLASTICS

ROUGH CARPENTRY	UNIT	MAT.	INST.	TOTAL
06110.30 FLOOR FRAMING				
4x8	L.F.	1.75	3.00	4.75
4x10	"	2.20	3.40	5.60
4x12	"	2.65	3.85	6.50
06110.40 FURRING				
Furring, wood strips				
Walls				
On masonry or concrete walls				
1x2 furring				
12" o.c.	S.F.	0.14	0.85	0.99
16" o.c.	"	0.12	0.77	0.89
1x3 furring				
12" o.c.	S.F.	0.19	0.85	1.04
16" o.c.	"	0.17	0.77	0.94
On wood walls				
1x2 furring				
12" o.c.	S.F.	0.14	0.60	0.74
16" o.c.	"	0.12	0.54	0.66
1x3 furring				
12" o.c.	S.F.	0.19	0.60	0.79
16" o.c.	"	0.17	0.54	0.71
Ceilings				
On masonry or concrete ceilings				
1x2 furring				
12" o.c.	S.F.	0.14	1.50	1.64
16" o.c.	"	0.12	1.35	1.47
1x3 furring				
12" o.c.	S.F.	0.19	1.50	1.69
16" o.c.	"	0.17	1.35	1.52
On wood ceilings				
1x2 furring				
12" o.c.	S.F.	0.14	1.00	1.14
16" o.c.	"	0.12	0.90	1.02
1x3				
12" o.c.	S.F.	0.19	1.00	1.19
16" o.c.	"	0.17	0.90	1.07
06110.50 ROOF FRAMING				
Roof framing				
Rafters, gable end				
0-2 pitch (flat to 2-in-12)				
12" o.c.				
2x4	S.F.	0.34	0.56	0.90
2x6	"	0.50	0.59	1.09
2x8	"	0.69	0.62	1.31
16" o.c.				
2x6	S.F.	0.42	0.48	0.90
2x8	"	0.58	0.50	1.08
Sister rafters				
2x4	L.F.	0.29	1.95	2.24

06 WOOD AND PLASTICS

ROUGH CARPENTRY

	UNIT	MAT.	INST.	TOTAL
06110.50 ROOF FRAMING				
2x6	L.F.	0.42	2.25	2.67
2x8	"	0.58	2.70	3.28
2x10	"	0.94	3.40	4.34
2x12	"	1.20	4.50	5.70
06110.60 SLEEPERS				
Sleepers, over concrete				
12" o.c.				
1x2	S.F.	0.15	0.62	0.77
1x3	"	0.20	0.65	0.85
2x4	"	0.34	0.75	1.09
2x6	"	0.50	0.80	1.30
16" o.c.				
1x2	S.F.	0.13	0.54	0.67
1x3	"	0.18	0.54	0.72
2x4	"	0.29	0.65	0.94
2x6	"	0.42	0.68	1.10
06110.65 SOFFITS				
Soffit framing				
2x3	L.F.	0.22	1.95	2.17
2x4	"	0.29	2.10	2.39
2x6	"	0.42	2.25	2.67
2x8	"	0.58	2.45	3.03
06110.70 WALL FRAMING				
Framing wall, studs				
12" o.c.				
2x3	S.F.	0.25	0.50	0.75
2x4	"	0.34	0.50	0.84
16" o.c.				
2x3	S.F.	0.22	0.42	0.64
2x4	"	0.29	0.42	0.71
24" o.c.				
2x3	S.F.	0.19	0.37	0.56
2x4	"	0.23	0.37	0.60
Plates, top or bottom				
2x3	L.F.	0.22	0.80	1.02
2x4	"	0.29	0.85	1.14
Headers, door or window				
2x6				
Single				
3' long	EA.	1.25	13.55	14.80
6' long	"	2.50	16.95	19.45
Double				
3' long	EA.	2.50	15.05	17.55
6' long	"	5.00	19.35	24.35
2x8				
Single				
4' long	EA.	2.30	16.95	19.25

06 WOOD AND PLASTICS

ROUGH CARPENTRY	UNIT	MAT.	INST.	TOTAL
06110.70 WALL FRAMING				
8' long	EA.	4.60	20.85	25.45
Double				
4' long	EA.	4.60	19.35	23.95
8' long	"	9.25	24.65	33.90
2x10				
Single				
5' long	EA.	4.65	20.85	25.50
10' long	"	9.30	27.10	36.40
Double				
5' long	EA.	9.30	22.60	31.90
10' long	"	18.55	27.10	45.65
06115.10 FLOOR SHEATHING				
Sub-flooring, plywood, CDX				
1/2" thick	S.F.	0.44	0.34	0.78
5/8" thick	"	0.53	0.39	0.92
3/4" thick	"	0.63	0.45	1.08
Structural plywood				
1/2" thick	S.F.	0.48	0.34	0.82
Board type subflooring				
1x6				
Minimum	S.F.	0.64	0.60	1.24
Maximum	"	0.77	0.68	1.45
1x8				
Minimum	S.F.	0.75	0.57	1.32
Maximum	"	0.83	0.64	1.47
1x10				
Minimum	S.F.	0.87	0.54	1.41
Maximum	"	0.94	0.60	1.54
Underlayment				
Hardboard, 1/4" tempered	S.F.	0.33	0.34	0.67
Plywood				
3/8" thick	S.F.	0.38	0.34	0.72
1/2" thick	"	0.44	0.36	0.80
5/8" thick	"	0.53	0.39	0.92
3/4" thick	"	0.63	0.42	1.05
06115.20 ROOF SHEATHING				
Sheathing				
3/8" thick	S.F.	0.38	0.35	0.73
1/2" thick	"	0.44	0.36	0.80
5/8" thick	"	0.53	0.39	0.92
06115.30 WALL SHEATHING				
Sheathing				
Plywood, CDX				
3/8" thick	S.F.	0.38	0.40	0.78
1/2" thick	"	0.44	0.42	0.86
5/8" thick	"	0.53	0.45	0.98
3/4" thick	"	0.63	0.49	1.12

06 WOOD AND PLASTICS

ROUGH CARPENTRY

06220.10 MILLWORK

ROUGH CARPENTRY	UNIT	MAT.	INST.	TOTAL
Countertop, laminated plastic				
25" x 7/8" thick				
Minimum	L.F.	10.45	6.80	17.25
Average	"	17.25	9.05	26.30
Maximum	"	27.50	10.85	38.35
25" x 1-1/4" thick				
Minimum	L.F.	11.55	9.05	20.60
Average	"	18.50	10.85	29.35
Maximum	"	31.40	13.55	44.95
Add for cutouts	EA.	0.00	16.95	16.95
Backsplash, 4" high, 7/8" thick	L.F.	9.00	5.40	14.40
Base cabinets, 34-1/2" high, 24" deep, hardwood, no tops				
Minimum	L.F.	38.50	10.85	49.35
Average	"	54.00	13.55	67.55
Maximum	"	88.00	18.10	106.10
Wall cabinets				
Minimum	L.F.	27.50	9.05	36.55
Average	"	41.80	10.85	52.65
Maximum	"	77.00	13.55	90.55

07 THERMAL AND MOISTURE

MOISTURE PROTECTION

	UNIT	MAT.	INST.	TOTAL
07150.10 DAMPPROOFING				
Silicone dampproofing, sprayed on				
Concrete surface				
1 coat	S.F.	0.30	0.12	0.42
2 coats	"	0.52	0.16	0.68
Concrete block				
1 coat	S.F.	0.32	0.14	0.46
2 coats	"	0.54	0.19	0.73
Brick				
1 coat	S.F.	0.35	0.16	0.51
2 coats	"	0.57	0.21	0.78
07160.10 BITUMINOUS DAMPPROOFING				
Building paper, asphalt felt				
15 lb	S.F.	0.07	0.85	0.92
30 lb	"	0.11	0.88	0.99
07190.10 VAPOR BARRIERS				
Vapor barrier, polyethylene				
2 mil	S.F.	0.01	0.11	0.12
6 mil	"	0.02	0.11	0.13
8 mil	"	0.03	0.12	0.15
10 mil	"	0.04	0.12	0.16

INSULATION

	UNIT	MAT.	INST.	TOTAL
07210.10 BATT INSULATION				
Ceiling, fiberglass, unfaced				
3-1/2" thick, R11	S.F.	0.24	0.25	0.49
6" thick, R19	"	0.41	0.28	0.69
9" thick, R30	"	0.63	0.33	0.96
Suspended ceiling, unfaced				
3-1/2" thick, R11	S.F.	0.24	0.24	0.48
6" thick, R19	"	0.41	0.26	0.67
9" thick, R30	"	0.63	0.30	0.93
Crawl space, unfaced				
3-1/2" thick, R11	S.F.	0.24	0.33	0.57
6" thick, R19	"	0.41	0.35	0.76
9" thick, R30	"	0.63	0.39	1.02
Wall, fiberglass				
Paper backed				
2" thick, R7	S.F.	0.17	0.22	0.39
3" thick, R8	"	0.18	0.24	0.42

07 THERMAL AND MOISTURE

INSULATION	UNIT	MAT.	INST.	TOTAL
07210.10 BATT INSULATION				
4" thick, R11	S.F.	0.28	0.25	0.53
6" thick, R19	"	0.42	0.26	0.68
Foil backed, 1 side				
2" thick, R7	S.F.	0.35	0.22	0.57
3" thick, R11	"	0.37	0.24	0.61
4" thick, R14	"	0.42	0.25	0.67
6" thick, R21	"	0.44	0.26	0.70
Foil backed, 2 sides				
2" thick, R7	S.F.	0.42	0.25	0.67
3" thick, R11	"	0.52	0.26	0.78
4" thick, R14	"	0.61	0.28	0.89
6" thick, R21	"	0.65	0.30	0.95
Unfaced				
2" thick, R7	S.F.	0.18	0.22	0.40
3" thick, R9	"	0.20	0.24	0.44
4" thick, R11	"	0.24	0.25	0.49
6" thick, R19	"	0.41	0.26	0.67
Mineral wool batts				
Paper backed				
2" thick, R6	S.F.	0.34	0.22	0.56
4" thick, R12	"	0.42	0.24	0.66
6" thick, R19	"	0.48	0.26	0.74
Fasteners, self adhering, attached to ceiling deck				
2-1/2" long	EA.	0.11	0.35	0.46
4-1/2" long	"	0.13	0.39	0.52
Capped, self-locking washers for fastening insulation	"	0.11	0.21	0.32
07210.20 BOARD INSULATION				
Insulation, rigid				
Fiberglass, roof				
0.75" thick, R2.78	S.F.	0.17	0.19	0.36
1.06" thick, R4.17	"	0.38	0.20	0.58
1.31" thick, R5.26	"	0.55	0.21	0.76
1.63" thick, R6.67	"	0.61	0.22	0.83
2.25" thick, R8.33	"	0.66	0.24	0.90
Perlite board, roof				
1.00" thick, R2.78	S.F.	0.44	0.18	0.62
1.50" thick, R4.17	"	0.68	0.18	0.86
2.00" thick, R5.92	"	0.84	0.19	1.03
2.50" thick, R6.67	"	1.00	0.20	1.20
3.00" thick, R8.33	"	1.30	0.21	1.51
4.00" thick, R10.00	"	1.45	0.22	1.67
5.25" thick, R14.29	"	1.60	0.24	1.84
1" thick, R3.0	"	0.33	0.18	0.51
07210.60 LOOSE FILL INSULATION				
Blown-in type				
Fiberglass				
5" thick, R11	S.F.	0.20	0.18	0.38
6" thick, R13	"	0.24	0.21	0.45
9" thick, R19	"	0.28	0.30	0.58
Rockwool, attic application				

INSULATION	UNIT	MAT.	INST.	TOTAL
07210.60 LOOSE FILL INSULATION				
6" thick, R13	S.F.	0.35	0.21	0.56
8" thick, R19	"	0.43	0.26	0.69
10" thick, R22	"	0.52	0.33	0.85
12" thick, R26	"	0.67	0.35	1.02
15" thick, R30	"	0.81	0.42	1.23
Poured type				
Fiberglass				
1" thick, R4	S.F.	0.12	0.13	0.25
2" thick, R8	"	0.23	0.15	0.38
3" thick, R12	"	0.35	0.18	0.53
4" thick, R16	"	0.46	0.21	0.67
Mineral wool				
1" thick, R3	S.F.	0.15	0.13	0.28
2" thick, R6	"	0.29	0.15	0.44
3" thick, R9	"	0.46	0.18	0.64
4" thick, R12	"	0.58	0.21	0.79
Vermiculite or perlite				
2" thick, R4.8	S.F.	0.28	0.15	0.43
3" thick, R7.2	"	0.44	0.18	0.62
4" thick, R9.6	"	0.57	0.21	0.78
Masonry, poured vermiculite or perlite				
4" block	S.F.	0.21	0.11	0.32
6" block	"	0.28	0.13	0.41
8" block	"	0.43	0.15	0.58
10" block	"	0.61	0.16	0.77
12" block	"	0.66	0.18	0.84
07210.70 SPRAYED INSULATION				
Foam, sprayed on				
Polystyrene				
1" thick, R4	S.F.	0.28	0.21	0.49
2" thick, R8	"	0.55	0.28	0.83
Urethane				
1" thick, R7.7	S.F.	0.33	0.21	0.54
2" thick, R15.4	"	0.55	0.28	0.83
07250.10 FIREPROOFING				
Sprayed on				
1" thick				
On beams	S.F.	0.36	0.47	0.83
On columns	"	0.38	0.42	0.80
On decks				
Flat surface	S.F.	0.36	0.21	0.57
Fluted surface	"	0.40	0.26	0.66
1-1/2" thick				
On beams	S.F.	0.55	0.61	1.16
On columns	"	0.61	0.53	1.14
On decks				
Flat surface	S.F.	0.55	0.26	0.81
Fluted surface	"	0.58	0.35	0.93

07 THERMAL AND MOISTURE

SHINGLES AND TILES	UNIT	MAT.	INST.	TOTAL
07310.10 ASPHALT SHINGLES				
Standard asphalt shingles, strip shingles				
210 lb/square	SQ.	36.30	26.20	62.50
235 lb/square	"	38.50	29.10	67.60
240 lb/square	"	39.60	32.70	72.30
260 lb/square	"	61.00	37.40	98.40
300 lb/square	"	63.50	43.60	107.10
385 lb/square	"	88.00	52.50	140.50
Roll roofing, mineral surface				
90 lb	SQ.	18.50	18.70	37.20
110 lb	"	30.00	21.80	51.80
140 lb	"	31.20	26.20	57.40
07310.30 METAL SHINGLES				
Aluminum, .020" thick				
Plain	SQ.	116.00	52.50	168.50
Colors	"	150.00	52.50	202.50
Steel, galvanized				
26 ga.				
Plain	SQ.	154.00	52.50	206.50
Colors	"	198.00	52.50	250.50
Porcelain enamel, 22 ga.				
Minimum	SQ.	473.00	65.50	538.50
Average	"	539.00	65.50	604.50
Maximum	"	605.00	65.50	670.50
Replacement shingles				
Small jobs	EA.	5.80	8.70	14.50
Large jobs	S.F.	3.85	4.35	8.20
07310.70 WOOD SHINGLES				
Wood shingles, on roofs				
White cedar, #1 shingles				
4" exposure	SQ.	154.00	87.00	241.00
5" exposure	"	138.00	65.50	203.50
On walls				
White cedar, #1 shingles				
4" exposure	SQ.	154.00	131.00	285.00
5" exposure	"	138.00	105.00	243.00
07310.80 WOOD SHAKES				
Shakes, hand split, 24" red cedar, on roofs				
5" exposure	SQ.	107.00	131.00	238.00
9" exposure	"	88.00	87.00	175.00

07 THERMAL AND MOISTURE

MEMBRANE ROOFING

MEMBRANE ROOFING	UNIT	MAT.	INST.	TOTAL
07510.10 BUILT-UP ASPHALT ROOFING				
Built-up roofing, asphalt felt, including gravel				
2 ply	SQ.	25.40	65.50	90.90
3 ply	"	35.80	87.00	122.80
4 ply	"	47.40	105.00	152.40
07530.10 SINGLE-PLY ROOFING				
Elastic sheet roofing				
Neoprene, 1/16" thick	S.F.	1.15	0.33	1.48
EPDM rubber				
45 mil	S.F.	0.76	0.33	1.09
60 mil	"	1.05	0.33	1.38
PVC				
45 mil	S.F.	1.05	0.33	1.38
60 mil	"	1.75	0.33	2.08
Flashing				
Pipe flashing, 90 mil thick				
1" pipe	EA.	13.20	6.55	19.75
2" pipe	"	14.85	6.55	21.40
3" pipe	"	15.05	6.90	21.95
4" pipe	"	16.15	6.90	23.05
5" pipe	"	17.50	7.25	24.75
6" pipe	"	18.90	7.25	26.15
8" pipe	"	20.90	7.70	28.60
10" pipe	"	24.75	8.70	33.45
12" pipe	"	29.70	8.70	38.40
Neoprene flashing, 60 mil thick strip				
6" wide	L.F.	0.85	2.20	3.05
12" wide	"	1.65	3.25	4.90
18" wide	"	2.40	4.35	6.75
24" wide	"	3.15	6.55	9.70
Adhesives				
1/4" bead	L.F.	0.04	0.13	0.17
Ballast, 3/4" through 1-1/2" dia. river gravel, 100lb/sf	S.F.	0.19	26.20	26.39
Walkway for membrane roofs, 1/2" thick	"	1.15	0.87	2.02

FLASHING AND SHEET METAL

FLASHING AND SHEET METAL	UNIT	MAT.	INST.	TOTAL
07610.10 METAL ROOFING				
Sheet metal roofing, copper, 16 oz, batten seam	SQ.	440.00	174.00	614.00
Standing seam	"	429.00	164.00	593.00
Aluminum roofing, natural finish				
Corrugated, on steel frame				
.0175" thick	SQ.	66.00	74.50	140.50
.032" thick	"	154.00	74.50	228.50

FLASHING AND SHEET METAL

	UNIT	MAT.	INST.	TOTAL
07610.10 METAL ROOFING				
Corrugated galvanized steel roofing, on steel frame				
28 ga.	SQ.	77.00	74.50	151.50
22 ga.	"	110.00	74.50	184.50
07620.10 FLASHING AND TRIM				
Counter flashing				
Aluminum, .032"	S.F.	0.88	2.60	3.48
Stainless steel, .015"	"	2.65	2.60	5.25
16 oz.	"	2.65	2.60	5.25
Base flashing				
Aluminum, .040"	S.F.	1.20	2.20	3.40
Stainless steel, .018"	"	2.85	2.20	5.05
16 oz.	"	2.65	2.20	4.85
24 oz.	"	3.95	2.20	6.15
Scupper outlets				
10" x 10" x 4"	EA.	15.75	6.55	22.30
22" x 4" x 4"	"	19.35	6.55	25.90
8" x 8" x 5"	"	15.75	6.55	22.30
Drainage boots, roof, cast iron				
2 x 3	L.F.	29.00	3.25	32.25
3 x 4	"	36.30	3.25	39.55
4 x 5	"	52.00	3.50	55.50
4 x 6	"	49.60	3.50	53.10
5 x 7	"	59.50	3.75	63.25
6 x 6	EA.	56.00	6.55	62.55
8 x 8	"	59.50	6.55	66.05
8 x 10	"	64.00	6.55	70.55
8 x 12	"	77.00	6.55	83.55
Reglets, copper 10 oz.	L.F.	2.00	1.75	3.75
Stainless steel, .020"	"	1.40	1.75	3.15
07620.20 GUTTERS AND DOWNSPOUTS				
Copper gutter and downspout				
Downspouts, 16 oz. copper				
Round				
3" dia.	L.F.	4.45	1.75	6.20
4" dia.	"	5.70	1.75	7.45
Rectangular, corrugated				
2" x 3"	L.F.	4.30	1.65	5.95
3" x 4"	"	5.60	1.65	7.25
Rectangular, flat surface				
2" x 3"	L.F.	4.70	1.75	6.45
3" x 4"	"	6.65	1.75	8.40
Lead-coated copper downspouts				
Round				
3" dia.	L.F.	5.50	1.65	7.15
4" dia.	"	6.60	1.85	8.45
Rectangular, corrugated				
2" x 3"	L.F.	4.05	1.75	5.80
3" x 4"	"	4.85	1.75	6.60

FLASHING AND SHEET METAL

07620.20 GUTTERS AND DOWNSPOUTS	UNIT	MAT.	INST.	TOTAL
Rectangular, plain				
2" x 3"	L.F.	3.50	1.75	5.25
3" x 4"	"	4.20	1.75	5.95
Gutters, 16 oz. copper				
Half round				
4" wide	L.F.	4.10	2.60	6.70
5" wide	"	4.95	2.90	7.85
Type K				
4" wide	L.F.	4.50	2.60	7.10
5" wide	"	4.95	2.90	7.85
Lead-coated copper gutters				
Half round				
4" wide	L.F.	3.85	2.60	6.45
6" wide	"	6.05	2.90	8.95
Type K				
4" wide	L.F.	5.30	2.60	7.90
5" wide	"	5.70	2.90	8.60
Aluminum gutter and downspout				
Downspouts				
2" x 3"	L.F.	0.63	1.75	2.38
3" x 4"	"	0.84	1.85	2.69
4" x 5"	"	0.94	2.00	2.94
Round				
3" dia.	L.F.	1.00	1.75	2.75
4" dia.	"	1.30	1.85	3.15
Gutters, stock units				
4" wide	L.F.	1.05	2.75	3.80
5" wide	"	1.15	2.90	4.05
Galvanized steel gutter and downspout				
Downspouts, round corrugated				
3" dia.	L.F.	0.88	1.75	2.63
4" dia.	"	0.99	1.75	2.74
5" dia.	"	1.20	1.85	3.05
6" dia.	"	1.55	1.85	3.40
Rectangular				
2" x 3"	L.F.	0.88	1.75	2.63
3" x 4"	"	1.40	1.65	3.05
4" x 4"	"	1.75	1.65	3.40
Gutters, stock units				
5" wide				
Plain	L.F.	0.99	2.90	3.89
Painted	"	1.10	2.90	4.00
6" wide				
Plain	L.F.	1.30	3.10	4.40
Painted	"	1.55	3.10	4.65

07700.10 MANUFACTURED SPECIALTIES	UNIT	MAT.	INST.	TOTAL
Ceiling access doors				
Swing up model, metal frame				
Steel door				
2'6" x 2'6"	EA.	275.00	26.20	301.20

07 THERMAL AND MOISTURE

FLASHING AND SHEET METAL

07700.10 MANUFACTURED SPECIALTIES	UNIT	MAT.	INST.	TOTAL
2'6" x 3'0"	EA.	297.00	26.20	323.20
Aluminum door				
2'6" x 2'6"	EA.	297.00	26.20	323.20
2'6" x 3'0"	"	319.00	26.20	345.20
Swing down model, metal frame				
Steel door				
2'6" x 2'6"	EA.	275.00	26.20	301.20
2'6" x 3'0"	"	297.00	26.20	323.20
Aluminum door				
2'6" x 2'6"	EA.	286.00	26.20	312.20
2'6" x 3'0"	"	308.00	26.20	334.20
Gravity ventilators, with curb, base, damper and screen				
Stationary siphon				
6" dia.	EA.	19.80	17.45	37.25
12" dia.	"	30.80	17.45	48.25
24" dia.	"	132.00	26.20	158.20
36" dia.	"	242.00	26.20	268.20
Wind driven spinner				
6" dia.	EA.	30.80	17.45	48.25
12" dia.	"	41.80	17.45	59.25
24" dia.	"	154.00	26.20	180.20
36" dia.	"	330.00	26.20	356.20
Stationary mushroom				
16" dia.	EA.	198.00	26.20	224.20
30" dia.	"	462.00	32.70	494.70
36" dia.	"	594.00	43.60	637.60
42" dia.	"	880.00	52.50	932.50

JOINT SEALERS

07920.10 CAULKING	UNIT	MAT.	INST.	TOTAL
Caulk exterior, two component				
1/4 x 1/2	L.F.	0.24	1.35	1.59
3/8 x 1/2	"	0.36	1.50	1.86
1/2 x 1/2	"	0.47	1.70	2.17
Caulk interior, single component				
1/4 x 1/2	L.F.	0.15	1.30	1.45
3/8 x 1/2	"	0.23	1.45	1.68
1/2 x 1/2	"	0.30	1.60	1.90

SPECIALTIES	UNIT	MAT.	INST.	TOTAL
10165.10 TOILET PARTITIONS				
Toilet partition, plastic laminate				
Ceiling mounted	EA.	606.00	90.50	696.50
Floor mounted	"	549.00	68.00	617.00
Metal				
Ceiling mounted	EA.	404.00	90.50	494.50
Floor mounted	"	370.00	68.00	438.00
Wheel chair partition, plastic laminate				
Ceiling mounted	EA.	764.00	90.50	854.50
Floor mounted	"	715.00	68.00	783.00
Painted metal				
Ceiling mounted	EA.	572.00	90.50	662.50
Floor mounted	"	528.00	68.00	596.00
Urinal screen, plastic laminate				
Wall hung	EA.	277.00	33.90	310.90
Floor mounted	"	243.00	33.90	276.90
Porcelain enameled steel, floor mounted	"	308.00	33.90	341.90
Painted metal, floor mounted	"	208.00	33.90	241.90
Stainless steel, floor mounted	"	385.00	33.90	418.90
Metal toilet partitions				
Toilet partitions, front door and side divider, floor mounted				
Porcelain enameled steel	EA.	660.00	68.00	728.00
Painted steel	"	393.00	68.00	461.00
Stainless steel	"	759.00	68.00	827.00
10185.10 SHOWER STALLS				
Shower receptors				
Precast, terrazzo				
32" x 32"	EA.	196.00	25.00	221.00
32" x 48"	"	264.00	30.00	294.00
Concrete				
32" x 32"	EA.	88.00	25.00	113.00
48" x 48"	"	132.00	33.30	165.30
Shower door, trim and hardware				
Economy, 24" wide, chrome frame, tempered glass	EA.	139.00	30.00	169.00
Porcelain enameled steel, flush	"	266.00	30.00	296.00
Baked enameled steel, flush	"	156.00	30.00	186.00
Aluminum frame, tempered glass, 48" wide, sliding	"	341.00	37.50	378.50
Folding	"	319.00	37.50	356.50
Aluminum frame and tempered glass, molded plastic				
Complete with receptor and door				
32" x 32"	EA.	455.00	75.00	530.00
36" x 36"	"	495.00	75.00	570.00
40" x 40"	"	572.00	85.50	657.50
Shower compartment, precast concrete receptor				
Single entry type				
Porcelain enameled steel	EA.	1,390	300.00	1,690
Baked enameled steel	"	808.00	300.00	1,108
Stainless steel	"	1,560	300.00	1,860
Double entry type				
Porcelain enameled steel	EA.	4,040	375.00	4,415
Baked enameled steel	"	2,540	375.00	2,915

10 SPECIALTIES

SPECIALTIES	UNIT	MAT.	INST.	TOTAL
10185.10 SHOWER STALLS				
Stainless steel	EA.	4,040	375.00	4,415
10270.40 ACCESS & PEDESTAL FLOOR				
Panels, no covering, 2'x2'				
Plain	S.F.	5.20	0.34	5.54
Perforated	"	7.80	13.55	21.35
Pedestals				
For 6" to 12" clearance	EA.	4.35	2.70	7.05
Stringers				
2'	L.F.	0.69	1.30	1.99
6'	"	0.69	0.90	1.59
Accessories				
Ramp assembly	S.F.	26.00	1.10	27.10
Elevated floor assembly	"	42.20	1.00	43.20
Handrail	L.F.	31.50	13.55	45.05
Fascia plate	"	14.75	6.80	21.55
For carpet tiles, add	S.F.			3.30
For vinyl flooring, add	"			4.10
RF shielding components, floor liner				
Hot rolled steel sheet				
14 ga.	S.F.	1.15	0.68	1.83
11 ga.	"	1.80	2.10	3.90
10290.10 PEST CONTROL				
Termite control				
Under slab spraying				
Minimum	S.F.	0.10	0.05	0.15
Average	"	0.12	0.11	0.23
Maximum	"	0.18	0.21	0.39
10500.10 LOCKERS				
Locker bench, floor mounted, laminated maple				
4'	EA.	63.00	22.60	85.60
6'	"	91.50	22.60	114.10
Wardrobe locker, single tier type				
12" x 15" x 72"	EA.	105.00	27.10	132.10
18" x 15" x 72"	"	132.00	28.50	160.50
12" x 18" x 72"	"	120.00	30.10	150.10
18" x 18" x 72"	"	143.00	31.90	174.90
10520.10 FIRE PROTECTION				
Portable fire extinguishers				
Water pump tank type				
2.5 gal.				
Red enameled galvanized	EA.	71.50	14.15	85.65
Red enameled copper	"	81.00	14.15	95.15
Polished copper	"	123.00	14.15	137.15

10 SPECIALTIES

SPECIALTIES	UNIT	MAT.	INST.	TOTAL
10520.10 FIRE PROTECTION				
Carbon dioxide type, red enamel steel				
Squeeze grip with hose and horn				
2.5 lb	EA.	37.90	14.15	52.05
5 lb	"	79.00	16.30	95.30
10 lb	"	132.00	21.20	153.20
15 lb	"	154.00	26.50	180.50
20 lb	"	176.00	26.50	202.50
Wheeled type				
125 lb	EA.	1,040	42.40	1,082
250 lb	"	974.00	42.40	1,016
500 lb	"	1,920	42.40	1,962
Dry chemical, pressurized type				
Red enameled steel				
2.5 lb	EA.	19.80	14.15	33.95
5 lb	"	30.80	16.30	47.10
10 lb	"	44.00	21.20	65.20
20 lb	"	77.00	26.50	103.50
30 lb	"	214.00	26.50	240.50
Chrome plated steel, 2.5 lb	"	44.00	14.15	58.15
Other type extinguishers				
2.5 gal, stainless steel, pressurized water tanks	EA.	44.00	14.15	58.15
Soda and acid type	"	87.50	14.15	101.65
Cartridge operated, water type	"	146.00	14.15	160.15
Loaded stream, water type	"	69.50	14.15	83.65
Foam type	"	57.50	14.15	71.65
40 gal, wheeled foam type	"	3,210	42.40	3,252
Fire extinguisher cabinets				
Enameled steel				
8" x 12" x 27"	EA.	114.00	42.40	156.40
8" x 16" x 38"	"	146.00	42.40	188.40
Aluminum				
8" x 12" x 27"	EA.	145.00	42.40	187.40
8" x 16" x 38"	"	174.00	42.40	216.40
8" x 12" x 27"	"	251.00	42.40	293.40
Stainless steel				
8" x 16" x 38"	EA.	321.00	42.40	363.40
10670.10 SHELVING				
Metal storage shelving, baked enamel				
7 shelf unit, 72" or 84" high				
10" shelf	L.F.	27.50	27.10	54.60
18" shelf	"	33.00	31.90	64.90
4 shelf unit, 40" high				
10" shelf	L.F.	23.10	22.60	45.70
18" shelf	"	31.20	28.50	59.70
3 shelf unit, 32" high				
10" shelf	L.F.	19.35	13.55	32.90
18" shelf	"	22.80	15.95	38.75
Single shelf unit, attached to masonry				
10" shelf	L.F.	7.05	4.50	11.55
18" shelf	"	9.25	5.55	14.80

10 SPECIALTIES

SPECIALTIES	UNIT	MAT.	INST.	TOTAL
10750.10 TELEPHONE ENCLOSURES				
Telephone enclosure, wall mounted, shelf, 28" x 30" x 15"	EA.	660.00	68.00	728.00
Directory shelf, stainless steel, 3 binders	"	550.00	45.20	595.20
10800.10 BATH ACCESSORIES				
Ash receiver, wall mounted, aluminum	EA.	112.00	13.55	125.55
Grab bar, 1-1/2" dia., stainless steel, wall mounted				
24" long	EA.	31.40	13.55	44.95
36" long	"	44.90	14.25	59.15
42" long	"	51.50	15.05	66.55
48" long	"	61.50	15.95	77.45
52" long	"	67.50	16.95	84.45
1" dia., stainless steel				
12" long	EA.	24.70	11.80	36.50
18" long	"	26.90	12.35	39.25
24" long	"	29.20	13.55	42.75
30" long	"	32.50	14.25	46.75
36" long	"	37.00	15.05	52.05
48" long	"	38.10	15.95	54.05
Hand dryer, surface mounted, 110 volt	"	352.00	33.90	385.90
Medicine cabinet, 16 x 22, baked enamel, steel, lighted	"	56.00	10.85	66.85
With mirror, lighted	"	101.00	18.10	119.10
Mirror, 1/4" plate glass, up to 10 sf	S.F.	5.40	2.70	8.10
Mirror, stainless steel frame				
18"x24"	EA.	84.00	9.05	93.05
18"x32"	"	90.00	10.85	100.85
18"x36"	"	95.50	13.55	109.05
24"x30"	"	112.00	13.55	125.55
24"x36"	"	129.00	15.05	144.05
24"x48"	"	224.00	22.60	246.60
24"x60"	"	236.00	27.10	263.10
30"x30"	"	129.00	27.10	156.10
30"x72"	"	359.00	33.90	392.90
48"x72"	"	376.00	45.20	421.20
With shelf, 18"x24"	"	129.00	10.85	139.85
Sanitary napkin dispenser, stainless steel, wall mounted	"	309.00	18.10	327.10
Shower rod, 1" diameter				
Chrome finish over brass	EA.	50.50	13.55	64.05
Stainless steel	"	48.80	13.55	62.35
Soap dish, stainless steel, wall mounted	"	61.50	18.10	79.60
Toilet tissue dispenser, stainless, wall mounted				
Single roll	EA.	30.30	6.80	37.10
Double roll	"	47.10	7.75	54.85
Towel dispenser, stainless steel				
Flush mounted	EA.	88.00	15.05	103.05
Surface mounted	"	82.50	13.55	96.05
Combination towel dispenser and waste receptacle	"	275.00	18.10	293.10
Towel bar, stainless steel				
18" long	EA.	25.80	10.85	36.65
24" long	"	30.30	12.35	42.65
30" long	"	32.50	13.55	46.05
36" long	"	35.90	15.05	50.95
Toothbrush and tumbler holder	"	21.90	9.05	30.95

10 SPECIALTIES

SPECIALTIES	UNIT	MAT.	INST.	TOTAL
10800.10 BATH ACCESSORIES				
Waste receptacle, stainless steel, wall mounted	EA.	185.00	22.60	207.60

11 EQUIPMENT

ARCHITECTURAL EQUIPMENT	UNIT	MAT.	INST.	TOTAL
11010.10 MAINTENANCE EQUIPMENT				
Vacuum cleaning system				
3 valves				
1.5 hp	EA.	813.00	301.00	1,114
2.5 hp	"	1,120	387.00	1,507
5 valves	"	1,510	542.00	2,052
7 valves	"	1,740	678.00	2,418
11060.10 THEATER EQUIPMENT				
Roll out stage, steel frame, wood floor				
Manual	S.F.	22.35	1.70	24.05
Electric	"	24.10	2.70	26.80
Portable stages				
8" high	S.F.	7.15	1.35	8.50
18" high	"	8.25	1.50	9.75
36" high	"	8.80	1.60	10.40
48" high	"	9.35	1.70	11.05
Band risers				
Minimum	S.F.	19.25	1.35	20.60
Maximum	"	32.20	1.35	33.55
Chairs for risers				
Minimum	EA.	55.00	0.96	55.96
Maximum	"	77.00	0.96	77.96
11080.10 POLICE EQUIPMENT				
Firing range equipment, rifle				
3 position	EA.	8,580	904.00	9,484
4 position	"	11,000	1,360	12,360
5 position	"	13,750	1,510	15,260
6 position	"	16,500	1,600	18,100
11090.10 CHECKROOM EQUIPMENT				
Motorized checkroom equipment				
No shelf system, 6'4" height				
7'6" length	EA.	1,650	271.00	1,921
14'6" length	"	1,870	271.00	2,141
28' length	"	2,530	271.00	2,801
One shelf, 6'8" height				
7'6" length	EA.	1,980	271.00	2,251
14'6" length	"	2,420	271.00	2,691
28' length	"	3,630	271.00	3,901
Two shelves, 7'5" height				
7'6" length	EA.	2,420	271.00	2,691
14'6" length	"	2,970	271.00	3,241
28' length	"	4,180	271.00	4,451
Three shelves, 8' height				
7'6" length	EA.	2,530	542.00	3,072
14'6" length	"	3,080	542.00	3,622
28' length	"	4,400	542.00	4,942
Four shelves, 8'7" height				
7'6" length	EA.	2,530	542.00	3,072
14'6" length	"	3,080	542.00	3,622

11 EQUIPMENT

ARCHITECTURAL EQUIPMENT

ARCHITECTURAL EQUIPMENT	UNIT	MAT.	INST.	TOTAL
11090.10 CHECKROOM EQUIPMENT				
28' length				
200 lb	EA.	4,400	542.00	4,942
11110.10 LAUNDRY EQUIPMENT				
High capacity, heavy duty				
Washer extractors				
135 lb				
Standard	EA.	37,920	226.00	38,146
Pass through	"	43,790	226.00	44,016
200 lb				
Standard	EA.	42,580	226.00	42,806
Pass through	"	48,410	226.00	48,636
110 lb dryer	"	6,620	226.00	6,846
Hand operated presser	"	5,160	301.00	5,461
Mushroom press	"	3,310	301.00	3,611
Spreader feeders				
2 station	EA.	49,590	301.00	49,891
4 station	"	54,870	542.00	55,412
Delivery carts				
12 bushel	EA.	330.00	3.40	333.40
Low capacity				
Pressers				
Air operated	EA.	4,100	108.00	4,208
Hand operated	"	3,340	108.00	3,448
Extractor, low capacity	"	4,770	108.00	4,878
Ironer, 48"	"	2,220	54.00	2,274
Coin washers				
10 lb capacity	EA.	1,000	54.00	1,054
20 lb capacity	"	2,520	54.00	2,574
Coin dryer	"	421.00	33.90	454.90
Coin dry cleaner, 20 lb	"	19,800	108.00	19,908
11161.10 LOADING DOCK EQUIPMENT				
Dock leveler, 10 ton capacity				
6' x 8'	EA.	3,370	271.00	3,641
7' x 8'	"	3,700	271.00	3,971
11170.10 WASTE HANDLING				
Incinerator, electric				
100 lb/hr				
Minimum	EA.	14,580	278.00	14,858
Maximum	"	20,900	278.00	21,178
400 lb/hr				
Minimum	EA.	26,240	555.00	26,795
Maximum	"	41,800	555.00	42,355
1000 lb/hr				
Minimum	EA.	44,890	841.00	45,731
Maximum	"	60,500	841.00	61,341
Incinerator, medical-waste				
25 lb/hr, 2-7 x 4-0	EA.	13,410	555.00	13,965

ARCHITECTURAL EQUIPMENT

ARCHITECTURAL EQUIPMENT	UNIT	MAT.	INST.	TOTAL
11170.10 WASTE HANDLING				
50 lb/hr, 2-11 x 4-11	EA.	14,690	555.00	15,245
75 lb/hr, 3-8 x 5-0	"	21,220	1,110	22,330
100 lb/hr, 3-8 x 6-0	"	23,320	1,110	24,430
Industrial compactor				
1 cy	EA.	7,520	308.00	7,828
3 cy	"	14,810	397.00	15,207
5 cy	"	17,130	555.00	17,685
11400.10 FOOD SERVICE EQUIPMENT				
Unit kitchens				
30" compact kitchen				
Refrigerator, with range, sink	EA.	2,410	139.00	2,549
Sink only	"	2,300	92.50	2,393
Range only	"	2,020	69.50	2,090
Cabinet for upper wall section	"	280.00	39.70	319.70
Stainless shield, for rear wall	"	95.50	11.10	106.60
Side wall	"	93.00	11.10	104.10
42" compact kitchen				
Refrigerator with range, sink	EA.	3,000	154.00	3,154
Sink only	"	2,580	139.00	2,719
Cabinet for upper wall section	"	404.00	46.30	450.30
Stainless shield, for rear wall	"	123.00	11.55	134.55
Side wall	"	50.50	11.55	62.05
54" compact kitchen				
Refrigerator, oven, range, sink	EA.	3,700	198.00	3,898
Cabinet for upper wall section	"	533.00	55.50	588.50
Stainless shield, for				
Rear wall	EA.	185.00	12.60	197.60
Side wall	"	50.50	12.60	63.10
60" compact kitchen				
Refrigerator, oven, range, sink	EA.	4,260	198.00	4,458
Cabinet for upper wall section	"	606.00	55.50	661.50
Stainless shield, for				
Rear wall	EA.	185.00	12.60	197.60
Side wall	"	50.50	12.60	63.10
72" compact kitchen				
Refrigerator, oven, range, sink	EA.	4,100	231.00	4,331
Cabinet for upper wall section	"	1,180	55.50	1,236
Stainless shield for				
Rear wall	EA.	219.00	13.90	232.90
Side wall	"	50.50	13.90	64.40
Bake oven				
Single deck				
Minimum	EA.	1,320	34.70	1,355
Maximum	"	3,080	69.50	3,150
Double deck				
Minimum	EA.	1,540	46.30	1,586
Maximum	"	5,500	69.50	5,570
Triple deck				
Minimum	EA.	2,200	46.30	2,246
Maximum	"	6,600	92.50	6,693

ARCHITECTURAL EQUIPMENT

11400.10 FOOD SERVICE EQUIPMENT

ARCHITECTURAL EQUIPMENT	UNIT	MAT.	INST.	TOTAL
Convection type oven, electric, 40" x 45" x 57"				
Minimum	EA.	2,200	34.70	2,235
Maximum	"	3,850	69.50	3,920
Broiler, without oven, 69" x 26" x 39"				
Minimum	EA.	1,980	34.70	2,015
Maximum	"	3,850	46.30	3,896
Coffee urns, 10 gallons				
Minimum	EA.	4,400	92.50	4,493
Maximum	"	7,150	139.00	7,289
Fryer, with submerger				
Single				
Minimum	EA.	660.00	55.50	715.50
Maximum	"	1,980	92.50	2,073
Double				
Minimum	EA.	1,100	69.50	1,170
Maximum	"	3,740	92.50	3,833
Griddle, counter				
3' long				
Minimum	EA.	1,100	46.30	1,146
Maximum	"	3,300	55.50	3,356
5' long				
Minimum	EA.	1,650	69.50	1,720
Maximum	"	4,290	92.50	4,383
Kettles, steam, jacketed				
20 gallons				
Minimum	EA.	1,760	69.50	1,830
Maximum	"	4,180	139.00	4,319
40 gallons				
Minimum	EA.	2,090	69.50	2,160
Maximum	"	4,620	139.00	4,759
60 gallons				
Minimum	EA.	2,750	69.50	2,820
Maximum	"	6,160	139.00	6,299
Range				
Heavy duty, single oven, open top				
Minimum	EA.	1,540	34.70	1,575
Maximum	"	3,960	92.50	4,053
Fry top				
Minimum	EA.	1,980	34.70	2,015
Maximum	"	4,290	92.50	4,383
Steamers, electric				
27 kw				
Minimum	EA.	5,950	69.50	6,020
Maximum	"	11,500	92.50	11,593
18 kw				
Minimum	EA.	4,380	69.50	4,450
Maximum	"	7,570	92.50	7,663
Dishwasher, rack type				
Single tank, 190 racks/hr	EA.	9,350	139.00	9,489
Double tank				
234 racks/hr	EA.	13,410	154.00	13,564
265 racks/hr	"	14,920	185.00	15,105

11 EQUIPMENT

ARCHITECTURAL EQUIPMENT

11400.10 FOOD SERVICE EQUIPMENT

ARCHITECTURAL EQUIPMENT	UNIT	MAT.	INST.	TOTAL
Dishwasher, automatic 100 meals/hr	EA.	44,210	92.50	44,303
Disposals				
100 gal/hr	EA.	982.00	92.50	1,075
120 gal/hr	"	1,420	95.50	1,516
250 gal/hr	"	3,700	99.00	3,799
Exhaust hood for dishwasher, gutter 4 sides, s-steel				
4'x4'x2'	EA.	1,320	103.00	1,423
4'x7'x2'	"	2,200	111.00	2,311
Food preparation machines				
Vertical cutter mixers				
25 quart	EA.	5,780	92.50	5,873
40 quart	"	6,960	92.50	7,053
80 quart	"	7,970	139.00	8,109
130 quart	"	8,860	231.00	9,091
Choppers				
5 lb	EA.	2,000	69.50	2,070
16 lb	"	2,750	92.50	2,843
40 lb	"	3,810	139.00	3,949
Mixers, floor models				
20 quart	EA.	2,200	34.70	2,235
60 quart	"	6,600	34.70	6,635
80 quart	"	8,800	39.70	8,840
140 quart	"	9,900	55.50	9,956
Ice cube maker				
50 lb per day				
Minimum	EA.	880.00	278.00	1,158
Maximum	"	1,760	278.00	2,038
500 lb per day				
Minimum	EA.	3,480	463.00	3,943
Maximum	"	6,400	463.00	6,863
Ice flakers				
300 lb per day	EA.	2,360	278.00	2,638
600 lb per day	"	2,800	463.00	3,263
1000 lb per day	"	3,310	617.00	3,927
2000 lb per day	"	7,520	694.00	8,214
Refrigerated cases				
Dairy products				490.00
Multi deck type	L.F.	499.00	18.50	517.50
For rear sliding doors, add	"			22.00
Delicatessen case, service deli				
Single deck	L.F.	462.00	139.00	601.00
Multi deck	"	506.00	174.00	680.00
Meat case				
Single deck	L.F.	330.00	163.00	493.00
Multi deck	"	363.00	174.00	537.00
Produce case				
Single deck	L.F.	348.00	163.00	511.00
Multi deck	"	396.00	174.00	570.00
Bottle coolers				
6' long				
Minimum	EA.	1,260	555.00	1,815
Maximum	"	2,660	555.00	3,215

11 EQUIPMENT

ARCHITECTURAL EQUIPMENT	UNIT	MAT.	INST.	TOTAL
11400.10 FOOD SERVICE EQUIPMENT				
10' long				
Minimum	EA.	1,800	925.00	2,725
Maximum	"	3,530	925.00	4,455
Frozen food cases				
Chest type	L.F.	359.00	163.00	522.00
Reach-in, glass door	"	748.00	174.00	922.00
Island case, single	"	365.00	163.00	528.00
Multi deck	"	808.00	174.00	982.00
Ice storage bins				
500 lb capacity	EA.	999.00	397.00	1,396
1000 lb capacity	"	2,000	793.00	2,793
11450.10 RESIDENTIAL EQUIPMENT				
Compactor, 4 to 1 compaction	EA.	990.00	69.50	1,060
Dishwasher, built-in				
2 cycles	EA.	825.00	139.00	964.00
4 or more cycles	"	1,210	139.00	1,349
Disposal				
Garbage disposer	EA.	330.00	92.50	422.50
Heaters, electric, built-in				
Ceiling type	EA.	165.00	92.50	257.50
Wall type		162.00	89.50	251.50
Minimum	EA.	110.00	69.50	179.50
Maximum	"	220.00	92.50	312.50
Hood for range, 2-speed, vented				
30" wide	EA.	118.00	92.50	210.50
42" wide	"	462.00	92.50	554.50
Ice maker, automatic				
30 lb per day	EA.	572.00	39.70	611.70
50 lb per day	"	1,210	139.00	1,349
Ranges electric				
Built-in, 30", 1 oven	EA.	800.00	92.50	892.50
2 oven	"	1,210	92.50	1,303
Counter top, 4 burner, standard	"	440.00	69.50	509.50
With grill	"	1,320	69.50	1,390
Free standing, 21", 1 oven	"	506.00	92.50	598.50
30", 1 oven	"	660.00	55.50	715.50
2 oven	"	1,980	55.50	2,036
Water softener				
30 grains per gallon	EA.	660.00	92.50	752.50
70 grains per gallon	"	1,210	139.00	1,349
11470.10 DARKROOM EQUIPMENT				
Dryers				
36" x 25" x 68"	EA.	6,600	150.00	6,750
48" x 25" x 68"	"	6,820	150.00	6,970
Processors, film				
Black and white	EA.	9,820	150.00	9,970
Color negatives	"	10,940	150.00	11,090
Prints	"	11,440	150.00	11,590
Transparencies	"	12,230	150.00	12,380

11 EQUIPMENT

ARCHITECTURAL EQUIPMENT	UNIT	MAT.	INST.	TOTAL
11470.10 DARKROOM EQUIPMENT				
Sinks with cabinet and/or stand				
5" sink with stand				
24" x 48"	EA.	722.00	75.00	797.00
32" x 64"	"	895.00	100.00	995.00
38" x 52"	"	895.00	100.00	995.00
42" x 132"	"	1,130	150.00	1,280
48" x 52"	"	1,090	150.00	1,240
5" sink with cabinet				
24" x 48"	EA.	1,440	75.00	1,515
32" x 64"	"	1,560	100.00	1,660
38" x 52"	"	1,530	100.00	1,630
42" x 132"	"	3,180	150.00	3,330
48" x 52"	"	1,960	150.00	2,110
10" sink with stand				
24" x 48"	EA.	884.00	75.00	959.00
32" x 64"	"	1,030	100.00	1,130
38" x 52"	"	1,020	100.00	1,120
10" sink with cabinet				
24" x 48"	EA.	1,440	75.00	1,515
38" x 52"	"	1,720	100.00	1,820
11500.10 INDUSTRIAL EQUIPMENT				
Vehicular paint spray booth, solid back, 14'4" x 9'6"				
24' deep	EA.	8,770	271.00	9,041
26'6" deep	"	8,940	271.00	9,211
28'6" deep	"	9,140	271.00	9,411
Drive through, 14'9" x 9'6"				
24' deep	EA.	9,200	271.00	9,471
26'6" deep	"	9,420	271.00	9,691
28'6" deep	"	9,540	271.00	9,811
Water wash, paint spray booth				
5' x 11'2" x 10'8"	EA.	4,600	271.00	4,871
6' x 11'2" x 10'8"	"	4,710	271.00	4,981
8' x 11'2" x 10'8"	"	5,950	271.00	6,221
10' x 11'2" x 11'2"	"	6,620	271.00	6,891
12' x 12'2" x 11'2"	"	8,080	271.00	8,351
14' x 12'2" x 11'2"	"	9,540	271.00	9,811
16' x 12'2" x 11'2"	"	10,770	271.00	11,041
20' x 12'2" x 11'2"	"	12,310	271.00	12,581
Dry type spray booth, with paint arrestors				
5'4" x 7'2" x 6'8"	EA.	1,930	271.00	2,201
6'4" x 7'2" x 6'8"	"	2,330	271.00	2,601
8'4" x 7'2" x 9'2"	"	2,740	271.00	3,011
10'4" x 7'2" x 9'2"	"	3,130	271.00	3,401
12'4" x 7'6" x 9'2"	"	3,700	271.00	3,971
14'4" x 7'6" x 9'8"	"	4,220	271.00	4,491
16'4" x 7'7" x 9'8"	"	4,820	271.00	5,091
20'4" x 7'7" x 10'8"	"	5,410	271.00	5,681
Air compressor, electric				
1 hp				
115 volt	EA.	2,410	181.00	2,591
5 hp				

ARCHITECTURAL EQUIPMENT

11500.10 INDUSTRIAL EQUIPMENT

ARCHITECTURAL EQUIPMENT	UNIT	MAT.	INST.	TOTAL
115 volt	EA.	4,990	271.00	5,261
230 volt	"	4,210	271.00	4,481
Hydraulic lifts				
8,000 lb capacity	EA.	370.00	678.00	1,048
11,000 lb capacity	"	5,270	1,080	6,350
24,000 lb capacity	"	7,410	1,810	9,220
Power tools				
Band saws				
10"	EA.	495.00	22.60	517.60
14"	"	880.00	27.10	907.10
Motorized shaper	"	473.00	20.85	493.85
Motorized lathe	"	550.00	22.60	572.60
Bench saws				
9" saw	EA.	1,010	18.10	1,028
10" saw	"	1,520	19.35	1,539
12" saw	"	1,960	22.60	1,983
Electric grinders				
1/3 hp	EA.	187.00	10.85	197.85
1/2 hp	"	264.00	11.80	275.80
3/4 hp	"	319.00	11.80	330.80

11600.10 LABORATORY EQUIPMENT

ARCHITECTURAL EQUIPMENT	UNIT	MAT.	INST.	TOTAL
Cabinets, base				
Minimum	L.F.	88.00	22.60	110.60
Maximum	"	154.00	22.60	176.60
Full storage, 7' high				
Minimum	L.F.	132.00	22.60	154.60
Maximum	"	198.00	22.60	220.60
Wall				
Minimum	L.F.	58.50	27.10	85.60
Maximum	"	99.00	27.10	126.10
Counter tops				78.50
Minimum	S.F.	12.10	3.40	15.50
Average	"	28.60	3.85	32.45
Maximum	"	47.30	4.50	51.80
Tables				
Open underneath	S.F.	66.00	13.55	79.55
Doors underneath	"	121.00	16.95	137.95
Medical laboratory equipment				
Analyzer				
Chloride	EA.	385.00	13.90	398.90
Blood	"	16,500	23.15	16,523
Bath, water, utility, countertop unit	"	275.00	27.80	302.80
Hot plate, lab, countertop	"	88.00	25.20	113.20
Stirrer	"	154.00	25.20	179.20
Incubator, anaerobic, 23x23x36"	"	4,400	139.00	4,539
Dry heat bath	"	330.00	46.30	376.30
Incinerator, for sterilizing	"	132.00	2.80	134.80
Meter, serum protein	"	594.00	3.45	597.45
Ph analog, general purpose	"	440.00	3.95	443.95
Refrigerator, blood bank, undercounter type 153 litres	"	3,190	46.30	3,236
5.4 cf, undercounter type	"	2,750	46.30	2,796

ARCHITECTURAL EQUIPMENT

ARCHITECTURAL EQUIPMENT	UNIT	MAT.	INST.	TOTAL
11600.10 LABORATORY EQUIPMENT				
Refrigerator/freezer, 4.4 cf, undercounter type	EA.	308.00	46.30	354.30
Sealer, impulse, free standing, 20x12x4"	"	165.00	9.25	174.25
Timer, electric, 1-60 minutes, bench or wall mounted	"	45.10	15.40	60.50
Glassware washer - dryer, undercounter	"	5,120	347.00	5,467
Balance, torsion suspension, tabletop, 4.5 lb capacity	"	352.00	15.40	367.40
Binocular microscope, with in base illuminator	"	1,380	10.70	1,391
Centrifuge, table model, 19x16x13"	"	836.00	11.10	847.10
Clinical model, with four place head	"	429.00	6.15	435.15
11700.10 MEDICAL EQUIPMENT				
Hospital equipment, lights				
Examination, portable	EA.	693.00	23.15	716.15
Meters				
Air flow meter	EA.	40.70	15.40	56.10
Oxygen flow meters	"	27.50	11.55	39.05
Physical therapy				
Chair, hydrotherapy	EA.	198.00	4.50	202.50
Diathermy, shortwave, portable, on casters	"	1,760	10.85	1,771
Exercise bicycle, floor standing, 35" x 15"	"	506.00	9.05	515.05
Hydrocollator, 4 pack, portable, 129 x 90 x 160"	"	198.00	3.85	201.85
Lamp, infrared, mobile with variable heat control	"	264.00	20.85	284.85
Ultra violet, base mounted	"	242.00	20.85	262.85
Stimulator, galvanic-faradic, hand held	"	57.00	1.80	58.80
Ultrasound, muscle stimulator, portable, 13x13x8"	"	1,320	2.25	1,322
Whirlpool, 85 gallon	"	1,760	136.00	1,896
65 gallon capacity	"	1,500	136.00	1,636
Radiology				
Radiographic table, motor driven tilting table	EA.	19,580	2,710	22,290
Fluoroscope image/tv system	"	55,000	5,420	60,420
Processor for washing and drying radiographs				
Water filter unit, 30" x 48-1/2" x 37-1/2"	EA.	17.60	463.00	480.60
Steam sterilizers				
For heat and moisture stable materials	EA.	1,980	27.80	2,008
For fast drying after sterilization	"	1,320	34.70	1,355
Compact unit	"	990.00	34.70	1,025
Semi-automatic	"	3,960	139.00	4,099
Floor loading				
Single door	EA.	74,800	231.00	75,031
Double door	"	94,600	278.00	94,878
Utensil washer, sanitizer	"	4,180	214.00	4,394
Automatic washer/sterilizer	"	15,400	555.00	15,955
16 x 16 x 26", including generator & accessories	"	14,850	925.00	15,775
Steam generator, elec., 10 kw to 180 kw	"	14,240	555.00	14,795
Surgical scrub				
Minimum	EA.	1,540	92.50	1,633
Maximum	"	3,250	92.50	3,343
Gas sterilizers				
Automatic, free standing, 21x19x29"	EA.	2,800	278.00	3,078
Surgical lights, ceiling mounted				
Minimum	EA.	2,750	463.00	3,213
Maximum	"	4,510	555.00	5,065
Water stills				

ARCHITECTURAL EQUIPMENT	UNIT	MAT.	INST.	TOTAL
11700.10 MEDICAL EQUIPMENT				
4 liters/hr	EA.	1,420	92.50	1,513
8 liters/hr	"	2,090	92.50	2,183
19 liters/hr	"	5,780	231.00	6,011
X-ray equipment				
Mobile unit				
Minimum	EA.	4,950	139.00	5,089
Maximum	"	24,200	278.00	24,478
Minimum	"	2,750	278.00	3,028
Maximum	"	5,500	278.00	5,778
Incubators				
15 cf	EA.	2,530	139.00	2,669
29 cf	"	3,080	231.00	3,311
Infant transport, portable	"	6,600	146.00	6,746
Headwall				
Aluminum, with back frame and console	EA.	2,090	139.00	2,229
Hospital ground detection system				
Power ground module	EA.	495.00	79.50	574.50
Ground slave module	"	330.00	60.50	390.50
Master ground module	"	214.00	52.50	266.50
Remote indicator	"	231.00	55.50	286.50
X-ray indicator	"	732.00	60.50	792.50
Micro ammeter	"	891.00	69.50	960.50
Supervisory module	"	682.00	60.50	742.50
Ground cords	"	55.00	10.30	65.30
Hospital isolation monitors, 5 ma				
120v	EA.	1,380	121.00	1,501
208v	"	1,320	121.00	1,441
240v	"	1,320	121.00	1,441
Digital clock-timers separate display	"	550.00	55.50	605.50
One display	"	517.00	55.50	572.50
Remote control	"	308.00	43.40	351.40
Battery pack	"	71.50	43.40	114.90
Surgical chronometer clock and 3 timers	"	1,100	87.00	1,187
Auxilary control	"	407.00	40.20	447.20
11700.20 DENTAL EQUIPMENT				
Dental care equipment				
Drill console with accessories	EA.	8,250	463.00	8,713
Amalgamator	"	206.00	13.90	219.90
Lathe	"	215.00	9.25	224.25
Finish polisher	"	162.00	18.50	180.50
Model trimer	"	254.00	12.60	266.60
Motor, wall mounted	"	283.00	12.60	295.60
Cleaner, ultrasonic	"	1,070	27.80	1,098
Curing unit, bench mounted	"	1,420	46.30	1,466
Oral evacuation system, dual pump	"	8,030	34.70	8,065
Sterilizer, table top, self contained	"	560.00	15.40	575.40
Dental lights				
Light, floor or ceiling mounted	EA.	1,320	139.00	1,459
X-ray unit				
Portable	EA.	1,980	69.50	2,050
Wall mounted with remote control	"	4,180	231.00	4,411

11 EQUIPMENT

ARCHITECTURAL EQUIPMENT

11700.20 DENTAL EQUIPMENT	UNIT	MAT.	INST.	TOTAL
Illuminator, single panel	EA.	6,010	397.00	6,407
X-ray film processor	"	990.00	231.00	1,221
Shield, portable x-ray, lead lined	"	209.00	18.50	227.50

12 FURNISHINGS

INTERIOR	UNIT	MAT.	INST.	TOTAL
12302.10 CASEWORK				
Kitchen base cabinet, prefinished, 24" deep, 35" high				
12"wide	EA.	108.00	27.10	135.10
18" wide	"	132.00	27.10	159.10
24" wide	"	144.00	30.10	174.10
27" wide	"	174.00	30.10	204.10
36" wide	"	196.00	33.90	229.90
48" wide	"	224.00	33.90	257.90
Corner cabinet, 36" wide	"	151.00	33.90	184.90
Wall cabinet, 12" deep, 12" high				
30" wide	EA.	86.00	27.10	113.10
36" wide	"	91.00	27.10	118.10
15" high				
30" wide	EA.	91.00	30.10	121.10
36" wide	"	98.50	30.10	128.60
24" high				
30" wide	EA.	108.00	30.10	138.10
36" wide	"	132.00	30.10	162.10
30" high				
12" wide	EA.	82.50	33.90	116.40
18" wide	"	94.00	33.90	127.90
24" wide	"	101.00	33.90	134.90
27" wide	"	120.00	33.90	153.90
30" wide	"	135.00	38.70	173.70
36" wide	"	142.00	38.70	180.70
Corner cabinet, 30" high				
24" wide	EA.	132.00	45.20	177.20
30" wide	"	159.00	45.20	204.20
36" wide	"	176.00	45.20	221.20
Wardrobe	"	407.00	68.00	475.00
Vanity with top, laminated plastic				
24" wide	EA.	252.00	68.00	320.00
30" wide	"	309.00	68.00	377.00
36" wide	"	337.00	90.50	427.50
48" wide	"	372.00	108.00	480.00
12390.10 COUNTER TOPS				
Stainless steel, counter top, with backsplash	S.F.	66.00	6.80	72.80
Acid-proof, kemrock surface	"	17.60	4.50	22.10

13 SPECIAL

CONSTRUCTION

13152.10 SWIMMING POOL EQUIPMENT	UNIT	MAT.	INST.	TOTAL
Diving boards				
14' long				
Aluminum	EA.	874.00	118.00	992.00
Fiberglass	"	539.00	118.00	657.00
Ladders, heavy duty				
2 steps				
Minimum	EA.	346.00	42.40	388.40
Maximum	"	556.00	42.40	598.40
4 steps				
Minimum	EA.	385.00	53.00	438.00
Maximum	"	737.00	53.00	790.00
Lifeguard chair				
Minimum	EA.	814.00	212.00	1,026
Maximum	"	1,280	212.00	1,492
Lights, underwater				
12 volt, with transformer	EA.	253.00	53.00	306.00
110 volt				
Minimum	EA.	231.00	53.00	284.00
Maximum	"	759.00	53.00	812.00
Ground fault interrupter for 110 volt, each light	"	95.00	17.65	112.65
Pool covers				
Reinforced polyethylene	S.F.	0.43	1.65	2.08
Vinyl water tube				
Minimum	S.F.	0.48	1.65	2.13
Maximum	"	0.74	1.65	2.39
Slides with water tube				
Minimum	EA.	253.00	177.00	430.00
Maximum	"	737.00	177.00	914.00

14 CONVEYING

ELEVATORS

14210.10 ELEVATORS

ELEVATORS	UNIT	MAT.	INST.	TOTAL
Passenger elevators, electric, geared				
Based on a shaft of 6 stops and 6 openings				
50 fpm, 2000 lb	EA.	58,300	1,150	59,450
100 fpm, 2000 lb	"	60,500	1,280	61,780
150 fpm				
2000 lb	EA.	66,000	1,440	67,440
3000 lb	"	84,150	1,640	85,790
4000 lb	"	88,640	1,910	90,550
Based on a shaft of 8 stops and 8 openings				
300 fpm				
3000 lb	EA.	100,530	2,300	102,830
3500 lb	"	102,660	2,300	104,960
4000 lb	"	108,270	2,550	110,820
5000 lb	"	122,070	2,740	124,810
4000 lb	"	11,880	2,550	14,430
Freight elevators, electric				
Based on a shaft of 6 stops and 6 openings				
50 fpm				
3500 lb	EA.	117,810	1,280	119,090
4000 lb	"	118,260	1,280	119,540
5000 lb	"	119,380	1,440	120,820
100 fpm				
3500 lb	EA.	123,980	1,440	125,420
4000 lb	"	124,540	1,440	125,980
5000 lb	"	121,510	1,640	123,150
For variable voltage control, add 20%				

14300.10 ESCALATORS

ESCALATORS	UNIT	MAT.	INST.	TOTAL
Escalators				
32" wide, floor to floor				
12' high	EA.	76,650	1,910	78,560
15' high	"	83,510	2,300	85,810
18' high	"	90,380	2,870	93,250
22' high	"	97,240	3,830	101,070
25' high	"	108,680	4,600	113,280

LIFTS

14410.10 PERSONNEL LIFTS

PERSONNEL LIFTS	UNIT	MAT.	INST.	TOTAL
Electrically operated, 1 or 2 person lift				
With attached foot platforms				
3 stops	EA.			13,280
5 stops	"			16,400

14 CONVEYING

LIFTS	UNIT	MAT.	INST.	TOTAL
14410.10 PERSONNEL LIFTS				
7 stops	EA.			19,520
For each additional stop, add $1250				
Residential stair climber, per story	EA.	3,930	231.00	4,161

MATERIAL HANDLING	UNIT	MAT.	INST.	TOTAL
14580.10 PNEUMATIC SYSTEMS				
Pneumatic message tube system				
Average, 20 station job				
3" round system	E.A.	203,940	252.00	204,192
4" round system	"	254,920	278.00	255,198
6" round system	"	385,220	308.00	385,528
4" x 7" oval system	"	402,210	555.00	402,765
Trash and linen tube system				
10 stations	EA.	130,290	5,740	136,030
15 stations	"	164,280	7,660	171,940
20 stations	"	220,390	8,840	229,230
30 stations	"	288,910	10,440	299,350

HOISTS AND CRANES	UNIT	MAT.	INST.	TOTAL
14600.10 INDUSTRIAL HOISTS				
Industrial hoists, electric, light to medium duty				4,350
500 lb	EA.	4,470	139.00	4,609
1000 lb	"	4,680	146.00	4,826
2000 lb	"	4,800	154.00	4,954
3000 lb	"	4,890	163.00	5,053
4000 lb	"	5,530	174.00	5,704
5000 lb	"	7,820	185.00	8,005
6000 lb	"	8,080	191.00	8,271
7500 lb	"	8,640	198.00	8,838
10,000 lb	"	14,590	206.00	14,796
15,000 lb	"	21,650	214.00	21,864
20,000 lb	"	25,580	231.00	25,811
25,000 lb	"	28,050	252.00	28,302

HOISTS AND CRANES

14600.10 INDUSTRIAL HOISTS	UNIT	MAT.	INST.	TOTAL
30,000 lb	EA.	27,500	278.00	27,778
Heavy duty				
500 lb	EA.	6,590	139.00	6,729
1000 lb	"	6,790	146.00	6,936
2000 lb	"	7,550	154.00	7,704
3000 lb	"	7,970	163.00	8,133
4000 lb	"	8,420	174.00	8,594
5000 lb	"	8,750	185.00	8,935
6000 lb	"	8,980	191.00	9,171
7500 lb	"	12,340	198.00	12,538
10,000 lb	"	12,900	206.00	13,106
15,000 lb	"	14,250	214.00	14,464
20,000 lb	"	17,620	231.00	17,851
25,000 lb	"	25,580	252.00	25,832
30,000 lb	"	26,700	278.00	26,978

15 MECHANICAL

BASIC MATERIALS	UNIT	MAT.	INST.	TOTAL
15100.10 SPECIALTIES				
Wall penetration				
Concrete wall, 6" thick				
2" dia.	EA.	0.00	7.05	7.05
4" dia.	"	0.00	10.60	10.60
8" dia.	"	0.00	15.15	15.15
12" thick				
2" dia.	EA.	0.00	9.65	9.65
4" dia.	"	0.00	15.15	15.15
8" dia.	"	0.00	23.55	23.55
Non-destructive testing, piping systems				
X-ray of welds				
3" dia. pipe	EA.	42.70	30.00	72.70
4" dia. pipe	"	37.00	30.00	67.00
6" dia. pipe	"	31.20	30.00	61.20
8" dia. pipe	"	33.50	37.50	71.00
10" dia. pipe	"	35.80	37.50	73.30
Liquid penetration of welds				
2" dia. pipe	EA.	21.25	18.75	40.00
3" dia. pipe	"	21.85	18.75	40.60
4" dia. pipe	"	24.25	18.75	43.00
6" dia. pipe	"	24.25	18.75	43.00
8" dia. pipe	"	30.00	18.75	48.75
10" dia. pipe	"	30.00	18.75	48.75
15120.10 BACKFLOW PREVENTERS				
Backflow preventer, flanged, cast iron, with valves				
3" pipe	EA.	1,810	150.00	1,960
4" pipe	"	2,420	167.00	2,587
6" pipe	"	3,640	250.00	3,890
8" pipe	"	6,530	300.00	6,830
Threaded				
3/4" pipe	EA.	330.00	18.75	348.75
2" pipe	"	660.00	30.00	690.00
Reduced pressure assembly, bronze, threaded				
3/4"	EA.	422.00	18.75	440.75
1"	"	531.00	21.45	552.45
1-1/4"	"	774.00	25.00	799.00
1-1/2"	"	843.00	30.00	873.00
15140.10 PIPE HANGERS, HEAVY				
Hangers				
1/2" pipe, clevis pipe hanger				
Black steel	EA.	1.75	10.00	11.75
Galvanized	"	2.00	10.00	12.00
U bolt	"	0.95	3.00	3.95
3/4" pipe, clevis pipe hanger				
Black steel	EA.	1.75	10.00	11.75
Galvanized	"	2.05	10.00	12.05
U bolt	"	0.95	3.00	3.95
1" pipe, clevis pipe hanger				

BASIC MATERIALS	UNIT	MAT.	INST.	TOTAL
15140.10 PIPE HANGERS, HEAVY				
Black steel	EA.	1.85	10.00	11.85
Galvanized	"	2.15	10.00	12.15
U bolt	"	0.99	3.00	3.99
1-1/4" pipe, clevis pipe hanger				
Black steel	EA.	2.05	10.00	12.05
Galvanized	"	2.45	10.00	12.45
U bolt	"	1.15	3.00	4.15
1-1/2" pipe, clevis pipe hanger				
Black steel	EA.	2.30	10.00	12.30
Galvanized	"	2.75	10.00	12.75
U bolt	"	1.15	3.00	4.15
2" pipe, clevis pipe hanger				
Black steel	EA.	2.75	10.00	12.75
Galvanized	"	3.10	10.00	13.10
Adjustable pipe roll stand	"	36.30	37.50	73.80
U bolt	"	1.20	3.00	4.20
Adjustable roller hanger	"	6.05	10.00	16.05
2-1/2" pipe, clevis pipe hanger				
Black steel	EA.	4.10	10.00	14.10
Galvanized	"	5.25	10.00	15.25
Adjustable pipe roll stand	"	36.50	10.00	46.50
Adjustable roller hanger	"	6.60	10.00	16.60
3" pipe, clevis pipe hanger				
Black steel	EA.	4.65	10.00	14.65
Galvanized	"	8.25	10.00	18.25
Adjustable pipe roll stand	"	36.50	10.00	46.50
U bolt	"	2.05	3.00	5.05
Adjustable roller hanger	"	7.60	10.00	17.60
3-1/2" pipe, clevis pipe hanger				
Black steel	EA.	4.05	10.00	14.05
Galvanized	"	6.15	10.00	16.15
Adjustable pipe roll stand	"	36.50	10.00	46.50
U bolt	"	2.15	3.00	5.15
Adjustable roller hanger	"	8.70	10.00	18.70
4" pipe, clevis pipe hanger				
Black steel	EA.	6.45	10.00	16.45
Galvanized	"	8.15	10.00	18.15
Adjustable pipe roll stand	"	42.50	10.00	52.50
U bolt	"	2.15	3.00	5.15
Adjustable roller hanger	"	10.80	10.00	20.80
5" pipe, clevis pipe hanger				
Black steel	EA.	7.15	12.00	19.15
Galvanized	"	9.35	12.00	21.35
Adjustable pipe roll stand	"	75.00	12.00	87.00
U bolt	"	2.30	3.00	5.30
Adjustable roller hanger	"	15.05	12.00	27.05
6" pipe, clevis pipe hanger				
Black steel	EA.	10.05	12.00	22.05
Galvanized	"	14.95	12.00	26.95
Adjustable pipe roll stand	"	42.00	12.00	54.00
U bolt	"	4.10	3.75	7.85
Adjustable roller hanger	"	17.50	12.00	29.50

BASIC MATERIALS	UNIT	MAT.	INST.	TOTAL
15140.10 PIPE HANGERS, HEAVY				
8" pipe, clevis pipe hanger				
Black steel	EA.	16.45	12.00	28.45
Galvanized	"	16.60	12.00	28.60
Adjustable pipe roll stand	"	65.50	12.00	77.50
U bolt	"	5.00	3.75	8.75
Adjustable roller hanger	"	21.25	12.00	33.25
10" clevis pipe hanger				
Black steel	EA.	17.40	12.00	29.40
Galvanized	"	23.80	12.00	35.80
Adjustable pipe roll stand	"	65.50	12.00	77.50
Adjustable roller hanger	"	26.10	12.00	38.10
12" pipe, clevis pipe hanger				
Black steel	EA.	16.05	12.00	28.05
Galvanized	"	22.75	12.00	34.75
Adjustable pipe roll stand	"	43.60	12.00	55.60
Adjustable roller hanger	"	28.50	12.00	40.50
14" pipe, clevis pipe hanger				
Black steel	EA.	19.15	13.65	32.80
Galvanized	"	27.70	13.65	41.35
Threaded rod, galvanized				
3/8"	L.F.			0.26
1/2"	"			0.45
5/8"	"			0.68
3/4"	"			1.20
7/8"	"			1.60
1"	"			2.20
Hex nuts, galvanized				
3/8"	EA.			0.03
1/2"	"			0.07
5/8"	"			0.13
3/4"	"			0.19
7/8"	"			0.31
1"	"			0.50
C-clamp, steel, with lock nut				
3/8"	EA.	1.20	3.75	4.95
1/2"	"	1.45	3.75	5.20
5/8"	"	2.30	3.75	6.05
3/4"	"	2.90	3.75	6.65
7/8"	"	6.40	3.75	10.15
Angle support, medium, welded steel				
12"x18"	EA.	65.00	30.00	95.00
18"x24"	"	78.00	30.00	108.00
24"x30"	"	102.00	30.00	132.00
Heavy, welded steel				
12"x18"	EA.	91.50	30.00	121.50
18"x24"	"	129.00	30.00	159.00
24"x30"	"	150.00	30.00	180.00
15140.11 PIPE HANGERS, LIGHT				
A band, black iron				
1/2"	EA.	0.42	2.15	2.57

15 MECHANICAL

BASIC MATERIALS	UNIT	MAT.	INST.	TOTAL
15140.11 PIPE HANGERS, LIGHT				
1"	EA.	0.53	2.20	2.73
1-1/4"	"	0.58	2.30	2.88
1-1/2"	"	0.67	2.50	3.17
2"	"	0.75	2.75	3.50
2-1/2"	"	1.00	3.00	4.00
3"	"	1.05	3.35	4.40
4"	"	1.55	3.75	5.30
5"	"	2.00	4.00	6.00
6"	"	2.80	4.30	7.10
8"	"	4.75	5.00	9.75
Copper				
1/2"	EA.	0.48	2.15	2.63
3/4"	"	0.48	2.20	2.68
1"	"	0.51	2.20	2.71
1-1/4"	"	0.53	2.30	2.83
1-1/2"	"	0.62	2.50	3.12
2"	"	0.73	2.75	3.48
2-1/2"	"	1.00	3.00	4.00
3"	"	1.25	3.35	4.60
4"	"	1.65	3.75	5.40
Black riser friction hangers				
3/4"	EA.	2.30	2.50	4.80
1"	"	2.30	2.60	4.90
1-1/4"	"	2.95	2.75	5.70
1-1/2"	"	3.15	2.85	6.00
2"	"	3.20	3.00	6.20
2-1/2"	"	3.50	3.35	6.85
3"	"	3.60	3.75	7.35
4"	"	4.55	4.30	8.85
5"	"	6.70	4.60	11.30
6"	"	7.65	5.00	12.65
8"	"	13.05	5.45	18.50
10"	"	16.80	6.00	22.80
Short pattern black riser clamps				
1-1/2"	EA.	1.95	2.75	4.70
2"	"	2.05	2.85	4.90
3"	"	2.20	3.00	5.20
4"	"	2.45	3.35	5.80
Copper riser friction hanger				
1/2"	EA.	2.20	2.30	4.50
3/4"	"	2.20	2.40	4.60
1"	"	2.45	2.50	4.95
1-1/4"	"	2.50	2.60	5.10
1-1/2"	"	2.65	2.75	5.40
2"	"	3.15	2.85	6.00
2-1/2"	"	4.65	3.00	7.65
3"	"	5.20	3.00	8.20
4"	"	5.80	3.35	9.15
Auto grip hangers, galvanized				
1/2"	EA.	0.55	2.15	2.70
3/4"	"	0.55	2.30	2.85
1"	"	0.55	2.40	2.95

BASIC MATERIALS	UNIT	MAT.	INST.	TOTAL
15140.11 PIPE HANGERS, LIGHT				
1-1/4"	EA.	0.59	2.50	3.09
1-1/2"	"	0.62	2.60	3.22
2"	"	0.65	2.75	3.40
2-1/2"	"	1.85	2.85	4.70
3"	"	2.80	3.00	5.80
4"	"	1.85	3.35	5.20
Copper				
1/2"	EA.	0.73	2.15	2.88
3/4"	"	0.73	2.30	3.03
1"	"	0.78	2.40	3.18
1-1/4"	"	0.83	2.50	3.33
1-1/2"	"	0.85	2.60	3.45
2"	"	0.91	2.75	3.66
2-1/2"	"	2.00	2.85	4.85
3"	"	2.10	3.00	5.10
4"	"	3.05	3.35	6.40
Split rings (F&M), galvanized				
3/8"	EA.	1.20	2.15	3.35
1/2"	"	1.45	2.30	3.75
3/4"	"	1.60	2.40	4.00
1"	"	1.95	2.50	4.45
1-1/4"	"	2.10	2.60	4.70
1-1/2"	"	2.45	2.75	5.20
2"	"	2.85	2.85	5.70
2-1/2"	"	8.00	3.00	11.00
3"	"	8.30	3.15	11.45
4"	"	10.10	3.35	13.45
Copper				
1/4"	EA.	1.15	2.15	3.30
3/8"	"	1.15	2.15	3.30
1/2"	"	1.20	2.30	3.50
1"	"	1.45	2.50	3.95
1-1/4"	"	1.70	2.60	4.30
1-1/2"	"	1.80	2.75	4.55
2"	"	2.10	2.85	4.95
2-1/2"	"	3.95	3.00	6.95
3"	"	5.05	3.15	8.20
4"	"	13.00	3.35	16.35
F&M plates, galvanized				
3/8"	EA.	0.92	2.15	3.07
1/2"	"	1.25	2.20	3.45
Copper				
3/8"	EA.	0.88	2.15	3.03
1/2"	"	0.99	2.20	3.19
2 hole clips, galvanized				
3/4"	EA.	0.14	2.00	2.14
1"	"	0.17	2.05	2.22
1-1/4"	"	0.21	2.15	2.36
1-1/2"	"	0.25	2.20	2.45
2"	"	0.33	2.30	2.63
2-1/2"	"	0.62	2.40	3.02
3"	"	0.89	2.50	3.39

BASIC MATERIALS	UNIT	MAT.	INST.	TOTAL
15140.11 PIPE HANGERS, LIGHT				
4"	EA.	1.90	2.75	4.65
Perforated strap				
3/4"				
Galvanized, 20 ga.	L.F.	0.21	1.50	1.71
Copper, 22 ga.	"	0.25	1.50	1.75
Threaded rod-couplings				
1/4"	EA.	0.69	1.85	2.54
3/4"	"	0.73	2.00	2.73
1/2"	"	0.81	2.15	2.96
5/8"	"	1.30	2.30	3.60
Reducing rod coupling, 1/2" x 3/8"	"	1.15	2.15	3.30
C-clamps				
3/4"	EA.	1.05	3.00	4.05
Top beam clamp				
3/8"	EA.	1.65	2.50	4.15
1/2"	"	2.05	2.75	4.80
Side beam connector				
3/8"	EA.	0.75	2.50	3.25
1/2"	"	1.60	2.75	4.35
Hex nuts, heavy				
1"	EA.			1.60
Heavy washers				
3/8"	EA.			0.07
1/2"	"			0.14
5/8"	"			0.28
3/4"	"			0.40
Lag rod, 3/8" x				
4"	EA.			0.55
4-1/2"	"			0.55
6"	"			0.65
8"	"			0.81
10"	"			1.10
12"	"			1.35
18"	"			1.90
6"	"			1.75
8"	"			2.00
10"	"			2.25
12"	"			2.35
Drive screws				
1-1/2" x 12"	EA.			0.14
2" x 12"	"			0.23
Wood screws				
3/4"				
#8	EA.			0.03
1"				
#8	EA.			0.04
#10	"			0.04
#12	"			0.08
1-1/4"				
#8	EA.			0.04
#10	"			0.06
#12	"			0.08

BASIC MATERIALS	UNIT	MAT.	INST.	TOTAL
15140.11 PIPE HANGERS, LIGHT				
1-1/2"				
#8	EA.			0.06
#10	"			0.07
2-1/2"				
#10	EA.			0.10
3"				
#12	EA.			0.17
4"				
#12	EA.			0.33
#14	"			0.41
J-Hooks				
1/2"	EA.	0.24	1.35	1.59
3/4"	"	0.25	1.35	1.60
1"	"	0.26	1.45	1.71
1-1/4"	"	0.28	1.45	1.73
1-1/2"	"	0.29	1.50	1.79
2"	"	0.30	1.50	1.80
3"	"	0.35	1.60	1.95
4"	"	0.35	1.60	1.95
PVC coated hangers, galvanized, 28 ga.				
1-1/2" x 12"	EA.	0.70	2.00	2.70
2" x 12"	"	0.78	2.15	2.93
3" x 12"	"	0.86	2.30	3.16
4" x 12"	"	0.94	2.50	3.44
Copper, 30 ga.				
1-1/2" x 12"	EA.	0.76	2.00	2.76
2" x 12"	"	0.90	2.15	3.05
3" x 12"	"	1.00	2.30	3.30
4" x 12"	"	1.10	2.50	3.60
2" x 24"	"	1.80	2.30	4.10
3" x 24"	"	2.05	2.50	4.55
4" x 24"	"	2.20	2.75	4.95
Milford hangers				
1/2" x 6"	EA.	0.63	2.15	2.78
1/2" x 12"	"	0.86	2.30	3.16
3/4" x 6"	"	0.69	2.20	2.89
3/4" x 12"	"	0.88	2.40	3.28
1" x 6"	"	0.74	2.30	3.04
1" x 12"	"	0.92	2.40	3.32
1-1/4" x 6"	"	0.84	2.40	3.24
1-1/4" x 12"	"	1.00	2.50	3.50
1-1/2" x 6"	"	0.92	2.50	3.42
1-1/2" x 12"	"	1.05	2.60	3.65
2" x 6"	"	1.00	2.60	3.60
2" x 12"	"	1.15	2.75	3.90
Wire hook hangers				
Black wire, 1/2" x				
4"	EA.	0.21	1.50	1.71
6"	"	0.24	1.60	1.84
8"	"	0.26	1.65	1.91
10"	"	0.34	1.65	1.99
12"	"	0.43	1.75	2.18

BASIC MATERIALS

15140.11 PIPE HANGERS, LIGHT

	UNIT	MAT.	INST.	TOTAL
3/4" x				
4"	EA.	0.25	1.60	1.85
6"	"	0.28	1.65	1.93
8"	"	0.29	1.75	2.04
10"	"	0.41	1.85	2.26
12"	"	0.42	2.00	2.42
1" x				
4"	EA.	0.25	1.65	1.90
6"	"	0.26	1.75	2.01
8"	"	0.29	1.85	2.14
10"	"	0.42	2.00	2.42
12"	"	0.45	2.15	2.60
1-1/4" x				
4"	EA.	0.28	1.75	2.03
6"	"	0.29	1.85	2.14
8"	"	0.33	2.00	2.33
10"	"	0.37	2.15	2.52
12"	"	0.45	2.30	2.75
1-1/2" x				
6"	EA.	0.32	2.00	2.32
8"	"	0.33	2.15	2.48
10"	"	0.43	2.30	2.73
12"	"	0.47	2.50	2.97
2" x				
6"	EA.	0.31	2.15	2.46
8"	"	0.41	2.30	2.71
10"	"	0.42	2.50	2.92
12"	"	0.48	2.75	3.23
Copper wire hooks				
1/2" x				
4"	EA.	0.26	1.50	1.76
6"	"	0.32	1.60	1.92
8"	"	0.37	1.65	2.02
10"	"	0.44	1.75	2.19
12"	"	0.52	1.85	2.37
3/4" x				
4"	EA.	0.29	1.60	1.89
6"	"	0.33	1.65	1.98
8"	"	0.38	1.75	2.13
10"	"	0.45	1.85	2.30
12"	"	0.54	2.00	2.54
1" x				
4"	EA.	0.31	1.65	1.96
6"	"	0.34	1.75	2.09
8"	"	0.40	1.85	2.25
10"	"	0.48	2.00	2.48
12"	"	0.56	2.15	2.71
1-1/4" x				
6"	EA.	0.40	1.75	2.15
8"	"	0.45	1.85	2.30
10"	"	0.53	2.00	2.53
12"	"	0.55	2.15	2.70

15 MECHANICAL

BASIC MATERIALS

BASIC MATERIALS	UNIT	MAT.	INST.	TOTAL
15140.11 PIPE HANGERS, LIGHT				
1-1/2" x				
6"	EA.	0.44	2.00	2.44
8"	"	0.48	2.15	2.63
10"	"	0.53	2.30	2.83
12"	"	0.58	2.50	3.08
2" x				
6"	EA.	0.47	2.15	2.62
8"	"	0.52	2.30	2.82
10"	"	0.56	2.50	3.06
12"	"	0.59	2.75	3.34
15175.60 EXPANSION TANKS				
Expansion tank, 125 psi, galvanized steel				
20 gallon	EA.	189.00	37.50	226.50
30 gallon	"	204.00	50.00	254.00
65 gallon	"	334.00	75.00	409.00
80 gallon	"	334.00	85.50	419.50
15240.10 VIBRATION CONTROL				
Vibration isolator, in-line, stainless connector, screwed				
1/2"	EA.	55.00	16.65	71.65
3/4"	"	64.00	17.65	81.65
1"	"	68.00	18.75	86.75
1-1/4"	"	77.00	20.00	97.00
1-1/2"	"	86.00	21.45	107.45
2"	"	108.00	23.10	131.10
2-1/2"	"	253.00	25.00	278.00
3"	"	286.00	27.30	313.30
4"	"	385.00	30.00	415.00
6"	"	572.00	33.30	605.30
Flanged				
8"	EA.	886.00	37.50	923.50
10"	"	1,490	42.90	1,533
12"	"	1,870	50.00	1,920

INSULATION

INSULATION	UNIT	MAT.	INST.	TOTAL
15260.10 FIBERGLASS PIPE INSULATION				
Fiberglass insulation on 1/2" pipe				
1" thick	L.F.	2.10	1.00	3.10
1-1/2" thick	"	2.40	1.25	3.65
3/4" pipe				

INSULATION	UNIT	MAT.	INST.	TOTAL
15260.10 FIBERGLASS PIPE INSULATION				
1" thick	L.F.	2.15	1.00	3.15
1-1/2" thick	"	2.55	1.25	3.80
1" pipe				
1" thick	L.F.	2.20	1.00	3.20
1-1/2" thick	"	2.75	1.25	4.00
2" thick	"	3.85	1.50	5.35
1-1/4" dia. pipe				
1" thick	L.F.	2.75	1.25	4.00
1-1/2" thick	"	3.30	1.35	4.65
2" thick	"	4.40	1.50	5.90
1-1/2" pipe				
1" thick	L.F.	2.85	1.25	4.10
1-1/2" thick	"	3.40	1.35	4.75
2" thick	"	5.05	1.45	6.50
2-1/2" thick	"	5.60	1.50	7.10
3-1/2" thick	"	8.90	1.60	10.50
2" pipe				
1" thick	L.F.	3.10	1.25	4.35
1-1/2" thick	"	3.65	1.35	5.00
2" thick	"	5.40	1.50	6.90
2-1/2" thick	"	5.95	1.65	7.60
3-1/2" thick	"	9.25	1.85	11.10
2-1/2" pipe				
1" thick	L.F.	3.40	1.25	4.65
1-1/2" thick	"	3.95	1.35	5.30
2" thick	"	5.85	1.50	7.35
2-1/2" thick	"	6.25	1.65	7.90
3" thick	"	7.80	1.85	9.65
3-1/2" thick	"	9.25	2.15	11.40
3" pipe				
1" thick	L.F.	3.50	1.45	4.95
1-1/2" thick	"	4.30	1.50	5.80
2" thick	"	5.95	1.65	7.60
2-1/2" thick	"	7.05	1.85	8.90
3" thick	"	8.35	2.15	10.50
3-1/2" thick	"	9.25	2.50	11.75
4" pipe				
1" thick	L.F.	4.40	1.45	5.85
1-1/2" thick	"	5.50	1.50	7.00
2" thick	"	7.70	1.65	9.35
2-1/2" thick	"	8.80	1.85	10.65
3" thick	"	10.45	2.15	12.60
3-1/2" thick	"	12.10	2.50	14.60
5" pipe				
1" thick	L.F.	4.40	1.45	5.85
2" thick	"	8.25	1.50	9.75
3" thick	"	11.00	1.75	12.75
4" thick	"	14.30	2.30	16.60
6" pipe				
1" thick	L.F.	5.50	1.60	7.10
2" thick	"	8.25	1.65	9.90
4" thick	"	18.70	2.00	20.70

INSULATION	UNIT	MAT.	INST.	TOTAL
15260.10 FIBERGLASS PIPE INSULATION				
6" thick	L.F.	30.80	2.75	33.55
8" pipe				
2" thick	L.F.	11.00	1.60	12.60
3" thick	"	15.40	1.65	17.05
4" thick	"	20.90	2.00	22.90
6" thick	"	36.30	2.75	39.05
10" pipe				
2" thick	L.F.	12.10	1.60	13.70
3" thick	"	17.60	1.65	19.25
4" thick	"	24.20	2.00	26.20
6" thick	"	40.70	2.75	43.45
12" pipe				
2" thick	L.F.	13.20	1.60	14.80
3" thick	"	19.80	1.65	21.45
4" thick	"	26.40	2.00	28.40
6" thick	"	46.20	2.75	48.95
15260.20 CALCIUM SILICATE				
Calcium silicate insulation, 6" pipe				
2" thick	L.F.	6.60	2.15	8.75
2-1/2" thick	"	8.45	2.30	10.75
3" thick	"	12.10	2.50	14.60
4" thick	"	16.50	2.75	19.25
6" thick	"	26.40	3.00	29.40
8" pipe				
2" thick	L.F.	7.70	2.30	10.00
2-1/2" thick	"	9.90	2.50	12.40
3" thick	"	14.85	2.75	17.60
4" thick	"	19.70	3.00	22.70
6" thick	"	31.90	3.35	35.25
10" pipe				
2" thick	L.F.	9.90	2.30	12.20
2-1/2" thick	"	12.65	2.50	15.15
3" thick	"	18.15	2.75	20.90
4" thick	"	24.20	3.00	27.20
6" thick	"	36.30	3.35	39.65
12" pipe				
2" thick	L.F.	11.00	2.30	13.30
2-1/2" thick	"	13.75	2.50	16.25
3" thick	"	19.80	2.75	22.55
4" thick	"	25.90	3.00	28.90
6" thick	"	45.10	3.35	48.45
15260.60 EXTERIOR PIPE INSULATION				
Fiberglass insulation, aluminum jacket				
1/2" pipe				
1" thick	L.F.	2.75	2.30	5.05
1-1/2" thick	"	3.25	2.50	5.75
3/4" pipe				
1" thick	L.F.	2.85	2.30	5.15
1-1/2" thick	"	3.35	2.50	5.85
1" pipe				

INSULATION	UNIT	MAT.	INST.	TOTAL
15260.60 EXTERIOR PIPE INSULATION				
1" thick	L.F.	2.95	2.30	5.25
1-1/2" thick	"	3.65	2.50	6.15
3-1/2" thick	"	9.90	3.35	13.25
1-1/4" pipe				
1" thick	L.F.	3.40	2.75	6.15
1-1/2" thick	"	3.85	2.85	6.70
3-1/2" thick	"	10.10	3.35	13.45
1-1/2" pipe				
1" thick	L.F.	3.50	2.75	6.25
1-1/2" thick	"	4.20	2.85	7.05
2" thick	"	5.85	3.00	8.85
2-1/2" thick	"	6.50	3.15	9.65
3-1/2" thick	"	10.10	3.35	13.45
2" pipe				
1" thick	L.F.	3.75	2.75	6.50
1-1/2" thick	"	4.50	2.85	7.35
2" thick	"	6.05	3.00	9.05
2-1/2" thick	"	6.70	3.15	9.85
3-1/2" thick	"	10.45	3.35	13.80
2-1/2" pipe				
1" thick	L.F.	3.95	2.75	6.70
1-1/2" thick	"	4.75	2.85	7.60
2" thick	"	6.70	3.00	9.70
2-1/2" thick	"	7.25	3.15	10.40
3" thick	"	8.80	3.35	12.15
3-1/2" thick	"	11.00	3.55	14.55
3" pipe				
1" thick	L.F.	4.20	3.00	7.20
1-1/2" thick	"	4.95	3.15	8.10
2" thick	"	7.00	3.35	10.35
2-1/2" thick	"	8.05	3.55	11.60
3" thick	"	9.50	3.75	13.25
3-1/2" thick	"	11.20	4.00	15.20
4" pipe				
1" thick	L.F.	4.95	3.00	7.95
1-1/2" thick	"	5.95	3.15	9.10
2" thick	"	7.50	3.35	10.85
2-1/2" thick	"	9.15	3.55	12.70
3" thick	"	11.20	3.75	14.95
3-1/2" thick	"	13.20	4.00	17.20
5" pipe				
1" thick	L.F.	5.15	3.00	8.15
2" thick	"	8.80	3.15	11.95
2-1/2" thick	"	10.00	3.35	13.35
3" thick	"	13.20	3.55	16.75
3-1/2" thick	"	13.85	3.75	17.60
6" pipe				
1" thick	L.F.	5.45	3.35	8.80
2" thick	"	9.00	3.55	12.55
2-1/2" thick	"	11.45	3.75	15.20
3" thick	"	13.40	4.00	17.40
3-1/2" thick	"	15.40	4.30	19.70

INSULATION	UNIT	MAT.	INST.	TOTAL
15260.60 EXTERIOR PIPE INSULATION				
4" thick	L.F.	20.90	4.60	25.50
6" thick	"	32.70	5.00	37.70
8" pipe				
2" thick	L.F.	11.00	3.35	14.35
3" thick	"	16.50	3.55	20.05
3-1/2" thick	"	22.00	3.75	25.75
4" thick	"	23.10	4.00	27.10
6" thick	"	38.50	4.30	42.80
10" pipe				
2" thick	L.F.	13.75	3.35	17.10
3" thick	"	19.80	3.55	23.35
3-1/2" thick	"	23.10	3.75	26.85
4" thick	"	27.50	4.00	31.50
6" thick	"	42.90	4.30	47.20
12" pipe				
2" thick	L.F.	15.40	3.35	18.75
3" thick	"	22.00	3.55	25.55
3-1/2" thick	"	27.50	3.75	31.25
4" thick	"	29.70	4.00	33.70
6" thick	"	47.30	4.30	51.60
Calcium silicate with aluminum jacket, 6" pipe				
2" thick	L.F.	8.05	3.75	11.80
2-1/2" thick	"	10.00	4.00	14.00
3" thick	"	13.65	4.30	17.95
4" thick	"	19.05	5.00	24.05
8" pipe				
2" thick	L.F.	9.35	3.75	13.10
2-1/2" thick	"	11.35	4.00	15.35
3" thick	"	15.95	4.30	20.25
4" thick	"	22.00	4.60	26.60
6" thick	"	33.00	5.00	38.00
10" pipe				
2" thick	L.F.	11.35	4.30	15.65
2-1/2" thick	"	13.20	4.60	17.80
3" thick	"	19.80	5.00	24.80
4" thick	"	27.50	5.45	32.95
6" thick	"	38.50	6.00	44.50
12" pipe				
2" thick	L.F.	12.75	4.30	17.05
2-1/2" thick	"	14.80	4.60	19.40
3" thick	"	20.90	5.00	25.90
4" thick	"	28.60	5.45	34.05
6" thick	"	49.50	6.00	55.50
15260.90 PIPE INSULATION FITTINGS				
Insulation protection saddle				
1" thick covering				
1/2" pipe	EA.	3.40	12.00	15.40
3/4" pipe	"	3.40	12.00	15.40
1" pipe	"	3.40	12.00	15.40
1-1/4" pipe	"	3.40	12.00	15.40

INSULATION	UNIT	MAT.	INST.	TOTAL
15260.90 PIPE INSULATION FITTINGS				
1-1/2" pipe	EA.	3.65	12.00	15.65
2" pipe	"	4.75	12.00	16.75
2-1/2" pipe	"	4.75	12.00	16.75
3" pipe	"	5.30	13.65	18.95
4" pipe	"	5.70	15.00	20.70
6" pipe	"	6.70	18.75	25.45
1-1/2" thick covering				
3/4" pipe	EA.	5.50	12.00	17.50
1" pipe	"	5.50	12.00	17.50
1-1/4" pipe	"	5.50	12.00	17.50
1-1/2" pipe	"	5.50	12.00	17.50
2" pipe	"	6.15	12.00	18.15
2-1/2" pipe	"	6.95	12.00	18.95
3" pipe	"	6.95	12.00	18.95
4" pipe	"	7.35	15.00	22.35
6" pipe	"	10.25	18.75	29.00
8" pipe	"	12.85	25.00	37.85
10" pipe	"	12.85	25.00	37.85
12" pipe	"	16.15	25.00	41.15
2" thick covering				
3/4" pipe	EA.	5.70	12.00	17.70
1" pipe	"	5.70	12.00	17.70
1-1/4" pipe	"	5.70	12.00	17.70
1-1/2" pipe	"	6.80	12.00	18.80
2" pipe	"	6.80	12.00	18.80
2-1/2" pipe	"	6.80	12.00	18.80
3" pipe	"	8.05	12.00	20.05
4" pipe	"	8.05	15.00	23.05
6" pipe	"	11.35	18.75	30.10
8" pipe	"	13.55	25.00	38.55
10" pipe	"	14.20	25.00	39.20
12" pipe	"	18.70	25.00	43.70
3" thick covering				
2" pipe	EA.	8.75	12.00	20.75
2-1/2" pipe	"	8.75	12.00	20.75
3" pipe	"	9.35	12.00	21.35
4" pipe	"	9.35	15.00	24.35
6" pipe	"	16.05	18.75	34.80
8" pipe	"	16.05	25.00	41.05
10" pipe	"	18.05	25.00	43.05
12" pipe	"	22.00	25.00	47.00
15280.10 EQUIPMENT INSULATION				
Equipment insulation, 2" thick, cellular glass	S.F.	1.45	1.85	3.30
Urethane, rigid, field applied jacket, plastered finish	"	1.60	3.75	5.35
Fiberglass, rigid, with vapor barrier	"	1.45	1.65	3.10
15290.10 DUCTWORK INSULATION				
Fiberglass duct insulation, plain blanket				
1-1/2" thick	S.F.	0.23	0.38	0.61

15 MECHANICAL

INSULATION

15290.10 DUCTWORK INSULATION	UNIT	MAT.	INST.	TOTAL
2" thick	S.F.	0.28	0.50	0.78
With vapor barrier				
1-1/2" thick	S.F.	0.30	0.38	0.68
2" thick	"	0.35	0.50	0.85
Rigid with vapor barrier				
2" thick	S.F.	1.10	1.00	2.10
3" thick	"	1.80	1.20	3.00
4" thick	"	2.35	1.50	3.85
6" thick	"	3.20	2.00	5.20
Weatherproof, polystyrene, 3" thick, w/vapor barrier	"	2.40	3.00	5.40
Urethane board with vapor barrier	"	3.35	3.75	7.10

FIRE PROTECTION

15330.10 WET SPRINKLER SYSTEM	UNIT	MAT.	INST.	TOTAL
Sprinkler head, 212 deg, brass, exposed piping	EA.	4.55	12.00	16.55
Chrome, concealed piping	"	4.45	16.65	21.10
Water motor alarm	"	61.50	50.00	111.50
Fire department inlet connection	"	94.50	60.00	154.50
Wall plate for fire dept connection	"	46.20	25.00	71.20
Swing check valve flanged iron body, 4"	"	102.00	100.00	202.00
Check valve, 6"	"	456.00	150.00	606.00
Wet pipe valve, flange to groove, 4"	"	402.00	33.30	435.30
Flange to flange				
6"	EA.	506.00	50.00	556.00
8"	"	666.00	100.00	766.00
Alarm valve, flange to flange, (wet valve)				
4"	EA.	330.00	33.30	363.30
8"	"	566.00	250.00	816.00
Inspector's test connection	"	28.10	25.00	53.10
Wall hydrant, polished brass, 2-1/2" x 2-1/2", single	"	264.00	21.45	285.45
2-way	"	726.00	21.45	747.45
3-way	"	1,070	21.45	1,091
Wet valve trim, includes retard chamber & guages, 4"-6"	"	368.00	25.00	393.00
Retard pressure switch for wet systems	"	671.00	60.00	731.00
Air maintenance device	"	189.00	25.00	214.00
Wall hydrant non-freeze, 8" thick wall, vacuum breaker	"	15.60	15.00	30.60
12" thick wall	"	16.85	15.00	31.85
15330.50 DRY SPRINKLER SYSTEM				
Dry pipe valve, flange to flange				
4"	EA.	764.00	60.00	824.00

15 MECHANICAL

FIRE PROTECTION	UNIT	MAT.	INST.	TOTAL
15330.50 DRY SPRINKLER SYSTEM				
6"	EA.	990.00	75.00	1,065
Trim, 4" and 6", includes gauges	"	358.00	25.00	383.00
Field testing and flushing	"	0.00	250.00	250.00
Disinfection	"	0.00	250.00	250.00
Pressure switch double circuit, open/close contacts	"	178.00	75.00	253.00
Low air				
Supervisory unit	EA.	605.00	50.00	655.00
Pressure switch	"	190.00	25.00	215.00
15330.70 CO2 SYSTEM				
CO2 system, high pressure, 75# cylinder with				
Valve assemblies	EA.	1,120	60.00	1,180
Storage rack	"	649.00	42.90	691.90
Manifold	"	478.00	214.00	692.00
Flexible loops	"	34.10	3.75	37.85
Beam scale for cylinders	"	324.00	50.00	374.00
Mechanically control head	"	248.00	20.00	268.00
Electrically control head	"	248.00	20.00	268.00
Stop valves	"	671.00	30.00	701.00
Check valves	"	297.00	37.50	334.50
Activation station	"	352.00	30.00	382.00
Nozzles	"	57.00	25.00	82.00
Hose reel with 75' of 3/4" hose	"	1,970	150.00	2,120
Main/reserve transfer switch	"	2,310	50.00	2,360
Pressure switch	"	209.00	30.00	239.00
Heat responsive device	"	341.00	50.00	391.00
Battery and charger	"	1,880	150.00	2,030
Low pressure				
Battery and charger	EA.	1,880	150.00	2,030
Pressure switch	"	198.00	33.30	231.30
Nozzles	"	57.00	27.30	84.30
Master selector valve	"	167.00	50.00	217.00
Selector valve	"	2,300	50.00	2,350
Low pressure hose reel with 75' of 3/4" hose	"	2,910	150.00	3,060
Tank fill lines	"	660.00	37.50	697.50
Activation stations	"	330.00	25.00	355.00
Electro manual pilot panels	"	660.00	37.50	697.50
15330.90 HALON SYSTEM				
Halon fire protection system, per computer room				
Minimum	EA.			2,200
Average	"			3,140
Maximum	"			4,950

PLUMBING

15410.05 C.I. PIPE, ABOVE GROUND

	UNIT	MAT.	INST.	TOTAL
No hub pipe				
1-1/2" pipe	L.F.	3.85	2.15	6.00
2" pipe	"	4.05	2.50	6.55
3" pipe	"	6.05	3.00	9.05
4" pipe	"	8.25	5.00	13.25
6" pipe	"	11.00	6.00	17.00
8" pipe	"	13.75	10.00	23.75
10" pipe	"	24.20	12.00	36.20
No hub fittings, 1-1/2" pipe				
1/4 bend	EA.	4.95	10.00	14.95
1/8 bend	"	3.30	10.00	13.30
Sanitary tee	"	6.05	15.00	21.05
Sanitary cross	"	7.50	15.00	22.50
Plug	"			1.10
Coupling	"			6.05
Wye	"	6.05	15.00	21.05
Tapped tee	"	6.15	10.00	16.15
P-trap	"	9.35	10.00	19.35
Tapped cross	"	9.55	10.00	19.55
2" pipe				
1/4 bend	EA.	5.05	12.00	17.05
1/8 bend	"	3.40	12.00	15.40
Sanitary tee	"	6.25	20.00	26.25
Sanitary cross	"	7.90	20.00	27.90
Plug	"			1.30
Coupling	"			6.40
Wye	"	6.25	25.00	31.25
Double wye	"	8.80	25.00	33.80
2x1-1/2" wye & 1/8 bend	"	8.60	18.75	27.35
Double wye & 1/8 bend	"	11.55	25.00	36.55
Test tee less 2" plug	"	5.85	12.00	17.85
Tapped tee				
2"x2"	EA.	6.80	12.00	18.80
2"x1-1/2"	"	6.70	12.00	18.70
P-trap				
2"x2"	EA.	9.55	12.00	21.55
Tapped cross				
2"x1-1/2"	EA.	9.90	12.00	21.90
3" pipe				
1/4 bend	EA.	6.40	15.00	21.40
1/8 bend	"	5.70	15.00	20.70
Sanitary tee	"	7.50	18.75	26.25
3"x2" sanitary tee	"	7.25	18.75	26.00
3"x1-1/2" sanitary tee	"	7.35	18.75	26.10
Sanitary cross	"	11.20	25.00	36.20
3x2" sanitary cross	"	12.35	25.00	37.35
Plug	"	2.00	15.55	17.55
Coupling	"			7.70
Wye	"	7.70	25.00	32.70
3x2" wye	"	7.90	25.00	32.90
Double wye	"	11.55	25.00	36.55
3x2" double wye	"	11.20	25.00	36.20

PLUMBING

15410.05 C.I. PIPE, ABOVE GROUND

PLUMBING	UNIT	MAT.	INST.	TOTAL
3x2" wye & 1/8 bend	EA.	9.90	21.45	31.35
3x1-1/2" wye & 1/8 bend	"	9.90	21.45	31.35
Double wye & 1/8 bend	"	18.15	25.00	43.15
3x2" double wye & 1/8 bend	"	13.75	25.00	38.75
3x2" reducer	"	4.20	13.65	17.85
Test tee, less 3" plug	"	8.80	15.00	23.80
Plug	"			2.75
3x3" tapped tee	"	13.75	15.00	28.75
3x2" tapped tee	"	12.30	15.00	27.30
3x1-1/2" tapped tee	"	12.10	15.00	27.10
P-trap	"	11.55	15.00	26.55
3x2" tapped cross	"	9.35	15.00	24.35
3x1-1/2" tapped cross	"	9.25	15.00	24.25
Closet flange, 3-1/2" deep	"	8.25	7.50	15.75
4" pipe				
1/4 bend	EA.	8.35	15.00	23.35
1/8 bend	"	7.50	15.00	22.50
Sanitary tee	"	9.35	25.00	34.35
4x3" sanitary tee	"	9.15	25.00	34.15
4x2" sanitary tee	"	8.60	25.00	33.60
Sanitary cross	"	23.10	30.00	53.10
4x3" sanitary cross	"	17.05	30.00	47.05
4x2" sanitary cross	"	14.95	30.00	44.95
Plug	"			2.85
Coupling	"			8.25
Wye	"	10.45	25.00	35.45
4x3" wye	"	8.80	25.00	33.80
4x2" wye	"	6.70	25.00	31.70
Double wye	"	24.20	30.00	54.20
4x3" double wye	"	14.30	30.00	44.30
4x2" double wye	"	12.65	30.00	42.65
Wye & 1/8 bend	"	13.20	25.00	38.20
4x3" wye & 1/8 bend	"	9.80	25.00	34.80
4x2" wye & 1/8 bend	"	7.90	25.00	32.90
Double wye & 1/8 bend	"	30.80	30.00	60.80
4x3" double wye & 1/8 bend	"	20.90	30.00	50.90
4x2" double wye & 1/8 bend	"	14.30	30.00	44.30
4x3" reducer	"	5.05	15.00	20.05
4x2" reducer	"	4.95	15.00	19.95
Test tee, less 4" plug	"	11.00	15.00	26.00
Plug	"			4.40
4x2" tapped tee	"	7.70	15.00	22.70
4x1-1/2" tapped tee	"	6.60	15.00	21.60
P-trap	"	19.70	15.00	34.70
4x2" tapped cross	"	12.10	15.00	27.10
4x1-1/2" tapped cross	"	9.80	15.00	24.80
Closet flange				
3" deep	EA.	7.50	15.00	22.50
8" deep	"	13.85	15.00	28.85
6" pipe				
1/4 bend	EA.	15.95	25.00	40.95
1/8 bend	"	14.30	25.00	39.30

PLUMBING

15410.05 C.I. PIPE, ABOVE GROUND

	UNIT	MAT.	INST.	TOTAL
Sanitary tee	EA.	26.40	30.00	56.40
6x4" sanitary tee	"	18.15	30.00	48.15
Coupling	"			17.60
Wye	"	25.30	30.00	55.30
6x4" wye	"	17.05	30.00	47.05
6x3" wye	"	15.20	30.00	45.20
6x2" wye	"	13.85	30.00	43.85
Double wye	"	36.30	37.50	73.80
6x4" double wye	"	30.80	37.50	68.30
Wye & 1/8 bend	"	32.50	30.00	62.50
6x4" wye & 1/8 bend	"	21.65	30.00	51.65
6x3" wye & 1/8 bend	"	21.10	30.00	51.10
6x2" wye & 1/8 bend	"	16.60	30.00	46.60
6x4" reducer	"	10.00	16.65	26.65
6x4" reducer	"	9.90	15.00	24.90
6x3" reducer	"	9.90	16.65	26.55
Test tee				
Less 6" plug	EA.	24.20	18.75	42.95
Plug	"	6.40	10.35	16.75
P-trap	"	39.60	18.75	58.35
8" pipe				
1/4 bend	EA.	37.40	25.00	62.40
1/8 bend	"	24.20	25.00	49.20
Sanitary tee	"	61.50	37.50	99.00
8x6" sanitary tee	"	44.00	37.50	81.50
Plug	"			11.00
Coupling	"			16.50
Wye	"	50.50	30.00	80.50
8x6" wye	"	39.60	25.00	64.60
8x4" wye	"	30.80	25.00	55.80
Double wye	"	72.50	25.00	97.50
Wye & 1/8 bend	"	75.00	25.00	100.00
8x6" wye & 1/8 bend	"	48.40	25.00	73.40
8x4" wye & 1/8 bend	"	39.60	25.00	64.60
8x6" reducer	"	14.30	25.00	39.30
8x4" reducer	"	13.75	25.00	38.75
8x3" reducer	"	12.00	25.00	37.00
8x2" reducer	"	11.90	16.65	28.55
Test tee				
Less 8" plug	EA.	48.40	25.00	73.40
Plug	"			11.20
10" pipe				
1/4 bend	EA.	75.00	25.00	100.00
1/8 bend	"	50.50	25.00	75.50
Plug	"			18.70
Coupling	"			25.30
Wye	"	103.00	50.00	153.00
10x8" wye	"	94.50	50.00	144.50
10x6" wye	"	83.50	50.00	133.50
10x4" wye	"	83.50	50.00	133.50
10x8" reducer	"	30.80	25.00	55.80
10x6" reducer	"	30.80	25.00	55.80

15 MECHANICAL

PLUMBING	UNIT	MAT.	INST.	TOTAL
15410.05 C.I. PIPE, ABOVE GROUND				
10x4" reducer	EA.	23.65	25.00	48.65
15410.06 C.I. PIPE, BELOW GROUND				
No hub pipe				
1-1/2" pipe	L.F.	3.85	1.50	5.35
2" pipe	"	4.05	1.65	5.70
3" pipe	"	6.05	1.85	7.90
4" pipe	"	8.25	2.50	10.75
6" pipe	"	11.00	2.75	13.75
8" pipe	"	13.75	3.35	17.10
10" pipe	"	24.20	3.75	27.95
Fittings, 1-1/2"				
1/4 bend	EA.	4.95	8.55	13.50
1/8 bend	"	3.30	8.55	11.85
Plug	"	1.10	4.45	5.55
Wye	"	6.05	12.00	18.05
Wye & 1/8 bend	"	6.35	8.55	14.90
P-trap	"	9.35	8.55	17.90
2"				
1/4 bend	EA.	5.05	10.00	15.05
1/8 bend	"	3.40	10.00	13.40
Plug	"			1.30
Double wye	"	8.80	18.75	27.55
Wye & 1/8 bend	"	8.60	15.00	23.60
Double wye & 1/8 bend	"	11.55	18.75	30.30
P-trap	"	9.55	10.00	19.55
3"				
1/4 bend	EA.	6.40	12.00	18.40
1/8 bend	"	5.70	12.00	17.70
Plug	"			2.00
Wye	"	7.70	18.75	26.45
3x2" wye	"	7.90	18.75	26.65
Wye & 1/8 bend	"	9.90	18.75	28.65
Double wye & 1/8 bend	"	18.15	18.75	36.90
3x2" double wye & 1/8 bend	"	13.75	18.75	32.50
3x2" reducer	"	4.20	12.00	16.20
P-trap	"	11.55	12.00	23.55
4"				
1/4 bend	EA.	8.35	12.00	20.35
1/8 bend	"	7.50	12.00	19.50
Plug	"			2.85
Wye	"	10.45	18.75	29.20
4x3" wye	"	8.80	18.75	27.55
4x2" wye	"	6.70	18.75	25.45
Double wye	"	24.20	25.00	49.20
4x3" double wye	"	14.30	25.00	39.30
4x2" double wye	"	12.65	25.00	37.65
Wye & 1/8 bend	"	13.20	18.75	31.95
4x3" wye & 1/8 bend	"	9.80	18.75	28.55
4x2" wye & 1/8 bend	"	7.90	18.75	26.65

PLUMBING

15410.06 C.I. PIPE, BELOW GROUND

	UNIT	MAT.	INST.	TOTAL
Double wye & 1/8 bend	EA.	30.80	25.00	55.80
4x3" double wye & 1/8 bend	"	20.90	25.00	45.90
4x2" double wye & 1/8 bend	"	14.30	25.00	39.30
4x3" reducer	"	5.05	12.00	17.05
4x2" reducer	"	4.95	12.00	16.95
6"				
1/4 bend	EA.	15.95	18.75	34.70
1/8 bend	"	14.30	18.75	33.05
Wye & 1/8 bend	"	32.50	25.00	57.50
6x4" wye & 1/8 bend	"	21.65	25.00	46.65
6x3" wye & 1/8 bend	"	21.10	25.00	46.10
6x2" wye & 1/8 bend	"	16.60	25.00	41.60
6x3" reducer	"	9.90	13.65	23.55
P-trap	"	39.60	15.00	54.60
8"				
1/4 bend	EA.	37.40	18.75	56.15
1/8 bend	"	24.20	18.75	42.95
Plug	"			11.00
Wye	"	50.50	25.00	75.50
8x6" wye	"	39.60	18.75	58.35
8x4" wye	"	30.80	18.75	49.55
8x6" wye & 1/8 bend	"	48.40	18.75	67.15
8x4" reducer	"	13.75	15.00	28.75
8x3" reducer	"	12.00	15.00	27.00
8x2" reducer	"	11.90	15.00	26.90
10"				
1/4 bend	EA.	75.00	18.75	93.75
1/8 bend	"	50.50	18.75	69.25
Plug	"			18.70
Wye	"	103.00	37.50	140.50
10x8" wye	"	94.50	37.50	132.00
10x6" wye	"	83.50	37.50	121.00
10x4" wye	"	83.50	37.50	121.00
10x8" reducer	"	30.80	18.75	49.55
10x6" reducer	"	30.80	18.75	49.55

15410.08 EXTRA HEAVY SOIL PIPE

	UNIT	MAT.	INST.	TOTAL
Extra heavy soil pipe, single hub				
2" x 5'	EA.	19.80	6.00	25.80
3" x 5'	"	25.30	6.40	31.70
4" x 5'	"	38.50	7.00	45.50
6" x 5'	"	57.00	7.50	64.50
Double hub				
2" x 5'	EA.	28.60	7.50	36.10
4" x 5'	"	39.60	7.90	47.50
5" x 5'	"	54.00	8.35	62.35
6" x 5'	"	66.00	8.55	74.55
Single hub				
3" x 10'	EA.	39.60	5.00	44.60
4" x 10'	"	50.50	5.15	55.65
6" x 10'	"	77.00	5.45	82.45

PLUMBING	UNIT	MAT.	INST.	TOTAL
15410.08 EXTRA HEAVY SOIL PIPE				
8" x 10'	EA.	121.00	5.75	126.75
"Mini", single hub				
2" x 42"	EA.	16.50	6.00	22.50
3" x 42"	"	19.80	6.40	26.20
4" x 42"	"	20.90	7.00	27.90
6" x 42"	"	28.60	7.50	36.10
8" x 42"	"	39.60	7.90	47.50
Fittings, 1/4" bend				
2"	EA.	6.15	10.00	16.15
3"	"	10.20	12.00	22.20
4"	"	11.95	12.00	23.95
1/8 bend				
2"	EA.	5.05	10.00	15.05
3"	"	8.60	12.00	20.60
4"	"	12.65	12.00	24.65
5"	"	16.50	15.00	31.50
6"	"	20.35	18.75	39.10
8"	"	46.20	18.75	64.95
Long sweep				
2"	EA.	11.00	10.00	21.00
3"	"	16.50	12.00	28.50
4"	"	23.10	12.00	35.10
Straight T				
4" x 2"	EA.	20.35	18.75	39.10
4" x 3"	"	22.00	18.75	40.75
4"	"	23.65	21.45	45.10
Sanitary T				
3" x 2"	EA.	17.60	18.75	36.35
3"	"	19.80	18.75	38.55
4" x 2"	"	20.35	21.45	41.80
4" x 3"	"	21.75	21.45	43.20
4"	"	23.65	23.10	46.75
4" x 2"	"	20.35	21.45	41.80
4" x 3"	"	23.65	21.45	45.10
4"	"	27.50	23.10	50.60
Combination Y and 1/8 bend, 4"	"	27.50	23.10	50.60
Double wye, 4"	"	30.80	23.10	53.90
Tapped sanitary T, 4" x 2"	"	15.40	25.00	40.40
P trap				
2"	EA.	11.55	12.00	23.55
4"	"	27.00	20.00	47.00
Dandy				
4", with 3" brass plug	EA.	21.25	16.65	37.90
6", 4" brass plug	"	38.50	18.75	57.25
15410.09 SERVICE WEIGHT PIPE				
Service weight pipe, single hub				
2" x 5'	EA.	18.15	6.00	24.15
3" x 5'	"	20.90	6.40	27.30
4" x 5'	"	25.30	6.65	31.95
5" x 5'	"	30.80	7.15	37.95
6" x 5'	"	42.90	7.50	50.40

PLUMBING	UNIT	MAT.	INST.	TOTAL
15410.09 SERVICE WEIGHT PIPE				
8" x 5'	EA.	61.50	7.90	69.40
10" x 5'	"	91.50	8.35	99.85
12" x 5'	"	121.00	9.35	130.35
Double hub				
2" x 5'	EA.	19.80	7.50	27.30
3" x 5'	"	22.00	8.10	30.10
4" x 5'	"	27.50	8.55	36.05
5" x 5'	"	38.50	9.35	47.85
6" x 5'	"	45.10	10.00	55.10
10" x 5'	"	92.50	10.70	103.20
12' x 5'	"	121.00	11.55	132.55
Single hub				
2" x 10'	EA.	23.10	7.50	30.60
3" x 10'	"	29.20	8.10	37.30
4" x 10'	"	36.30	8.55	44.85
5" x 10'	"	49.00	9.35	58.35
6" x 10'	"	58.50	10.00	68.50
8" x 10'	"	92.50	10.70	103.20
10" x 10'	"	153.00	11.55	164.55
12" x 10'	"	222.00	12.50	234.50
Shorty				
2" x 42"	EA.	12.10	6.00	18.10
3" x 42"	"	13.20	6.40	19.60
4" x 42"	"	16.50	6.65	23.15
5" x 42"	"	23.10	7.15	30.25
6" x 42"	"	25.30	7.50	32.80
8" x 42"	"	36.30	7.90	44.20
10" x 42"	"	50.50	8.35	58.85
Soil plug				
2"	EA.	2.65	10.00	12.65
3"	"	3.75	10.70	14.45
4"	"	4.30	11.55	15.85
5"	"	6.60	12.00	18.60
6"	"	6.95	12.50	19.45
1/5 bend				
3"	EA.	8.80	12.00	20.80
4"	"	12.10	13.65	25.75
1/6 bend				
2"	EA.	6.60	10.00	16.60
3"	"	9.00	12.00	21.00
4"	"	12.45	13.65	26.10
1/8 bend				
2"	EA.	3.50	10.00	13.50
3"	"	5.50	12.00	17.50
4"	"	7.70	13.65	21.35
5"	"	10.45	14.30	24.75
6"	"	12.10	15.00	27.10
8"	"	38.50	18.75	57.25
10"	"	53.00	21.45	74.45
12"	"	86.00	25.00	111.00
1/16 bend				
2"	EA.	3.75	10.00	13.75

15 MECHANICAL

PLUMBING

15410.09 SERVICE WEIGHT PIPE

	UNIT	MAT.	INST.	TOTAL
3"	EA.	5.60	12.00	17.60
4"	"	7.70	13.65	21.35
5"	"	10.45	14.30	24.75
6"	"	11.55	15.00	26.55
1/4 bend				
2"	EA.	4.40	10.00	14.40
3"	"	6.95	12.00	18.95
4"	"	9.80	13.65	23.45
5"	"	12.65	14.30	26.95
6"	"	15.95	15.00	30.95
8"	"	47.90	18.75	66.65
10"	"	67.00	21.45	88.45
12"	"	90.00	25.00	115.00
1/4 bend, long				
2" x 12"	EA.	8.25	10.70	18.95
4" x 18"	"	17.80	15.00	32.80
4" x 12"	"	16.50	15.00	31.50
Sweep				
2"	EA.	6.60	10.00	16.60
3"	"	9.35	12.00	21.35
4"	"	14.20	13.65	27.85
5"	"	20.80	14.30	35.10
6"	"	25.30	15.00	40.30
8"	"	61.50	18.75	80.25
Reducing long sweep				
3" x 2"	EA.	11.00	12.00	23.00
4" x 3"	"	13.20	15.00	28.20
Straight T				
2"	EA.	8.25	18.75	27.00
3"	"	12.65	20.00	32.65
4" x 2"	"	12.65	21.45	34.10
4" x 3"	"	14.85	23.10	37.95
4"	"	16.05	25.00	41.05
Sanitary T				
2"	EA.	6.70	18.75	25.45
3" x 2"	"	9.70	20.00	29.70
3"	"	10.95	21.45	32.40
4" x 2"	"	11.30	23.10	34.40
4" x 3"	"	13.10	25.00	38.10
4"	"	13.85	25.00	38.85
5"	"	24.00	27.30	51.30
6"	"	27.50	27.30	54.80
Wye				
2"	EA.	6.60	15.00	21.60
3" x 2"	"	9.35	16.65	26.00
3"	"	11.65	16.65	28.30
4" x 2"	"	11.40	17.65	29.05
4" x 3"	"	13.40	17.65	31.05
4"	"	14.85	17.65	32.50
5" x 2"	"	15.85	18.75	34.60
5" x 3"	"	16.95	18.75	35.70
5" x 4"	"	19.25	18.75	38.00

PLUMBING	UNIT	MAT.	INST.	TOTAL
15410.09 SERVICE WEIGHT PIPE				
5"	EA.	23.10	18.75	41.85
6" x 2"	"	20.90	18.75	39.65
6" x 3"	"	21.10	18.75	39.85
6" x 4"	"	22.55	20.00	42.55
6" x 5"	"	29.20	20.00	49.20
6"	"	31.90	21.45	53.35
8" x 4"	"	39.50	21.45	60.95
8" x 6"	"	48.40	23.10	71.50
8"	"	80.50	25.00	105.50
10" x 4"	"	70.50	25.00	95.50
10" x 6"	"	76.00	27.30	103.30
10" x 8"	"	106.00	30.00	136.00
Service wye				
10"	EA.	121.00	37.50	158.50
12" x 4"	"	107.00	37.50	144.50
12" x 6"	"	111.00	37.50	148.50
12" x 8"	"	132.00	42.90	174.90
12"	"	198.00	50.00	248.00
Combination wye and 1/8 bend				
2"	EA.	8.80	20.00	28.80
3" x 2"	"	10.00	21.45	31.45
3"	"	13.20	21.45	34.65
4" x 2"	"	12.65	23.10	35.75
4" x 3"	"	14.30	23.10	37.40
4"	"	16.50	25.00	41.50
5" x 4"	"	28.10	25.00	53.10
6" x 4"	"	28.40	30.00	58.40
8"	"	92.50	37.50	130.00
Straight cross, 4"	"	18.60	25.00	43.60
Sanitary cross				
2"	EA.	9.70	21.45	31.15
3"	"	15.40	23.10	38.50
3" x 2"	"	13.20	23.10	36.30
4"	"	20.15	25.00	45.15
4" x 3"	"	17.95	25.00	42.95
4" x 2"	"	16.85	27.30	44.15
Double wye				
2"	EA.	9.90	21.45	31.35
3" x 2"	"	13.75	23.10	36.85
3"	"	15.30	23.10	38.40
4" x 2"	"	16.40	25.00	41.40
4" x 3"	"	18.70	25.00	43.70
4"	"	22.00	25.00	47.00
5"	"	32.50	30.00	62.50
6" x 4"	"	36.90	37.50	74.40
Combinatioin double wye and 1/8 bend				
2"	EA.	15.40	21.45	36.85
3" x 2"	"	17.05	23.10	40.15
3"	"	20.25	23.10	43.35
4" x 3"	"	23.10	25.00	48.10
4"	"	27.50	25.00	52.50
Tapped sanitary T				

PLUMBING

15410.09 SERVICE WEIGHT PIPE

	UNIT	MAT.	INST.	TOTAL
2" x 1-1/2"	EA.	8.80	21.45	30.25
2" x 2"	"	8.90	21.45	30.35
3" x 1-1/2"	"	9.70	23.10	32.80
3" x 2"	"	9.70	23.10	32.80
4" x 1-1/2"	"	11.00	25.00	36.00
4" x 2"	"	11.20	25.00	36.20
Tapped straight T				
3" x 1-1/2"	EA.	8.80	23.10	31.90
3" x 2"	"	8.80	23.10	31.90
4" x 2"	"	11.00	25.00	36.00
4" x 1-1/2"	"	14.85	25.00	39.85
Tapped Y				
4" x 2"	EA.	14.95	25.00	39.95
Tapped sanitary cross				
2" x 1-1/2"	EA.	7.60	21.45	29.05
2" x 2"	"	7.60	23.10	30.70
3" x 1-1/2"	"	9.90	23.10	33.00
3" x 2"	"	9.90	23.10	33.00
3" x 3"	"	14.95	23.10	38.05
4" x 1-1/2"	"	13.20	25.00	38.20
4" x 2"	"	13.20	25.00	38.20
Cleanout, dandy, with brass plug				
2", 1-1/2" plug	EA.	10.90	21.45	32.35
3", 2" plug	"	21.45	23.10	44.55
4", 3" plug	"	25.30	25.00	50.30
5", 4" plug	"	30.80	27.30	58.10
6", 4" plug	"	38.50	30.00	68.50
8", 6" plug	"	79.00	37.50	116.50
Reducer, 3" x 2"	"	5.70	21.45	27.15
4" x				
2"	EA.	5.95	23.10	29.05
3"	"	6.95	23.10	30.05
5" x				
2"	EA.	8.45	25.00	33.45
3"	"	8.60	25.00	33.60
4"	"	8.70	25.00	33.70
6" x				
2"	EA.	8.80	30.00	38.80
3"	"	9.90	30.00	39.90
4"	"	10.90	30.00	40.90
5"	"	11.35	30.00	41.35
8" x				
4"	EA.	19.25	37.50	56.75
6"	"	19.60	37.50	57.10
10" x				
4"	EA.	27.50	42.90	70.40
6"	"	28.60	42.90	71.50
8"	"	31.90	42.90	74.80
12" x				
4"	EA.	43.50	50.00	93.50
6"	"	46.80	50.00	96.80
8"	"	47.30	50.00	97.30

PLUMBING	UNIT	MAT.	INST.	TOTAL
15410.09 SERVICE WEIGHT PIPE				
10"	EA.	49.50	50.00	99.50
P trap				
2"	EA.	7.70	21.45	29.15
3"	"	10.45	23.10	33.55
4"	"	15.40	25.00	40.40
5"	"	30.80	30.00	60.80
6"	"	43.50	37.50	81.00
15410.10 COPPER PIPE				
Type "K" copper				
1/2"	L.F.	2.10	0.94	3.04
3/4"	"	4.30	1.00	5.30
1"	"	5.40	1.05	6.45
1-1/4"	"	6.50	1.15	7.65
1-1/2"	"	8.25	1.25	9.50
2"	"	11.55	1.35	12.90
2-1/2"	"	13.75	1.50	15.25
3"	"	16.30	1.60	17.90
4"	"	22.00	1.65	23.65
DWV, copper				
1-1/4"	L.F.	5.15	1.25	6.40
1-1/2"	"	6.50	1.35	7.85
2"	"	8.60	1.50	10.10
3"	"	11.90	1.65	13.55
4"	"	20.35	1.85	22.20
6"	"	55.00	2.15	57.15
Refrigeration tubing, copper, sealed				
1/8"	L.F.	0.61	1.20	1.81
3/16"	"	0.66	1.25	1.91
1/4"	"	0.77	1.30	2.07
5/16"	"	0.99	1.35	2.34
3/8"	"	1.10	1.45	2.55
1/2"	"	1.45	1.50	2.95
7/8"	"	2.75	1.70	4.45
1-1/8"	"	4.20	2.00	6.20
1-3/8"	"	6.95	2.30	9.25
Type "L" copper				
1/4"	L.F.	0.66	0.88	1.54
3/8"	"	0.99	0.88	1.87
1/2"	"	1.75	0.94	2.69
3/4"	"	2.20	1.00	3.20
1"	"	3.85	1.05	4.90
1-1/4"	"	5.30	1.15	6.45
1-1/2"	"	6.50	1.25	7.75
2"	"	8.80	1.35	10.15
2-1/2"	"	11.55	1.50	13.05
3"	"	13.20	1.60	14.80
3-1/2"	"	15.05	1.60	16.65
4"	"	22.00	1.65	23.65
Type "M" copper				
1/2"	L.F.	1.10	0.94	2.04

PLUMBING	UNIT	MAT.	INST.	TOTAL
15410.10 COPPER PIPE				
3/4"	L.F.	1.65	1.00	2.65
1"	"	2.20	1.05	3.25
1-1/4"	"	3.30	1.15	4.45
2"	"	7.70	1.35	9.05
2-1/2"	"	9.90	1.50	11.40
3"	"	11.00	1.60	12.60
4"	"	13.75	1.65	15.40
Type "K" tube, coil				
1/4" x 60'	EA.			48.40
1/2" x 60'	"			92.50
1/2" x 100'	"			132.00
3/4" x 60'	"			143.00
3/4" x 100'	"			220.00
1" x 60'	"			220.00
1" x 100'	"			330.00
1-1/4" x 60'	"			330.00
1-1/2" x 60'	"			418.00
2" x 40'	"			440.00
Type "L" tube, coil				
1/4" x 60'	EA.			44.00
3/8" x 60'	"			66.00
1/2" x 60'	"			88.00
1/2" x 100'	"			132.00
3/4" x 60'	"			132.00
3/4" x 100'	"			198.00
1" x 60'	"			220.00
1" x 100'	"			330.00
1-1/4" x 60'	"			330.00
1-1/2" x 60'	"			352.00
15410.11 COPPER FITTINGS				
Coupling, with stop				
1/4"	EA.	0.33	10.00	10.33
3/8"	"	0.64	12.00	12.64
1/2"	"	0.77	13.05	13.82
5/8"	"	0.99	15.00	15.99
3/4"	"	1.10	16.65	17.75
1"	"	2.20	17.65	19.85
3"	"	17.60	30.00	47.60
4"	"	39.60	37.50	77.10
Reducing coupling				
1/4" x 1/8"	EA.	1.10	12.00	13.10
3/8" x 1/4"	"	1.20	13.05	14.25
1/2" x				
3/8"	EA.	1.15	15.00	16.15
1/4"	"	1.15	15.00	16.15
1/8"	"	1.15	15.00	16.15
3/4" x				
3/8"	EA.	1.55	16.65	18.20
1/2"	"	1.60	16.65	18.25
1" x				

PLUMBING	UNIT	MAT.	INST.	TOTAL
15410.11 COPPER FITTINGS				
3/8"	EA.	3.30	18.75	22.05
1" x 1/2"	"	3.40	18.75	22.15
1" x 3/4"	"	3.50	18.75	22.25
1-1/4" x				
1/2"	EA.	3.95	20.00	23.95
3/4"	"	4.05	20.00	24.05
1"	"	4.30	20.00	24.30
1-1/2" x				
1/2"	EA.	7.15	21.45	28.60
3/4"	"	7.25	21.45	28.70
1"	"	7.35	21.45	28.80
1-1/4"	"	7.50	21.45	28.95
2" x				
1/2"	EA.	10.45	25.00	35.45
3/4"	"	10.65	25.00	35.65
1"	"	10.90	25.00	35.90
1-1/4"	"	11.10	25.00	36.10
1-1/2"	"	11.20	25.00	36.20
2-1/2" x				
1"	EA.	22.00	30.00	52.00
1-1/4"	"	22.55	30.00	52.55
1-1/2"	"	22.55	30.00	52.55
2"	"	22.75	30.00	52.75
3" x				
1-1/2"	EA.	28.60	37.50	66.10
2"	"	29.70	37.50	67.20
2-1/2"	"	30.20	37.50	67.70
4" x				
2"	EA.	62.50	42.90	105.40
2-1/2"	"	64.00	42.90	106.90
3"	"	65.00	42.90	107.90
Slip coupling				
1/8"	EA.	0.33	10.00	10.33
1/2"	"	0.50	12.00	12.50
3/4"	"	1.10	15.00	16.10
1"	"	2.10	16.65	18.75
1-1/4"	"	4.40	18.75	23.15
1-1/2"	"	5.70	20.00	25.70
2"	"	8.70	25.00	33.70
2-1/2"	"	9.90	25.00	34.90
3"	"	17.60	30.00	47.60
4"	"	37.40	37.50	74.90
Coupling with drain				
1/2"	EA.	3.95	15.00	18.95
3/4"	"	5.50	16.65	22.15
1"	"	7.15	18.75	25.90
Reducer				
3/8" x 1/4"	EA.	1.20	12.00	13.20
1/2" x 3/8"	"	1.30	12.00	13.30
3/4" x				
1/4"	EA.	1.85	13.65	15.50
3/8"	"	2.00	13.65	15.65

PLUMBING	UNIT	MAT.	INST.	TOTAL
15410.11 COPPER FITTINGS				
1/2"	EA.	2.10	13.65	15.75
1" x				
1/2"	EA.	3.20	15.00	18.20
3/4"	"	3.30	15.00	18.30
1-1/4" x				
1/2"	EA.	4.50	16.65	21.15
3/4"	"	5.50	16.65	22.15
1"	"	5.60	16.65	22.25
1-1/2" x				
1/2"	EA.	6.50	18.75	25.25
3/4"	"	6.55	18.75	25.30
1"	"	6.70	18.75	25.45
1-1/4"	"	6.80	18.75	25.55
2" x				
1/2"	EA.	9.90	21.45	31.35
3/4"	"	10.00	21.45	31.45
1"	"	10.00	21.45	31.45
1-1/4"	"	10.35	21.45	31.80
1-1/2"	"	10.35	21.45	31.80
2-1/2" x				
1"	EA.	20.90	25.00	45.90
1-1/4"	"	22.00	25.00	47.00
1-1/2"	"	22.00	25.00	47.00
2"	"	23.10	25.00	48.10
3" x				
1-1/4"	EA.	27.50	30.00	57.50
1-1/2"	"	27.50	30.00	57.50
2"	"	28.60	30.00	58.60
2-1/2"	"	28.60	30.00	58.60
4" x				
2"	EA.	53.00	37.50	90.50
3"	"	55.00	37.50	92.50
Female adapters				
1/4"	EA.	4.40	12.00	16.40
3/8"	"	4.60	13.65	18.25
1/2"	"	2.20	15.00	17.20
3/4"	"	3.30	16.65	19.95
1"	"	5.50	16.65	22.15
1-1/4"	"	8.70	18.75	27.45
1-1/2"	"	13.20	18.75	31.95
2"	"	16.50	20.00	36.50
2-1/2"	"	55.00	21.45	76.45
3"	"	77.00	25.00	102.00
4"	"	121.00	30.00	151.00
Increasing female adapters				
1/8" x				
3/8"	EA.	3.30	12.00	15.30
1/2"	"	3.30	12.00	15.30
1/4" x 1/2"	"	3.10	13.05	16.15
3/8" x 1/2"	"	3.10	13.65	16.75
1/2" X				
3/4"	EA.	4.40	15.00	19.40

15 MECHANICAL

PLUMBING	UNIT	MAT.	INST.	TOTAL
15410.11 COPPER FITTINGS				
1"	EA.	8.25	15.00	23.25
3/4" X				
1"	EA.	7.70	16.65	24.35
1-1/4"	"	13.20	16.65	29.85
1" x				
1-1/4"	EA.	13.20	16.65	29.85
1-1/2"	"	15.40	16.65	32.05
1-1/4" x				
1-1/2"	EA.	15.40	18.75	34.15
2"	"	20.90	18.75	39.65
1-1/2" x 2"	"	21.45	20.00	41.45
Reducing female adapters				
3/8" x 1/4"	EA.	3.30	13.65	16.95
1/2" x				
1/4"	EA.	3.30	15.00	18.30
3/8"	"	3.40	15.00	18.40
3/4" x 1/2"	"	4.40	16.65	21.05
1" x				
1/2"	EA.	7.60	16.65	24.25
3/4"	"	7.60	16.65	24.25
1-1/4" x				
1/2"	EA.	14.30	18.75	33.05
3/4"	"	14.30	18.75	33.05
1"	"	14.30	18.75	33.05
1-1/2" x				
1"	EA.	15.40	20.00	35.40
1-1/4"	"	15.40	20.00	35.40
2" x				
1"	EA.	22.00	21.45	43.45
1-1/4"	"	23.10	21.45	44.55
1-1/2"	"	24.20	21.45	45.65
Female fitting adapters				
1/2"	EA.	3.30	15.00	18.30
3/4"	"	4.40	15.00	19.40
3/4" x 1/2"	"	7.70	15.80	23.50
1"	"	7.15	16.65	23.80
1-1/4"	"	11.00	17.65	28.65
1-1/2"	"	14.30	18.75	33.05
2"	"	15.40	20.00	35.40
Male adapters				
1/4"	EA.	4.20	13.65	17.85
3/8"	"	2.40	13.65	16.05
3"	"	53.00	25.00	78.00
4"	"	79.00	30.00	109.00
Increasing male adapters				
3/8" x 1/2"	EA.	2.20	13.65	15.85
1/2" x				
3/4"	EA.	3.30	15.00	18.30
1"	"	6.60	15.00	21.60
3/4" x				
1"	EA.	6.60	15.80	22.40
1-1/4"	"	9.35	15.80	25.15

PLUMBING	UNIT	MAT.	INST.	TOTAL
15410.11 COPPER FITTINGS				
1" x 1-1/4"	EA.	9.35	16.65	26.00
1-1/2" x				
3/4"	EA.	12.10	17.65	29.75
1"	"	12.30	17.65	29.95
1-1/4"	"	12.55	17.65	30.20
2" x				
1"	EA.	20.90	18.75	39.65
1-1/4"	"	20.90	18.75	39.65
1-1/2"	"	21.45	18.75	40.20
2" x 2-1/2"	"	68.00	20.00	88.00
Reducing male adapters				
1/2" x				
1/4"	EA.	2.40	15.00	17.40
3/8"	"	2.40	15.00	17.40
3/4" x 1/2"	"	4.20	15.80	20.00
1" x				
1/2"	EA.	6.60	16.65	23.25
3/4"	"	6.60	16.65	23.25
1-1/4" x				
3/4"	EA.	8.80	17.65	26.45
1"	"	9.00	17.65	26.65
1-1/2" x				
3/4"	EA.	12.65	18.75	31.40
1"	"	12.85	18.75	31.60
1-1/4'	"	13.20	18.75	31.95
2" x				
3/4"	EA.	21.45	20.00	41.45
1"	"	21.45	20.00	41.45
1-1/4"	"	22.55	20.00	42.55
1-1/2"	"	22.55	20.00	42.55
2-1/2" x				
2"	EA.	75.00	23.10	98.10
Fitting x male adapters				
1/2"	EA.	5.50	15.00	20.50
3/4"	"	6.80	15.80	22.60
1"	"	8.25	16.65	24.90
1-1/4"	"	19.80	17.65	37.45
1-1/2"	"	23.10	18.75	41.85
2"	"	39.60	20.00	59.60
90 ells				
1/8"	EA.	1.45	12.00	13.45
1/4"	"	1.45	12.00	13.45
3/8"	"	1.20	13.65	14.85
1/2"	"	0.83	15.00	15.83
3/4"	"	1.35	15.80	17.15
1"	"	3.40	16.65	20.05
1-1/4"	"	5.40	17.65	23.05
1-1/2"	"	7.90	18.75	26.65
2"	"	16.30	20.00	36.30
2-1/2"	"	23.10	23.10	46.20
3"	"	28.60	25.00	53.60
4"	"	70.50	30.00	100.50

15 MECHANICAL

PLUMBING	UNIT	MAT.	INST.	TOTAL
15410.11 COPPER FITTINGS				
Reducing 90 ell				
3/8" x 1/4"	EA.	2.65	13.65	16.30
1/2" x				
1/4"	EA.	3.95	15.00	18.95
3/8"	"	3.95	15.00	18.95
3/4" x 1/2"	"	3.50	15.80	19.30
1" x				
1/2"	EA.	5.50	16.65	22.15
3/4"	"	5.50	16.65	22.15
1-1/4" x				
1/2"	EA.	13.20	17.65	30.85
3/4"	"	13.20	17.65	30.85
1"	"	13.40	17.65	31.05
1-1/2" x				
1/2"	EA.	19.80	18.75	38.55
3/4"	"	20.00	18.75	38.75
1"	"	20.25	18.75	39.00
x 1-1/4"	"	20.25	18.75	39.00
2" x				
3/4"	EA.	24.20	20.00	44.20
x 1"	"	24.20	20.00	44.20
x 1-1/4"	"	24.20	20.00	44.20
x 1-1/2"	"	24.20	20.00	44.20
2-1/2" x				
1-1/2"	EA.	77.00	23.10	100.10
x 2"	"	78.00	23.10	101.10
3" x				
2"	EA.	101.00	25.00	126.00
x 2-1/2"	"	103.00	25.00	128.00
Street ells, copper				
1/4"	EA.	3.30	12.00	15.30
3/8"	"	2.30	13.65	15.95
1/2"	"	0.77	15.00	15.77
3/4"	"	1.75	15.80	17.55
1"	"	4.40	16.65	21.05
1-1/4"	"	6.50	17.65	24.15
1-1/2"	"	8.15	18.75	26.90
2"	"	16.50	20.00	36.50
2-1/2"	"	28.60	23.10	51.70
3"	"	39.60	25.00	64.60
4"	"	87.00	30.00	117.00
Female, 90 ell				
1/2"	EA.	2.30	15.00	17.30
3/4"	"	4.20	15.80	20.00
1"	"	11.00	16.65	27.65
1-1/4"	"	18.60	17.65	36.25
1-1/2"	"	20.90	18.75	39.65
2"	"	31.90	20.00	51.90
Female increasing, 90 ell				
3/8" x 1/2"	EA.	3.20	13.65	16.85
1/2" x				
3/4"	EA.	6.60	15.00	21.60

PLUMBING	UNIT	MAT.	INST.	TOTAL
15410.11 COPPER FITTINGS				
1"	EA.	10.45	15.00	25.45
3/4" x 1"	"	10.45	15.80	26.25
1" x 1-1/4"	"	17.05	16.65	33.70
1-1/4" x 1-1/2"	"	17.05	17.65	34.70
Female reducing, 90 ell				
1/2" x 3/8"	EA.	4.40	15.00	19.40
3/4" x 1/2"	"	6.60	15.80	22.40
1" x				
1/2"	EA.	10.45	16.65	27.10
x 3/4"	"	10.45	16.65	27.10
1-1/4" x				
3/4"	EA.	17.05	17.65	34.70
x 1"	"	17.05	17.65	34.70
x 1/2"	"	17.05	17.65	34.70
1-1/2" x				
1"	EA.	15.40	18.75	34.15
x 1-1/4"	"	15.40	18.75	34.15
2" x 1-1/2"	"	23.10	20.00	43.10
Male, 90 ell				
1/4"	EA.	3.85	12.00	15.85
3/8"	"	3.95	13.65	17.60
1/2"	"	2.40	15.00	17.40
3/4"	"	5.30	15.80	21.10
1"	"	8.80	16.65	25.45
1-1/4"	"	14.30	17.65	31.95
1-1/2"	"	18.60	18.75	37.35
2"	"	29.70	20.00	49.70
Male, increasing 90 ell				
1/2" x				
3/4"	EA.	4.95	15.00	19.95
1"	"	9.35	15.00	24.35
3/4" x 1"	"	9.35	15.80	25.15
1" x 1-1/4"	"	17.05	16.65	33.70
1-1/4" x 1-1/2"	"	17.05	17.65	34.70
Male, reducing 90 ell				
1/2" x 3/8"	EA.	4.05	15.00	19.05
3/4" x 1/2"	"	4.95	15.80	20.75
1" x				
1/2"	EA.	9.80	16.65	26.45
3/4"	"	9.80	16.65	26.45
1-1/4" x 1"	"	17.05	17.65	34.70
Drop ear ells				
1/2"	EA.	3.75	15.00	18.75
Female drop ear ells				
1/2"	EA.	3.20	15.00	18.20
1/2" x 3/8"	"	5.40	15.00	20.40
3/4"	"	6.05	15.80	21.85
Female flanged sink ell				
1/2"	EA.	5.40	15.00	20.40
45 ells				
1/4"	EA.	3.30	12.00	15.30
3/8"	"	3.10	13.65	16.75

PLUMBING	UNIT	MAT.	INST.	TOTAL
15410.11 COPPER FITTINGS				
3"	EA.	20.80	25.00	45.80
4"	"	71.50	30.00	101.50
45 street ell				
1/4"	EA.	3.45	12.00	15.45
3/8"	"	2.65	13.65	16.30
1/2"	"	0.88	15.00	15.88
3/4"	"	1.55	15.80	17.35
1"	"	5.30	16.65	21.95
1-1/2"	"	7.60	18.75	26.35
2"	"	11.55	20.00	31.55
2-1/2"	"	25.30	23.10	48.40
3"	"	37.40	25.00	62.40
4"	"	126.00	30.00	156.00
Tee				
1/8"	EA.	4.20	12.00	16.20
1/4"	"	4.20	12.00	16.20
3/8"	"	2.65	13.65	16.30
3"	"	52.50	25.00	77.50
4"	"	132.00	30.00	162.00
Caps				
1/4"	EA.	0.44	12.00	12.44
3/8"	"	0.66	13.65	14.31
3"	"	21.45	25.00	46.45
4"	"	39.60	30.00	69.60
Test caps				
1/2"	EA.	0.44	15.00	15.44
3/4"	"	0.50	15.80	16.30
1"	"	1.10	16.65	17.75
1-1/4"	"	1.20	17.65	18.85
1-1/2"	"	1.30	18.75	20.05
2"	"	2.20	20.00	22.20
3"	"	4.40	25.00	29.40
Flush bushing				
1/4" x 1/8"	EA.	1.10	12.00	13.10
1/2" x				
1/4"	EA.	1.30	15.00	16.30
3/8"	"	1.30	15.00	16.30
3/4" x				
3/8"	EA.	2.30	15.80	18.10
1/2"	"	2.30	15.80	18.10
1" x				
1/2"	EA.	3.85	16.65	20.50
3/4"	"	3.85	16.65	20.50
1-1/4" x				
1/2"	EA.	4.60	17.65	22.25
3/4"	"	4.60	17.65	22.25
x 1"	"	4.60	17.65	22.25
1-1/2" x				
1/2"	EA.	5.50	18.75	24.25
1-1/4"	"	5.50	18.75	24.25
2" x				
1"	EA.	13.20	20.00	33.20

PLUMBING	UNIT	MAT.	INST.	TOTAL
15410.11 COPPER FITTINGS				
1-1/4"	EA.	13.20	20.00	33.20
1-1/2"	"	13.40	20.00	33.40
Female flush bushing				
1/2" x				
1/2" x 1/8"	EA.	3.10	15.00	18.10
1/4"	"	3.10	15.00	18.10
Union				
1/4"	EA.	7.50	12.00	19.50
3/8"	"	7.50	13.65	21.15
3"	"	231.00	25.00	256.00
Female				
1/2"	EA.	9.90	15.00	24.90
3/4"	"	13.20	15.80	29.00
Male				
1/2"	EA.	8.80	15.00	23.80
3/4"	"	11.55	15.80	27.35
1"	"	25.90	16.65	42.55
45 degree wye				
1/2"	EA.	6.60	15.00	21.60
3/4"	"	10.90	15.80	26.70
1"	"	14.30	16.65	30.95
1" x 3/4" x 3/4"	"	29.70	16.65	46.35
1-1/4"	"	29.70	17.65	47.35
1-1/4" x 1" x 1"	"	30.80	17.65	48.45
Twin ells				
1" x 3/4" x 3/4"	EA.	7.70	16.65	24.35
1" x 1" x 1"	"	7.70	16.65	24.35
1-1/4" x 1" x 1"	"	12.00	17.65	29.65
Companion flanges, 125#				
2" x 6"	EA.	75.00	20.00	95.00
2-1/2"	"	86.00	23.10	109.10
3" x 7-1/2"	"	83.50	25.00	108.50
4" x 9"	"	138.00	30.00	168.00
90 union ells, male				
1/2"	EA.	11.55	15.00	26.55
3/4"	"	15.40	15.80	31.20
1"	"	35.20	16.65	51.85
DWV fittings, coupling with stop				
1-1/4"	EA.	3.30	17.65	20.95
1-1/2"	"	3.85	18.75	22.60
1-1/2" x 1-1/4"	"	4.40	18.75	23.15
2"	"	4.30	20.00	24.30
2" x 1-1/4"	"	5.50	20.00	25.50
2" x 1-1/2"	"	5.50	20.00	25.50
3"	"	7.15	25.00	32.15
3" x 1-1/2"	"	14.30	25.00	39.30
3" x 2"	"	14.30	25.00	39.30
4"	"	14.85	30.00	44.85
Slip coupling				
1-1/2"	EA.	4.30	18.75	23.05
2"	"	4.60	20.00	24.60
3"	"	6.50	25.00	31.50

15 MECHANICAL

PLUMBING

15410.11 COPPER FITTINGS

	UNIT	MAT.	INST.	TOTAL
90 ells				
1-1/2"	EA.	7.70	18.75	26.45
1-1/2" x 1-1/4"	"	9.80	18.75	28.55
2"	"	8.25	20.00	28.25
2" x 1-1/2"	"	12.65	20.00	32.65
3"	"	24.20	25.00	49.20
4"	"	99.00	30.00	129.00
Street, 90 elbows				
1-1/2"	EA.	6.60	18.75	25.35
2"	"	11.55	20.00	31.55
3"	"	24.20	25.00	49.20
4"	"	106.00	30.00	136.00
Female, 90 elbows				
1-1/2"	EA.	11.00	18.75	29.75
2"	"	24.20	20.00	44.20
Male, 90 elbows				
1-1/2"	EA.	10.90	18.75	29.65
2"	"	24.20	20.00	44.20
90 with side inlet				
3" x 3" x 1"	EA.	36.30	25.00	61.30
3" x 3" x 1-1/2"	"	37.40	25.00	62.40
3" x 3" x 2"	"	37.40	25.00	62.40
45 ells				
1-1/4"	EA.	4.60	17.65	22.25
1-1/2"	"	4.30	18.75	23.05
2"	"	8.60	20.00	28.60
3"	"	15.40	25.00	40.40
4"	"	45.10	30.00	75.10
Street, 45 ell				
1-1/2"	EA.	5.50	18.75	24.25
2"	"	9.90	20.00	29.90
3"	"	23.10	25.00	48.10
60 ell				
1-1/2"	EA.	7.80	18.75	26.55
2"	"	13.75	20.00	33.75
3"	"	27.50	25.00	52.50
22-1/2 ell				
1-1/2"	EA.	7.70	18.75	26.45
2"	"	10.65	20.00	30.65
3"	"	16.40	25.00	41.40
11-1/4 ell				
1-1/2"	EA.	7.90	18.75	26.65
2"	"	9.35	20.00	29.35
3"	"	19.25	25.00	44.25
Wye				
1-1/4"	EA.	15.40	17.65	33.05
1-1/2"	"	13.20	18.75	31.95
2"	"	20.90	20.00	40.90
2" x 1-1/2" x 1-1/2"	"	20.35	20.00	40.35
2" x 1-1/2" x 2"	"	20.35	20.00	40.35
2" x 1-1/2" x 2"	"	20.45	20.00	40.45
3"	"	53.00	25.00	78.00

PLUMBING	UNIT	MAT.	INST.	TOTAL
15410.11 COPPER FITTINGS				
3" x 3" x 1-1/2"	EA.	54.00	25.00	79.00
3" x 3" x 2"	"	50.50	25.00	75.50
4"	"	112.00	30.00	142.00
4" x 4" x 2"	"	103.00	30.00	133.00
4" x 4" x 3"	"	103.00	30.00	133.00
Sanitary tee				
1-1/4"	EA.	10.45	17.65	28.10
1-1/2"	"	9.90	18.75	28.65
2"	"	15.40	20.00	35.40
2" x 1-1/2" x 1-1/2"	"	14.85	20.00	34.85
2" x 1-1/2" x 2"	"	14.85	20.00	34.85
2" x 2" x 1-1/2"	"	14.85	20.00	34.85
3"	"	37.40	25.00	62.40
3" x 3" x 1-1/2"	"	35.20	25.00	60.20
3" x 3" x 2"	"	35.20	25.00	60.20
4"	"	93.50	30.00	123.50
4" x 4" x 3"	"	82.50	30.00	112.50
Female sanitary tee				
1-1/2"	EA.	26.40	18.75	45.15
Long turn tee				
1-1/2"	EA.	17.60	18.75	36.35
2"	"	27.50	20.00	47.50
3" x 1-1/2"	"	39.60	25.00	64.60
Double wye				
1-1/2"	EA.	25.30	18.75	44.05
2"	"	38.50	20.00	58.50
2" x 2" x 1-1/2" x 1-1/2"	"	31.90	20.00	51.90
3"	"	54.00	25.00	79.00
3" x 3" x 1-1/2" x 1-1/2"	"	54.00	25.00	79.00
3" x 3" x 2" x 2"	"	54.00	25.00	79.00
4" x 4" x 1-1/2" x 1-1/2"	"	76.00	30.00	106.00
Double sanitary tee				
1-1/2"	EA.	17.60	18.75	36.35
2"	"	31.80	20.00	51.80
2" x 2" x 1-1/2"	"	23.10	20.00	43.10
3"	"	48.40	25.00	73.40
3" x 3" x 1-1/2" x 1-1/2"	"	34.10	25.00	59.10
3" x 3" x 2" x 2"	"	34.10	25.00	59.10
4" x 4" x 1-1/2" x 1-1/2"	"	154.00	30.00	184.00
Long				
2" x 1-1/2"	EA.	41.80	20.00	61.80
Twin elbow				
1-1/2"	EA.	19.80	18.75	38.55
2"	"	30.80	20.00	50.80
2" x 1-1/2" x 1-1/2"	"	19.80	20.00	39.80
Spigot adapter, manoff				
1-1/2" x 2"	EA.	17.60	18.75	36.35
1-1/2" x 3"	"	24.20	18.75	42.95
2"	"	9.90	20.00	29.90
2" x 3"	"	24.20	20.00	44.20
2" x 4"	"	34.10	20.00	54.10
3"	"	18.70	25.00	43.70

15 MECHANICAL

PLUMBING	UNIT	MAT.	INST.	TOTAL
15410.11 COPPER FITTINGS				
3" x 4"	EA.	34.10	25.00	59.10
4"	"	34.10	30.00	64.10
No-hub adapters				
1-1/2" x 2"	EA.	9.90	18.75	28.65
2"	"	9.90	20.00	29.90
2" x 3"	"	28.60	20.00	48.60
3"	"	18.15	25.00	43.15
3" x 4"	"	28.60	25.00	53.60
4"	"	30.80	30.00	60.80
Fitting reducers				
1-1/2" x 1-1/4"	EA.	4.40	18.75	23.15
2" x 1-1/2"	"	4.30	20.00	24.30
3" x 1-1/2"	"	13.75	25.00	38.75
3" x 2"	"	14.10	25.00	39.10
Slip joint (Desanco)				
1-1/4"	EA.	7.60	17.65	25.25
1-1/2"	"	7.90	18.75	26.65
1-1/2" x 1-1/4"	"	7.90	18.75	26.65
Street x slip joint (Desanco)				
1-1/2"	EA.	9.90	18.75	28.65
1-1/2" x 1-1/4"	"	10.65	18.75	29.40
Flush bushing				
1-1/2" x 1-1/4"	EA.	5.50	18.75	24.25
2" x 1-1/2"	"	8.80	20.00	28.80
3" x 1-1/2"	"	23.10	25.00	48.10
3" x 2"	"	23.10	25.00	48.10
Male hex trap bushing				
1-1/4" x 1-1/2"	EA.	7.15	17.65	24.80
1-1/2"	"	5.50	18.75	24.25
1-1/2" x 2"	"	8.80	18.75	27.55
2"	"	6.50	20.00	26.50
Round trap bushing				
1-1/2"	EA.	6.60	18.75	25.35
2"	"	7.15	20.00	27.15
Female adapter				
1-1/4"	EA.	6.60	17.65	24.25
1-1/2"	"	6.60	18.75	25.35
1-1/2" x 2"	"	11.55	18.75	30.30
2"	"	11.55	20.00	31.55
2" x 1-1/2"	"	11.55	20.00	31.55
3"	"	26.40	25.00	51.40
Fitting x female adapter				
1-1/2"	EA.	7.60	18.75	26.35
2"	"	11.00	20.00	31.00
Male adapters				
1-1/4"	EA.	6.60	17.65	24.25
1-1/4" x 1-1/2"	"	7.70	17.65	25.35
1-1/2"	"	7.60	18.75	26.35
1-1/2" x 2"	"	9.35	18.75	28.10
2"	"	8.80	20.00	28.80
2" x 1-1/2"	"	9.35	20.00	29.35
3"	"	24.20	25.00	49.20

PLUMBING

15410.11 COPPER FITTINGS

	UNIT	MAT.	INST.	TOTAL
Male x slip joint adapters				
1-1/2" x 1-1/4"	EA.	11.00	18.75	29.75
Dandy cleanout				
1-1/2"	EA.	18.15	18.75	36.90
2"	"	29.70	20.00	49.70
3"	"	69.50	25.00	94.50
End cleanout, flush pattern				
1-1/2" x 1"	EA.	13.20	18.75	31.95
2" x 1-1/2"	"	19.80	20.00	39.80
3" x 2-1/2"	"	64.00	25.00	89.00
Copper caps				
1-1/2"	EA.	7.70	18.75	26.45
2"	"	11.55	20.00	31.55
Closet flanges				
3"	EA.	19.80	25.00	44.80
4"	"	19.80	30.00	49.80
Drum traps, with cleanout				
1-1/2" x 3" x 6"	EA.	46.20	18.75	64.95
P-trap, swivel, with cleanout				
1-1/2"	EA.	29.00	18.75	47.75
P-trap, solder union				
1-1/2"	EA.	19.80	18.75	38.55
2"	"	35.20	20.00	55.20
With cleanout				
1-1/2"	EA.	21.45	18.75	40.20
2"	"	38.50	20.00	58.50
2" x 1-1/2"	"	38.50	20.00	58.50
Swivel joint, with cleanout				
1-1/2" x 1-1/4"	EA.	26.40	18.75	45.15
1-1/2"	"	35.20	18.75	53.95
2" x 1-1/2"	"	42.90	20.00	62.90
Estabrook TY, with inlets				
3", with 1-1/2" inlet	EA.	57.00	25.00	82.00
Fine thread adapters				
1/2"	EA.	1.55	15.00	16.55
1/2" x 1/2" IPS	"	1.75	15.00	16.75
1/2" x 3/4" IPS	"	2.85	15.00	17.85
1/2" x male	"	1.05	15.00	16.05
1/2" x female	"	2.10	15.00	17.10

15410.14 BRASS I.P.S. FITTINGS

	UNIT	MAT.	INST.	TOTAL
Fittings, iron pipe size, 45 deg ell				
1/8"	EA.	4.60	12.00	16.60
1/4"	"	4.60	12.00	16.60
3/8"	"	4.60	13.65	18.25
1/2"	"	4.75	15.00	19.75
3/4"	"	6.05	15.80	21.85
1"	"	10.45	16.65	27.10
1-1/4"	"	15.40	17.65	33.05
1-1/2"	"	19.80	18.75	38.55
2"	"	26.40	20.00	46.40

PLUMBING	UNIT	MAT.	INST.	TOTAL
15410.14 BRASS I.P.S. FITTINGS				
90 deg ell				
1/8"	EA.	4.40	12.00	16.40
1/4"	"	4.40	12.00	16.40
3/8"	"	4.40	13.65	18.05
1/2"	"	4.40	15.00	19.40
3/4"	"	5.50	15.80	21.30
1"	"	9.90	16.65	26.55
1-1/4"	"	13.75	17.65	31.40
1-1/2"	"	18.15	18.75	36.90
2"	"	24.20	20.00	44.20
2-1/2"	"	44.00	23.10	67.10
90 deg ell, reducing				
1/4" x 1/8"	EA.	5.50	12.00	17.50
3/8" x 1/8"	"	5.50	13.65	19.15
3/8" x 1/4"	"	5.50	13.65	19.15
1/2" x 1/4"	"	5.50	15.00	20.50
1/2" x 3/8"	"	5.50	15.00	20.50
3/4" x 1/2"	"	8.25	15.80	24.05
1" x 3/8"	"	10.90	16.65	27.55
1" x 1/2"	"	10.90	16.65	27.55
1" x 3/4"	"	10.90	16.65	27.55
1-1/4" x 3/4"	"	17.60	17.65	35.25
1-1/4" x 1"	"	17.60	17.65	35.25
1-1/2" x 1/2"	"	20.90	18.75	39.65
1-1/2" x 3/4"	"	20.90	18.75	39.65
1-1/2" x 1"	"	20.90	18.75	39.65
2" x 1"	"	30.80	20.00	50.80
2" x 1-1/4"	"	30.80	20.00	50.80
2" x 1-1/2"	"	30.80	20.00	50.80
Street ell, 45 deg				
1/2"	EA.	5.50	15.00	20.50
3/4"	"	7.60	15.80	23.40
2"	"	37.40	20.00	57.40
90 deg				
1/8"	EA.	5.40	12.00	17.40
1/4"	"	5.40	12.00	17.40
3/8"	"	5.40	13.65	19.05
1/2"	"	5.40	15.00	20.40
3/4"	"	6.60	15.80	22.40
1"	"	8.80	16.65	25.45
1-1/4"	"	17.60	17.65	35.25
1-1/2"	"	18.70	18.75	37.45
2"	"	30.80	20.00	50.80
Tee, 1/8"	"	4.50	12.00	16.50
1/4"	"	4.50	12.00	16.50
3/8"	"	4.50	13.65	18.15
1/2"	"	4.50	15.00	19.50
3/4"	"	6.15	15.80	21.95
1"	"	9.25	16.65	25.90
1-1/4"	"	15.40	17.65	33.05
1-1/2"	"	16.50	18.75	35.25
2"	"	26.40	20.00	46.40

PLUMBING

15410.14 BRASS I.P.S. FITTINGS

PLUMBING	UNIT	MAT.	INST.	TOTAL
2-1/2"	EA.	59.50	23.10	82.60
Tee, reducing, 3/8" x				
1/4"	EA.	5.70	13.65	19.35
1/2"	"	5.70	13.65	19.35
1/2" x				
1/4"	EA.	5.70	15.00	20.70
3/8"	"	5.70	15.00	20.70
3/4"	"	6.60	15.00	21.60
3/4" x				
1/4"	EA.	6.60	15.80	22.40
1/2"	"	6.60	15.80	22.40
1"	"	13.20	15.80	29.00
1" x				
1/2"	EA.	13.20	16.65	29.85
3/4"	"	13.20	16.65	29.85
1-1/4" x				
1/2"	EA.	17.60	17.65	35.25
1"	"	17.60	17.65	35.25
1-1/2" x				
1/2"	EA.	24.20	18.75	42.95
3/4"	"	24.20	18.75	42.95
1"	"	24.20	18.75	42.95
1-1/2"	"	26.40	18.75	45.15
2" x				
1/2"	EA.	37.40	20.00	57.40
3/4"	"	37.40	20.00	57.40
1"	"	37.40	20.00	57.40
1-1/4"	"	37.40	20.00	57.40
1-1/2"	"	37.40	20.00	57.40
2-1/2" x 2"	"	83.50	23.10	106.60
Tee, reducing				
1/2" x 3/8" x 1/2"	EA.	5.70	15.00	20.70
3/4" x 1/2" x 1/2"	"	8.80	15.80	24.60
3/4" x 1/2" x 3/4"	"	8.25	15.80	24.05
1" x 1/2" x 1/2"	"	14.30	16.65	30.95
1" x 1/2" x 3/4"	"	14.30	16.65	30.95
1" x 3/4" x 1/2"	"	14.30	16.65	30.95
1" x 3/4" x 3/4"	"	14.30	16.65	30.95
1-1/4" x 1/2" x 1-1/4"	"	17.60	17.65	35.25
1-1/4" x 1" x 1"	"	18.70	17.65	36.35
1-1/4" x 1-1/4" x 3/4"	"	17.60	17.65	35.25
1-1/2" x 1-1/4" x 1-1/4"	"	24.20	18.75	42.95
Union				
1/8"	EA.	9.90	12.00	21.90
1/4"	"	9.90	12.00	21.90
3/8"	"	9.90	13.65	23.55
1/2"	"	9.90	15.00	24.90
3/4"	"	12.10	15.80	27.90
1"	"	15.40	16.65	32.05
1-1/4"	"	20.90	17.65	38.55
1-1/2"	"	25.30	18.75	44.05
2"	"	37.40	20.00	57.40

PLUMBING

15410.14 BRASS I.P.S. FITTINGS

	UNIT	MAT.	INST.	TOTAL
Brass face bushing				
3/8" x 1/4"	EA.	4.50	13.65	18.15
1/2" x 3/8"	"	4.50	15.00	19.50
3/4" x 1/2"	"	5.70	15.80	21.50
1" x 3/4"	"	9.80	16.65	26.45
1-1/4" x 1"	"	12.10	17.65	29.75
Hex bushing, 1/4" x 1/8"	"	3.20	12.00	15.20
1/2" x				
1/4"	EA.	2.85	15.00	17.85
3/8"	"	2.85	15.00	17.85
5/8" x				
1/8"	EA.	2.85	15.00	17.85
1/4"	"	2.85	15.00	17.85
3/4" x				
1/8"	EA.	4.30	15.80	20.10
1/4"	"	4.30	15.80	20.10
3/8"	"	3.20	15.80	19.00
1/2"	"	3.20	15.80	19.00
1" x				
1/4"	EA.	5.60	16.65	22.25
3/8"	"	5.60	16.65	22.25
1/2"	"	5.05	16.65	21.70
3/4"	"	5.05	16.65	21.70
1-1/4" x				
3/8"	EA.	8.80	17.65	26.45
1/2"	"	8.80	17.65	26.45
3/4"	"	6.50	17.65	24.15
1"	"	6.50	17.65	24.15
1-1/2" x				
1/4"	EA.	11.55	18.75	30.30
1/2"	"	11.55	18.75	30.30
3/4"	"	11.55	18.75	30.30
1"	"	9.90	18.75	28.65
1-1/4"	"	9.90	18.75	28.65
2" x				
1/2"	EA.	15.40	20.00	35.40
3/4"	"	15.40	20.00	35.40
1"	"	15.40	20.00	35.40
1-1/4"	"	10.55	20.00	30.55
1-1/2"	"	10.55	20.00	30.55
2-1/2" x				
1"	EA.	30.80	23.10	53.90
2"	"	26.40	23.10	49.50
3" x				
1-1/2"	EA.	45.10	25.00	70.10
2"	"	37.40	25.00	62.40
2-1/2"	"	37.40	25.00	62.40
4" x				
2"	EA.	109.00	30.00	139.00
3"	"	79.00	30.00	109.00
Caps				
1/8"	EA.	3.10	12.00	15.10

PLUMBING	UNIT	MAT.	INST.	TOTAL
15410.14 BRASS I.P.S. FITTINGS				
1/4"	EA.	3.10	12.00	15.10
3/8"	"	3.10	13.65	16.75
1/2"	"	3.10	15.00	18.10
3/4"	"	3.50	15.80	19.30
1"	"	5.50	16.65	22.15
1-1/4"	"	8.25	17.65	25.90
1-1/2"	"	10.45	18.75	29.20
2"	"	17.60	20.00	37.60
Couplings				
1/8"	EA.	3.40	12.00	15.40
1/4"	"	3.40	12.00	15.40
3/8"	"	3.40	13.65	17.05
1/2"	"	3.40	15.00	18.40
3/4"	"	4.40	15.80	20.20
1"	"	6.60	16.65	23.25
1-1/4"	"	10.45	17.65	28.10
1-1/2"	"	13.20	18.75	31.95
2"	"	20.90	20.00	40.90
2-1/2"	"	31.90	23.10	55.00
Couplings, reducing, 1/4" x 1/8"	"	3.95	12.00	15.95
3/8" x				
1/8"	EA.	3.95	13.65	17.60
1/4"	"	3.95	13.65	17.60
1/2" x				
1/8"	EA.	4.40	15.00	19.40
1/4"	"	4.30	15.00	19.30
3/8"	"	4.30	15.00	19.30
3/4" x				
1/4"	EA.	6.60	15.80	22.40
3/8"	"	5.60	15.80	21.40
1/2"	"	5.60	15.80	21.40
1/2"	"	7.60	15.80	23.40
3/4"	"	7.60	15.80	23.40
1-1/4" x				
1/2"	EA.	14.30	17.65	31.95
3/4"	"	13.20	17.65	30.85
1"	"	13.20	17.65	30.85
1-1/2" x				
3/4"	EA.	19.80	18.75	38.55
1"	"	17.60	18.75	36.35
1-1/4"	"	17.60	18.75	36.35
2" x				
3/4"	EA.	26.40	20.00	46.40
1"	"	26.40	20.00	46.40
1-1/4"	"	25.30	20.00	45.30
1-1/2"	"	25.30	20.00	45.30
2-1/2" x 1-1/2"	"	39.60	23.10	62.70
Square head plug, solid				
1/8"	EA.	3.30	12.00	15.30
1/4"	"	3.30	12.00	15.30
3/8"	"	3.30	13.65	16.95
1/2"	"	3.30	15.00	18.30

PLUMBING	UNIT	MAT.	INST.	TOTAL
15410.14 BRASS I.P.S. FITTINGS				
3/4"	EA.	3.85	15.80	19.65
Cored				
1/2"	EA.	3.20	15.00	18.20
3/4"	"	3.30	15.80	19.10
1"	"	5.50	16.65	22.15
1-1/4"	"	6.60	17.65	24.25
1-1/2"	"	6.60	18.75	25.35
2"	"	11.00	20.00	31.00
3"	"	28.60	25.00	53.60
4"	"	50.50	30.00	80.50
Countersunk				
1/2"	EA.	3.30	15.00	18.30
3/4"	"	3.50	15.80	19.30
1-1/2"	"	9.45	18.75	28.20
2"	"	12.65	20.00	32.65
Locknut				
3/4"	EA.	3.30	15.80	19.10
1"	"	3.75	16.65	20.40
1-1/4"	"	7.35	17.65	25.00
2"	"	11.55	20.00	31.55
Close standard red nipple, 1/8"	"	1.10	12.00	13.10
1/8" x				
1-1/2"	EA.	1.20	12.00	13.20
2"	"	1.30	12.00	13.30
2-1/2"	"	1.55	12.00	13.55
3"	"	1.75	12.00	13.75
3-1/2"	"	2.10	12.00	14.10
4"	"	2.20	12.00	14.20
4-1/2"	"	2.30	12.00	14.30
5"	"	2.40	12.00	14.40
5-1/2"	"	2.85	12.00	14.85
6"	"	3.10	12.00	15.10
1/4" x close	"	1.30	12.00	13.30
1/4" x				
1-1/2"	EA.	2.00	12.00	14.00
2"	"	2.10	12.00	14.10
2-1/2"	"	2.20	12.00	14.20
3"	"	2.30	12.00	14.30
3-1/2"	"	2.55	12.00	14.55
4"	"	2.65	12.00	14.65
4-1/2"	"	2.75	12.00	14.75
5"	"	2.85	12.00	14.85
5-1/2"	"	3.20	12.00	15.20
6"	"	3.30	12.00	15.30
3/8" x close	"	1.30	13.65	14.95
3/8" x				
1-1/2"	EA.	1.45	13.65	15.10
2"	"	1.55	13.65	15.20
2-1/2"	"	1.85	13.65	15.50
3"	"	2.30	13.65	15.95
3-1/2"	"	2.55	13.65	16.20
4"	"	3.20	13.65	16.85

PLUMBING	UNIT	MAT.	INST.	TOTAL
15410.14 BRASS I.P.S. FITTINGS				
4-1/2"	EA.	3.30	13.65	16.95
5"	"	3.50	13.65	17.15
5-1/2"	"	3.65	13.65	17.30
6"	"	4.30	13.65	17.95
1/2" x close	"	1.65	15.00	16.65
1/2" x				
1-1/2"	EA.	1.85	15.00	16.85
2"	"	2.30	15.00	17.30
2-1/2"	"	2.55	15.00	17.55
3"	"	2.75	15.00	17.75
3-1/2"	"	3.10	15.00	18.10
4"	"	3.30	15.00	18.30
4-1/2"	"	3.50	15.00	18.50
5"	"	3.75	15.00	18.75
5-1/2"	"	3.95	15.00	18.95
6"	"	4.20	15.00	19.20
7-1/2"	"	4.40	15.00	19.40
8"	"	4.40	15.00	19.40
3/4" x close	"	2.40	15.80	18.20
3/4" x				
1-1/2"	EA.	2.55	15.80	18.35
2"	"	2.85	15.80	18.65
2-1/2"	"	3.10	15.80	18.90
3"	"	3.40	15.80	19.20
3-1/2"	"	3.75	15.80	19.55
4"	"	4.05	15.80	19.85
4-1/2"	"	4.30	15.80	20.10
5"	"	4.50	15.80	20.30
5-1/2"	"	5.05	15.80	20.85
6"	"	5.30	15.80	21.10
1" x close	"	3.50	16.65	20.15
1" x				
2"	EA.	4.95	16.65	21.60
2-1/2"	"	5.05	16.65	21.70
3"	"	5.30	16.65	21.95
3-1/2"	"	5.50	16.65	22.15
4"	"	6.05	16.65	22.70
4-1/2"	"	6.15	16.65	22.80
5"	"	7.15	16.65	23.80
5-1/2"	"	7.35	16.65	24.00
6"	"	7.90	16.65	24.55
1-1/4" x close	"	5.05	17.65	22.70
1-1/4" x				
2"	EA.	5.15	17.65	22.80
2-1/2"	"	5.70	17.65	23.35
3"	"	7.35	17.65	25.00
3-1/2"	"	8.05	17.65	25.70
4"	"	8.05	17.65	25.70
4-1/2"	"	8.35	17.65	26.00
5"	"	9.80	17.65	27.45
5-1/2"	"	10.90	17.65	28.55
6"	"	11.00	17.65	28.65

PLUMBING	UNIT	MAT.	INST.	TOTAL
15410.14 BRASS I.P.S. FITTINGS				
1-1/2" x close	EA.	5.50	18.75	24.25
1-1/2" x				
2"	EA.	5.70	18.75	24.45
2-1/2"	"	5.95	18.75	24.70
3"	"	6.50	18.75	25.25
3-1/2"	"	7.05	18.75	25.80
4-1/2"	"	9.80	18.75	28.55
5"	"	10.65	18.75	29.40
5-1/2"	"	11.65	18.75	30.40
6"	"	11.90	18.75	30.65
2" x close	"	7.70	20.00	27.70
2"				
2-1/2"	EA.	8.80	20.00	28.80
3"	"	9.25	20.00	29.25
3-1/2"	"	10.45	20.00	30.45
4"	"	10.65	20.00	30.65
4-1/2"	"	12.00	20.00	32.00
5"	"	12.65	20.00	32.65
5-1/2"	"	15.40	20.00	35.40
6"	"	16.50	20.00	36.50
2-1/2" x close	"	17.60	23.10	40.70
2-1/2" x				
3"	EA.	18.15	23.10	41.25
3-1/2"	"	19.25	23.10	42.35
4"	"	23.10	23.10	46.20
2-1/2" x 5"	"	26.40	23.10	49.50
3"				
Close	EA.	25.30	25.00	50.30
15410.15 BRASS FITTINGS				
Compression fittings, union				
3/8"	EA.	2.10	5.00	7.10
1/2"	"	3.20	5.00	8.20
5/8"	"	3.95	5.00	8.95
Union elbow				
3/8"	EA.	3.50	5.00	8.50
1/2"	"	5.50	5.00	10.50
5/8"	"	7.05	5.00	12.05
Union tee				
3/8"	EA.	4.60	5.00	9.60
1/2"	"	7.60	5.00	12.60
5/8"	"	9.90	5.00	14.90
Male connector				
3/8"	EA.	1.55	5.00	6.55
1/2"	"	1.10	5.00	6.10
5/8"	"	2.40	5.00	7.40
Female connector				
3/8"	EA.	2.85	5.00	7.85
1/2"	"	3.75	5.00	8.75
5/8"	"	3.75	5.00	8.75
Brass flare fittings, union				
3/8"	EA.	1.65	4.85	6.50

PLUMBING	UNIT	MAT.	INST.	TOTAL
15410.15 BRASS FITTINGS				
1/2"	EA.	2.45	4.85	7.30
5/8"	"	3.45	4.85	8.30
90 deg elbow union				
3/8"	EA.	3.70	4.85	8.55
1/2"	"	5.20	4.85	10.05
5/8"	"	7.50	4.85	12.35
Three way tee				
3/8"	EA.	4.95	8.10	13.05
1/2"	"	6.25	8.10	14.35
5/8"	"	9.90	8.10	18.00
Cross				
3/8"	EA.	6.35	10.70	17.05
1/2"	"	9.25	10.70	19.95
5/8"	"	18.50	10.70	29.20
Male connector, half union				
3/8"	EA.	2.10	4.85	6.95
1/2"	"	2.80	4.85	7.65
5/8"	"	5.70	4.85	10.55
Female connector, half union				
3/8"	EA.	2.10	4.85	6.95
1/2"	"	4.30	4.85	9.15
5/8"	"	7.25	4.85	12.10
Long forged nut				
3/8"	EA.	1.55	4.85	6.40
1/2"	"	2.65	4.85	7.50
5/8"	"	3.95	4.85	8.80
Short forged nut				
3/8"	EA.	1.00	4.85	5.85
1/2"	"	1.45	4.85	6.30
5/8"	"	1.85	4.85	6.70
Compression elbow				
1/4"	EA.	0.73	4.85	5.58
5/16"	"	1.30	4.85	6.15
3/4"	"	6.60	4.85	11.45
Nut				
1/8"	EA.			0.10
1/4"	"			0.10
5/16"	"			0.15
3/8"	"			0.18
1/2"	"			0.29
5/8"	"			0.48
3/4"	"			0.87
7/8"	"			0.84
Sleeve				
1/8"	EA.	0.06	6.00	6.06
1/4"	"	0.04	6.00	6.04
5/16"	"	0.07	6.00	6.07
3/8"	"	0.06	6.00	6.06
1/2"	"	0.11	6.00	6.11
5/8"	"	0.15	6.00	6.15
3/4"	"	0.31	6.00	6.31
7/8"	"	0.48	6.00	6.48

15 MECHANICAL

PLUMBING	UNIT	MAT.	INST.	TOTAL
15410.15 BRASS FITTINGS				
Tee				
1/4"	EA.	1.20	8.55	9.75
5/16"	"	1.75	8.55	10.30
Male tee				
5/16" x 1/8"	EA.	1.30	8.55	9.85
Female union				
1/8" x 1/8"	EA.	0.46	7.50	7.96
1/4" x 3/8"	"	0.84	7.50	8.34
3/8" x 1/4"	"	0.69	7.50	8.19
3/8" x 1/2"	"	0.86	7.50	8.36
5/8" x 1/2"	"	1.35	8.55	9.90
Male union, 1/4"				
1/4" x 1/4"	EA.	0.50	7.50	8.00
3/8"	"	0.76	7.50	8.26
1/2"	"	1.30	7.50	8.80
5/16" x				
1/8"	EA.	0.46	7.50	7.96
1/4"	"	0.51	7.50	8.01
3/8"	"	0.81	7.50	8.31
3/8" x				
1/8"	EA.	0.53	7.50	8.03
1/4"	"	0.62	7.50	8.12
1/2"	"	0.84	7.50	8.34
3/4"	"	1.00	7.50	8.50
1/2" x				
1/4"	EA.	0.79	8.55	9.34
3/8"	"	0.81	8.55	9.36
5/8" x				
3/8"	EA.	1.20	8.55	9.75
1/2"	"	1.05	8.55	9.60
3/4"	"	1.65	8.55	10.20
1/2"	"	2.65	8.55	11.20
7/8" x				
1/2"	EA.	2.55	8.55	11.10
3/4"	"	3.95	8.55	12.50
Female elbow, 1/4" x 1/4"	"	1.20	8.55	9.75
5/16" x				
1/8"	EA.	1.15	8.55	9.70
1/4"	"	1.45	8.55	10.00
3/8" x				
3/8"	EA.	1.35	8.55	9.90
1/2"	"	1.45	8.55	10.00
Male elbow, 1/8" x 1/8"	"	0.73	8.55	9.28
3/16" x 1/4"	"	1.05	8.55	9.60
1/4" x				
1/8"	EA.	0.62	8.55	9.17
1/4"	"	0.69	8.55	9.24
3/8"	"	1.05	8.55	9.60
5/16" x				
1/8"	EA.	0.84	8.55	9.39
1/4"	"	0.91	8.55	9.46
3/8"	"	1.15	8.55	9.70

PLUMBING	UNIT	MAT.	INST.	TOTAL
15410.15 BRASS FITTINGS				
3/8" x				
1/8"	EA.	0.86	8.55	9.41
1/4"	"	0.88	8.55	9.43
3/8"	"	1.20	8.55	9.75
1/2"	"	1.50	8.55	10.05
1/2" x				
1/4"	EA.	1.50	10.00	11.50
3/8"	"	1.55	10.00	11.55
1/2"	"	1.85	10.00	11.85
5/8" x				
3/8"	EA.	2.05	10.00	12.05
1/2"	"	2.00	10.00	12.00
3/4"	"	3.10	10.00	13.10
3/4" x				
1/2"	EA.	3.45	10.00	13.45
3/4"	"	3.55	10.00	13.55
7/8" x				
3/4"	EA.	4.75	10.00	14.75
Union				
1/8"	EA.	0.58	8.55	9.13
3/16"	"	0.58	8.55	9.13
1/4"	"	0.59	8.55	9.14
5/16"	"	0.75	8.55	9.30
3/8"	"	1.85	8.55	10.40
3/4"	"	4.05	8.55	12.60
7/8"	"	4.20	8.55	12.75
Reducing union				
3/8" x 1/4"	EA.	1.00	10.00	11.00
5/8" x				
3/8"	EA.	1.30	10.00	11.30
1/2"	"	2.00	10.00	12.00
15410.17 CHROME PLATED FITTINGS				
Fittings				
90 ell				
1/2"	EA.	10.45	7.50	17.95
3/4"	"	13.75	7.50	21.25
45 ell				
3/8"	EA.	13.75	7.50	21.25
1/2"	"	18.15	7.50	25.65
Tee				
3/8"	EA.	13.20	10.00	23.20
1/2"	"	15.40	10.00	25.40
Coupling				
3/8"	EA.	8.25	7.50	15.75
1/2"	"	8.25	7.50	15.75
Union				
3/8"	EA.	13.20	7.50	20.70
1/2"	"	13.75	7.50	21.25
Tee				
1/2" x 3/8" x 3/8"	EA.	15.40	10.00	25.40

15 MECHANICAL

PLUMBING

	UNIT	MAT.	INST.	TOTAL
15410.17 CHROME PLATED FITTINGS				
1/2" x 3/8" x 1/2"	EA.	15.60	10.00	25.60
15410.18 GLASS PIPE				
Glass pipe				
1-1/2" dia.	L.F.	10.45	6.00	16.45
2" dia.	"	12.10	6.65	18.75
3" dia.	"	21.45	7.50	28.95
4" dia.	"	30.80	8.55	39.35
6" dia.	"	54.00	10.00	64.00
15410.30 PVC/CPVC PIPE				
PVC schedule 40				
1/2" pipe	L.F.	0.55	1.25	1.80
3/4" pipe	"	0.66	1.35	2.01
1" pipe	"	0.86	1.50	2.36
1-1/4" pipe	"	0.97	1.65	2.62
1-1/2" pipe	"	0.99	1.85	2.84
2" pipe	"	1.30	2.15	3.45
2-1/2" pipe	"	2.40	2.50	4.90
3" pipe	"	2.75	3.00	5.75
4" pipe	"	3.85	3.75	7.60
6" pipe	"	7.70	7.50	15.20
8" pipe	"	12.65	10.00	22.65
Fittings, 1/2"				
90 deg ell	EA.	0.29	3.75	4.04
45 deg ell	"	0.44	3.75	4.19
Tee	"	0.37	4.30	4.67
Reducing insert	"	0.29	5.00	5.29
Threaded	"	0.70	3.75	4.45
Male adapter	"	0.28	5.00	5.28
Female adapter	"	0.30	3.75	4.05
Coupling	"	0.22	3.75	3.97
Union	"	4.40	6.00	10.40
Cap	"	0.26	5.00	5.26
Flange	"	5.05	6.00	11.05
3/4"				
90 deg elbow	EA.	0.30	5.00	5.30
45 deg elbow	"	0.68	5.00	5.68
Tee	"	0.38	6.00	6.38
Reducing insert	"	0.28	4.30	4.58
Threaded	"	0.42	5.00	5.42
1"				
90 deg elbow	EA.	0.50	6.00	6.50
45 deg elbow	"	0.75	6.00	6.75
Tee	"	0.68	6.65	7.33
Reducing insert	"	0.50	6.00	6.50
Threaded	"	0.67	6.65	7.32
Male adapter	"	0.44	7.50	7.94
Female adapter	"	0.40	7.50	7.90
Coupling	"	0.37	7.50	7.87

PLUMBING	UNIT	MAT.	INST.	TOTAL
15410.30 PVC/CPVC PIPE				
Union	EA.	3.60	10.00	13.60
Cap	"	0.41	6.00	6.41
Flange	"	5.05	10.00	15.05
1-1/4"				
90 deg elbow	EA.	0.87	8.55	9.42
45 deg elbow	"	1.05	8.55	9.60
Tee	"	1.05	10.00	11.05
Reducing insert	"	0.62	10.00	10.62
Threaded	"	1.00	10.00	11.00
Male adapter	"	0.54	10.00	10.54
Female adapter	"	0.62	10.00	10.62
Coupling	"	0.53	10.00	10.53
Union	"	8.20	12.00	20.20
Cap	"	0.56	10.00	10.56
Flange	"	5.20	12.00	17.20
1-1/2"				
90 deg elbow	EA.	0.97	8.55	9.52
45 deg elbow	"	1.45	8.55	10.00
Tee	"	1.30	10.00	11.30
Reducing insert	"	0.67	10.00	10.67
Threaded	"	1.20	10.00	11.20
Male adapter	"	0.76	10.00	10.76
Female adapter	"	0.76	10.00	10.76
Coupling	"	0.57	10.00	10.57
Union	"	11.00	15.00	26.00
Cap	"	0.64	10.00	10.64
Flange	"	8.70	15.00	23.70
2"				
90 deg elbow	EA.	1.45	10.00	11.45
45 deg elbow	"	1.85	10.00	11.85
Tee	"	2.00	12.00	14.00
Reducing insert	"	1.20	12.00	13.20
Threaded	"	1.65	12.00	13.65
Male adapter	"	0.99	12.00	12.99
Female adapter	"	1.00	12.00	13.00
Coupling	"	0.90	12.00	12.90
Union	"	15.40	18.75	34.15
Cap	"	0.79	12.00	12.79
Flange	"	8.80	18.75	27.55
2-1/2"				
90 deg elbow	EA.	4.45	18.75	23.20
45 deg elbow	"	6.60	18.75	25.35
Tee	"	5.95	20.00	25.95
Reducing insert	"	1.75	20.00	21.75
Threaded	"	2.60	20.00	22.60
Male adapter	"	2.75	20.00	22.75
Female adapter	"	2.45	20.00	22.45
Coupling	"	2.75	20.00	22.75
Union	"	19.80	25.00	44.80
Cap	"	2.40	18.75	21.15
Flange	"	11.55	25.00	36.55
3"				

15 MECHANICAL

PLUMBING	UNIT	MAT.	INST.	TOTAL
15410.30 PVC/CPVC PIPE				
90 deg elbow	EA.	5.30	25.00	30.30
45 deg elbow	"	6.90	25.00	31.90
Tee	"	8.50	27.30	35.80
Reducing insert	"	2.60	25.00	27.60
Threaded	"	3.35	25.00	28.35
Male adapter	"	4.05	25.00	29.05
Female adapter	"	3.25	25.00	28.25
Coupling	"	3.05	25.00	28.05
Union	"	23.15	30.00	53.15
Cap	"	2.65	25.00	27.65
Flange	"	12.10	30.00	42.10
4"				
90 deg elbow	EA.	9.55	30.00	39.55
45 deg elbow	"	12.40	30.00	42.40
Tee	"	14.20	33.30	47.50
Reducing insert	"	5.80	30.00	35.80
Threaded	"	7.45	30.00	37.45
Male adapter	"	5.15	30.00	35.15
Female adapter	"	4.85	30.00	34.85
Coupling	"	4.35	30.00	34.35
Union	"	28.30	37.50	65.80
Cap	"	5.95	30.00	35.95
Flange	"	15.35	37.50	52.85
PVC schedule 80 pipe				
1-1/2" pipe	L.F.	0.85	1.85	2.70
2" pipe	"	1.10	2.15	3.25
3" pipe	"	1.75	3.00	4.75
4" pipe	"	2.75	3.75	6.50
Fittings, 1-1/2"				
90 deg elbow	EA.	2.85	10.00	12.85
45 deg elbow	"	5.05	10.00	15.05
Tee	"	7.90	15.00	22.90
Reducing insert	"	2.75	10.00	12.75
Threaded	"	3.30	10.00	13.30
Male adapter	"	5.25	10.00	15.25
Female adapter	"	5.70	10.00	15.70
Coupling	"	4.40	10.00	14.40
Union	"	11.00	15.00	26.00
Cap	"	3.10	10.00	13.10
Flange	"	7.15	15.00	22.15
2"				
90 deg elbow	EA.	3.40	12.00	15.40
45 deg elbow	"	6.50	12.00	18.50
Tee	"	9.85	18.75	28.60
Reducing insert	"	3.95	12.00	15.95
Threaded	"	4.05	12.00	16.05
Male adapter	"	7.25	12.00	19.25
Female adapter	"	9.95	12.00	21.95
2-1/2"				
90 deg elbow	EA.	6.45	18.75	25.20
45 deg elbow	"	13.60	18.75	32.35
Tee	"	10.70	25.00	35.70

PLUMBING	UNIT	MAT.	INST.	TOTAL
15410.30 PVC/CPVC PIPE				
Reducing insert	EA.	6.85	18.75	25.60
Threaded	"	8.55	18.75	27.30
Male adapter	"	8.65	18.75	27.40
Female adapter	"	15.70	18.75	34.45
Coupling	"	8.45	18.75	27.20
Union	"	23.45	25.00	48.45
Cap	"	9.90	18.75	28.65
Flange	"	12.65	25.00	37.65
3"				
90 deg elbow	EA.	8.80	25.00	33.80
45 deg elbow	"	16.70	25.00	41.70
Tee	"	13.40	30.00	43.40
Reducing insert	"	10.85	25.00	35.85
Threaded	"	15.80	25.00	40.80
Male adapter	"	9.60	25.00	34.60
Female adapter	"	17.75	25.00	42.75
Coupling	"	9.70	25.00	34.70
Union	"	30.00	30.00	60.00
Cap	"	12.65	25.00	37.65
Flange	"	14.30	30.00	44.30
4"				
90 deg elbow	EA.	11.10	30.00	41.10
45 deg elbow	"	30.00	30.00	60.00
Tee	"	14.75	37.50	52.25
Reducing insert	"	15.10	30.00	45.10
Threaded	"	24.35	30.00	54.35
Male adapter	"	17.05	30.00	47.05
Coupling	"	12.20	30.00	42.20
Union	"	28.30	37.50	65.80
Cap	"	15.40	30.00	45.40
Flange	"	17.60	37.50	55.10
CPVC schedule 40				
1/2" pipe	L.F.	0.44	1.25	1.69
3/4" pipe	"	0.67	1.35	2.02
1" pipe	"	0.98	1.50	2.48
1-1/4" pipe	"	1.35	1.65	3.00
1-1/2" pipe	"	1.65	1.85	3.50
2" pipe	"	2.10	2.15	4.25
Fittings, CPVC, schedule 80				
1/2", 90 deg ell	EA.	2.20	3.00	5.20
Tee	"	3.30	5.00	8.30
3/4", 90 deg ell	"	3.40	3.00	6.40
Tee	"	3.85	5.00	8.85
1", 90 deg ell	"	4.20	3.35	7.55
Tee	"	4.40	5.45	9.85
1-1/4", 90 deg ell	"	4.60	3.35	7.95
Tee	"	5.30	5.45	10.75
1-1/2", 90 deg ell	"	5.05	6.00	11.05
Tee	"	7.25	7.50	14.75
2", 90 deg ell	"	7.70	6.00	13.70
Tee	"	9.35	7.50	16.85
Polypropylene, acid resistant, DWV pipe				

PLUMBING	UNIT	MAT.	INST.	TOTAL
15410.30 PVC/CPVC PIPE				
Schedule 40				
1-1/2" pipe	L.F.	3.30	2.15	5.45
2" pipe	"	4.45	2.50	6.95
3" pipe	"	8.25	3.00	11.25
4" pipe	"	11.65	3.75	15.40
6" pipe	"	20.95	7.50	28.45
Fittings				
1-1/2"				
1/4 bend	EA.	12.00	7.50	19.50
1/8 bend	"	11.85	7.50	19.35
Sanitary tee	"	15.00	15.00	30.00
Cleanout with plug	"	30.60	15.00	45.60
Wye	"	28.00	15.00	43.00
Combination wye & 1/8 bend	"	27.10	15.00	42.10
P-trap	"	26.20	18.75	44.95
Hub adapter	"	14.30	7.50	21.80
Coupling	"	9.45	7.50	16.95
2"				
1/4 bend	EA.	15.00	8.55	23.55
1/8 bend	"	14.40	8.55	22.95
Sanitary tee	"	17.95	17.65	35.60
Cleanout with plug	"	35.80	17.65	53.45
Wye	"	37.90	17.65	55.55
Combination wye & 1/8 bend	"	36.10	17.65	53.75
P-trap	"	35.90	25.00	60.90
Hub adapter	"	15.35	10.00	25.35
Mechanical joint adapter	"	15.15	10.00	25.15
Coupling	"	11.75	10.00	21.75
3"				
1/4 bend	EA.	25.40	10.00	35.40
1/8 bend	"	26.40	10.00	36.40
Sanitary tee	"	35.90	20.00	55.90
Cleanout with plug	"	59.00	20.00	79.00
Wye	"	38.20	20.00	58.20
Combination wye & 1/8 bend	"	64.50	20.00	84.50
P-trap	"	83.00	30.00	113.00
Hub adapter	"	19.20	10.00	29.20
Mechanical joint adapter	"	18.75	10.00	28.75
Coupling	"	15.10	10.00	25.10
4"				
1/4 bend	EA.	40.40	15.00	55.40
1/8 bend	"	30.00	15.00	45.00
Sanitary tee	"	53.50	30.00	83.50
Cleanout with plug	"	79.50	30.00	109.50
Wye	"	55.50	30.00	85.50
Combination wye & 1/8 bend	"	86.00	30.00	116.00
P-trap	"	106.00	50.00	156.00
Hub adapter	"	28.60	15.00	43.60
Mechanical joint adapter	"	27.60	15.00	42.60
Coupling	"	21.35	15.00	36.35
6"				
1/4 bend	EA.	99.50	25.00	124.50

PLUMBING	UNIT	MAT.	INST.	TOTAL
15410.30 PVC/CPVC PIPE				
1/8 bend	EA.	83.00	25.00	108.00
Sanitary tee	"	151.00	50.00	201.00
Cleanout with plug	"	109.00	50.00	159.00
Wye	"	143.00	50.00	193.00
Hub adapter	"	37.90	25.00	62.90
Coupling	"	33.90	25.00	58.90
Polyethylene pipe and fittings				
SDR-21				
3" pipe	L.F.	0.90	3.75	4.65
4" pipe	"	1.75	5.00	6.75
6" pipe	"	3.50	7.50	11.00
8" pipe	"	5.30	8.55	13.85
10" pipe	"	8.25	10.00	18.25
12" pipe	"	11.55	12.00	23.55
14" pipe	"	14.30	15.00	29.30
16" pipe	"	18.70	18.75	37.45
18" pipe	"	23.10	23.10	46.20
20" pipe	"	28.60	30.00	58.60
22" pipe	"	34.10	33.30	67.40
24" pipe	"	39.60	37.50	77.10
Fittings, 3"				
90 deg elbow	EA.	58.50	15.00	73.50
45 deg elbow	"	37.10	15.00	52.10
Tee	"	33.90	25.00	58.90
45 deg wye	"	87.00	25.00	112.00
Reducer	"	19.45	18.75	38.20
Flange assembly	"	33.10	15.00	48.10
4"				
90 deg elbow	EA.	86.50	18.75	105.25
45 deg elbow	"	52.50	18.75	71.25
Tee	"	74.50	30.00	104.50
45 deg wye	"	137.00	30.00	167.00
Reducer	"	46.90	25.00	71.90
Flange assembly	"	46.90	18.75	65.65
8"				
90 deg elbow	EA.	235.00	37.50	272.50
45 deg elbow	"	130.00	37.50	167.50
Tee	"	221.00	60.00	281.00
45 deg wye	"	340.00	60.00	400.00
Reducer	"	114.00	50.00	164.00
Flange assembly	"	124.00	37.50	161.50
10"				
90 deg elbow	EA.	324.00	50.00	374.00
45 deg elbow	"	176.00	50.00	226.00
Tee	"	291.00	75.00	366.00
45 deg wye	"	464.00	75.00	539.00
Reducer	"	154.00	60.00	214.00
Flange assembly	"	149.00	50.00	199.00
12"				
90 deg elbow	EA.	526.00	60.00	586.00
45 deg elbow	"	320.00	60.00	380.00
Tee	"	412.00	100.00	512.00

PLUMBING

15410.30 PVC/CPVC PIPE

	UNIT	MAT.	INST.	TOTAL
45 deg wye	EA.	641.00	100.00	741.00
Reducer	"	246.00	75.00	321.00
Flange assembly	"	198.00	60.00	258.00
14"				
90 deg elbow	EA.	715.00	75.00	790.00
45 deg elbow	"	406.00	75.00	481.00
Tee	"	522.00	120.00	642.00
45 deg wye	"	961.00	120.00	1,081
Reducer	"	206.00	100.00	306.00
Flange assembly	"	240.00	75.00	315.00
16"				
90 deg elbow	EA.	892.00	75.00	967.00
45 deg elbow	"	530.00	75.00	605.00
Tee	"	646.00	120.00	766.00
45 deg wye	"	1,080	120.00	1,200
Reducer	"	418.00	100.00	518.00
Flange assembly	"	300.00	75.00	375.00
18"				
90 deg elbow	EA.	1,360	100.00	1,460
45 deg elbow	"	881.00	100.00	981.00
Tee	"	1,050	150.00	1,200
45 deg wye	"	1,820	150.00	1,970
Reducer	"	389.00	100.00	489.00
Flange assembly	"	532.00	100.00	632.00
20"				
90 deg elbow	EA.	1,100	100.00	1,200
45 deg elbow	"	652.00	100.00	752.00

15410.33 ABS DWV PIPE

	UNIT	MAT.	INST.	TOTAL
Schedule 40 ABS				
1-1/2" pipe	L.F.	0.77	1.50	2.27
2" pipe	"	0.99	1.65	2.64
3" pipe	"	2.10	2.15	4.25
4" pipe	"	3.10	3.00	6.10
6" pipe	"	5.05	3.75	8.80
Fittings				
1/8 bend				
1-1/2"	EA.	0.99	6.00	6.99
2"	"	1.10	7.50	8.60
3"	"	3.30	10.00	13.30
4"	"	5.50	12.00	17.50
6"	"	35.20	15.00	50.20
Tee, sanitary				
1-1/2"	EA.	1.55	10.00	11.55
2"	"	2.30	12.00	14.30
3"	"	5.40	15.00	20.40
4"	"	9.90	18.75	28.65
6"	"	57.00	25.00	82.00
Tee, sanitary reducing				
2 x 1-1/2 x 1-1/2	EA.	2.95	12.00	14.95
2 x 1-1/2 x 2	"	3.65	12.50	16.15

PLUMBING	UNIT	MAT.	INST.	TOTAL
15410.33 ABS DWV PIPE				
2 x 2 x 1-1/2	EA.	2.20	13.65	15.85
3 x 3 x 1-1/2	"	4.40	15.00	19.40
3 x 3 x 2	"	4.40	16.65	21.05
4 x 4 x 1-1/2	"	12.10	18.75	30.85
4 x 4 x 2	"	9.90	21.45	31.35
4 x 4 x 3	"	13.20	23.10	36.30
6 x 6 x 4	"	56.00	25.00	81.00
Wye				
1-1/2"	EA.	2.35	8.55	10.90
2"	"	2.75	12.00	14.75
3"	"	5.50	15.00	20.50
4"	"	9.90	18.75	28.65
6"	"	57.00	25.00	82.00
Reducer				
2 x 1-1/2	EA.	1.55	7.50	9.05
3 x 1-1/2	"	3.50	10.00	13.50
3 x 2	"	3.20	10.00	13.20
4 x 2	"	6.80	12.00	18.80
4 x 3	"	6.70	12.00	18.70
6 x 4	"	10.10	15.00	25.10
P-trap				
1-1/2"	EA.	3.50	10.00	13.50
2"	"	4.60	11.10	15.70
3"	"	20.90	13.05	33.95
4"	"	36.60	15.00	51.60
6"	"	60.00	18.75	78.75
Double sanitary, tee				
1-1/2"	EA.	4.40	12.00	16.40
2"	"	5.95	15.00	20.95
3"	"	13.20	18.75	31.95
4"	"	22.00	25.00	47.00
Long sweep, 1/4 bend				
1-1/2"	EA.	2.20	6.00	8.20
2"	"	2.30	7.50	9.80
3"	"	6.15	10.00	16.15
4"	"	12.65	15.00	27.65
Wye, standard				
1-1/2"	EA.	2.65	10.00	12.65
2"	"	2.40	12.00	14.40
3"	"	5.95	15.00	20.95
4"	"	10.45	18.75	29.20
Wye, reducing				
2 x 1-1/2 x 1-1/2	EA.	4.95	10.00	14.95
2 x 2 x 1-1/2	"	3.95	12.00	15.95
4 x 4 x 2	"	10.45	18.75	29.20
4 x 4 x 3	"	9.45	20.00	29.45
Double wye				
1-1/2"	EA.	4.70	12.00	16.70
2"	"	6.05	15.00	21.05
3"	"	19.80	18.75	38.55
4"	"	35.20	25.00	60.20
2 x 2 x 1-1/2 x 1-1/2	"	6.60	15.00	21.60

PLUMBING	UNIT	MAT.	INST.	TOTAL
15410.33 ABS DWV PIPE				
3 x 3 x 2 x 2	EA.	13.20	18.75	31.95
4 x 4 x 3 x 3	"	30.80	25.00	55.80
Combination wye and 1/8 bend				
1-1/2"	EA.	5.05	10.00	15.05
2"	"	5.70	12.00	17.70
3"	"	7.90	15.00	22.90
4"	"	15.40	18.75	34.15
2 x 2 x 1-1/2	"	5.10	12.00	17.10
3 x 3 x 1-1/2	"	9.90	15.00	24.90
3 x 3 x 2	"	6.15	15.00	21.15
4 x 4 x 2	"	12.25	18.75	31.00
4 x 4 x 3	"	14.90	18.75	33.65
15410.35 PLASTIC PIPE				
Fiberglass reinforced pipe				
2" pipe	L.F.	4.05	2.30	6.35
3" pipe	"	5.60	2.50	8.10
4" pipe	"	7.05	2.75	9.80
6" pipe	"	9.45	3.00	12.45
8" pipe	"	18.70	5.00	23.70
10" pipe	"	30.80	6.00	36.80
12" pipe	"	35.20	7.50	42.70
Fittings				
90 deg elbow, flanged				
2"	EA.	70.50	30.00	100.50
3"	"	84.50	33.30	117.80
4"	"	110.00	37.50	147.50
6"	"	196.00	50.00	246.00
8"	"	358.00	60.00	418.00
10"	"	474.00	75.00	549.00
12"	"	635.00	100.00	735.00
45 deg elbow, flanged				
2"	EA.	70.50	25.00	95.50
3"	"	84.50	30.00	114.50
4"	"	110.00	37.50	147.50
6"	"	196.00	50.00	246.00
8"	"	300.00	60.00	360.00
10"	"	404.00	75.00	479.00
12"	"	508.00	100.00	608.00
Tee, flanged				
2"	EA.	90.50	37.50	128.00
3"	"	127.00	42.90	169.90
4"	"	142.00	50.00	192.00
6"	"	238.00	60.00	298.00
8"	"	425.00	75.00	500.00
10"	"	691.00	100.00	791.00
12"	"	952.00	150.00	1,102
Wye, flanged				
2"	EA.	177.00	37.50	214.50
3"	"	244.00	42.90	286.90
4"	"	317.00	50.00	367.00
6"	"	402.00	60.00	462.00

PLUMBING	UNIT	MAT.	INST.	TOTAL
15410.35 PLASTIC PIPE				
8"	EA.	544.00	75.00	619.00
10"	"	929.00	100.00	1,029
12"	"	1,190	150.00	1,340
Concentric reducer, flanged				
2"	EA.	68.00	25.00	93.00
4"	"	102.00	30.00	132.00
6"	"	136.00	42.90	178.90
8"	"	220.00	60.00	280.00
10"	"	340.00	75.00	415.00
12"	"	487.00	100.00	587.00
Adapter, bell x male or female				
2"	EA.	8.70	25.00	33.70
3"	"	17.45	27.30	44.75
4"	"	18.45	30.00	48.45
6"	"	39.40	42.90	82.30
8"	"	57.50	60.00	117.50
10"	"	82.50	75.00	157.50
12"	"	146.00	100.00	246.00
Nipples				
2" x 6"	EA.	3.70	3.00	6.70
2" x 12"	"	5.55	3.75	9.30
3" x 8"	"	5.55	4.60	10.15
3" x 12"	"	5.55	5.00	10.55
4" x 8"	"	5.55	5.00	10.55
4" x 12"	"	6.25	6.00	12.25
6" x 12"	"	14.30	7.50	21.80
8" x 18"	"	40.80	7.50	48.30
8" x 24"	"	47.60	8.55	56.15
10" x 18"	"	47.60	10.00	57.60
10" x 24"	"	64.50	12.00	76.50
12" x 18"	"	64.50	13.65	78.15
12" x 24"	"	76.00	15.00	91.00
Sleeve coupling				
2"	EA.	8.05	25.00	33.05
3"	"	8.85	30.00	38.85
4"	"	12.25	42.90	55.15
6"	"	29.50	60.00	89.50
8"	"	48.70	75.00	123.70
10"	"	72.50	100.00	172.50
Flanges				
2"	EA.	12.00	25.00	37.00
3"	"	16.20	30.00	46.20
4"	"	21.55	42.90	64.45
6"	"	36.80	60.00	96.80
8"	"	64.50	75.00	139.50
10"	"	92.00	100.00	192.00
12"	"	113.00	100.00	213.00
15410.70 STAINLESS STEEL PIPE				
Stainless steel, schedule 40, threaded				
1/2" pipe	L.F.	3.50	4.30	7.80

PLUMBING	UNIT	MAT.	INST.	TOTAL
15410.70 STAINLESS STEEL PIPE				
3/4" pipe	L.F.	4.75	4.40	9.15
1" pipe	"	6.15	4.60	10.75
1-1/2" pipe	"	8.80	5.00	13.80
2" pipe	"	13.20	5.45	18.65
2-1/2" pipe	"	17.60	6.00	23.60
3" pipe	"	26.40	6.65	33.05
4" pipe	"	33.00	7.50	40.50
Fittings, 1/2"				
90 deg ell	EA.	4.95	37.50	42.45
45 deg ell	"	5.85	37.50	43.35
Tee	"	6.60	50.00	56.60
Cap	"	2.20	18.75	20.95
Reducer	"	7.15	25.00	32.15
Union	"	11.55	37.50	49.05
Flange	"	19.80	37.50	57.30
3/4"				
90 deg ell	EA.	6.60	37.50	44.10
45 deg ell	"	7.60	37.50	45.10
Tee	"	8.90	50.00	58.90
Cap	"	3.95	18.75	22.70
Reducer	"	8.80	25.00	33.80
Union	"	14.30	37.50	51.80
Flange	"	24.20	37.50	61.70
1"				
90 deg ell	EA.	8.25	37.50	45.75
45 deg ell	"	8.25	37.50	45.75
Tee	"	9.80	50.00	59.80
Cap	"	5.40	18.75	24.15
Reducer	"	8.60	37.50	46.10
Union	"	17.05	37.50	54.55
Flange	"	24.20	37.50	61.70
1-1/4"				
90 deg ell	EA.	11.55	37.50	49.05
45 deg ell	"	11.55	37.50	49.05
Tee	"	16.50	50.00	66.50
Cap	"	10.45	18.75	29.20
Union	"	31.40	37.50	68.90
Flange	"	27.50	37.50	65.00
1-1/2"				
90 deg ell	EA.	14.30	50.00	64.30
45 deg ell	"	14.85	50.00	64.85
Tee	"	20.80	60.00	80.80
Cap	"	13.20	25.00	38.20
Reducer	"	15.95	30.00	45.95
Union	"	37.40	42.90	80.30
Flange	"	34.10	42.90	77.00
2"				
90 deg ell	EA.	21.45	60.00	81.45
45 deg ell	"	21.45	60.00	81.45
Tee	"	25.30	100.00	125.30
Cap	"	16.50	30.00	46.50
Reducer	"	19.25	37.50	56.75

PLUMBING	UNIT	MAT.	INST.	TOTAL
15410.70 STAINLESS STEEL PIPE				
Union	EA.	41.80	60.00	101.80
Flange	"	31.90	60.00	91.90
Type 304, sch 10 pipe				
1" pipe	L.F.	5.50	3.75	9.25
1-1/4" pipe	"	7.15	5.00	12.15
1-1/2" pipe	"	7.70	5.45	13.15
2" pipe	"	8.25	6.65	14.90
2-1/2" pipe	"	11.00	7.50	18.50
3" pipe	"	13.10	8.55	21.65
4" pipe	"	17.15	10.00	27.15
6" pipe	"	27.50	12.00	39.50
Fittings, 1"				
90 deg elbow	EA.	5.70	60.00	65.70
45 deg elbow	"	8.05	60.00	68.05
Tee	"	17.60	100.00	117.60
Reducer	"	6.05	60.00	66.05
Flange	"	22.00	37.50	59.50
Cap	"	17.60	37.50	55.10
1-1/4"				
90 deg elbow	EA.	9.35	60.00	69.35
45 deg elbow	"	11.20	60.00	71.20
Tee	"	34.10	100.00	134.10
Reducer	"	8.80	60.00	68.80
Flange	"	27.50	37.50	65.00
Cap	"	27.50	37.50	65.00
1-1/2"				
90 deg elbow	EA.	10.45	50.00	60.45
45 deg elbow	"	11.55	50.00	61.55
Tee	"	24.20	100.00	124.20
Reducer	"	13.20	75.00	88.20
Flange	"	24.20	50.00	74.20
Cap	"	22.00	50.00	72.00
2"				
90 deg elbow	EA.	9.90	60.00	69.90
45 deg elbow	"	9.00	60.00	69.00
Tee	"	24.20	150.00	174.20
Reducer	"	13.20	100.00	113.20
Flange	"	30.80	75.00	105.80
Cap	"	22.00	75.00	97.00
2-1/2"				
90 deg elbow	EA.	16.50	60.00	76.50
45 deg elbow	"	16.50	60.00	76.50
Tee	"	53.00	150.00	203.00
Reducer	"	17.60	100.00	117.60
Flange	"	53.00	75.00	128.00
Cap	"	22.00	75.00	97.00
3"				
90 deg elbow	EA.	16.70	75.00	91.70
45 deg elbow	"	13.85	75.00	88.85
Tee	"	41.80	200.00	241.80
Reducer	"	18.15	150.00	168.15
Flange	"	49.50	100.00	149.50

PLUMBING	UNIT	MAT.	INST.	TOTAL
15410.70 STAINLESS STEEL PIPE				
Cap	EA.	23.10	100.00	123.10
4"				
90 deg elbow	EA.	35.20	75.00	110.20
45 deg elbow	"	26.40	75.00	101.40
Reducer	"	33.00	150.00	183.00
Flange	"	69.50	100.00	169.50
Cap	"	24.20	100.00	124.20
6"				
90 deg elbow	EA.	76.00	100.00	176.00
45 deg elbow	"	49.50	100.00	149.50
Tee	"	154.00	429.00	583.00
Reducer	"	46.20	150.00	196.20
Flange	"	106.00	150.00	256.00
Cap	"	50.50	150.00	200.50
Type 304 tubing				
.035 wall				
1/4"	L.F.	0.66	1.65	2.31
3/8"	"	0.79	1.85	2.64
1/2"	"	0.92	2.15	3.07
5/8"	"	1.30	2.50	3.80
3/4"	"	1.55	3.00	4.55
7/8"	"	1.60	3.35	4.95
1"	"	1.65	3.75	5.40
.049 wall				
1/4"	L.F.	0.83	1.75	2.58
3/8"	"	0.91	2.00	2.91
1/2"	"	0.99	2.30	3.29
5/8"	"	1.30	2.75	4.05
3/4"	"	1.55	3.35	4.90
7/8"	"	1.95	3.75	5.70
1"	"	2.00	4.30	6.30
.065 wall				
1/4"	L.F.	1.25	2.00	3.25
3/8"	"	1.55	2.50	4.05
1/2"	"	1.65	2.75	4.40
5/8"	"	1.75	3.35	5.10
3/4"	"	1.15	4.30	5.45
7/8"	"	2.55	5.00	7.55
1"	"	2.40	6.00	8.40
Type 316 tubing				
.035 wall				
1/4"	L.F.	0.64	1.65	2.29
3/8"	"	0.79	1.85	2.64
1/2"	"	0.64	2.15	2.79
5/8"	"	1.55	2.50	4.05
3/4"	"	1.75	3.00	4.75
7/8"	"	2.40	3.35	5.75
1"	"	2.75	3.75	6.50
.049 wall				
1/4"	L.F.	1.10	2.00	3.10
3/8"	"	1.15	2.50	3.65
1/2"	"	1.45	2.75	4.20

PLUMBING	UNIT	MAT.	INST.	TOTAL
15410.70 STAINLESS STEEL PIPE				
5/8"	L.F.	1.85	3.35	5.20
3/4"	"	2.00	4.30	6.30
7/8"	"	2.85	5.00	7.85
1"	"	1.80	6.00	7.80
.065 wall				
1/4"	L.F.	1.55	2.00	3.55
3/8"	"	2.10	2.50	4.60
1/2"	"	2.10	2.75	4.85
5/8"	"	2.20	3.35	5.55
3/4"	"	2.55	4.30	6.85
7/8"	"	5.45	5.00	10.45
1"	"	3.05	6.00	9.05
Fittings, 1/4"				
90 deg elbow	EA.	12.10	6.00	18.10
Union tee	"	20.35	10.00	30.35
Union	"	9.90	10.00	19.90
Male connector	"	6.60	7.50	14.10
3/8"				
90 deg elbow	EA.	17.60	7.50	25.10
Union tee	"	26.40	11.55	37.95
Union	"	13.75	11.55	25.30
Male connector	"	9.90	7.50	17.40
1/2"				
90 deg elbow	EA.	26.40	7.90	34.30
Union tee	"	39.60	12.50	52.10
Union	"	20.90	12.50	33.40
Male connector	"	14.30	7.50	21.80
5/8"				
90 deg elbow	EA.	64.00	10.00	74.00
Union tee	"	84.50	15.00	99.50
Union	"	41.80	15.00	56.80
Male connector	"	20.90	10.00	30.90
3/4"				
90 deg elbow	EA.	94.50	10.00	104.50
Union tee	"	81.50	15.00	96.50
Union	"	68.00	15.00	83.00
Male connector	"	29.70	10.00	39.70
7/8"				
90 deg elbow	EA.	88.00	10.70	98.70
Union tee	"	132.00	16.65	148.65
Union	"	68.00	16.65	84.65
Male connector	"	54.00	10.70	64.70
1"				
90 deg elbow	EA.	160.00	13.65	173.65
Union tee	"	198.00	18.75	216.75
Union	"	106.00	18.75	124.75
Male conector	"	71.50	15.00	86.50
Type 316 valves				
Gate valves				
1/4"	EA.	242.00	10.00	252.00
3/8"	"	242.00	12.00	254.00
1/2"	"	154.00	13.05	167.05

PLUMBING	UNIT	MAT.	INST.	TOTAL
15410.70 STAINLESS STEEL PIPE				
3/4"	EA.	192.00	15.00	207.00
1"	"	211.00	20.00	231.00
Globe valves				
1/4"	EA.	93.50	10.00	103.50
3/8"	"	148.00	12.00	160.00
1/2"	"	187.00	13.05	200.05
3/4"	"	214.00	15.00	229.00
1"	"	220.00	20.00	240.00
Check valves				
1/4"	EA.	71.50	10.00	81.50
3/8"	"	138.00	12.00	150.00
1/2"	"	160.00	13.05	173.05
3/4"	"	209.00	15.00	224.00
1"	"	220.00	20.00	240.00
Test and balance	"	0.00	25.00	25.00
Pipe identification	"	0.11	6.00	6.11
Disinfect	"	0.00	25.00	25.00
15410.80 STEEL PIPE				
Black steel, extra heavy pipe, threaded				
1/2" pipe	L.F.	1.10	1.20	2.30
3/4" pipe	"	1.30	1.20	2.50
1" pipe	"	1.75	1.50	3.25
1-1/2" pipe	"	2.85	1.65	4.50
2-1/2" pipe	"	6.05	3.75	9.80
3" pipe	"	8.25	5.00	13.25
4" pipe	"	13.10	6.00	19.10
5" pipe	"	17.60	7.50	25.10
6" pipe	"	22.00	7.50	29.50
8" pipe	"	24.75	10.00	34.75
10" pipe	"	33.00	12.00	45.00
12" pipe	"	41.80	15.00	56.80
Fittings, malleable iron, threaded, 1/2" pipe				
90 deg ell	EA.	1.30	10.00	11.30
45 deg ell	"	2.10	10.00	12.10
Tee	"	1.75	15.00	16.75
Reducing tee	"	3.85	15.00	18.85
Cap	"	1.30	6.00	7.30
Coupling	"	1.60	12.00	13.60
Union	"	7.45	10.00	17.45
Nipple, 4" long	"	1.35	10.00	11.35
3/4" pipe				
90 deg ell	EA.	1.75	10.00	11.75
45 deg ell	"	2.85	15.00	17.85
Tee	"	2.90	15.00	17.90
Reducing tee	"	4.30	10.00	14.30
Cap	"	1.75	6.00	7.75
Coupling	"	1.90	10.00	11.90
Union	"	8.45	10.00	18.45
Nipple, 4" long	"	1.95	10.00	11.95
1" pipe				

PLUMBING	UNIT	MAT.	INST.	TOTAL
15410.80 STEEL PIPE				
90 deg ell	EA.	2.85	12.00	14.85
45 deg ell	"	3.20	12.00	15.20
Tee	"	4.30	16.65	20.95
Reducing tee	"	5.50	16.65	22.15
Cap	"	2.40	6.00	8.40
Coupling	"	3.30	12.00	15.30
Union	"	10.10	12.00	22.10
Nipple, 4" long	"	2.30	12.00	14.30
1-1/2" pipe				
90 deg ell	EA.	5.45	15.00	20.45
45 deg ell	"	6.35	15.00	21.35
Tee	"	7.85	21.45	29.30
Reducing tee	"	12.15	21.45	33.60
Cap	"	3.50	7.50	11.00
Coupling	"	4.85	15.00	19.85
Union	"	13.20	15.00	28.20
Nipple, 4" long	"	3.40	15.00	18.40
2-1/2" pipe				
90 deg ell	EA.	22.00	37.50	59.50
45 deg ell	"	27.50	37.50	65.00
Tee	"	29.70	50.00	79.70
Reducing tee	"	38.50	50.00	88.50
Cap	"	11.55	18.75	30.30
Coupling	"	18.15	50.00	68.15
Union	"	49.50	50.00	99.50
Nipple, 4" long	"	11.55	50.00	61.55
3" pipe				
90 deg ell	EA.	31.90	50.00	81.90
45 deg ell	"	36.30	50.00	86.30
Tee	"	40.70	75.00	115.70
Reducing tee	"	65.00	75.00	140.00
Cap	"	17.60	25.00	42.60
Coupling	"	25.30	50.00	75.30
Union	"	81.50	50.00	131.50
Nipple, 4" long	"	14.30	50.00	64.30
4" pipe				
90 deg ell	EA.	64.00	60.00	124.00
45 deg ell	"	65.00	60.00	125.00
Tee	"	93.50	100.00	193.50
Reducing tee	"	99.00	100.00	199.00
Cap	"	26.40	100.00	126.40
Coupling	"	49.50	30.00	79.50
Union	"	86.00	100.00	186.00
Nipple, 4" long	"	17.60	100.00	117.60
6" pipe				
90 deg ell	EA.	182.00	60.00	242.00
45 deg ell	"	220.00	60.00	280.00
Tee	"	253.00	100.00	353.00
Reducing tee	"	248.00	100.00	348.00
Cap	"	104.00	30.00	134.00
8" pipe				
90 deg ell	EA.	189.00	120.00	309.00

15 MECHANICAL

PLUMBING	UNIT	MAT.	INST.	TOTAL
15410.80 STEEL PIPE				
45 deg ell	EA.	231.00	120.00	351.00
Tee	"	275.00	188.00	463.00
Reducing tee	"	253.00	158.00	411.00
Cap	"	30.80	60.00	90.80
10" pipe				
90 deg ell	EA.	211.00	150.00	361.00
45 deg ell	"	248.00	250.00	498.00
Tee	"	314.00	188.00	502.00
Reducing tee	"	297.00	75.00	372.00
Cap	"	30.80	75.00	105.80
12" pipe				
90 deg ell	EA.	248.00	188.00	436.00
45 deg ell	"	286.00	188.00	474.00
Tee	"	363.00	250.00	613.00
Reducing tee	"	336.00	250.00	586.00
Cap	"	31.90	100.00	131.90
Butt welded, 1/2" pipe				
90 deg ell	EA.	8.70	10.00	18.70
45 deg ell	"	12.65	10.00	22.65
Tee	"	35.20	15.00	50.20
3/4" pipe				
90 deg ell	EA.	8.70	10.00	18.70
45 deg. ell	"	13.20	10.00	23.20
Tee	"	36.30	15.00	51.30
1" pipe				
90 deg ell	EA.	8.70	12.00	20.70
45 deg ell	"	13.20	12.00	25.20
Tee	"	36.30	16.65	52.95
1-1/2" pipe				
90 deg ell	EA.	12.30	15.00	27.30
45 deg. ell	"	13.10	15.00	28.10
Tee	"	32.50	21.45	53.95
Reducing tee	"	21.65	21.45	43.10
Cap	"	11.55	12.00	23.55
2-1/2" pipe				
90 deg. ell	EA.	25.30	30.00	55.30
45 deg. ell	"	25.30	30.00	55.30
Tee	"	97.00	42.90	139.90
Reducing tee	"	20.90	42.90	63.80
Cap	"	26.40	15.00	41.40
3" pipe				
90 deg ell	EA.	24.20	37.50	61.70
45 deg. ell	"	59.50	37.50	97.00
Tee	"	24.20	50.00	74.20
Reducing tee	"	25.30	50.00	75.30
Cap	"	13.20	25.00	38.20
4" pipe				
90 deg ell	EA.	48.40	50.00	98.40
45 deg. ell	"	39.60	50.00	89.60
Tee	"	87.00	75.00	162.00
Reducing tee	"	31.90	75.00	106.90
Cap	"	14.30	25.00	39.30

PLUMBING

15410.80 STEEL PIPE

PLUMBING	UNIT	MAT.	INST.	TOTAL
6" pipe				
90 deg. ell	EA.	148.00	60.00	208.00
45 deg. ell	"	110.00	60.00	170.00
Tee	"	236.00	100.00	336.00
Reducing tee	"	214.00	100.00	314.00
Cap	"	25.30	30.00	55.30
8" pipe				
90 deg. ell	EA.	236.00	100.00	336.00
45 deg. ell	"	176.00	100.00	276.00
Tee	"	539.00	150.00	689.00
Reducing tee	"	446.00	150.00	596.00
Cap	"	38.50	60.00	98.50
10" pipe				
90 deg ell	EA.	517.00	100.00	617.00
45 deg. ell	"	358.00	100.00	458.00
Tee	"	952.00	150.00	1,102
Reducing tee	"	187.00	150.00	337.00
Cap	"	60.50	75.00	135.50
12" pipe				
90 deg. ell	EA.	869.00	120.00	989.00
45 deg. ell	"	566.00	120.00	686.00
Tee	"	1,210	214.00	1,424
Reducing tee	"	429.00	214.00	643.00
Cap	"	73.50	75.00	148.50
90 deg. ell	"	2.20	10.00	12.20
45 deg. ell	"	4.40	10.00	14.40
Tee	"	2.85	15.00	17.85
Reducing tee	"	5.40	15.00	20.40
3/4" pipe				
90 deg. ell	EA.	2.30	10.00	12.30
45 deg. ell	"	2.85	10.00	12.85
Tee	"	3.60	15.00	18.60
Reducing tee	"	4.70	15.00	19.70
1" pipe				
90 deg. ell	EA.	2.80	12.00	14.80
45 deg. ell	"	3.85	12.00	15.85
Tee	"	3.50	16.65	20.15
Reducing tee	"	4.60	16.65	21.25
1-1/2" pipe				
90 deg. ell	EA.	5.45	15.00	20.45
45 deg. ell	"	7.80	15.00	22.80
Tee	"	7.95	21.45	29.40
Reducing tee	"	10.75	21.45	32.20
2-1/2" pipe				
90 deg. ell	EA.	17.40	30.00	47.40
45 deg. ell	"	20.65	30.00	50.65
Tee	"	24.80	42.90	67.70
Reducing tee	"	28.80	42.90	71.70
3" pipe				
90 deg. ell	EA.	28.20	37.50	65.70
45 deg. ell	"	32.30	37.50	69.80
Tee	"	37.40	60.00	97.40

PLUMBING

15410.80 STEEL PIPE	UNIT	MAT.	INST.	TOTAL
Reducing tee	EA.	43.30	60.00	103.30
4" pipe				
90 deg. ell	EA.	50.50	50.00	100.50
45 deg. ell	"	62.50	50.00	112.50
Tee	"	72.00	75.00	147.00
Reducing tee	"	86.50	75.00	161.50
6" pipe				
90 deg. ell	EA.	118.00	50.00	168.00
45 deg. ell	"	136.00	50.00	186.00
Tee	"	165.00	75.00	240.00
Reducing tee	"	194.00	75.00	269.00
8" pipe				
90 deg. ell	EA.	244.00	100.00	344.00
45 deg. ell	"	271.00	100.00	371.00
Tee	"	344.00	150.00	494.00
Reducing tee	"	375.00	150.00	525.00

15410.82 GALVANIZED STEEL PIPE	UNIT	MAT.	INST.	TOTAL
Galvanized pipe				
1/2" pipe	L.F.	1.20	3.00	4.20
3/4" pipe	"	1.55	3.75	5.30
1" pipe	"	2.20	4.30	6.50
1-1/4" pipe	"	2.65	5.00	7.65
1-1/2" pipe	"	3.20	6.00	9.20
2" pipe	"	4.60	7.50	12.10
2-1/2" pipe	"	5.70	10.00	15.70
3" pipe	"	9.90	10.70	20.60
4" pipe	"	15.40	12.50	27.90
6" pipe	"	21.45	25.00	46.45
90 degree ell, 150 lb malleable iron, galvanized				
1/2"	EA.	1.10	6.00	7.10
3/4"	"	1.30	7.50	8.80
1"	"	2.65	7.90	10.55
1-1/4"	"	4.40	8.80	13.20
1-1/2"	"	4.60	10.00	14.60
2"	"	6.60	12.00	18.60
2-1/2"	"	11.00	18.75	29.75
3"	"	15.40	23.10	38.50
4"	"	24.20	25.00	49.20
5"	"	57.00	30.00	87.00
6"	"	69.50	30.00	99.50
45 degree ell, 150 lb m.i., galv.				
1/2"	EA.	1.30	6.00	7.30
3/4"	"	2.20	7.50	9.70
1"	"	3.30	7.90	11.20
1-1/4"	"	4.40	8.80	13.20
1-1/2"	"	5.50	10.00	15.50
2"	"	6.60	12.00	18.60
2-1/2"	"	11.00	18.75	29.75
3"	"	16.50	23.10	39.60
4"	"	25.30	30.00	55.30
5"	"	66.00	30.00	96.00

PLUMBING	UNIT	MAT.	INST.	TOTAL
15410.82 GALVANIZED STEEL PIPE				
6"	EA.	77.00	37.50	114.50
Tees, straight, 150 lb m.i., galv.				
1/2"	EA.	1.55	7.50	9.05
3/4"	"	2.65	8.55	11.20
1"	"	3.75	10.00	13.75
1-1/4"	"	4.40	12.00	16.40
1-1/2"	"	5.50	15.00	20.50
2"	"	8.80	18.75	27.55
2-1/2"	"	15.40	25.00	40.40
3"	"	16.50	30.00	46.50
4"	"	33.00	37.50	70.50
5"	"	66.00	42.90	108.90
6"	"	88.00	50.00	138.00
Tees, reducing, out, 150 lb m.i., galv.				
1/2"	EA.	3.10	7.50	10.60
3/4"	"	3.30	8.55	11.85
1"	"	4.40	10.00	14.40
1-1/4"	"	5.50	12.00	17.50
1-1/2"	"	6.60	15.00	21.60
2"	"	13.75	18.75	32.50
2-1/2"	"	15.40	25.00	40.40
3"	"	19.80	30.00	49.80
4"	"	28.60	37.50	66.10
5"	"	97.00	42.90	139.90
6"	"	75.00	50.00	125.00
Couplings, straight, 150 lb m.i., galv.				
1/2"	EA.	1.75	6.00	7.75
3/4"	"	2.20	6.65	8.85
1"	"	3.30	7.50	10.80
1-1/4"	"	4.40	8.55	12.95
1-1/2"	"	4.95	10.00	14.95
2"	"	6.05	12.00	18.05
2-1/2"	"	7.70	18.75	26.45
3"	"	9.35	25.00	34.35
4"	"	20.90	27.30	48.20
5"	"	35.20	30.00	65.20
6"	"	48.40	30.00	78.40
Couplings, reducing, 150 lb m.i., galv				
1/2"	EA.	1.75	6.00	7.75
3/4"	"	2.20	6.65	8.85
1"	"	3.30	7.50	10.80
1-1/4"	"	4.40	8.55	12.95
1-1/2"	"	4.95	10.00	14.95
2"	"	6.05	12.00	18.05
2-1/2"	"	7.70	18.75	26.45
3"	"	9.35	25.00	34.35
4"	"	20.90	27.30	48.20
5"	"	35.20	30.00	65.20
6"	"	48.40	30.00	78.40
Caps, 150 lb m.i., galv.				
1/2"	EA.	1.10	3.00	4.10
3/4"	"	1.30	3.15	4.45

PLUMBING	UNIT	MAT.	INST.	TOTAL
15410.82 GALVANIZED STEEL PIPE				
1"	EA.	2.00	3.35	5.35
1-1/4"	"	2.40	3.55	5.95
1-1/2"	"	3.30	3.75	7.05
2"	"	3.50	4.30	7.80
2-1/2"	"	4.40	5.45	9.85
3"	"	6.60	7.50	14.10
4"	"	11.00	9.35	20.35
5"	"	23.10	11.55	34.65
6"	"	42.90	15.00	57.90
Unions, 150 lb m.i., galv.				
1/2"	EA.	3.30	7.50	10.80
3/4"	"	3.50	8.55	12.05
1"	"	4.40	10.00	14.40
1-1/4"	"	5.50	12.00	17.50
1-1/2"	"	6.60	15.00	21.60
2"	"	8.80	16.65	25.45
2-1/2"	"	15.40	20.00	35.40
3"	"	24.20	25.00	49.20
Nipples, galvanized steel, 4" long				
1/2"	EA.	1.30	3.75	5.05
3/4"	"	1.75	4.00	5.75
1"	"	2.40	4.30	6.70
1-1/4"	"	2.75	4.60	7.35
1-1/2"	"	3.30	5.00	8.30
2"	"	3.85	5.45	9.30
2-1/2"	"	4.60	6.00	10.60
3"	"	6.05	7.50	13.55
4"	"	7.60	10.00	17.60
90 degree reducing ell, 150 lb m.i., galv.				
3/4" x 1/2"	EA.	2.00	6.00	8.00
1" x 3/4"	"	2.75	6.65	9.40
1-1/4" x 1"	"	4.40	7.50	11.90
1-1/4" x 3/4"	"	4.60	8.55	13.15
1-1/4" x 1/2"	"	4.60	10.00	14.60
1-1/2" x 1-1/4"	"	4.85	10.00	14.85
1-1/2" x 1"	"	5.05	10.00	15.05
1-1/2" x 3/4"	"	5.30	10.00	15.30
2" x 1-1/2"	"	6.60	13.05	19.65
2" x 1-1/4"	"	6.95	13.05	20.00
2" x 1"	"	6.95	13.05	20.00
2" x 3/4"	"	6.95	13.05	20.00
2-1/2" x 2"	"	13.20	18.75	31.95
2-1/2" x 1-1/2"	"	13.75	18.75	32.50
3" x 2-1/2"	"	19.80	25.00	44.80
3" x 2"	"	19.80	25.00	44.80
4" x 3"	"	35.20	27.30	62.50
Square head plug (C.I.)				
1/2"	EA.	0.99	3.35	4.34
3/4"	"	2.20	3.75	5.95
1"	"	2.30	4.00	6.30
1-1/4"	"	2.40	4.30	6.70
1-1/2"	"	3.05	4.60	7.65

15 MECHANICAL

PLUMBING	UNIT	MAT.	INST.	TOTAL
15410.82 GALVANIZED STEEL PIPE				
2"	EA.	3.85	5.00	8.85
2-1/2"	"	5.05	6.65	11.70
3"	"	7.35	7.50	14.85
4"	"	10.75	10.00	20.75
5"	"	17.60	12.00	29.60
6"	"	24.20	15.00	39.20
Screwed flanges, galv.				
1"	EA.	9.00	15.00	24.00
1-1/4"	"	9.90	16.65	26.55
1-1/2"	"	11.20	18.75	29.95
2"	"	12.00	18.75	30.75
2-1/2"	"	12.65	20.00	32.65
3"	"	16.50	27.30	43.80
4"	"	22.00	37.50	59.50
5"	"	27.50	50.00	77.50
6"	"	35.20	50.00	85.20
15430.23 CLEANOUTS				
Cleanout, wall				
2"	EA.	51.00	20.00	71.00
3"	"	57.00	20.00	77.00
4"	"	76.00	25.00	101.00
6"	"	156.00	30.00	186.00
8"	"	176.00	37.50	213.50
Floor				
2"	EA.	76.50	25.00	101.50
3"	"	81.50	25.00	106.50
4"	"	108.00	30.00	138.00
6"	"	185.00	37.50	222.50
8"	"	231.00	42.90	273.90
15430.24 GREASE TRAPS				
Grease traps, cast iron, 3" pipe				
35 gpm, 70 lb capacity	EA.	1,810	300.00	2,110
50 gpm, 100 lb capacity	"	2,190	375.00	2,565
15430.25 HOSE BIBBS				
Hose bibb				
1/2"	EA.	4.45	10.00	14.45
3/4"	"	4.70	10.00	14.70
15430.60 VALVES				
Gate valve, 125 lb, bronze, soldered				
1/2"	EA.	8.25	7.50	15.75
3/4"	"	10.00	7.50	17.50
1"	"	14.20	10.00	24.20
1-1/2"	"	24.60	12.00	36.60
2"	"	31.90	15.00	46.90
2-1/2"	"	65.00	18.75	83.75

15 MECHANICAL

PLUMBING	UNIT	MAT.	INST.	TOTAL
15430.60 VALVES				
Threaded				
1/4", 125 lb	EA.	5.50	12.00	17.50
1/2"				
125 lb	EA.	8.25	12.00	20.25
150 lb	"	11.55	12.00	23.55
300 lb	"	27.50	12.00	39.50
3/4"				
125 lb	EA.	10.00	12.00	22.00
150 lb	"	16.50	12.00	28.50
300 lb	"	34.10	12.00	46.10
1"				
125 lb	EA.	14.20	12.00	26.20
150 lb	"	20.90	12.00	32.90
300 lb	"	49.50	15.00	64.50
1-1/2"				
125 lb	EA.	24.20	15.00	39.20
150 lb	"	36.30	15.00	51.30
300 lb	"	93.50	16.65	110.15
2"				
125 lb	EA.	31.90	21.45	53.35
150 lb	"	49.50	21.45	70.95
300 lb	"	132.00	25.00	157.00
Cast iron, flanged				
2", 150 lb	EA.	183.00	25.00	208.00
2-1/2"				
125 lb	EA.	176.00	25.00	201.00
150 lb	"	275.00	25.00	300.00
250 lb	"	352.00	25.00	377.00
3"				
125 lb	EA.	209.00	30.00	239.00
150 lb	"	286.00	30.00	316.00
250 lb	"	440.00	30.00	470.00
4"				
125 lb	EA.	253.00	42.90	295.90
150 lb	"	341.00	42.90	383.90
250 lb	"	594.00	42.90	636.90
6"				
125 lb	EA.	385.00	60.00	445.00
250 lb	"	990.00	60.00	1,050
8"				
125 lb	EA.	770.00	75.00	845.00
250 lb	"	1,870	75.00	1,945
OS&Y, flanged				
2"				
125 lb	EA.	132.00	25.00	157.00
250 lb	"	308.00	25.00	333.00
2-1/2"				
125 lb	EA.	143.00	25.00	168.00
250 lb	"	352.00	30.00	382.00
3"				
125 lb	EA.	143.00	30.00	173.00
250 lb	"	385.00	30.00	415.00

PLUMBING

15430.60 VALVES

PLUMBING	UNIT	MAT.	INST.	TOTAL
4"				
125 lb	EA.	204.00	50.00	254.00
250 lb	"	594.00	50.00	644.00
6"				
125 lb	EA.	324.00	60.00	384.00
250 lb	"	957.00	60.00	1,017
Check valve, bronze, soldered, 125 lb				
1/2"	EA.	13.15	7.50	20.65
3/4"	"	15.40	7.50	22.90
1"	"	19.80	10.00	29.80
1-1/4"	"	23.10	12.00	35.10
1-1/2"	"	31.90	12.00	43.90
2"	"	44.00	15.00	59.00
Threaded				
1/2"				
125 lb	EA.	10.70	10.00	20.70
150 lb	"	17.60	10.00	27.60
200 lb	"	17.85	10.00	27.85
3/4"				
125 lb	EA.	13.60	12.00	25.60
150 lb	"	22.20	12.00	34.20
200 lb	"	22.20	12.00	34.20
1"				
125 lb	EA.	18.05	15.00	33.05
150 lb	"	29.10	15.00	44.10
200 lb	"	29.10	15.00	44.10
Flow check valve, cast iron, threaded				
1"	EA.	30.90	12.00	42.90
1-1/4"	"	32.80	15.00	47.80
1-1/2"				
125 lb	EA.	30.00	15.00	45.00
150 lb	"	49.70	15.00	64.70
200 lb	"	49.70	16.65	66.35
2"				
125 lb	EA.	44.00	16.65	60.65
150 lb	"	73.00	16.65	89.65
200 lb	"	73.00	18.75	91.75
2-1/2"				
125 lb	EA.	103.00	25.00	128.00
250 lb	"	323.00	30.00	353.00
3"				
125 lb	EA.	116.00	30.00	146.00
250 lb	"	400.00	37.50	437.50
4"				
125 lb	EA.	176.00	42.90	218.90
250 lb	"	515.00	50.00	565.00
6"				
125 lb	EA.	252.00	60.00	312.00
250 lb	"	892.00	60.00	952.00
Vertical check valve, bronze, 125 lb, threaded				
1/2"	EA.	21.05	12.00	33.05
3/4"	"	28.60	13.65	42.25

15 MECHANICAL

PLUMBING	UNIT	MAT.	INST.	TOTAL
15430.60 VALVES				
1"	EA.	34.30	15.00	49.30
1-1/4"	"	43.20	16.65	59.85
1-1/2"	"	49.70	18.75	68.45
2"	"	81.00	21.45	102.45
Cast iron, flanged				
2-1/2"	EA.	292.00	30.00	322.00
3"	"	324.00	37.50	361.50
4"	"	440.00	50.00	490.00
6	"	770.00	60.00	830.00
8"	"	1,430	75.00	1,505
10"	"	2,200	100.00	2,300
12"	"	2,750	120.00	2,870
Globe valve, bronze, soldered, 125 lb				
1/2"	EA.	11.00	8.55	19.55
3/4"	"	13.20	9.35	22.55
1"	"	24.20	10.00	34.20
1-1/4"	"	39.60	10.70	50.30
1-1/2"	"	53.00	12.50	65.50
2"	"	88.00	15.00	103.00
Threaded				
1/2"				
125 lb	EA.	11.00	10.00	21.00
150 lb	"	33.00	10.00	43.00
300 lb	"	30.80	10.00	40.80
3/4"				
125 lb	EA.	13.20	12.00	25.20
150 lb	"	35.20	12.00	47.20
300 lb	"	30.80	12.00	42.80
1"				
125 lb	EA.	24.20	15.00	39.20
150 lb	"	77.00	15.00	92.00
300 lb	"	71.50	15.00	86.50
1-1/4"				
125 lb	EA.	44.00	15.00	59.00
150 lb	"	104.00	15.00	119.00
300 lb	"	98.00	15.00	113.00
1-1/2"				
125 lb	EA.	53.00	16.65	69.65
150 lb	"	116.00	16.65	132.65
300 lb	"	110.00	16.65	126.65
2"				
125 lb	EA.	88.00	20.00	108.00
150 lb	"	220.00	20.00	240.00
300 lb	"	209.00	20.00	229.00
Cast iron flanged				
2-1/2"				
125 lb	EA.	492.00	30.00	522.00
250 lb	"	904.00	30.00	934.00
3"				
125 lb	EA.	583.00	37.50	620.50
250 lb	"	972.00	37.50	1,010
4"				

PLUMBING

15430.60 VALVES

	UNIT	MAT.	INST.	TOTAL
125 lb	EA.	829.00	50.00	879.00
250 lb	"	1,360	50.00	1,410
6"				
125 lb	EA.	1,490	60.00	1,550
250 lb	"	1,830	60.00	1,890
8"				
125 lb	EA.	2,450	75.00	2,525
250 lb	"	2,860	75.00	2,935
Butterfly valve, cast iron, wafer type				
2"				
150 lb	EA.	71.50	21.45	92.95
200 lb	"	77.00	25.00	102.00
2-1/2"				
150 lb	EA.	77.00	25.00	102.00
200 lb	"	82.50	27.30	109.80
3"				
150 lb	EA.	77.00	30.00	107.00
200 lb	"	88.00	33.30	121.30
4"				
150 lb	EA.	93.50	42.90	136.40
200 lb	"	99.00	50.00	149.00
6"				
150 lb	EA.	165.00	60.00	225.00
200 lb	"	176.00	60.00	236.00
8"				
150 lb	EA.	198.00	66.50	264.50
200 lb	"	231.00	75.00	306.00
10"				
150 lb	EA.	242.00	75.00	317.00
200 lb	"	330.00	100.00	430.00
Ball valve, bronze, 250 lb, threaded				
1/2"	EA.	6.75	12.00	18.75
3/4"	"	10.45	12.00	22.45
1"	"	11.55	15.00	26.55
1-1/4"	"	19.80	16.65	36.45
1-1/2"	"	30.80	18.75	49.55
2"	"	35.20	21.45	56.65
Angle valve, bronze, 150 lb, threaded				
1/2"	EA.	22.00	10.70	32.70
3/4"	"	30.80	12.00	42.80
1"	"	44.00	12.00	56.00
1-1/4"	"	55.50	15.00	70.50
1-1/2"	"	74.50	16.65	91.15
Balancing valve, with meter connections, circuit setter				
1/2"	EA.	37.20	12.00	49.20
3/4"	"	39.60	13.65	53.25
1"	"	50.50	15.00	65.50
1-1/4"	"	70.00	16.65	86.65
1-1/2"	"	82.50	20.00	102.50
2"	"	121.00	25.00	146.00
2-1/2"	"	231.00	30.00	261.00
3"	"	348.00	37.50	385.50

15 MECHANICAL

PLUMBING	UNIT	MAT.	INST.	TOTAL
15430.60 VALVES				
4"	EA.	492.00	50.00	542.00
Balancing valve, straight type				
1/2"	EA.	9.70	12.00	21.70
3/4"	"	11.55	12.00	23.55
Angle type				
1/2"	EA.	13.15	12.00	25.15
3/4"	"	17.35	12.00	29.35
Square head cock, 125 lb, bronze body				
1/2"	EA.	7.80	10.00	17.80
3/4"	"	9.05	12.00	21.05
1"	"	12.95	13.65	26.60
1-1/4"	"	17.60	15.00	32.60
Radiator temp control valve, with control and sensor				
1/2" valve	EA.	51.50	18.75	70.25
1" valve	"	60.50	18.75	79.25
Pressure relief valve, 1/2", bronze				
Low pressure	EA.	11.75	12.00	23.75
High pressure	"	12.95	12.00	24.95
Press and temperature relief valve				
Bronze, 3/4"	EA.	43.10	12.00	55.10
Cast iron, 3/4"				
High pressure	EA.	24.70	12.00	36.70
Temperature relief	"	32.00	12.00	44.00
Pressure & temp relief valve	"	32.00	12.00	44.00
Pressure reducing valve, bronze, threaded, 250 lb				
1/2"	EA.	207.00	18.75	225.75
3/4"	"	207.00	18.75	225.75
1"	"	238.00	18.75	256.75
1-1/4"	"	276.00	21.45	297.45
1-1/2"	"	311.00	25.00	336.00
Pressure regulating valve, bronze, class 300				
1"	EA.	446.00	18.75	464.75
1-1/2"	"	583.00	23.10	606.10
2"	"	686.00	30.00	716.00
3"	"	755.00	42.90	797.90
4"	"	1,010	60.00	1,070
5"	"	1,410	75.00	1,485
6"	"	1,420	100.00	1,520
Solar water temperature regulating valve				
3/4"	EA.	265.00	25.00	290.00
1"	"	271.00	30.00	301.00
1-1/4"	"	297.00	33.30	330.30
1-1/2"	"	332.00	37.50	369.50
2"	"	412.00	42.90	454.90
2-1/2"	"	824.00	75.00	899.00
Tempering valve, threaded				
3/4"	EA.	45.80	10.00	55.80
1"	"	114.00	12.00	126.00
1-1/4"	"	137.00	15.00	152.00
1-1/2"	"	183.00	15.00	198.00
2"	"	229.00	18.75	247.75
2-1/2"	"	286.00	25.00	311.00

PLUMBING

PLUMBING	UNIT	MAT.	INST.	TOTAL
15430.60 VALVES				
3"	EA.	355.00	30.00	385.00
4"	"	812.00	42.90	854.90
Thermostatic mixing valve, threaded				
1/2"	EA.	54.50	10.70	65.20
3/4"	"	57.50	12.00	69.50
1"	"	202.00	13.05	215.05
1-1/2"	"	220.00	15.00	235.00
2"	"	287.00	18.75	305.75
Sweat connection				
1/2"	EA.	64.50	10.70	75.20
3/4"	"	74.00	12.00	86.00
Mixing valve, sweat connection				
1/2"	EA.	25.20	10.70	35.90
3/4"	"	28.00	12.00	40.00
Liquid level gauge, aluminum body				
3/4"	EA.	205.00	12.00	217.00
125 psi, pvc body				
3/4"	EA.	205.00	12.00	217.00
150 psi, crs body				
3/4"	EA.	156.00	12.00	168.00
1"	"	177.00	12.00	189.00
175 psi, bronze body, 1/2"	"	327.00	10.70	337.70
15430.65 VACUUM BREAKERS				
Vacuum breaker, atmospheric, threaded connection				
3/4"	EA.	18.25	12.00	30.25
1"	"	28.10	12.00	40.10
Anti-siphon, brass				
3/4"	EA.	17.60	12.00	29.60
1"	"	27.00	12.00	39.00
1-1/4"	"	45.80	15.00	60.80
1-1/2"	"	54.00	16.65	70.65
2"	"	74.50	18.75	93.25
Air eliminators, purger, cast iron, threaded				
1"	EA.	16.50	12.00	28.50
1-1/4"	"	16.50	15.00	31.50
1-1/2"	"	35.00	15.00	50.00
2"	"	38.10	18.75	56.85
2-1/2"	"	101.00	37.50	138.50
3"	"	104.00	50.00	154.00
Airtrol fitting, 3/4"	"	24.20	12.00	36.20
Air eliminator, air vents, 1/4"	"	14.75	12.00	26.75
Air vent for hot water	"	12.20	10.70	22.90
15430.68 STRAINERS				
Strainer, Y pattern, 125 psi, cast iron body, threaded				
3/4"	EA.	9.95	10.70	20.65
1"	"	12.45	12.00	24.45
1-1/4"	"	17.05	15.00	32.05
1-1/2"	"	21.35	15.00	36.35
2"	"	32.50	18.75	51.25

PLUMBING	UNIT	MAT.	INST.	TOTAL
15430.68 STRAINERS				
250 psi, brass body, threaded				
3/4"	EA.	24.30	12.00	36.30
1"	"	34.50	12.00	46.50
1-1/4"	"	43.70	15.00	58.70
1-1/2"	"	63.00	15.00	78.00
2"	"	104.00	18.75	122.75
Cast iron body, threaded				
3/4"	EA.	16.05	12.00	28.05
1"	"	20.45	12.00	32.45
1-1/4"	"	27.30	15.00	42.30
1-1/2"	"	36.20	15.00	51.20
2"	"	45.90	18.75	64.65
15430.70 DRAINS, ROOF & FLOOR				
Floor drain, cast iron, with cast iron top				
2"	EA.	44.00	25.00	69.00
3"	"	67.00	25.00	92.00
4"	"	88.00	25.00	113.00
6"	"	162.00	30.00	192.00
Roof drain, cast iron				
2"	EA.	121.00	25.00	146.00
3"	"	132.00	25.00	157.00
4"	"	143.00	25.00	168.00
5"	"	204.00	30.00	234.00
6"	"	209.00	30.00	239.00
15430.80 TRAPS				
Bucket trap, threaded				
3/4"	EA.	64.00	18.75	82.75
1"	"	275.00	20.00	295.00
1-1/4"	"	326.00	23.10	349.10
1-1/2"	"	487.00	27.30	514.30
Inverted bucket steam trap, threaded				
3/4"	EA.	55.00	18.75	73.75
1"	"	212.00	18.75	230.75
1-1/4"	"	286.00	16.65	302.65
1-1/2"	"	297.00	25.00	322.00
With stainless interior				
1/2"	EA.	43.40	18.75	62.15
3/4"	"	58.50	18.75	77.25
1"	"	133.00	18.75	151.75
1-1/4"	"	194.00	21.45	215.45
Brass interior				
3/4"	EA.	76.00	18.75	94.75
1"	"	193.00	20.00	213.00
1-1/4"	"	257.00	21.45	278.45
Cast steel body, threaded, high temperature				
3/4"	EA.	332.00	18.75	350.75
1"	"	461.00	21.45	482.45
1-1/4	"	699.00	23.10	722.10
1-1/2"	"	1,300	25.00	1,325

PLUMBING	UNIT	MAT.	INST.	TOTAL
15430.80 TRAPS				
2"	EA.	1,300	30.00	1,330
Float trap, 15 psi				
3/4"	EA.	58.50	18.75	77.25
1"	"	70.50	20.00	90.50
1-1/4"	"	87.50	21.45	108.95
1-1/2"	"	122.00	25.00	147.00
2"	"	237.00	30.00	267.00
30 psi				
3/4"	EA.	74.50	18.75	93.25
1"	"	75.50	20.00	95.50
1-1/4"	"	149.00	25.00	174.00
1-1/2"	"	260.00	30.00	290.00
75 psi				
3/4"	EA.	89.00	18.75	107.75
1"	"	99.00	20.00	119.00
1-1/4"	"	151.00	21.45	172.45
1-1/2"	"	263.00	25.00	288.00
125 psi				
3/4"	EA.	89.00	18.75	107.75
1"	"	99.00	20.00	119.00
1-1/4	"	151.00	21.45	172.45
1-1/2	"	263.00	25.00	288.00
Float and thermostatic trap, 15 psi				
3/4"	EA.	61.50	18.75	80.25
1"	"	42.90	20.00	62.90
1-1/4"	"	90.50	21.45	111.95
1-1/2"	"	125.00	25.00	150.00
2"	"	237.00	30.00	267.00
30 psi				
3/4"	EA.	81.50	18.75	100.25
1"	"	76.00	20.00	96.00
1-1/4"	"	105.00	21.45	126.45
1-1/2"	"	157.00	25.00	182.00
75 psi				
3/4"	EA.	95.00	18.75	113.75
1"	"	105.00	20.00	125.00
1-1/4"	"	158.00	21.45	179.45
1-1/2"	"	271.00	25.00	296.00
Steam trap, cast iron body, threaded, 125 psi				
3/4"	EA.	94.50	18.75	113.25
1"	"	104.00	20.00	124.00
1-1/4"	"	157.00	21.45	178.45
1-1/2"	"	269.00	25.00	294.00
Thermostatic trap, low pressure, angle type, 25 psi				
1/2"	EA.	25.50	18.75	44.25
3/4"	"	42.20	18.75	60.95
1"	"	58.00	20.00	78.00
50 psi				
1/2"	EA.	40.30	18.75	59.05
3/4"	"	54.00	18.75	72.75
1"	"	69.00	20.00	89.00
Cast iron body, threaded, 125 psi				

PLUMBING	UNIT	MAT.	INST.	TOTAL
15430.80 TRAPS				
3/4"	EA.	75.50	18.75	94.25
1"	"	91.00	21.45	112.45
1-1/4"	"	109.00	23.10	132.10
1-1/2"	"	151.00	25.00	176.00
Low pressure, 25 psi, swivel type, 1/2"	"	33.10	18.75	51.85
Straightway type, 3/4"	"	45.80	18.75	64.55
Vertical type, 1/2"	"	30.80	18.75	49.55
Medium pressure, 50 psi, angle type, 1/2"	"	42.00	18.75	60.75
High pressure, 125 psi, angle type				
1/2"	EA.	55.00	18.75	73.75
3/4"	"	71.50	18.75	90.25
1"	"	107.00	21.45	128.45
Straightway type				
1/2"	EA.	57.00	18.75	75.75
Thermo disc trap				
3/4"	EA.	86.50	18.75	105.25
1"	"	98.00	20.00	118.00
Drip pan ell, cast iron				
2-1/2"	EA.	91.50	37.50	129.00
3"	"	98.50	42.90	141.40
4"	"	133.00	50.00	183.00
5"	"	174.00	60.00	234.00
6"	"	271.00	75.00	346.00
8"	"	303.00	100.00	403.00
Steel				
2-1/2"	EA.	469.00	37.50	506.50
3"	"	493.00	42.90	535.90
4"	"	641.00	50.00	691.00
5"	"	1,030	60.00	1,090
6"	"	1,030	75.00	1,105
8"	"	1,480	100.00	1,580

PLUMBING FIXTURES	UNIT	MAT.	INST.	TOTAL
15440.10 BATHS				
Bath tub, 5' long				
Minimum	EA.	323.00	100.00	423.00
Average	"	681.00	150.00	831.00
Maximum	"	1,020	300.00	1,320
6' long				
Minimum	EA.	370.00	100.00	470.00
Average	"	716.00	150.00	866.00
Maximum	"	1,120	300.00	1,420

15 MECHANICAL

PLUMBING FIXTURES	UNIT	MAT.	INST.	TOTAL
15440.10 BATHS				
Square tub, whirlpool, 4'x4'				
Minimum	EA.	1,120	150.00	1,270
Average	"	1,390	300.00	1,690
Maximum	"	2,990	375.00	3,365
5'x5'				
Minimum	EA.	1,140	150.00	1,290
Average	"	1,620	300.00	1,920
Maximum	"	3,120	375.00	3,495
6'x6'				
Minimum	EA.	1,330	150.00	1,480
Average	"	1,820	300.00	2,120
Maximum	"	3,810	375.00	4,185
For trim and rough-in				
Minimum	EA.	98.00	100.00	198.00
Average	"	139.00	150.00	289.00
Maximum	"	202.00	300.00	502.00
15440.12 DISPOSALS & ACCESSORIES				
Continuous feed				
Minimum	EA.	66.00	60.00	126.00
Average	"	104.00	75.00	179.00
Maximum	"	219.00	100.00	319.00
Batch feed, 1/2 hp				
Minimum	EA.	242.00	60.00	302.00
Average	"	330.00	75.00	405.00
Maximum	"	550.00	100.00	650.00
Hot water dispenser				
Minimum	EA.	99.00	60.00	159.00
Average	"	132.00	75.00	207.00
Maximum	"	275.00	100.00	375.00
Epoxy finish faucet	"	169.00	60.00	229.00
Lock stop assembly	"	36.20	37.50	73.70
Mounting gasket	"	4.15	25.00	29.15
Tailpipe gasket	"	0.62	25.00	25.62
Stopper assembly	"	14.35	30.00	44.35
Switch assembly, on/off	"	16.45	50.00	66.45
Tailpipe gasket washer	"	0.31	15.00	15.31
Stop gasket	"	1.45	16.65	18.10
Tailpipe flange	"	0.17	15.00	15.17
Tailpipe	"	1.90	18.75	20.65
15440.15 FAUCETS				
Kitchen				
Minimum	EA.	86.50	50.00	136.50
Average	"	139.00	60.00	199.00
Maximum	"	167.00	75.00	242.00
Bath				
Minimum	EA.	139.00	50.00	189.00
Average	"	203.00	60.00	263.00
Maximum	"	219.00	75.00	294.00
Lavatory, domestic				
Minimum	EA.	89.00	50.00	139.00

PLUMBING FIXTURES	UNIT	MAT.	INST.	TOTAL
15440.15 FAUCETS				
Average	EA.	185.00	60.00	245.00
Maximum	"	277.00	75.00	352.00
Hospital, patient rooms				
Minimum	EA.	219.00	75.00	294.00
Average	"	312.00	100.00	412.00
Maximum	"	393.00	150.00	543.00
Operating room				
Minimum	EA.	254.00	75.00	329.00
Average	"	335.00	100.00	435.00
Maximum	"	474.00	150.00	624.00
Washroom				
Minimum	EA.	127.00	50.00	177.00
Average	"	208.00	60.00	268.00
Maximum	"	300.00	75.00	375.00
Handicapped				
Minimum	EA.	162.00	60.00	222.00
Average	"	225.00	75.00	300.00
Maximum	"	335.00	100.00	435.00
Shower				
Minimum	EA.	104.00	50.00	154.00
Average	"	185.00	60.00	245.00
Maximum	"	289.00	75.00	364.00
For trim and rough-in				
Minimum	EA.	46.20	60.00	106.20
Average	"	69.50	75.00	144.50
Maximum	"	104.00	150.00	254.00
15440.18 HYDRANTS				
Wall hydrant				
8" thick	EA.	44.00	50.00	94.00
12" thick	"	55.00	60.00	115.00
18" thick	"	66.00	66.50	132.50
24" thick	"	77.00	75.00	152.00
Ground hydrant				
2' deep	EA.	132.00	37.50	169.50
4' deep	"	143.00	42.90	185.90
6' deep	"	176.00	50.00	226.00
8' deep	"	209.00	75.00	284.00
15440.20 LAVATORIES				
Lavatory, counter top, porcelain enamel on cast iron				
Minimum	EA.	113.00	60.00	173.00
Average	"	167.00	75.00	242.00
Maximum	"	243.00	100.00	343.00
Wall hung, china				
Minimum	EA.	127.00	60.00	187.00
Average	"	173.00	75.00	248.00
Maximum	"	462.00	100.00	562.00
Handicapped				
Minimum	EA.	254.00	75.00	329.00
Average	"	300.00	100.00	400.00
Maximum	"	450.00	150.00	600.00

PLUMBING FIXTURES

PLUMBING FIXTURES	UNIT	MAT.	INST.	TOTAL
15440.20 LAVATORIES				
For trim and rough-in				
Minimum	EA.	110.00	75.00	185.00
Average	"	133.00	100.00	233.00
Maximum	"	231.00	150.00	381.00
15440.30 SHOWERS				
Shower, fiberglass, 36"x34"x84"				
Minimum	EA.	450.00	214.00	664.00
Average	"	624.00	300.00	924.00
Maximum	"	895.00	300.00	1,195
Steel, 1 piece, 36"x36"				
Minimum	EA.	416.00	214.00	630.00
Average	"	612.00	300.00	912.00
Maximum	"	705.00	300.00	1,005
Receptor, molded stone, 36"x36"				
Minimum	EA.	139.00	100.00	239.00
Average	"	225.00	150.00	375.00
Maximum	"	346.00	250.00	596.00
For trim and rough-in				
Minimum	EA.	106.00	136.00	242.00
Average	"	133.00	167.00	300.00
Maximum	"	162.00	300.00	462.00
15440.40 SINKS				
Service sink, 24"x29"				
Minimum	EA.	416.00	75.00	491.00
Average	"	520.00	100.00	620.00
Maximum	"	693.00	150.00	843.00
Kitchen sink, single, stainless steel, single bowl				
Minimum	EA.	121.00	60.00	181.00
Average	"	162.00	75.00	237.00
Maximum	"	229.00	100.00	329.00
Double bowl				
Minimum	EA.	162.00	75.00	237.00
Average	"	185.00	100.00	285.00
Maximum	"	248.00	150.00	398.00
Porcelain enamel, cast iron, single bowl				
Minimum	EA.	122.00	60.00	182.00
Average	"	146.00	75.00	221.00
Maximum	"	200.00	100.00	300.00
Double bowl				
Minimum	EA.	167.00	75.00	242.00
Average	"	191.00	100.00	291.00
Maximum	"	254.00	150.00	404.00
Mop sink, 24"x36"x10"				
Minimum	EA.	300.00	60.00	360.00
Average	"	393.00	75.00	468.00
Maximum	"	485.00	100.00	585.00
Washing machine box				
Minimum	EA.	104.00	75.00	179.00
Average	"	139.00	100.00	239.00
Maximum	"	185.00	150.00	335.00

PLUMBING FIXTURES

PLUMBING FIXTURES	UNIT	MAT.	INST.	TOTAL
15440.40 SINKS				
For trim and rough-in				
Minimum	EA.	162.00	100.00	262.00
Average	"	231.00	150.00	381.00
Maximum	"	289.00	200.00	489.00
15440.50 URINALS				
Urinal, flush valve, floor mounted				
Minimum	EA.	433.00	75.00	508.00
Average	"	508.00	100.00	608.00
Maximum	"	693.00	150.00	843.00
Wall mounted				
Minimum	EA.	127.00	75.00	202.00
Average	"	312.00	100.00	412.00
Maximum	"	416.00	150.00	566.00
For trim and rough-in				
Minimum	EA.	81.00	75.00	156.00
Average	"	121.00	150.00	271.00
Maximum	"	144.00	200.00	344.00
15440.60 WATER CLOSETS				
Water closet flush tank, floor mounted				
Minimum	EA.	165.00	75.00	240.00
Average	"	420.00	100.00	520.00
Maximum	"	825.00	150.00	975.00
Handicapped				
Minimum	EA.	245.00	100.00	345.00
Average	"	431.00	150.00	581.00
Maximum	"	660.00	300.00	960.00
Bowl, with flush valve, floor mounted				
Minimum	EA.	280.00	75.00	355.00
Average	"	315.00	100.00	415.00
Maximum	"	632.00	150.00	782.00
Wall mounted				
Minimum	EA.	292.00	75.00	367.00
Average	"	338.00	100.00	438.00
Maximum	"	660.00	150.00	810.00
For trim and rough-in				
Minimum	EA.	99.00	75.00	174.00
Average	"	117.00	100.00	217.00
Maximum	"	146.00	150.00	296.00
15440.70 WATER HEATERS				
Water heater, electric				
6 gal	EA.	187.00	50.00	237.00
10 gal	"	198.00	50.00	248.00
15 gal	"	209.00	50.00	259.00
20 gal	"	231.00	60.00	291.00
30 gal	"	242.00	60.00	302.00
40 gal	"	253.00	60.00	313.00
52 gal	"	286.00	75.00	361.00
66 gal	"	440.00	75.00	515.00
80 gal	"	484.00	75.00	559.00

15 MECHANICAL

PLUMBING FIXTURES	UNIT	MAT.	INST.	TOTAL
15440.70 WATER HEATERS				
100 gal	EA.	594.00	100.00	694.00
120 gal	"	759.00	100.00	859.00
Oil fired				
20 gal	EA.	715.00	150.00	865.00
50 gal	"	1,040	214.00	1,254
15440.90 MISCELLANEOUS FIXTURES				
Electric water cooler				
Floor mounted	EA.	451.00	100.00	551.00
Wall mounted	"	550.00	100.00	650.00
Wash fountain				
Wall mounted	EA.	759.00	150.00	909.00
Circular, floor supported	"	1,430	300.00	1,730
Deluge shower and eye wash	"	539.00	150.00	689.00
15440.95 FIXTURE CARRIERS				
Water fountain, wall carrier				
Minimum	EA.	88.00	30.00	118.00
Average	"	110.00	37.50	147.50
Maximum	"	132.00	50.00	182.00
Lavatory, wall carrier				
Minimum	EA.	88.00	30.00	118.00
Average	"	110.00	37.50	147.50
Maximum	"	132.00	50.00	182.00
Sink, industrial, wall carrier				
Minimum	EA.	93.50	30.00	123.50
Average	"	116.00	37.50	153.50
Maximum	"	138.00	50.00	188.00
Toilets, water closets, wall carrier				
Minimum	EA.	116.00	30.00	146.00
Average	"	132.00	37.50	169.50
Maximum	"	176.00	50.00	226.00
Floor support				
Minimum	EA.	33.00	25.00	58.00
Average	"	44.00	30.00	74.00
Maximum	"	55.00	37.50	92.50
Urinals, wall carrier				
Minimum	EA.	71.50	30.00	101.50
Average	"	93.50	37.50	131.00
Maximum	"	116.00	50.00	166.00
Floor support				
Minimum	EA.	49.50	25.00	74.50
Average	"	66.00	30.00	96.00
Maximum	"	82.50	37.50	120.00
15450.30 PUMPS				
In-line pump, bronze, centrifugal				
5 gpm, 20' head	EA.	422.00	18.75	440.75
20 gpm, 40' head	"	781.00	18.75	799.75
50 gpm				

PLUMBING FIXTURES

15450.30 PUMPS

	UNIT	MAT.	INST.	TOTAL
50' head	EA.	1,040	37.50	1,078
100' head	"	1,560	37.50	1,598
70 gpm, 100' head	"	1,560	50.00	1,610
100 gpm, 80' head	"	1,560	50.00	1,610
250 gpm, 150' head	"	4,900	75.00	4,975
Cast iron, centrifugal				
50 gpm, 200' head	EA.	2,270	37.50	2,308
100 gpm				
100' head	EA.	2,100	50.00	2,150
200' head	"	2,450	50.00	2,500
200 gpm				
100' head	EA.	2,330	75.00	2,405
200' head	"	3,150	75.00	3,225
Centrifugal, close coupled, c.i., single stage				
50 gpm, 100' head	EA.	921.00	37.50	958.50
100 gpm, 100' head	"	1,120	50.00	1,170
Base mounted				
50 gpm, 100' head	EA.	1,500	37.50	1,538
100 gpm, 50' head	"	1,660	50.00	1,710
200 gpm, 100' head	"	2,120	75.00	2,195
300 gpm, 175' head	"	2,750	75.00	2,825
Suction diffuser, flanged, strainer				
3" inlet, 2-1/2" outlet	EA.	239.00	37.50	276.50
3" outlet	"	245.00	37.50	282.50
4" inlet				
3" outlet	EA.	286.00	50.00	336.00
4" outlet	"	332.00	50.00	382.00
6" inlet				
4" outlet	EA.	385.00	60.00	445.00
5" outlet	"	468.00	60.00	528.00
6" Outlet	"	485.00	60.00	545.00
8" inlet				
6" outlet	EA.	525.00	75.00	600.00
8" outlet	"	909.00	75.00	984.00
10" inlet				
8" outlet	EA.	1,220	100.00	1,320
Vertical turbine				
Single stage, C.I., 3550 rpm, 200 gpm, 50'head	EA.	2,970	100.00	3,070
Multi stage, 3550 rpm				
50 gpm, 100' head	EA.	2,890	75.00	2,965
100 gpm				
100' head	EA.	3,110	75.00	3,185
200 gpm				
50' head	EA.	3,320	100.00	3,420
100' head	"	3,380	100.00	3,480
Bronze				
Single stage, 3550 rpm, 100 gpm, 50' head	EA.	2,820	75.00	2,895
Multi stage, 3550 rpm, 50 gpm, 100' head	"	3,000	75.00	3,075
100 gpm				
100' head	EA.	3,000	75.00	3,075
200 gpm				
50' head	EA.	3,000	100.00	3,100

PLUMBING FIXTURES	UNIT	MAT.	INST.	TOTAL
15450.30 PUMPS				
100' head	EA.	3,260	100.00	3,360
Sump pump, bronze, 1750 rpm, 25 gpm				
20' head	EA.	2,870	375.00	3,245
150' head	"	4,150	500.00	4,650
50 gpm				
100' head	EA.	3,000	375.00	3,375
100 gpm				
50' head	EA.	3,000	375.00	3,375
Condensate pump, simplex				
1000 sf EDR, 2 gpm	EA.	723.00	250.00	973.00
2000 sf EDR, 3 gpm	"	735.00	250.00	985.00
4000 sf EDR, 6 gpm	"	740.00	273.00	1,013
6000 sf EDR, 9 gpm	"	746.00	273.00	1,019
Duplex, bronze				
8000 sf EDR, 12 gpm	EA.	1,000	273.00	1,273
10,000 sf EDR, 15 gpm	"	1,010	375.00	1,385
15,000 sf EDR, 23 gpm	"	1,270	429.00	1,699
20,000 sf EDR, 30 gpm	"	1,500	600.00	2,100
25,000 sf EDR, 38 gpm	"	1,520	600.00	2,120
30,000 sf EDR, 45 gpm	"	2,270	667.00	2,937
40,000 sf EDR, 60 gpm	"	2,270	375.00	2,645
50,000 sf EDR, 75 gpm	"	2,270	429.00	2,699
75,000 sf EDR, 112 gpm	"	3,260	500.00	3,760
100,000 sf EDR, 150 gpm	"	3,370	750.00	4,120
15450.40 STORAGE TANKS				
Hot water storage tank, cement lined				
10 gallon	EA.	2,630	100.00	2,730
70 gallon	"	2,620	150.00	2,770
200 gallon	"	2,710	214.00	2,924
900 gallon	"	3,110	375.00	3,485
1100 gallon	"	3,580	375.00	3,955
2000 gallon	"	7,160	375.00	7,535
15480.10 SPECIAL SYSTEMS				
Air compressor, air cooled, two stage				
5.0 cfm, 175 psi	EA.	2,250	600.00	2,850
10 cfm, 175 psi	"	2,750	667.00	3,417
20 cfm, 175 psi	"	3,800	714.00	4,514
50 cfm, 125 psi	"	5,530	789.00	6,319
80 cfm, 125 psi	"	7,920	857.00	8,777
Single stage, 125 psi				
1.0 cfm	EA.	2,080	429.00	2,509
1.5 cfm	"	2,110	429.00	2,539
2.0 cfm	"	2,160	429.00	2,589
Automotive compressor, hose reel, air and water, 50' hose	"	968.00	250.00	1,218
Lube equipment, 3 reel, with pumps	"	5,060	1,200	6,260
Tire changer				
Truck	EA.	7,920	429.00	8,349
Passenger car	"	1,870	231.00	2,101
Air hose reel, includes, 50' hose	"	495.00	231.00	726.00

PLUMBING FIXTURES

15480.10 SPECIAL SYSTEMS

PLUMBING FIXTURES	UNIT	MAT.	INST.	TOTAL
Hose reel, 5 reel, motor oil, gear oil, lube, air & water	EA.	4,620	1,200	5,820
Water hose reel, 50' hose	"	495.00	231.00	726.00
Pump, air operated, for motor or gear oil, fits 55 gal drum	"	649.00	30.00	679.00
For chassis lube	"	1,060	30.00	1,090
Fuel dispensing pump, lighted dial, one product				
One hose	EA.	2,310	250.00	2,560
Two hose	"	4,070	250.00	4,320
Two products, two hose	"	4,290	250.00	4,540

HEATING & VENTILATING

15555.10 BOILERS

HEATING & VENTILATING	UNIT	MAT.	INST.	TOTAL
Cast iron, gas fired, hot water				
115 mbh	EA.	1,100	957.00	2,057
175 mbh	"	1,320	1,040	2,360
235 mbh	"	1,980	1,150	3,130
940 mbh	"	5,720	2,300	8,020
1600 mbh	"	8,360	2,870	11,230
3000 mbh	"	17,600	3,830	21,430
6000 mbh	"	38,500	5,740	44,240
Steam				
115 mbh	EA.	1,910	957.00	2,867
175 mbh	"	2,430	1,040	3,470
235 mbh	"	2,660	1,150	3,810
940 mbh	"	6,710	2,300	9,010
1600 mbh	"	12,100	2,870	14,970
3000 mbh	"	23,100	3,830	26,930
6000 mbh	"	42,900	5,740	48,640
Electric, hot water				
115 mbh	EA.	3,120	574.00	3,694
175 mbh	"	3,850	574.00	4,424
235 mbh	"	5,170	574.00	5,744
940 mbh	"	11,000	1,150	12,150
1600 mbh	"	16,500	2,300	18,800
3000 mbh	"	28,600	2,870	31,470
6000 mbh	"	42,900	3,830	46,730
Steam				
115 mbh	EA.	3,580	574.00	4,154
175 mbh	"	4,740	574.00	5,314
235 mbh	"	6,010	574.00	6,584
940 mbh	"	16,500	1,150	17,650
1600 mbh	"	27,500	2,300	29,800
3000 mbh	"	31,900	2,870	34,770
6000 mbh	"	49,500	3,830	53,330

HEATING & VENTILATING

15555.10 BOILERS

	UNIT	MAT.	INST.	TOTAL
Oil fired, hot water				
115 mbh	EA.	1,730	766.00	2,496
175 mbh	"	2,190	884.00	3,074
235 mbh	"	2,540	1,040	3,580
940 mbh	"	8,140	1,910	10,050
1600 mbh	"	13,400	2,300	15,700
3000 mbh	"	23,560	2,870	26,430
6000 mbh	"	44,000	5,740	49,740
Steam				
115 mbh	EA.	1,960	766.00	2,726
175 mbh	"	2,370	884.00	3,254
235 mbh	"	3,120	1,040	4,160
940 mbh	"	10,050	1,910	11,960
1600 mbh	"	16,400	2,300	18,700
3000 mbh	"	28,410	2,870	31,280
6000 mbh	"	46,750	5,740	52,490

15610.10 FURNACES

	UNIT	MAT.	INST.	TOTAL
Electric, hot air				
40 mbh	EA.	520.00	150.00	670.00
60 mbh	"	606.00	158.00	764.00
80 mbh	"	664.00	167.00	831.00
100 mbh	"	901.00	176.00	1,077
125 mbh	"	993.00	182.00	1,175
160 mbh	"	1,440	188.00	1,628
200 mbh	"	2,660	194.00	2,854
400 mbh	"	4,740	200.00	4,940
Gas fired hot air				
40 mbh	EA.	520.00	150.00	670.00
60 mbh	"	549.00	158.00	707.00
80 mbh	"	635.00	167.00	802.00
100 mbh	"	664.00	176.00	840.00
125 mbh	"	722.00	182.00	904.00
160 mbh	"	866.00	188.00	1,054
200 mbh	"	1,540	194.00	1,734
400 mbh	"	6,600	200.00	6,800
Oil fired hot air				
40 mbh	EA.	635.00	150.00	785.00
60 mbh	"	791.00	158.00	949.00
80 mbh	"	866.00	167.00	1,033
100 mbh	"	1,010	176.00	1,186
125 mbh	"	1,130	182.00	1,312
160 mbh	"	1,210	188.00	1,398
200 mbh	"	2,770	194.00	2,964
400 mbh	"	9,590	200.00	9,790

REFRIGERATION

	UNIT	MAT.	INST.	TOTAL
15670.10 CONDENSING UNITS				
Air cooled condenser, single circuit				
3 ton	EA.	1,320	50.00	1,370
5 ton	"	2,200	50.00	2,250
7.5 ton	"	3,520	143.00	3,663
20 ton	"	7,420	150.00	7,570
25 ton	"	8,660	150.00	8,810
30 ton	"	9,980	150.00	10,130
40 ton	"	13,750	214.00	13,964
50 ton	"	16,720	214.00	16,934
60 ton	"	19,250	188.00	19,438
With low ambient dampers				
3 ton	EA.	1,540	75.00	1,615
5 ton	"	2,420	75.00	2,495
7.5 ton	"	3,740	150.00	3,890
20 ton	"	7,700	200.00	7,900
25 ton	"	8,910	200.00	9,110
30 ton	"	10,230	200.00	10,430
40 ton	"	14,410	250.00	14,660
50 ton	"	17,380	273.00	17,653
60 ton	"	19,910	273.00	20,183
Dual circuit				
10 ton	EA.	3,300	150.00	3,450
15 ton	"	4,840	214.00	5,054
20 ton	"	8,250	214.00	8,464
25 ton	"	9,460	214.00	9,674
30 ton	"	11,000	214.00	11,214
40 ton	"	15,730	250.00	15,980
50 ton	"	17,380	250.00	17,630
60 ton	"	18,040	250.00	18,290
80 ton	"	22,550	333.00	22,883
100 ton	"	26,400	333.00	26,733
120 ton	"	31,350	333.00	31,683
With low ambient dampers				
15 ton	EA.	5,390	214.00	5,604
20 ton	"	8,800	214.00	9,014
25 ton	"	10,010	214.00	10,224
30 ton	"	10,780	214.00	10,994
40 ton	"	16,280	250.00	16,530
50 ton	"	17,930	250.00	18,180
60 ton	"	18,590	250.00	18,840
80 ton	"	23,650	333.00	23,983
100 ton	"	27,500	333.00	27,833
120 ton	"	32,780	333.00	33,113
15680.10 CHILLERS				
Chiller, reciprocal				
Air cooled, remote condenser, starter				
20 ton	EA.	15,400	383.00	15,783
25 ton	"	17,380	383.00	17,763
30 ton	"	18,370	383.00	18,753
40 ton	"	27,170	574.00	27,744

REFRIGERATION	UNIT	MAT.	INST.	TOTAL
15680.10 CHILLERS				
50 ton	EA.	29,590	638.00	30,228
60 ton	"	33,880	676.00	34,556
80 ton	"	41,800	1,040	42,840
100 ton	"	49,500	1,150	50,650
120 ton	"	56,100	1,280	57,380
150 ton	"	71,500	1,440	72,940
180 ton	"	81,950	1,640	83,590
200 ton	"	90,200	1,910	92,110
Water cooled, with starter				
20 ton	EA.	13,200	383.00	13,583
25 ton	"	14,630	383.00	15,013
30 ton	"	17,600	574.00	18,174
40 ton	"	24,200	574.00	24,774
50 ton	"	26,400	638.00	27,038
60 ton	"	28,600	676.00	29,276
80 ton	"	33,000	1,040	34,040
100 ton	"	38,500	1,150	39,650
120 ton	"	45,100	1,280	46,380
150 ton	"	56,100	1,440	57,540
180 ton	"	58,300	1,640	59,940
200 ton	"	61,600	1,910	63,510
Packaged, air cooled, with starter				
20 ton	EA.	14,740	287.00	15,027
25 ton	"	15,950	287.00	16,237
30 ton	"	18,590	287.00	18,877
40 ton	"	21,340	287.00	21,627
50 ton	"	24,090	383.00	24,473
60 ton	"	28,380	383.00	28,763
80 ton	"	33,110	574.00	33,684
100 ton	"	38,830	574.00	39,404
120 ton	"	44,110	574.00	44,684
Heat recovery, air cooled, with starter				
40 ton	EA.	24,750	574.00	25,324
50 ton	"	28,600	574.00	29,174
60 ton	"	31,350	766.00	32,116
75 ton	"	35,860	1,150	37,010
100 ton	"	40,920	1,150	42,070
Water cooled, with starter				
40 ton	EA.	24,420	574.00	24,994
50 ton	"	29,150	574.00	29,724
60 ton	"	32,120	766.00	32,886
75 ton	"	39,050	1,150	40,200
100 ton	"	44,110	1,280	45,390
Centrifugal, single bundle condenser, with starter				
80 ton	EA.	57,750	1,640	59,390
130 ton	"	60,830	1,910	62,740
160 ton	"	61,820	2,090	63,910
180 ton	"	66,220	2,300	68,520
230 ton	"	69,190	2,550	71,740
280 ton	"	76,230	2,870	79,100
360 ton	"	85,030	2,870	87,900
460 ton	"	104,830	3,830	108,660

15 MECHANICAL

REFRIGERATION

REFRIGERATION	UNIT	MAT.	INST.	TOTAL
15680.10 CHILLERS				
560 ton	EA.	11,470	4,100	15,570
670 ton	"	13,980	4,600	18,580
15710.10 COOLING TOWERS				
Cooling tower, propeller type				
100 ton	EA.	6,730	383.00	7,113
200 ton	"	11,220	574.00	11,794
300 ton	"	16,830	957.00	17,787
400 ton	"	22,440	1,150	23,590
600 ton	"	33,660	1,640	35,300
800 ton	"	44,880	2,300	47,180
1000 ton	"	53,860	2,870	56,730
Centrifugal				
100 ton	EA.	6,730	383.00	7,113
200 ton	"	13,460	574.00	14,034
300 ton	"	20,200	957.00	21,157
400 ton	"	26,930	1,150	28,080
600 ton	"	40,390	1,640	42,030
800 ton	"	53,860	2,300	56,160
1000 ton	"	62,830	2,870	65,700

HEAT TRANSFER

HEAT TRANSFER	UNIT	MAT.	INST.	TOTAL
15780.10 COMPUTER ROOM A/C				
Air cooled, alarm, high efficiency filter, elec. heat				
3 ton	EA.	7,970	231.00	8,201
5 ton	"	8,550	250.00	8,800
7.5 ton	"	14,550	300.00	14,850
10 ton	"	15,010	375.00	15,385
15 ton	"	16,630	429.00	17,059
Steam heat				
3 ton	EA.	9,240	231.00	9,471
5 ton	"	9,820	250.00	10,070
7.5 ton	"	15,590	300.00	15,890
10 ton	"	16,050	375.00	16,425
15 ton	"	17,790	429.00	18,219
Hot water heat				
3 ton	EA.	9,240	231.00	9,471
5 ton	"	9,820	250.00	10,070
7.5 ton	"	15,590	300.00	15,890
10 ton	"	16,050	375.00	16,425
15 ton	"	17,790	429.00	18,219
Air cooled condenser, low ambient damper				
3 ton	EA.	2,080	60.00	2,140

HEAT TRANSFER	UNIT	MAT.	INST.	TOTAL
15780.10 COMPUTER ROOM A/C				
5 ton	EA.	2,310	75.00	2,385
7.5 ton	"	3,350	150.00	3,500
10 ton	"	3,350	214.00	3,564
15 ton	"	3,700	176.00	3,876
Water cooled, high efficiency filter, alarm, elec. heat				
3 ton	EA.	9,120	214.00	9,334
5 ton	"	9,820	250.00	10,070
7.5 ton	"	15,590	375.00	15,965
10 ton	"	16,170	429.00	16,599
15 ton	"	18,940	500.00	19,440
Steam heat				
3 ton	EA.	10,390	214.00	10,604
5 ton	"	11,880	250.00	12,130
7.5 ton	"	16,750	375.00	17,125
10 ton	"	17,320	429.00	17,749
15 ton	"	20,100	500.00	20,600
Hot water heat				
3 ton	EA.	10,390	214.00	10,604
5 ton	"	11,090	250.00	11,340
7.5 ton	"	16,750	375.00	17,125
10 ton	"	17,320	429.00	17,749
15 ton	"	20,100	500.00	20,600
Chilled water, alarm, high eff. filter, elec. heat				
7.5 ton	EA.	7,280	273.00	7,553
10 ton	"	7,620	333.00	7,953
15 ton	"	8,660	375.00	9,035
Steam heat				
7.5 ton	EA.	8,320	273.00	8,593
10 ton	"	8,660	333.00	8,993
15 ton	"	9,700	375.00	10,075
Hot water heat				
7.5 ton	EA.	8,320	273.00	8,593
10 ton	"	8,660	333.00	8,993
15 ton	"	9,700	375.00	10,075
15780.20 ROOFTOP UNITS				
Packaged, single zone rooftop unit, with roof curb				
2 ton	EA.	2,090	300.00	2,390
3 ton	"	2,310	300.00	2,610
4 ton	"	3,190	375.00	3,565
5 ton	"	3,520	500.00	4,020
7.5 ton	"	4,950	600.00	5,550
15820.10 DEHUMIDIFIERS				
Dessicant dehumidifier, 1125 cfm	EA.			11,000
15830.10 RADIATION UNITS				
Baseboard radiation unit				
1.7 mbh/lf	L.F.	39.90	12.00	51.90
2.1 mbh/lf	"	52.50	15.00	67.50

HEAT TRANSFER

HEAT TRANSFER	UNIT	MAT.	INST.	TOTAL
15830.10 RADIATION UNITS				
Enclosure only				
Two tier	L.F.	20.65	5.00	25.65
Three tier	"	27.00	5.00	32.00
Copper element only, 3/4" dia.				
Two tier	L.F.	24.85	7.50	32.35
Three tier	"	37.50	10.00	47.50
Fin-tube, 16 ga, sloping cover, 1-1/4" steel				
One tier	L.F.	29.50	10.00	39.50
Two tier	"	45.70	12.00	57.70
2" steel				
Two tier	L.F.	49.00	12.00	61.00
Three tier	"	67.50	15.00	82.50
1-1/4" copper				
Two tier	L.F.	49.70	10.00	59.70
18 ga flat cover, 1-1/4" steel				
One tier	L.F.	16.80	10.00	26.80
Two tier	"	30.20	12.00	42.20
Three tier	"	49.40	15.00	64.40
2" steel				
One tier	L.F.	21.35	10.00	31.35
Two tier	"	34.20	12.00	46.20
Three tier	"	49.70	15.00	64.70
1-1/4" copper				
One tier	L.F.	18.80	10.00	28.80
Two tier	"	34.00	12.00	46.00
Three tier	"	43.50	15.00	58.50
15830.20 FAN COIL UNITS				
Fan coil unit, 2 pipe, complete				
200 cfm ceiling hung	EA.	554.00	100.00	654.00
Floor mounted	"	520.00	75.00	595.00
300 cfm, ceiling hung	"	589.00	120.00	709.00
Floor mounted	"	560.00	100.00	660.00
400 cfm, ceiling hung	"	617.00	143.00	760.00
Floor mounted	"	594.00	100.00	694.00
500 cfm, ceiling hung	"	722.00	150.00	872.00
Floor mounted	"	689.00	115.00	804.00
600 cfm, ceiling hung	"	912.00	166.00	1,078
Floor mounted	"	843.00	136.00	979.00
800 cfm, ceiling hung	"	1,060	188.00	1,248
Floor mounted	"	843.00	143.00	986.00
1000 cfm, ceiling hung	"	1,210	214.00	1,424
Floor mounted	"	1,330	158.00	1,488
1200 cfm ceiling hung	"	1,370	250.00	1,620
Floor mounted	"	1,440	188.00	1,628
15830.70 UNIT HEATERS				
Steam unit heater, horizontal				
12,500 btuh, 200 cfm	EA.	185.00	50.00	235.00
17,000 btuh, 300 cfm	"	214.00	50.00	264.00
40,000 btuh, 500 cfm	"	260.00	50.00	310.00

HEAT TRANSFER	UNIT	MAT.	INST.	TOTAL
15830.70 UNIT HEATERS				
60,000 btuh, 700 cfm	EA.	300.00	50.00	350.00
70,000 btuh, 1000 cfm	"	404.00	75.00	479.00
Vertical				
12,500 btuh, 200 cfm	EA.	202.00	50.00	252.00
17,000 btuh, 300 cfm	"	329.00	50.00	379.00
40,000 btuh, 500 cfm	"	433.00	50.00	483.00
60,000 btuh, 700 cfm	"	491.00	50.00	541.00
70,000 btuh, 1000 cfm	"	537.00	50.00	587.00
Gas unit heater, horizontal				
27,400 btuh	EA.	404.00	120.00	524.00
38,000 btuh	"	462.00	120.00	582.00
56,000 btuh	"	520.00	120.00	640.00
82,200 btuh	"	578.00	120.00	698.00
103,900 btuh	"	606.00	188.00	794.00
125,700 btuh	"	635.00	188.00	823.00
133,200 btuh	"	578.00	188.00	766.00
149,000 btuh	"	751.00	188.00	939.00
172,000 btuh	"	797.00	188.00	985.00
190,000 btuh	"	866.00	188.00	1,054
225,000 btuh	"	953.00	188.00	1,141
Hot water unit heater, horizontal				
12,500 btuh, 200 cfm	EA.	208.00	50.00	258.00
17,000 btuh, 300 cfm	"	266.00	50.00	316.00
25,000 btuh, 500 cfm	"	289.00	50.00	339.00
30,000 btuh, 700 cfm	"	335.00	50.00	385.00
50,000 btuh, 1000 cfm	"	375.00	75.00	450.00
60,000 btuh, 1300 cfm	"	474.00	75.00	549.00
Vertical				
12,500 btuh, 200 cfm	EA.	243.00	50.00	293.00
17,000 btuh, 300 cfm	"	318.00	50.00	368.00
25,000 btuh, 500 cfm	"	346.00	50.00	396.00
30,000 btuh, 700 cfm	"	381.00	50.00	431.00
50,000 btuh, 1000 cfm	"	433.00	50.00	483.00
60,000 btuh, 1300 cfm	"	474.00	50.00	524.00
Cabinet unit heaters, ceiling, exposed, hot water				
200 cfm	EA.	653.00	100.00	753.00
300 cfm	"	681.00	120.00	801.00
400 cfm	"	743.00	143.00	886.00
600 cfm	"	762.00	158.00	920.00
800 cfm	"	1,070	188.00	1,258
1000 cfm	"	1,160	214.00	1,374
1200 cfm	"	1,250	250.00	1,500
2000 cfm	"	2,250	333.00	2,583

AIR HANDLING

15855.10 AIR HANDLING UNITS	UNIT	MAT.	INST.	TOTAL
Air handling unit, medium pressure, single zone				
1500 cfm	EA.	1,980	188.00	2,168
3000 cfm	"	2,640	333.00	2,973
4000 cfm	"	2,970	375.00	3,345
5000 cfm	"	3,520	400.00	3,920
6000 cfm	"	3,960	429.00	4,389
7000 cfm	"	4,400	462.00	4,862
8500 cfm	"	5,170	500.00	5,670
10,500 cfm	"	5,830	600.00	6,430
12,500 cfm	"	7,480	667.00	8,147
15,500 cfm	"	12,100	857.00	12,957
17,500 cfm	"	14,300	1,000	15,300
20,500 cfm	"	16,500	1,200	17,700
25,000 cfm	"	18,700	1,500	20,200
31,500 cfm	"	23,100	2,000	25,100
Rooftop air handling units				
4950 cfm	EA.	6,930	333.00	7,263
7370 cfm	"	7,920	429.00	8,349
9790 cfm	"	9,350	500.00	9,850
14,300 cfm	"	13,200	429.00	13,629
21,725 cfm	"	18,700	429.00	19,129
33,000 cfm	"	26,400	500.00	26,900

15870.20 EXHAUST FANS	UNIT	MAT.	INST.	TOTAL
Belt drive roof exhaust fans				
640 cfm, 2618 fpm	EA.	660.00	37.50	697.50
940 cfm, 2604 fpm	"	852.00	37.50	889.50
1050 cfm, 3325 fpm	"	759.00	37.50	796.50
1170 cfm, 2373 fpm	"	1,100	37.50	1,138
2440 cfm, 4501 fpm	"	869.00	37.50	906.50
2760 cfm, 4950 fpm	"	957.00	37.50	994.50
3890 cfm, 6769 fpm	"	1,090	37.50	1,128
2380 cfm, 3382 fpm	"	1,210	37.50	1,248
2880 cfm, 3859 fpm	"	1,260	37.50	1,298
3200 cfm, 4173 fpm	"	1,280	50.00	1,330
3660 cfm, 3437 fpm	"	1,300	50.00	1,350
4070 cfm, 3694 fpm	"	1,650	50.00	1,700
5030 cfm, 3251 fpm	"	1,160	50.00	1,210
5830 cfm, 6932 fpm	"	1,590	60.00	1,650
6380 cfm, 3817 fpm	"	1,590	60.00	1,650
8460 cfm, 6721 fpm	"	1,540	60.00	1,600
10,970 cfm, 5906 fpm	"	1,890	75.00	1,965
12,470 cfm, 6620 fpm	"	2,160	100.00	2,260
7000 cfm, 3449 fpm	"	1,540	75.00	1,615
13,000 cfm, 5456 fpm	"	2,280	75.00	2,355
11,250 cfm, 4854 fpm	"	2,090	75.00	2,165
18,490 cfm, 7405 fpm	"	3,050	136.00	3,186
11,300 cfm, 3232 fpm	"	1,980	130.00	2,110
18,330 cfm, 4488 fpm	"	3,300	130.00	3,430
21,720 cfm, 5131 fpm	"	3,470	130.00	3,600
31,110 cfm, 6965 fpm	"	3,820	150.00	3,970
Direct drive fans				

AIR HANDLING

15870.20 EXHAUST FANS	UNIT	MAT.	INST.	TOTAL
60 to 390 cfm	EA.	539.00	37.50	576.50
145 to 590 cfm	"	649.00	37.50	686.50
295 to 860 cfm	"	786.00	37.50	823.50
235 to 1300 cfm	"	847.00	37.50	884.50
415 to 1630 cfm	"	957.00	37.50	994.50
590 to 2045 cfm	"	1,100	37.50	1,138
805 cfm, 3235 fpm	"	726.00	37.50	763.50
1455 cfm, 4360 fpm	"	781.00	37.50	818.50
1385 cfm, 3655 fpm	"	803.00	37.50	840.50
2260 cfm, 4930 fpm	"	847.00	37.50	884.50
1720 cfm, 3870 fpm	"	814.00	37.50	851.50
2700 cfm, 5220 fpm	"	858.00	37.50	895.50
Terminal blenders and cooling				
400 cfm	EA.	297.00	60.00	357.00
800 cfm	"	319.00	60.00	379.00
1200 cfm	"	385.00	75.00	460.00
2000 cfm	"	440.00	75.00	515.00

AIR DISTRIBUTION

15890.10 METAL DUCTWORK	UNIT	MAT.	INST.	TOTAL
Rectangular duct				
Galvanized steel				
Minimum	Lb.	1.75	2.75	4.50
Average	"	2.00	3.35	5.35
Maximum	"	2.10	5.00	7.10
Aluminum				
Minimum	Lb.	4.60	6.00	10.60
Average	"	5.05	7.50	12.55
Maximum	"	5.50	10.00	15.50
Fittings				
Minimum	EA.	2.65	10.00	12.65
Average	"	5.30	15.00	20.30
Maximum	"	13.20	30.00	43.20
For work				
10-20' high, add per pound, $.30				
30-50', add per pound, $.50				

15890.30 FLEXIBLE DUCTWORK	UNIT	MAT.	INST.	TOTAL
Flexible duct, 1.25" fiberglass				
5" dia.	L.F.	1.30	1.50	2.80
6" dia.	"	1.45	1.65	3.10
7" dia.	"	1.65	1.75	3.40

AIR DISTRIBUTION	UNIT	MAT.	INST.	TOTAL
15890.30 FLEXIBLE DUCTWORK				
8" dia.	L.F.	2.00	1.85	3.85
10" dia.	"	2.20	2.15	4.35
12" dia.	"	2.75	2.30	5.05
14" dia.	"	3.30	2.50	5.80
16" dia.	"	3.85	2.75	6.60
Flexible duct connector, 3" wide fabric	"	1.35	5.00	6.35
15895.10 ROOF CURBS				
8" high, insulated, with liner and raised can				
15" x 15"	EA.	56.00	15.00	71.00
17" x 17"	"	60.50	15.00	75.50
19" x 19"	"	64.00	15.00	79.00
21" x 21"	"	70.50	15.00	85.50
25" x 25"	"	77.00	18.75	95.75
28" x 28"	"	81.50	20.00	101.50
32" x 32"	"	92.50	21.45	113.95
36" x 36"	"	104.00	21.45	125.45
40" x 40"	"	119.00	21.45	140.45
44" x 44"	"	130.00	23.10	153.10
48" x 48"	"	286.00	23.10	309.10
52" x 52"	"	352.00	25.00	377.00
56" x 56"	"	440.00	25.00	465.00
60" x 60"	"	528.00	30.00	558.00
64" x 64"	"	638.00	30.00	668.00
68" x 68"	"	742.00	33.30	775.30
72" x 72"	"	847.00	37.50	884.50
15910.10 DAMPERS				
Horizontal parallel aluminum backdraft damper				
12" x 12"	EA.	28.90	7.50	36.40
16" x 16"	"	40.40	8.55	48.95
20" x 20"	"	52.00	10.70	62.70
24" x 24"	"	63.50	15.00	78.50
28" x 28"	"	88.00	16.65	104.65
32" x 32"	"	120.00	18.75	138.75
36" x 36"	"	146.00	21.45	167.45
40" x 40"	"	182.00	25.00	207.00
44" x 44"	"	217.00	27.30	244.30
48" x 48"	"	260.00	30.00	290.00
"Up", parallel dampers				
12" x 12"	EA.	51.00	7.50	58.50
16" x 16"	"	71.50	8.55	80.05
20" x 20"	"	78.50	10.70	89.20
24" x 24"	"	88.00	15.00	103.00
28" x 28"	"	128.00	16.65	144.65
32" x 32"	"	144.00	18.75	162.75
36" x 36"	"	150.00	21.45	171.45
40" x 40"	"	196.00	25.00	221.00
44" x 44"	"	243.00	27.30	270.30
48" x 48"	"	293.00	30.00	323.00
"Down", parallel dampers				
12" x 12"	EA.	51.00	7.50	58.50

15 MECHANICAL

AIR DISTRIBUTION	UNIT	MAT.	INST.	TOTAL
15910.10 DAMPERS				
16" x 16"	EA.	71.50	8.55	80.05
20" x 20"	"	78.50	10.70	89.20
24" x 24"	"	88.00	15.00	103.00
28" x 28"	"	128.00	16.65	144.65
32" x 32"	"	144.00	18.75	162.75
36" x 36"	"	150.00	21.45	171.45
40" x 40"	"	202.00	25.00	227.00
44" x 44"	"	243.00	27.30	270.30
48" x 48"	"	293.00	30.00	323.00
Fire damper, 1.5 hr rating				
12" x 12"	EA.	17.35	15.00	32.35
16" x 16"	"	27.70	15.00	42.70
20" x 20"	"	30.00	15.00	45.00
24" x 24"	"	34.70	15.00	49.70
28" x 28"	"	41.60	21.45	63.05
32" x 32"	"	48.50	25.00	73.50
36" x 36"	"	58.00	30.00	88.00
40" x 40"	"	69.50	33.30	102.80
44" x 44"	"	83.00	37.50	120.50
48" x 48"	"	104.00	42.90	146.90
15940.10 DIFFUSERS				
Ceiling diffusers, round, baked enamel finish				
6" dia.	EA.	28.90	10.00	38.90
8" dia.	"	34.70	12.50	47.20
10" dia.	"	40.40	12.50	52.90
12" dia.	"	52.00	12.50	64.50
14" dia.	"	64.50	13.65	78.15
16" dia.	"	75.00	13.65	88.65
18" dia.	"	86.50	15.00	101.50
20" dia.	"	110.00	15.00	125.00
Rectangular				
6x6"	EA.	28.60	10.00	38.60
9x9"	"	34.10	15.00	49.10
12x12"	"	50.50	15.00	65.50
15x15"	"	64.00	15.00	79.00
18x18"	"	81.00	15.00	96.00
21x21"	"	104.00	18.75	122.75
24x24"	"	133.00	18.75	151.75
Lay in, flush mounted, perforated face, with grid				
6x6/24x24	EA.	46.20	12.00	58.20
8x8/24x24	"	46.20	12.00	58.20
9x9/24x24	"	46.20	12.00	58.20
10x10/24x24	"	48.40	12.00	60.40
12x12/24x24	"	50.50	12.00	62.50
15x15/24x24	"	50.50	12.00	62.50
18x6/24x24	"	50.50	12.00	62.50
18x18/24x24	"	53.00	12.00	65.00
Two-way slot diffuser with balancing damper, 4'	"	55.00	30.00	85.00

AIR DISTRIBUTION

AIR DISTRIBUTION	UNIT	MAT.	INST.	TOTAL
15940.20 RELIEF VENTILATORS				
Intake ventilator, aluminum, with screen, no curbs				
12" x 12"	EA.	104.00	25.00	129.00
16" x 16"	"	143.00	30.00	173.00
20" x 20"	"	242.00	30.00	272.00
30" x 30"	"	385.00	42.90	427.90
36" x 36"	"	578.00	50.00	628.00
42" x 42"	"	798.00	50.00	848.00
48" x 48"	"	946.00	60.00	1,006
15940.40 REGISTERS AND GRILLES				
Lay in flush mounted, perforated face, return				
6x6/24x24	EA.	24.25	12.00	36.25
8x8/24x24	"	24.25	12.00	36.25
9x9/24x24	"	24.25	12.00	36.25
10x10/24x24	"	24.25	12.00	36.25
12x12/24x24	"	24.25	12.00	36.25
Rectangular, ceiling return, single deflection				
10x10	EA.	11.00	15.00	26.00
12x12	"	13.20	15.00	28.20
14x14	"	16.50	15.00	31.50
16x8	"	13.20	15.00	28.20
16x16	"	13.20	15.00	28.20
18x8	"	15.40	15.00	30.40
20x20	"	26.40	15.00	41.40
24x12	"	37.40	15.00	52.40
24x18	"	49.50	15.00	64.50
36x24	"	99.00	16.65	115.65
36x30	"	143.00	16.65	159.65
Wall, return air register				
12x12	EA.	21.95	7.50	29.45
16x16	"	32.30	7.50	39.80
18x18	"	38.10	7.50	45.60
20x20	"	45.00	7.50	52.50
24x24	"	62.50	7.50	70.00
Ceiling, return air grille				
6x6	EA.	8.10	10.00	18.10
8x8	"	9.25	12.00	21.25
10x10	"	10.40	12.00	22.40
Ceiling, exhaust grille, aluminum egg crate				
6x6	EA.	10.40	10.00	20.40
8x8	"	10.40	12.00	22.40
10x10	"	11.55	12.00	23.55
12x12	"	15.00	15.00	30.00
14x14	"	19.65	15.00	34.65
16x16	"	23.10	15.00	38.10
18x18	"	27.70	15.00	42.70
15940.80 PENTHOUSE LOUVERS				
Penthouse louvers				
12" high, extruded aluminum, 4" louver				
6' perimeter	EA.	264.00	75.00	339.00
8' perimeter	"	352.00	75.00	427.00

AIR DISTRIBUTION	UNIT	MAT.	INST.	TOTAL
15940.80 PENTHOUSE LOUVERS				
10' perimeter	EA.	440.00	75.00	515.00
12' perimeter	"	638.00	75.00	713.00
14' perimeter	"	781.00	100.00	881.00
16' perimeter	"	902.00	120.00	1,022
18' perimeter	"	1,060	167.00	1,227
20' perimeter	"	1,320	200.00	1,520
16" high x 4' perimeter	"	220.00	75.00	295.00
6' perimeter	"	308.00	75.00	383.00
8' perimeter	"	396.00	75.00	471.00
10' perimeter	"	484.00	75.00	559.00
12' perimeter	"	715.00	75.00	790.00
14' perimeter	"	880.00	100.00	980.00
16' perimeter	"	990.00	120.00	1,110
18' perimeter	"	1,210	167.00	1,377
20' perimeter	"	1,540	200.00	1,740
22' perimeter	"	1,650	250.00	1,900
24' perimeter	"	1,760	333.00	2,093
20" high x 4' perimeter	"	286.00	75.00	361.00
6' perimeter	"	330.00	75.00	405.00
8' perimeter	"	440.00	75.00	515.00
10' perimeter	"	550.00	75.00	625.00
12' perimeter	"	803.00	75.00	878.00
14' perimeter	"	990.00	100.00	1,090
16' perimeter	"	1,210	120.00	1,330
18' perimeter	"	1,380	167.00	1,547
20' perimeter	"	1,650	200.00	1,850
22' perimeter	"	1,770	250.00	2,020
24' perimeter	"	1,970	333.00	2,303
24" high x 4' perimeter	"	308.00	75.00	383.00
6' perimeter	"	368.00	75.00	443.00
8' perimeter	"	490.00	75.00	565.00
10' perimeter	"	610.00	75.00	685.00
12' perimeter	"	896.00	75.00	971.00
16' perimeter	"	1,320	120.00	1,440
18' perimeter	"	1,540	167.00	1,707
20' perimeter	"	1,860	200.00	2,060
22' perimeter	"	2,060	250.00	2,310
24' perimeter	"	2,200	333.00	2,533

CONTROLS	UNIT	MAT.	INST.	TOTAL
15950.10 HVAC CONTROLS				
Pressure gauge, direct reading gage cock and syphon				
Control valve, 1", modulating	EA.	47.30	18.75	66.05

CONTROLS	UNIT	MAT.	INST.	TOTAL
15950.10 HVAC CONTROLS				
2-way	EA.	484.00	25.00	509.00
3-way	"	550.00	37.50	587.50
Self contained control valve with sensing element, 3/4"	"	88.00	18.75	106.75
Instrument air system 2-2 hp compressors, rcvr refrigerant dryer	"			4,070
Thermostat primary control device	"			88.00
Humidistat primary control device	"			66.00
Timers primary control device, indoor/outdoor, 24 hour	"			132.00
Thermometer, dir. reading, 3 dial	"			66.00
Control dampers, round				
6" dia.	EA.	48.40	12.00	60.40
8" dia	"	71.50	12.00	83.50
10" dia	"	94.50	12.00	106.50
12" dia	"	121.00	12.00	133.00
12" dia	"	176.00	15.00	191.00
18" dia	"	198.00	15.00	213.00
20" dia	"	253.00	15.00	268.00
Rectangular, parallel blade standard leakage				
12" x 12"	EA.	36.30	15.00	51.30
16" x 16"	"	54.00	15.00	69.00
20" x 20"	"	64.00	15.00	79.00
28" x 28"	"	82.50	18.75	101.25
32" x 32"	"	94.50	18.75	113.25
36" x 36"	"	106.00	25.00	131.00
40" x 40"	"	132.00	30.00	162.00
44" x 44"	"	165.00	37.50	202.50
48" x 48"	"	187.00	42.90	229.90
48" x 52"	"	204.00	50.00	254.00
48" x 56"	"	226.00	50.00	276.00
48" x 60"	"	236.00	50.00	286.00
48" x 64"	"	238.00	50.00	288.00
48" x 68"	"	258.00	50.00	308.00
48" x 72"	"	286.00	50.00	336.00
Low leakage				
12" x 12"	EA.	71.50	15.00	86.50
16" x 16"	"	91.50	15.00	106.50
20" x 20"	"	116.00	15.00	131.00
24" x 24"	"	158.00	15.00	173.00
28" x 28"	"	182.00	18.75	200.75
32" x 32"	"	204.00	21.45	225.45
36" x 36"	"	231.00	25.00	256.00
40" x 40"	"	363.00	30.00	393.00
44" x 44"	"	385.00	37.50	422.50
48" x 48"	"	407.00	42.90	449.90
48" x 56"	"	473.00	50.00	523.00
48" x 60"	"	495.00	50.00	545.00
48" x 64"	"	517.00	50.00	567.00
48" x 68"	"	539.00	50.00	589.00
48" x 72"	"	638.00	50.00	688.00
Rectangular, opposed horizontal blade				
12" x 12"	EA.	47.30	15.00	62.30
16" x 16"	"	66.00	15.00	81.00
20" x 20"	"	77.00	15.00	92.00

CONTROLS	UNIT	MAT.	INST.	TOTAL
15950.10 HVAC CONTROLS				
24" x 24"	EA.	88.00	15.00	103.00
28" x 28"	"	121.00	18.75	139.75
32" x 32"	"	132.00	20.00	152.00
36" x 36"	"	143.00	25.00	168.00
40" x 40"	"	187.00	30.00	217.00
44" x 44"	"	236.00	37.50	273.50
48" x 48"	"	286.00	42.90	328.90
48" x 52"	"	292.00	42.90	334.90
48" x 56"	"	308.00	50.00	358.00
48" x 60"	"	324.00	50.00	374.00
48" x 64"	"	341.00	50.00	391.00
48" x 68"	"	380.00	50.00	430.00
48" x 72"	"	407.00	50.00	457.00

BASIC MATERIALS

16050.30 BUS DUCT

	UNIT	MAT.	INST.	TOTAL
Bus duct, 100a, plug-in				
10', 600v	EA.	154.00	95.50	249.50
With ground	"	204.00	146.00	350.00
10', 277/480v	"	198.00	95.50	293.50
With ground	"	243.00	146.00	389.00
Cable tap box	"	83.50	87.00	170.50
End closure	"	19.80	13.90	33.70
Edgewise hanger	"	8.40	25.20	33.60
Flatwise hanger	"	8.90	25.20	34.10
Outside elbow	"	123.00	27.80	150.80
Inside elbow	"	123.00	27.80	150.80
Outside tee	"	158.00	38.20	196.20
Inside tee	"	158.00	38.20	196.20
Outlet cover	"	8.90	13.90	22.80
Wall flange	"	25.30	13.90	39.20
Circuit breakers, with enclosure				
1 pole				
15a-60a	EA.	155.00	34.70	189.70
70a-100a	"	178.00	43.40	221.40
2 pole				
15a-60a	EA.	230.00	38.20	268.20
70a-100a	"	277.00	45.10	322.10
3 pole				
15a-60a	EA.	263.00	40.20	303.20
70a-100a	"	304.00	52.50	356.50
Bus duct, copper feeder duct, 277/480v, 4 wire				
800a	L.F.	154.00	13.90	167.90
1000a	"	179.00	17.35	196.35
1200a	"	207.00	18.50	225.50
1350a	"	243.00	21.35	264.35
1600a	"	293.00	25.20	318.20
2000a	"	362.00	27.80	389.80
2500a	"	444.00	29.50	473.50
3000a	"	568.00	33.00	601.00
Weatherproof				
800a	L.F.	189.00	15.40	204.40
1000a	"	218.00	18.50	236.50
1350a	"	292.00	23.15	315.15
1600a	"	353.00	25.20	378.20
2000a	"	436.00	28.90	464.90
2500a	"	534.00	31.20	565.20
3000a	"	678.00	33.90	711.90
4000a	"	858.00	52.50	910.50
5000a	"	1,040	63.00	1,103
Plug-in feeder duct, 277/480v, 4 wire				
400a	L.F.	103.00	13.90	116.90
600a	"	114.00	15.40	129.40
800a	"	168.00	17.35	185.35
1000a	"	192.00	17.35	209.35
1200a	"	231.00	18.50	249.50
1350a	"	256.00	21.35	277.35
1600a	"	310.00	25.20	335.20

BASIC MATERIALS

16050.30 BUS DUCT	UNIT	MAT.	INST.	TOTAL
2000a	L.F.	372.00	27.80	399.80
2500a	"	459.00	29.50	488.50
3000a	"	578.00	33.00	611.00
Ground bus				
225a	L.F.			17.50
400a	"			18.30
600a	"			18.85
800a	"			20.20
1000a	"			21.10
1200a	"			26.10
1350a	"			27.20
1600a	"			36.50
2000a	"			44.40
2500a	"			62.00
3000a	"			82.00
4000a	"			100.00
5000a	"			129.00
Copper flanged ends, 277/480v, 4 wire				
225a	EA.	17.15	87.00	104.15
400a	"	206.00	95.50	301.50
600a	"	266.00	103.00	369.00
800a	"	308.00	107.00	415.00
1000a	"	352.00	111.00	463.00
1200a	"	378.00	111.00	489.00
1350a	"	395.00	116.00	511.00
1600a	"	460.00	121.00	581.00
2000a	"	531.00	121.00	652.00
2500a	"	536.00	126.00	662.00
3000a	"	762.00	132.00	894.00
4000a	"	950.00	154.00	1,104
5000a	"	1,190	163.00	1,353
Bus duct, copper elbows, 277/480v-4w				
225a-1000a	EA.	329.00	73.00	402.00
1200a-3000a	"	459.00	87.00	546.00
4000a-5000a	"	540.00	103.00	643.00
Tees, 277/480v-4w				
225a-1000a	EA.	385.00	77.00	462.00
1200a-3000a	"	540.00	89.50	629.50
4000a-5000a	"	631.00	103.00	734.00
Crosses, 277/480v-4w				
225a-1000a	EA.	448.00	77.00	525.00
1200a-3000a	"	631.00	89.50	720.50
4000a-5000a	"	725.00	103.00	828.00
Copper end closures, 277/480v-4w				
225a-1000a	EA.	81.50	31.20	112.70
1200a-3000a	"	107.00	41.40	148.40
4000a-5000a	"	133.00	58.00	191.00
Tap boxes, 277/480v-4w				
225a	EA.	368.00	121.00	489.00
400a	"	569.00	154.00	723.00
600a	"	813.00	252.00	1,065
800a	"	825.00	278.00	1,103

BASIC MATERIALS

16050.30 BUS DUCT

	UNIT	MAT.	INST.	TOTAL
1000a	EA.	834.00	347.00	1,181
1200a	"	847.00	382.00	1,229
1350a	"	856.00	451.00	1,307
1600a	"	867.00	486.00	1,353
2000a	"	928.00	589.00	1,517
2500a	"	1,010	798.00	1,808
3000a	"	1,050	971.00	2,021
4000a	"	1,110	1,320	2,430
5000a	"	1,270	1,560	2,830
Circuit breaker, adapter cubicle				
225a	EA.	1,880	52.50	1,933
200a	"	2,220	55.50	2,276
600a	"	3,290	59.00	3,349
800a	"	3,750	63.00	3,813
1000a	"	4,350	66.00	4,416
1200a	"	5,210	69.50	5,280
1600a	"	6,340	73.00	6,413
2000a	"	7,420	77.00	7,497
Transformer taps, 1 phase 277/480v				
600a	EA.	323.00	252.00	575.00
800a	"	343.00	278.00	621.00
1000a	"	412.00	347.00	759.00
1200a	"	451.00	382.00	833.00
1350a	"	476.00	451.00	927.00
1600a	"	544.00	486.00	1,030
2000a	"	620.00	589.00	1,209
2500a	"	750.00	798.00	1,548
3000a	"	886.00	971.00	1,857
4000a	"	1,040	1,320	2,360
5000a	"	1,310	1,600	2,910
3 phase, 227/480v, 3 wire				
600a	EA.	735.00	252.00	987.00
800a	"	790.00	278.00	1,068
1000a	"	903.00	347.00	1,250
1200a	"	978.00	382.00	1,360
1350a	"	1,030	451.00	1,481
1600a	"	1,240	486.00	1,726
2000a	"	1,390	589.00	1,979
2500a	"	1,620	798.00	2,418
3000a	"	1,880	971.00	2,851
4000a	"	2,160	1,320	3,480
5000a	"	2,500	1,600	4,100
3 phase, 4 wire, 277/480v				
600a	EA.	798.00	252.00	1,050
800a	"	852.00	278.00	1,130
1000a	"	984.00	347.00	1,331
1200a	"	1,080	382.00	1,462
1350a	"	1,130	451.00	1,581
1600a	"	1,320	486.00	1,806
2000a	"	1,510	589.00	2,099
2500a	"	1,770	798.00	2,568
3000a	"	2,070	971.00	3,041

BASIC MATERIALS

16050.30 BUS DUCT

BASIC MATERIALS	UNIT	MAT.	INST.	TOTAL
4000a	EA.			2,400
5000a	"	2,710	1,600	4,310
Transformer connection, 4 wire, 277/480v				
600a	EA.	1,650	95.50	1,746
800a	"	1,720	99.00	1,819
1000a	"	1,800	103.00	1,903
1200a	"	1,850	107.00	1,957
1350a	"	1,880	111.00	1,991
1600a	"	2,020	116.00	2,136
2000a	"	2,110	121.00	2,231
2500a	"	2,380	126.00	2,506
3000a	"	2,610	132.00	2,742
4000a	"	3,450	154.00	3,604
5000a	"	4,620	163.00	4,783
Unfused reducers, 4 wire, 277/480v				
400a	EA.	230.00	87.00	317.00
600a	"	250.00	132.00	382.00
800a	"	309.00	163.00	472.00
1000a	"	362.00	174.00	536.00
1200a	"	620.00	185.00	805.00
1350a	"	791.00	198.00	989.00
1600a	"	859.00	214.00	1,073
2000a	"	1,160	214.00	1,374
2500a	"	1,430	231.00	1,661
3000a	"	1,720	252.00	1,972
4000a	"	2,240	304.00	2,544
5000a	"	2,700	375.00	3,075
400a	"	389.00	95.50	484.50
600a	"	502.00	139.00	641.00
800a	"	618.00	174.00	792.00
1000a	"	725.00	185.00	910.00
1200a	"	1,240	198.00	1,438
1350a	"	1,580	214.00	1,794
1600a	"	1,720	231.00	1,951
2000a	"	2,310	252.00	2,562
2500a	"	2,850	252.00	3,102
3000a	"	3,450	286.00	3,736
4000a	"	4,480	338.00	4,818
5000a	"	5,400	416.00	5,816
Circuit breaker reducers, 4 wire, 277/480v				
400a	EA.	402.00	77.00	479.00
600a	"	436.00	121.00	557.00
800a	"	537.00	146.00	683.00
1000a	"	629.00	163.00	792.00
1350a	"	1,380	185.00	1,565
1600a	"	1,490	198.00	1,688
2000a	"	2,010	214.00	2,224
2500a	"	2,490	231.00	2,721
3000a	"	3,000	198.00	3,198
4000a	"	3,900	252.00	4,152
5000a	"	4,700	347.00	5,047
Expansion fittings, 4 wire, 277/480v				

BASIC MATERIALS

16050.30 BUS DUCT

BASIC MATERIALS	UNIT	MAT.	INST.	TOTAL
225a	EA.	525.00	87.00	612.00
400a	"	571.00	132.00	703.00
600a	"	650.00	163.00	813.00
800a	"	795.00	174.00	969.00
1000a	"	892.00	185.00	1,077
1200a	"	1,070	198.00	1,268
1350a	"	1,180	214.00	1,394
1600a	"	1,430	214.00	1,644
2000a	"	1,520	231.00	1,751
2500a	"	1,640	252.00	1,892
3000a	"	2,150	304.00	2,454
4000a	"	2,490	375.00	2,865
5000a	"	2,610	416.00	3,026
Wall flanges				
225a-2500a	EA.	100.00	139.00	239.00
3000a-5000a	"	154.00	214.00	368.00
Weather seals	"	196.00	34.70	230.70
Roof flanges	"	421.00	139.00	560.00
Fire barriers	"	201.00	52.50	253.50
Spring hangers	"	58.50	60.50	119.00
Sway brace collars	"	11.90	43.40	55.30
Hook sticks				
8'	EA.			106.00
14'	"			177.00
Fusible switches, 240v, 3 phase				
30a	EA.	154.00	34.70	188.70
60a	"	47.30	43.40	90.70
100a	"	249.00	52.50	301.50
200a	"	432.00	73.00	505.00
400a	"	1,190	139.00	1,329
600a	"	1,710	214.00	1,924
208v, 4 wire				
30a	EA.	184.00	41.40	225.40
60a	"	195.00	47.10	242.10
100a	"	273.00	63.00	336.00
200a	"	482.00	95.50	577.50
400a	"	1,280	174.00	1,454
600a	"	1,870	278.00	2,148
600v				
30a	EA.	168.00	34.70	202.70
60a	"	179.00	43.40	222.40
100a	"	254.00	52.50	306.50
200a	"	451.00	73.00	524.00
400a	"	1,190	139.00	1,329
600a	"	1,720	214.00	1,934
800a	"	2,970	231.00	3,201
1000a	"	3,510	278.00	3,788
1200a	"	5,580	382.00	5,962
1600a	"	5,650	416.00	6,066
480v, 4 wire				
30a	EA.	195.00	41.40	236.40
60a	"	204.00	47.10	251.10

BASIC MATERIALS

16050.30 BUS DUCT

BASIC MATERIALS	UNIT	MAT.	INST.	TOTAL
100a	EA.	296.00	63.00	359.00
200a	"	508.00	95.50	603.50
400a	"	1,280	174.00	1,454
600a	"	1,870	278.00	2,148
800a	"	3,080	286.00	3,366
1000a	"	3,640	382.00	4,022
1200a	"	5,670	416.00	6,086
1600a	"	5,920	486.00	6,406
Fusible combination starters, 600v, 3 phase				
Size 0	EA.	571.00	44.80	615.80
Size 1	"	644.00	55.50	699.50
Size 2	"	814.00	63.00	877.00
Size 3	"	1,330	89.50	1,420
Circuit breaker combination starters, 600v, 3 phase				
Size 0	EA.	622.00	44.80	666.80
Size 1	"	648.00	55.50	703.50
Size 2	"	938.00	63.00	1,001
Size 3	"	1,220	89.50	1,310
Fusible combination contactors, 600v, 3 phase				
30a	EA.	584.00	44.80	628.80
60a	"	722.00	55.50	777.50
100a	"	773.00	63.00	836.00
200a	"	1,260	89.50	1,350
Circuit breaker, combination contactors, 600v, 3 phase				
30a	EA.	601.00	44.80	645.80
60a	"	626.00	55.50	681.50
100a	"	873.00	63.00	936.00
200a	"	1,150	89.50	1,240
Fusible contactor electrically held, 480v, 4 wire				
30a	EA.	588.00	44.80	632.80
60a	"	754.00	55.50	809.50
100a	"	1,070	63.00	1,133
200a	"	2,240	95.50	2,336
Mechanically held				
30a	EA.	635.00	44.80	679.80
60a	"	906.00	55.50	961.50
100a	"	1,260	64.50	1,325
200a	"	2,590	95.50	2,686
Circuit breakers, 240v, 3 phase				
15a-60a	EA.	305.00	40.20	345.20
70a-100a	"	336.00	55.50	391.50
600v, 3 phase				
15a-60a	EA.	332.00	40.20	372.20
125a-225a	"	837.00	79.50	916.50
250a-400a	"	1,730	146.00	1,876
500a-600a	"	2,460	185.00	2,645
700a-800a	"	2,960	278.00	3,238
900a-1000a	"	3,520	347.00	3,867
1200a-1600a	"	5,580	382.00	5,962
120/208v, 4 wire				
15a-60a	EA.	279.00	44.80	323.80
70a-100a	"	326.00	64.50	390.50

BASIC MATERIALS	UNIT	MAT.	INST.	TOTAL
16050.30 BUS DUCT				
277/480v, 4 wire				
15a-60a	EA.	342.00	44.80	386.80
70a-100a	"	368.00	66.00	434.00
125a-225a	"	896.00	95.50	991.50
250a-400a	"	1,810	174.00	1,984
500a-600a	"	2,580	278.00	2,858
700a-800a	"	3,080	278.00	3,358
900a-1000a	"	3,640	382.00	4,022
1200a-1600a	"	5,550	486.00	6,036
600v, 3 phase, 65,000 aic.				
60a	EA.	425.00	40.20	465.20
70a-100a	"	460.00	55.50	515.50
125a-225a	"	1,490	79.50	1,570
250a-400a	"	2,390	146.00	2,536
500a-600a	"	2,830	214.00	3,044
700a-800a	"	3,350	278.00	3,628
900a-1000a	"	3,810	347.00	4,157
277/480v, 4 wire, 65,000 aic.				
15a-60a	EA.	460.00	44.80	504.80
70a-100a	"	497.00	66.00	563.00
125a-225a	"	1,550	95.50	1,646
250a-400a	"	2,480	174.00	2,654
500a-600a	"	2,960	278.00	3,238
700a-800a	"	3,470	278.00	3,748
900a-1000a	"	5,670	359.00	6,029
600v, 3 phase, current limiting				
15a-60a	EA.	1,100	40.20	1,140
70a-100a	"	1,360	55.50	1,416
125a-225a	"	2,540	79.50	2,620
250a-400a	"	3,020	146.00	3,166
500a-600a	"	4,530	214.00	4,744
700a-800a	"	5,110	278.00	5,388
900a-1000a	"	5,690	347.00	6,037
277/480v, 4 wire, current limiting				
15a-60a	EA.	1,130	44.80	1,175
70a-100a	"	1,460	1,390	2,850
125a-225a	"	2,610	95.50	2,706
250a-400a	"	3,100	174.00	3,274
500a-600a	"	3,460	278.00	3,738
700a-800a	"	5,230	278.00	5,508
900a-1000a	"	5,800	382.00	6,182
Capacitors, 3 phase, 240v				
5 kvar	EA.	1,270	174.00	1,444
7.5 kvar	"	1,600	214.00	1,814
10 kvar	"	1,860	281.00	2,141
15 kvar	"	2,450	319.00	2,769
480v				
2.5 kvar	EA.	551.00	95.50	646.50
5 kvar	"	823.00	163.00	986.00
7.5 kvar	"	1,000	198.00	1,198
10 kvar	"	1,120	281.00	1,401
15 kvar	"	1,340	312.00	1,652

BASIC MATERIALS

BASIC MATERIALS	UNIT	MAT.	INST.	TOTAL
16050.30 BUS DUCT				
20 kvar	EA.	1,670	399.00	2,069
25 kvar	"	2,080	451.00	2,531
30 kvar	"	2,460	493.00	2,953
Transformers, 3 phase, 480v				
1.0 kva	EA.	525.00	60.50	585.50
1.5 kva	"	571.00	69.50	640.50
2 kva	"	614.00	87.00	701.00
3 kva	"	730.00	95.50	825.50
5 kva	"	1,010	139.00	1,149
7.5 kva	"	1,210	174.00	1,384
10 kva	"	1,430	185.00	1,615
16110.12 CABLE TRAY				
Cable tray, 6"	L.F.	10.25	2.05	12.30
Ventilated cover	"	4.15	1.05	5.20
Solid cover	"	3.25	1.05	4.30
Flat 90	EA.	48.20	17.35	65.55
Outside 90	"	40.70	17.35	58.05
Inside 90	"	44.50	17.35	61.85
Flat 45	"	36.80	17.35	54.15
Outside 45	"	30.50	17.35	47.85
Inside 45	"	31.80	17.35	49.15
Adjustable elbow	"	45.70	17.35	63.05
Support riser	"	61.00	17.35	78.35
Adjustable riser	"	6.80	17.35	24.15
Tee	"	62.00	60.50	122.50
Cross	"	84.00	63.00	147.00
Blind end	"	4.95	10.30	15.25
Expansion joint	"	22.85	17.35	40.20
Box connector	"	15.25	87.00	102.25
Standard dropout	"	2.65	17.35	20.00
2"	"	13.00	21.35	34.35
3"	"	22.85	27.80	50.65
4"	"	36.80	34.70	71.50
Cable tray, 9"	L.F.	10.85	2.40	13.25
Ventilated cover	"	4.65	1.40	6.05
Solid cover	"	3.75	1.40	5.15
Flat 90	EA.	53.50	18.50	72.00
Outside 90	"	43.90	18.50	62.40
Inside 90	"	46.20	18.50	64.70
Flat 45	"	38.10	18.50	56.60
Outside 45	"	32.30	18.50	50.80
Inside 45	"	32.30	18.50	50.80
Adjustable elbow	"	48.50	18.50	67.00
Support riser	"	66.00	18.50	84.50
Adjustable riser	"	6.80	18.50	25.30
Tee	"	68.00	63.00	131.00
Cross	"	88.00	63.00	151.00
Blind end	"	5.35	11.10	16.45
Expansion joint	"	32.30	18.50	50.80
Box connector	"	15.45	89.50	104.95

BASIC MATERIALS

16110.12 CABLE TRAY

	UNIT	MAT.	INST.	TOTAL
Standard dropout	EA.	2.90	18.50	21.40
2"	"	13.15	23.15	36.30
3"	"	21.90	23.15	45.05
4"	"	36.80	38.00	74.80
Cable tray, 12"	L.F.	11.15	2.80	13.95
Ventilated cover	"	4.30	1.75	6.05
Solid cover	"	3.95	1.75	5.70
Flat 90	EA.	55.50	18.50	74.00
Outside 90	"	44.00	18.50	62.50
Inside 90	"	48.40	18.50	66.90
Flat 45	"	40.40	18.50	58.90
Outside 45	"	63.50	18.50	82.00
Inside 45	"	52.00	18.50	70.50
Adjustable elbow	"	51.00	18.50	69.50
Support riser	"	66.00	18.50	84.50
Adjustable riser	"	7.60	18.50	26.10
Tee	"	71.50	69.50	141.00
Cross	"	90.00	69.50	159.50
Blind end	"	6.05	12.05	18.10
Expansion joint	"	34.70	21.35	56.05
Box connector	"	16.15	95.50	111.65
Standard dropout	"	3.15	21.35	24.50
2"	"	13.75	25.20	38.95
3"	"	22.85	31.20	54.05
4"	"	38.10	40.20	78.30
Cable tray, 18"	L.F.	11.70	3.45	15.15
Ventilated cover	"	6.05	2.05	8.10
Solid cover	"	5.10	2.05	7.15
Flat 90	EA.	65.00	25.20	90.20
Outside 90	"	69.50	25.20	94.70
Inside 90	"	53.00	25.20	78.20
Flat 45	"	46.20	25.20	71.40
Outside 45	"	65.00	25.20	90.20
Inside 45	"	54.00	25.20	79.20
Adjustable elbow	"	69.50	25.20	94.70
Support riser	"	81.50	25.20	106.70
Adjustable riser	"	6.80	25.20	32.00
Tee	"	86.00	73.00	159.00
Cross	"	129.00	73.00	202.00
Blind end	"	8.95	13.90	22.85
Expansion joint	"	45.10	25.20	70.30
Box connector	"	19.05	103.00	122.05
Standard dropout	"	4.80	23.15	27.95
2"	"	15.25	25.20	40.45
3"	"	24.15	33.00	57.15
4"	"	39.30	41.40	80.70
Cable tray, 24"	L.F.	13.35	4.25	17.60
Ventilated cover	"	8.25	2.40	10.65
Solid cover	"	5.70	2.40	8.10
Flat 90	EA.	80.50	25.20	105.70
Outside 90	"	71.50	25.20	96.70
Inside 90	"	56.00	25.20	81.20

16 ELECTRICAL

BASIC MATERIALS

16110.12 CABLE TRAY	UNIT	MAT.	INST.	TOTAL
Flat 45	EA.	54.00	25.20	79.20
Outside 45	"	66.00	25.20	91.20
Inside 45	"	56.00	25.20	81.20
Adjustable elbow	"	70.50	25.20	95.70
Support riser	"	84.50	25.20	109.70
Adjustable riser	"	7.55	25.20	32.75
Tee	"	99.00	77.00	176.00
Cross	"	147.00	77.00	224.00
Blind end	"	9.85	15.40	25.25
Expansion joint	"	46.20	25.20	71.40
Box connector	"	20.35	121.00	141.35
Standard dropout	"	5.10	23.15	28.25
2"	"	15.85	27.80	43.65
3"	"	25.10	33.90	59.00
4"	"	40.40	43.40	83.80
Cable tray, 36"	L.F.	15.40	5.05	20.45
Ventilated cover	"	9.55	2.80	12.35
Solid cover	"	6.60	2.80	9.40
Flat 90	EA.	94.50	29.50	124.00
Outside 90	"	81.50	29.50	111.00
Inside 90	"	71.50	29.50	101.00
Flat 45	"	62.50	29.50	92.00
Outside 45	"	72.50	29.50	102.00
Inside 45	"	62.50	29.50	92.00
Adjustable elbow	"	72.50	29.50	102.00
Support riser	"	88.00	29.50	117.50
Adjustable riser	"	7.95	29.50	37.45
Tee	"	113.00	79.50	192.50
Cross	"	170.00	103.00	273.00
Blind end	"	10.15	16.35	26.50
Expansion joint	"	48.40	25.20	73.60
Box connector	"	22.85	132.00	154.85
Standard dropout	"	5.70	25.20	30.90
2"	"	16.80	27.80	44.60
3"	"	26.00	34.70	60.70
4"	"	40.70	44.80	85.50
Reducers				
9" - 6"	EA.	36.50	17.35	53.85
12" - 9"	"	38.10	17.35	55.45
18" - 12"	"	40.40	21.35	61.75
24" - 18"	"	45.10	25.20	70.30
36" - 18"	"	55.00	27.80	82.80
36" - 24"	"	56.00	31.20	87.20
Conduit dropouts				
3/4"	EA.	4.85	12.05	16.90
1"	"	5.00	12.05	17.05
1-1/4"	"	5.85	13.90	19.75
1-1/2"	"	6.70	17.35	24.05
2"	"	7.70	18.50	26.20
2-1/2"	"	8.80	25.20	34.00
3"	"	10.20	27.80	38.00
Wall brackets				

BASIC MATERIALS

BASIC MATERIALS	UNIT	MAT.	INST.	TOTAL
16110.12 CABLE TRAY				
6"	EA.	19.00	5.05	24.05
9"	"	21.90	5.05	26.95
12"	"	23.10	6.95	30.05
18"	"	27.70	8.70	36.40
24"	"	32.30	10.30	42.60
36"	"	40.70	13.90	54.60
16110.15 FIBERGLASS CABLE TRAY				
Fiberglass cable tray, 6"	L.F.	16.30	1.40	17.70
Tray cover	"	4.55	1.05	5.60
Horizontal				
90	EA.	189.00	9.55	198.55
45	"	80.50	9.55	90.05
30	"	82.50	9.55	92.05
Inside				
90	EA.	72.50	9.55	82.05
45	"	58.50	9.55	68.05
30	"	78.00	9.55	87.55
Horizontal tee	"	126.00	25.20	151.20
Horizontal cross	"	147.00	38.00	185.00
Splice plate	"	7.00	5.05	12.05
Floor flange	"	7.75	12.05	19.80
Panel flange	"	8.00	43.40	51.40
End plate	"	5.25	5.05	10.30
Nylon rivet	"	0.19	1.75	1.94
Barrier strip	"	2.65	1.75	4.40
Hold down clamp	"	1.15	1.75	2.90
Drop out	"	8.35	18.50	26.85
Cover stand off	"	2.20	1.75	3.95
Wall bracket	"	31.30	6.95	38.25
Sealer	"			11.40
Outside				
90	EA.	72.50	9.55	82.05
45	"	58.50	9.55	68.05
30	"	79.00	9.55	88.55
Fiberglass cable tray, 9"	L.F.	17.10	1.75	18.85
Tray cover	"	4.75	1.40	6.15
Horizontal				
90	EA.	198.00	9.55	207.55
45	"	83.50	9.55	93.05
30	"	84.50	9.55	94.05
Inside				
90	EA.	76.00	9.55	85.55
45	"	60.50	9.55	70.05
30	"	81.50	9.55	91.05
Horizontal tee	"	132.00	25.20	157.20
Horizontal cross	"	154.00	38.00	192.00
Splice plate	"	7.35	5.05	12.40
Floor flange	"	8.15	12.05	20.20
Panel flange	"	8.45	43.40	51.85
End plate	"	5.55	5.05	10.60
Nylon rivet	"	0.20	1.75	1.95

BASIC MATERIALS

16110.15 FIBERGLASS CABLE TRAY

	UNIT	MAT.	INST.	TOTAL
Barrier strap	EA.	2.75	1.75	4.50
Hold down clamp	"	1.25	1.75	3.00
Drop out	"	8.75	18.50	27.25
Cover stand off	"	2.35	1.75	4.10
Wall bracket	"	32.80	6.95	39.75
Sealer	"			11.40
Outside				
90	EA.	76.50	9.55	86.05
45	"	62.50	9.55	72.05
30	"	83.50	9.55	93.05
Fiberglass cable tray, 12"	L.F.	17.95	2.05	20.00
Tray cover	"	5.05	1.75	6.80
Horizontal				
90	EA.	208.00	12.05	220.05
45	"	88.00	12.05	100.05
30	"	90.00	12.05	102.05
Inside				
90	EA.	79.00	12.05	91.05
45	"	62.50	12.05	74.55
Inside 30	"	84.50	12.05	96.55
Horizontal tee	"	139.00	33.00	172.00
Horizontal cross	"	163.00	44.80	207.80
Splice plate	"	7.70	5.55	13.25
Floor flange	"	8.50	12.60	21.10
Panel flange	"	8.85	55.50	64.35
End plate	"	5.80	5.55	11.35
Nylon rivet	"	0.21	1.75	1.96
Barrier strip	"	2.90	1.75	4.65
Hold down clamp	"	1.30	1.75	3.05
Drop out	"	13.05	18.50	31.55
Cover stand off	"	2.45	1.75	4.20
Wall bracket	"	34.50	6.95	41.45
Sealer	"			11.40
Outside				
90	EA.	79.00	12.05	91.05
45	"	64.00	12.05	76.05
30	"	88.00	12.05	100.05
Fiberglass cable tray, 18"	L.F.	18.85	2.45	21.30
Tray cover	"	9.30	2.05	11.35
Horizontal				
90	EA.	218.00	13.90	231.90
45	"	92.50	13.90	106.40
30	"	94.50	13.90	108.40
Inside				
90	EA.	82.50	13.90	96.40
45	"	66.00	13.90	79.90
30	"	89.00	13.90	102.90
Horizontal tee	"	145.00	53.50	198.50
Horizontal cross	"	169.00	47.10	216.10
Splice plate	"	8.10	5.55	13.65
Floor flange	"	8.95	13.20	22.15
Panel flange	"	9.30	59.00	68.30

BASIC MATERIALS

16110.15 FIBERGLASS CABLE TRAY

	UNIT	MAT.	INST.	TOTAL
End plate	EA.	6.10	5.90	12.00
Nylon rivet	"	0.22	1.75	1.97
Barrier strip	"	3.05	1.75	4.80
Hold down clamp	"	1.35	1.75	3.10
Drop out	"	19.05	18.50	37.55
Cover stand off	"	2.55	1.75	4.30
Wall bracket	"	36.20	6.95	43.15
Sealer	"			11.40
Outside				
90	EA.	83.50	13.90	97.40
45	"	67.00	13.90	80.90
30	"	92.50	13.90	106.40
Fiberglass cable tray, 24"	L.F.	19.80	2.80	22.60
Tray cover	"	11.00	2.40	13.40
Horizontal				
90	EA.	221.00	17.35	238.35
45	"	94.50	17.35	111.85
30	"	97.00	17.35	114.35
Inside				
90	EA.	86.00	17.35	103.35
45	"	69.50	17.35	86.85
30	"	92.50	17.35	109.85
Horizontal tee	"	152.00	34.70	186.70
Horizontal cross	"	178.00	52.50	230.50
Splice plate	"	8.50	5.55	14.05
Floor flange	"	9.40	13.90	23.30
Panel flange	"	9.75	63.00	72.75
End plate	"	6.40	5.90	12.30
Nylon Rivet	"	0.23	1.75	1.98
Barrier strip	"	3.20	1.75	4.95
Hold down clamp	"	1.40	1.75	3.15
Drop out	"	24.85	18.50	43.35
Cover stand off	"	2.70	1.75	4.45
Wall bracket	"	39.50	6.95	46.45
Sealer	"			11.40
Outside				
90	EA.	86.00	17.35	103.35
45	"	70.50	17.35	87.85
30	"	93.50	17.35	110.85
Fiberglass cable tray, 30"	L.F.	22.60	3.10	25.70
Tray cover	"	15.65	2.80	18.45
Horizontal				
90	EA.	223.00	21.35	244.35
45	"	97.00	21.35	118.35
30	"	99.00	21.35	120.35
Inside				
90	EA.	88.00	21.35	109.35
45	"	71.50	21.35	92.85
30	"	94.50	21.35	115.85
Horizontal tee	"	169.00	38.00	207.00
Horizontal cross	"	199.00	55.50	254.50
Splice plate	"	8.90	5.55	14.45

16 ELECTRICAL

BASIC MATERIALS

16110.15 FIBERGLASS CABLE TRAY

BASIC MATERIALS	UNIT	MAT.	INST.	TOTAL
Floor flange	EA.	9.90	13.90	23.80
Panel flange	"	10.25	66.00	76.25
End plate	"	6.75	6.95	13.70
Nylon rivet	"	0.24	1.75	1.99
Barrier strip	"	3.35	1.75	5.10
Hold down clamp	"	1.45	1.75	3.20
Dropout	"	28.50	21.35	49.85
Cover stand off	"	2.85	1.75	4.60
Wall bracket	"	44.50	6.95	51.45
Sealer	"			11.40
Outside				
90	EA.	88.00	21.35	109.35
45	"	72.50	21.35	93.85
30	"	95.50	21.35	116.85
Fibeglass cable tray, 36"	L.F.	26.10	3.45	29.55
Tray cover	"	20.85	3.10	23.95
Horizontal				
90	EA.	224.00	23.15	247.15
45	"	98.00	23.15	121.15
30	"	100.00	23.15	123.15
Inside				
90	EA.	90.00	23.15	113.15
45	"	72.50	23.15	95.65
30	"	95.50	23.15	118.65
Horizontal tee	"	190.00	41.40	231.40
horizontal cross	"	219.00	59.00	278.00
Splice plate	"	9.35	5.55	14.90
Floor flange	"	10.35	15.40	25.75
Panel flange	"	10.75	73.00	83.75
End plate	"	7.05	6.95	14.00
Nylon rivet	"	0.26	1.75	2.01
Barrier strip	"	3.55	1.75	5.30
Hold down clamp	"	1.55	1.75	3.30
Drop out	"	32.70	23.15	55.85
5Cover stand off	"	2.95	1.75	4.70
Wall bracket	"	49.40	6.95	56.35
Sealer	"			11.40
Outside				
90	EA.	90.00	23.15	113.15
45	"	75.00	23.15	98.15
30	"	98.00	23.15	121.15
Reducers				
12" - 6"	EA.	116.00	10.30	126.30
12" - 9"	"	117.00	103.00	220.00
18" - 6"	"	103.00	12.05	115.05
18" - 9"	"	107.00	12.05	119.05
18" - 12"	"	109.00	13.90	122.90
24" - 6"	"	126.00	13.90	139.90
24" - 9"	"	129.00	13.90	142.90
24" - 12"	"	131.00	13.90	144.90
24" - 18"	"	133.00	15.40	148.40
30" - 9"	"	129.00	13.90	142.90

16 ELECTRICAL

BASIC MATERIALS

BASIC MATERIALS	UNIT	MAT.	INST.	TOTAL
16110.15 FIBERGLASS CABLE TRAY				
30" - 12"	EA.	131.00	13.90	144.90
30" - 18"	"	133.00	15.40	148.40
30" - 24"	"	142.00	17.35	159.35
36" - 9"	"	134.00	17.35	151.35
36" - 12"	"	142.00	17.35	159.35
36" - 18"	"	145.00	18.50	163.50
36" - 24"	"	148.00	21.35	169.35
36" - 30"	"	152.00	25.20	177.20
16110.20 CONDUIT SPECIALTIES				
Rod beam clamp, 1/2"	EA.	3.15	1.75	4.90
Hanger rod				
3/8"	L.F.	0.68	1.40	2.08
1/2"	"	1.10	1.75	2.85
All thread rod				
1/4"	L.F.	0.38	1.05	1.43
3/8"	"	0.65	1.40	2.05
1/2"	"	1.10	1.75	2.85
5/8"	"	1.70	2.80	4.50
Hanger channel, 1-1/2"				
No holes	EA.	2.55	1.05	3.60
Holes	"	2.85	1.05	3.90
Channel strap				
1/2"	EA.	0.70	1.75	2.45
3/4"	"	0.83	1.75	2.58
1"	"	0.89	1.75	2.64
1-1/4"	"	1.00	2.80	3.80
1-1/2"	"	1.10	2.80	3.90
2"	"	1.30	2.80	4.10
2-1/2"	"	1.45	4.25	5.70
3"	"	1.55	4.25	5.80
3-1/2"	"	1.90	4.25	6.15
4"	"	2.15	5.05	7.20
5"	"	2.95	5.05	8.00
6"	"	3.55	5.05	8.60
Conduit penetrations, roof and wall, 8" thick				
1/2"	EA.	0.00	21.35	21.35
3/4"	"	0.00	21.35	21.35
1"	"	0.00	27.80	27.80
1-1/4"	"	0.00	27.80	27.80
1-1/2"	"	0.00	27.80	27.80
2"	"	0.00	55.50	55.50
2-1/2"	"	0.00	55.50	55.50
3"	"	0.00	55.50	55.50
3-1/2"	"	0.00	69.50	69.50
4"	"	0.00	69.50	69.50
Plastic duct bank conduit spacer, 3" separation				
2"	EA.	0.33	1.75	2.08
3"	"	0.35	1.75	2.10
4"	"	0.38	1.75	2.13
5"	"	0.44	1.75	2.19
6"	"	0.50	1.75	2.25

BASIC MATERIALS	UNIT	MAT.	INST.	TOTAL
16110.20 CONDUIT SPECIALTIES				
Intermediate, 3" separation				
2"	EA.	0.37	1.75	2.12
3"	"	0.40	1.75	2.15
4"	"	0.41	1.75	2.16
5"	"	0.46	1.75	2.21
6"	"	0.54	1.75	2.29
Base with 1-1/2" separation				
2"	EA.	0.31	1.75	2.06
3"	"	0.33	5.55	5.88
4"	"	0.38	5.55	5.93
5"	"	0.42	5.55	5.97
6"	"	0.45	5.55	6.00
Intermediate, 1-1/2" separation				
2"	EA.	0.33	5.55	5.88
3"	"	0.35	5.55	5.90
3-1/2"	"	0.40	5.55	5.95
4"	"	0.40	5.55	5.95
5"	"	0.44	6.00	6.44
6"	"	0.48	5.55	6.03
OD beam clamp, 1/4"	"	0.59	6.95	7.54
Threaded rod couplings				
1/4"	EA.	0.50	1.75	2.25
3/8"	"	0.57	1.75	2.32
1/2"	"	0.75	1.75	2.50
5/8"	"	1.85	1.75	3.60
3/4"	"	2.70	1.75	4.45
Hex nuts				
1/4"	EA.	0.06	1.75	1.81
3/8"	"	0.13	1.75	1.88
1/2"	"	0.22	1.75	1.97
5/8"	"	0.66	1.75	2.41
3/4"	"	0.73	1.75	2.48
Square nuts				
1/4"	EA.	0.06	1.75	1.81
3/8"	"	0.13	1.75	1.88
3/8"	"	0.22	1.75	1.97
5/8"	"	0.66	1.75	2.41
3/4"	"	0.73	1.75	2.48
Flat washers				
1/4"	EA.			0.06
3/8"	"			0.08
1/2"	"			0.13
5/8"	"			0.21
3/4"	"			0.30
Lockwashers				
1/4"	EA.			0.06
3/8"	"			0.06
1/2"	"			0.07
5/8"	"			0.09
3/4"	"			0.18
Channel closure strip	L.F.	1.20	4.65	5.85
Channel end cap	EA.	0.53	4.65	5.18

16 ELECTRICAL

BASIC MATERIALS	UNIT	MAT.	INST.	TOTAL
16110.20 CONDUIT SPECIALTIES				
Li-channel trapeze hangers				
12" long	EA.	9.20	5.05	14.25
18" long	"	10.45	5.05	15.50
24" long	"	12.95	5.05	18.00
30" long	"	14.60	8.70	23.30
36" long	"	17.05	8.70	25.75
42" long	"	20.25	10.30	30.55
Channel spring nuts				
1/4"	EA.	0.71	2.05	2.76
3/8"	"	0.81	2.80	3.61
1/2"	"	0.89	3.45	4.34
Fireproofing, for conduit penetrations				
1/2"	EA.	1.45	17.35	18.80
3/4"	"	1.45	17.35	18.80
1"	"	1.50	17.35	18.85
1-1/4"	"	2.10	27.30	29.40
1-1/2"	"	2.10	25.20	27.30
2"	"	2.10	25.20	27.30
2-1/2"	"	4.20	31.20	35.40
3"	"	4.20	33.70	37.90
3-1/2"	"	5.05	43.40	48.45
4"	"	6.05	52.50	58.55
16110.21 ALUMINUM CONDUIT				
Aluminum conduit				
1/2"	L.F.	0.79	1.05	1.84
3/4"	"	1.05	1.40	2.45
1"	"	1.05	1.75	2.80
1-1/4"	"	1.95	2.05	4.00
1-1/2"	"	2.45	2.80	5.25
2"	"	3.30	3.10	6.40
2-1/2"	"	5.20	3.45	8.65
3"	"	6.80	3.70	10.50
3-1/2"	"	8.15	4.25	12.40
4"	"	9.65	5.05	14.70
5"	"	13.85	6.30	20.15
6"	"	18.25	6.95	25.20
90 deg. elbow				
1/2"	EA.	2.75	6.60	9.35
3/4"	"	3.75	8.70	12.45
1"	"	5.25	10.70	15.95
1-1/4"	"	8.35	13.20	21.55
1-1/2"	"	11.00	13.90	24.90
2"	"	16.35	15.40	31.75
2-1/2"	"	27.50	19.85	47.35
3"	"	42.60	23.15	65.75
3-1/2"	"	66.50	27.80	94.30
4"	"	79.00	30.80	109.80
5"	"	214.00	39.70	253.70
6"	"	297.00	77.00	374.00
Coupling				
1/2"	EA.	0.88	1.75	2.63

BASIC MATERIALS	UNIT	MAT.	INST.	TOTAL
16110.21 ALUMINUM CONDUIT				
3/4"	EA.	1.35	2.05	3.40
1"	"	1.75	2.80	4.55
1-1/4"	"	2.20	3.10	5.30
1-1/2"	"	2.50	3.45	5.95
2"	"	3.60	3.70	7.30
2-1/2"	"	8.10	4.25	12.35
3"	"	10.55	4.25	14.80
3-1/2"	"	14.50	5.05	19.55
4"	"	17.45	5.55	23.00
5"	"	44.20	5.55	49.75
6"	"	69.00	6.60	75.60
16110.22 EMT CONDUIT				
EMT conduit				
1/2"	L.F.	0.17	1.05	1.22
3/4"	"	0.23	1.40	1.63
1"	"	0.37	1.75	2.12
1-1/4"	"	0.53	2.05	2.58
1-1/2"	"	0.62	2.80	3.42
2"	"	0.78	3.10	3.88
2-1/2"	"	1.75	3.45	5.20
3"	"	2.20	4.25	6.45
3-1/2"	"	3.10	5.05	8.15
4"	"	3.60	6.30	9.90
90 deg. elbow				
1/2"	EA.	2.00	3.10	5.10
3/4"	"	2.05	3.45	5.50
1"	"	2.20	3.70	5.90
1-1/4"	"	3.00	4.25	7.25
1-1/2"	"	3.75	5.05	8.80
2"	"	6.05	6.60	12.65
2-1/2"	"	17.20	7.30	24.50
3"	"	26.80	8.40	35.20
3-1/2"	"	35.90	9.70	45.60
4"	"	42.30	9.90	52.20
Connector, steel compression				
1/2"	EA.	0.55	3.10	3.65
3/4"	"	0.78	3.10	3.88
1"	"	1.30	3.10	4.40
1-1/4"	"	2.50	3.70	6.20
1-1/2"	"	3.65	5.05	8.70
2"	"	5.25	6.60	11.85
2-1/2"	"	16.50	8.70	25.20
3"	"	22.75	9.90	32.65
3-1/2"	"	34.40	10.70	45.10
4"	"	35.10	11.55	46.65
Coupling, steel, compression				
1/2"	EA.	0.67	2.05	2.72
3/4"	"	0.95	2.05	3.00
1"	"	1.50	2.05	3.55
1-1/4"	"	2.75	3.10	5.85

16 ELECTRICAL

BASIC MATERIALS

16110.22 EMT CONDUIT

	UNIT	MAT.	INST.	TOTAL
1-1/2"	EA.	3.95	3.70	7.65
2"	"	5.40	5.05	10.45
2-1/2"	"	20.60	7.70	28.30
3"	"	25.80	8.70	34.50
3-1/2"	"	40.30	9.90	50.20
4"	"	41.20	10.70	51.90
1 hole strap, steel				
1/2"	EA.	0.09	1.40	1.49
3/4"	"	0.13	1.40	1.53
1"	"	0.18	1.40	1.58
1-1/4"	"	0.28	1.75	2.03
1-1/2"	"	0.43	1.75	2.18
2"	"	0.56	1.75	2.31
2-1/2"	"	1.10	2.05	3.15
3"	"	1.25	2.05	3.30
3-1/2"	"	1.85	2.05	3.90
4"	"	2.30	2.05	4.35
Connector, steel set screw				
1/2"	EA.	0.34	3,970	3,970
3/4"	"	0.52	2.40	2.92
1"	"	0.86	2.40	3.26
1-1/4"	"	1.75	3.70	5.45
1-1/2"	"	2.50	5.05	7.55
2"	"	3.45	6.30	9.75
2-1/2"	"	10.35	8.40	18.75
3"	"	12.35	9.25	21.60
3-1/2"	"	16.45	10.30	26.75
4"	"	19.05	12.05	31.10
Insulated throat				
1/2"	EA.	0.47	2.40	2.87
3/4"	"	0.73	2.40	3.13
1"	"	1.20	2.40	3.60
1-1/4"	"	2.20	3.70	5.90
1-1/2"	"	3.15	5.05	8.20
2"	"	4.50	6.30	10.80
2-1/2"	"	20.00	8.40	28.40
3"	"	23.50	9.25	32.75
3-1/2"	"	31.40	10.30	41.70
4"	"	34.80	12.05	46.85
Connector, die cast set screw				
1/2"	EA.	0.18	2.05	2.23
3/4"	"	0.29	2.05	2.34
1"	"	0.46	2.05	2.51
1-1/4"	"	0.80	3.10	3.90
1-1/2"	"	1.25	3.70	4.95
2"	"	1.60	5.05	6.65
2-1/2"	"	5.20	6.95	12.15
3"	"	6.35	7.70	14.05
3-1/2"	"	8.95	8.70	17.65
4"	"	10.25	9.90	20.15
Insulated throat				
1/2"	EA.	0.30	2.05	2.35

BASIC MATERIALS

16110.22 EMT CONDUIT

BASIC MATERIALS	UNIT	MAT.	INST.	TOTAL
3/4"	EA.	0.48	2.05	2.53
1"	"			0.73
1-1/4"	"	1.50	3.10	4.60
1-1/2"	"	2.05	3.70	5.75
2"	"	2.85	5.05	7.90
2-1/2"	"	10.80	6.95	17.75
3"	"	13.80	7.70	21.50
3-1/2"	"	17.85	9.35	27.20
4"	"	19.75	9.90	29.65
Coupling, steel set screw				
1/2"	EA.	0.46	1.40	1.86
3/4"	"	0.69	1.40	2.09
1"	"	1.05	1.40	2.45
1-1/4"	"	1.95	1.75	3.70
1-1/2"	"	2.90	2.80	5.70
2"	"	3.85	3.70	7.55
2-1/2"	"	8.65	5.55	14.20
3"	"	9.70	6.60	16.30
3-1/2"	"	11.45	7.70	19.15
4"	"	13.05	8.70	21.75
Diecast set screw				
1/2"	EA.	0.18	1.40	1.58
3/4"	"	0.29	1.40	1.69
1"	"	0.50	1.40	1.90
1-1/4"	"	0.86	1.75	2.61
1-1/2"	"	1.25	2.80	4.05
2"	"	1.65	3.70	5.35
2-1/2"	"	4.25	5.55	9.80
3"	"	4.75	6.45	11.20
3-1/2"	"	5.80	7.70	13.50
4"	"	6.45	8.70	15.15
1 hole malleable straps				
1/2"	EA.	0.19	1.40	1.59
3/4"	"	0.24	1.40	1.64
1"	"	0.36	1.40	1.76
1-1/4"	"	0.58	1.75	2.33
1-1/2"	"	0.70	1.75	2.45
2"	"	1.35	1.75	3.10
2-1/2"	"	2.50	2.05	4.55
3"	"	3.50	2.05	5.55
3-1/2"	"	5.05	2.05	7.10
4"	"	11.30	2.05	13.35
EMT to rigid compression coupling				
1/2"	EA.	1.30	3.45	4.75
3/4"	"	1.90	3.45	5.35
1"	"	2.75	5.20	7.95
Set screw couplings				
1/2"	EA.	0.94	3.45	4.39
3/4"	"	1.35	3.45	4.80
1"	"	2.05	5.05	7.10
Set screw offset connectors				
1/2"	EA.	1.30	3.45	4.75

16 ELECTRICAL

BASIC MATERIALS	UNIT	MAT.	INST.	TOTAL
16110.22 EMT CONDUIT				
3/4"	EA.	1.75	3.45	5.20
1"	"	2.65	5.05	7.70
Compression offset connectors				
1/2"	EA.	1.45	3.45	4.90
3/4"	"	2.10	3.45	5.55
1"	"	3.25	5.05	8.30
Type "LB" set screw condulets				
1/2"	EA.	2.75	7.95	10.70
3/4"	"	3.40	13.20	16.60
1"	"	5.25	13.20	18.45
1-1/4"	"	7.80	15.40	23.20
1-1/2"	"	10.05	18.50	28.55
2"	"	16.50	21.35	37.85
2-1/2"	"	51.50	25.20	76.70
3"	"	68.00	34.70	102.70
3-1/2"	"	105.00	46.30	151.30
4"	"	127.00	55.50	182.50
Type "T" set screw condulets				
1/2"	EA.	3.80	10.30	14.10
3/4"	"	4.75	15.40	20.15
1"	"	6.40	15.40	21.80
1-1/4"	"	11.20	18.50	29.70
1-1/2"	"	15.15	21.35	36.50
2"	"	25.10	23.15	48.25
Type "C" set screw condulets				
1/2"	EA.	3.30	8.70	12.00
3/4"	"	4.05	10.30	14.35
1"	"	6.00	13.20	19.20
1-1/4"	"	9.05	15.40	24.45
1-1/2"	"	11.90	18.50	30.40
2"	"	23.25	13.20	36.45
Type "LL" setscrew condulets				
1/2"	EA.	3.30	8.70	12.00
3/4"	"	4.05	10.30	14.35
1"	"	6.00	13.20	19.20
1-1/4"	"	9.05	15.40	24.45
1-1/2"	"	11.85	18.50	30.35
2"	"	19.90	21.35	41.25
Type "LR" set screw condulets				
1/2"	EA.	2.80	8.70	11.50
3/4"	"	4.05	10.30	14.35
1"	"	6.00	13.20	19.20
1-1/4"	"	9.05	15.40	24.45
1-1/2"	"	11.85	18.50	30.35
2"	"	19.90	21.35	41.25
Type "LB" compression condulets				
1/2"	EA.	6.95	10.30	17.25
3/4"	"	10.25	17.35	27.60
1"	"	13.15	17.35	30.50
Type "T" compression condulets				
1/2"	EA.	9.30	13.90	23.20
3/4"	"	12.25	15.40	27.65

BASIC MATERIALS

16110.22 EMT CONDUIT

	UNIT	MAT.	INST.	TOTAL
1"	EA.	18.95	21.35	40.30
Condulet covers				
1/2"	EA.	0.78	4.25	5.03
3/4"	"	1.05	4.25	5.30
1"	"	1.35	4.25	5.60
1-1/4"	"	1.65	4.25	5.90
1-1/2"	"	1.75	5.05	6.80
2"	"	2.95	5.05	8.00
2-1/2"	"	4.80	5.05	9.85
3"	"	5.00	6.30	11.30
3-1/2"	"	6.75	6.30	13.05
4"	"	7.45	6.30	13.75
Clamp type entrance caps				
1/2"	EA.	3.55	8.70	12.25
3/4"	"	4.25	10.30	14.55
1"	"	5.20	0.79	5.99
1-1/4"	"	6.10	18.50	24.60
1-1/2"	"	10.95	21.35	32.30
2"	"	20.20	31.20	51.40
2-1/2"	"	53.50	34.70	88.20
3"	"	54.00	52.50	106.50
3-1/2"	"	67.50	60.50	128.00
4"	"	91.50	77.00	168.50
Slip fitter type entrance caps				
1/2"	EA.	2.40	8.70	11.10
3/4"	"	3.40	10.30	13.70
1"	"	4.35	13.90	18.25
1-1/4"	"	4.60	18.50	23.10
1-1/2"	"	6.40	21.35	27.75
2"	"	7.90	31.20	39.10
2-1/2"	"	29.10	34.70	63.80
3"	"	48.40	52.50	100.90
3-1/2"	"	65.50	65.00	130.50
4"	"	73.00	77.00	150.00

16110.23 FLEXIBLE CONDUIT

	UNIT	MAT.	INST.	TOTAL
Flexible conduit, steel				
3/8"	L.F.	0.19	1.05	1.24
1/2	"	0.20	1.05	1.25
3/4"	"	0.28	1.40	1.68
1"	"	0.55	1.40	1.95
1-1/4"	"	0.70	1.75	2.45
1-1/2"	"	0.92	2.05	2.97
2"	"	1.20	2.80	4.00
2-1/2"	"	1.40	3.10	4.50
3"	"	1.75	3.70	5.45
Flexible conduit, liquid tight				
3/8"	L.F.	0.79	1.05	1.84
1/2"	"	0.86	1.05	1.91
3/4"	"	0.95	1.40	2.35
1"	"	1.80	1.40	3.20
1-1/4"	"	2.50	1.75	4.25

16 ELECTRICAL

BASIC MATERIALS	UNIT	MAT.	INST.	TOTAL
16110.23 FLEXIBLE CONDUIT				
1-1/2"	L.F.	3.45	2.05	5.50
2"	EA.	4.35	2.80	7.15
2-1/2"	"	8.00	3.10	11.10
3"	"	11.05	3.70	14.75
4"	"	16.05	5.05	21.10
Connector, straight				
3/8"	EA.	0.50	2.80	3.30
1/2"	"	1.20	2.80	4.00
3/4"	"	0.99	3.10	4.09
1"	"	1.95	3.45	5.40
1-1/4"	"	3.05	3.70	6.75
1-1/2"	"	5.30	4.25	9.55
2"	"	7.30	5.05	12.35
2-1/2"	"	13.75	6.30	20.05
3"	"	19.15	6.60	25.75
Straight insulated throat connectors				
3/8"	EA.	2.10	4.25	6.35
1/2"	"	2.15	4.25	6.40
3/4"	"	3.10	5.05	8.15
1"	"	4.80	5.05	9.85
1-1/4"	"	7.50	6.30	13.80
1-1/2"	"	10.70	7.30	18.00
2"	"	20.10	7.95	28.05
2-1/2"	"	102.00	9.25	111.25
3"	"	113.00	11.55	124.55
4"	"	134.00	14.60	148.60
90 deg connectors				
3/8"	EA.	2.95	5.15	8.10
1/2"	"	3.00	5.15	8.15
3/4"	"	4.45	5.90	10.35
1"	"	9.05	6.30	15.35
1-1/4"	"	14.05	7.95	22.00
1-1/2"	"	16.85	8.70	25.55
2"	"	24.65	9.25	33.90
2-1/2"	"	109.00	11.55	120.55
3"	"	131.00	13.20	144.20
4"	"	168.00	15.40	183.40
90 degree insulated throat connectors				
3/8"	EA.	3.50	5.05	8.55
1/2"	"	3.60	5.05	8.65
3/4"	"	5.30	5.90	11.20
1"	"	10.10	6.15	16.25
1-1/4"	"	14.65	7.95	22.60
1-1/2"	"	18.00	8.70	26.70
2"	"	26.90	9.25	36.15
2-1/2"	"	125.00	11.55	136.55
3"	"	152.00	13.20	165.20
4"	"	198.00	15.40	213.40
Flexible aluminum conduit				
3/8"	L.F.	0.17	1.05	1.22
1/2"	"	0.25	1.05	1.30
3/4"	"	0.34	1.40	1.74

BASIC MATERIALS	UNIT	MAT.	INST.	TOTAL
16110.23 FLEXIBLE CONDUIT				
1"	L.F.	0.75	1.40	2.15
1-1/4"	"	0.98	1.75	2.73
1-1/2"	"	1.20	2.05	3.25
2"	"	1.60	2.80	4.40
2-1/2"	"	2.05	3.10	5.15
3"	"	2.35	3.70	6.05
3-1/2"	"	3.05	4.25	7.30
4"	"	3.65	5.05	8.70
Connector, straight				
3/8"	EA.	1.90	3.45	5.35
1/2"	"	1.95	3.45	5.40
3/4"	"	2.70	3.70	6.40
1"	"	3.95	4.25	8.20
1-1/4"	"	6.80	5.05	11.85
1-1/2"	"	9.70	6.30	16.00
2"	"	17.80	6.60	24.40
2-1/2"	"	84.50	7.70	92.20
3"	"	95.50	9.55	105.05
4"	"	113.00	12.05	125.05
Straight insulated throat connectors				
3/8"	EA.	0.53	3.10	3.63
1/2"	"	1.05	2.80	3.85
3/4"	"	1.20	3.10	4.30
1"	"	2.55	3.45	6.00
1-1/4"	"	4.15	3.70	7.85
1-1/2"	"	6.15	4.25	10.40
2"	"	9.30	5.05	14.35
2-1/2"	"	18.60	6.30	24.90
3"	"	24.10	6.60	30.70
3-1/2"	"	78.00	7.70	85.70
4"	"	101.00	9.55	110.55
90 deg connectors				
3/8"	EA.	0.97	5.05	6.02
1/2"	"	1.50	5.05	6.55
3/4"	"	2.35	5.05	7.40
1"	"	3.95	5.90	9.85
1-1/4"	"	8.00	6.30	14.30
1-1/2"	"	13.20	6.95	20.15
2"	"	17.10	7.30	24.40
2-1/2"	"	49.50	7.95	57.45
3"	"	60.50	9.25	69.75
90 deg insulated throat connectors				
3/8"	EA.	1.05	5.05	6.10
1/2"	"	1.70	5.05	6.75
3/4"	"	2.55	5.05	7.60
1"	"	4.40	5.90	10.30
1-1/4"	"	8.45	6.30	14.75
1-1/2"	"	15.10	6.95	22.05
2"	"	19.75	7.30	27.05
2-1/2"	"	53.00	7.95	60.95
3"	"	66.00	9.25	75.25
3-1/2"	"	195.00	11.55	206.55

BASIC MATERIALS	UNIT	MAT.	INST.	TOTAL
16110.23 FLEXIBLE CONDUIT				
4"	EA.	294.00	14.60	308.60
3/4"	"	0.86	1.75	2.61
1"	"	1.35	1.75	3.10
1-1/4"	"	1.75	2.05	3.80
1-1/2"	"	2.45	2.40	4.85
16110.24 GALVANIZED CONDUIT				
Galvanized rigid steel conduit				
1/2"	L.F.	0.76	1.40	2.16
3/4"	"	0.95	1.75	2.70
1"	"	1.30	2.05	3.35
1-1/4"	"	1.70	2.80	4.50
1-1/2"	"	2.05	3.10	5.15
2"	"	2.65	3.45	6.10
2-1/2"	"	4.20	5.05	9.25
3"	"	5.60	6.30	11.90
3-1/2"	"	5.65	6.60	12.25
4"	"	8.30	7.30	15.60
5"	"	17.65	9.90	27.55
6"	"	25.40	13.20	38.60
90 degree ell				
1/2"	EA.	2.70	8.70	11.40
3/4"	"	3.25	10.70	13.95
1"	"	4.85	13.20	18.05
1-1/4"	"	7.00	15.40	22.40
1-1/2"	"	8.35	17.35	25.70
2"	"	12.30	18.50	30.80
2-1/2"	"	20.70	23.15	43.85
3"	"	31.00	30.80	61.80
3-1/2"	"	53.50	34.70	88.20
4"	"	62.00	46.30	108.30
5"	"	162.00	77.00	239.00
6"	"	218.00	116.00	334.00
Couplings, with set screws				
1/2"	EA.	1.35	1.75	3.10
3/4"	"	1.75	2.05	3.80
1"	"	2.85	2.80	5.65
1-1/4"	"	4.85	3.45	8.30
1-1/2"	"	6.25	4.25	10.50
2"	"	14.05	5.05	19.10
2-1/2"	"	34.80	6.60	41.40
3"	"	41.80	8.70	50.50
3-1/2"	"	59.00	9.90	68.90
4"	"	77.00	10.70	87.70
5"	"	123.00	15.40	138.40
6"	"	158.00	17.35	175.35
Split couplings				
1/2"	EA.	1.20	6.60	7.80
3/4"	"	1.50	8.70	10.20
1"	"	2.10	9.55	11.65
1-1/4"	"	4.25	10.70	14.95

16 ELECTRICAL

BASIC MATERIALS	UNIT	MAT.	INST.	TOTAL
16110.24 GALVANIZED CONDUIT				
1-1/2"	EA.	5.35	13.20	18.55
2"	"	24.70	19.85	44.55
2-1/2"	"	25.10	19.85	44.95
3"	"	37.30	25.20	62.50
3-1/2"	"	59.50	34.70	94.20
4"	"	70.50	46.30	116.80
5"	"	121.00	56.50	177.50
6"	"	162.00	71.00	233.00
Erickson couplings				
1/2"	EA.	1.95	15.40	17.35
3/4"	"	2.40	17.35	19.75
1"	"	4.80	21.35	26.15
1-1/4"	"	8.65	30.80	39.45
1-1/2"	"	11.25	34.70	45.95
2"	"	22.55	46.30	68.85
2-1/2"	"	44.00	64.50	108.50
3"	"	65.00	73.00	138.00
3-1/2"	"	117.00	87.00	204.00
4"	"	139.00	92.50	231.50
5"	"	277.00	103.00	380.00
6"	"	363.00	111.00	474.00
Seal fittings				
1/2"	EA.	8.25	23.15	31.40
3/4"	"	9.50	27.80	37.30
1"	"	11.65	34.70	46.35
1-1/4"	"	14.10	39.70	53.80
1-1/2"	"	20.90	46.30	67.20
2"	"	27.30	55.50	82.80
2-1/2"	"	42.00	66.00	108.00
3"	"	50.50	73.00	123.50
3-1/2"	"	139.00	87.00	226.00
4"	"	211.00	103.00	314.00
5"	"	320.00	154.00	474.00
6"	"	436.00	174.00	610.00
Entrance fitting, (weather head), threaded				
1/2"	EA.	2.30	15.40	17.70
3/4"	"	2.80	17.35	20.15
1"	"	3.65	19.85	23.50
1-1/4"	"	4.70	25.20	29.90
1-1/2"	"	6.90	27.80	34.70
2"	"	12.65	30.80	43.45
2-1/2"	"	43.60	34.70	78.30
3"	"	62.50	46.30	108.80
3-1/2"	"	83.50	60.50	144.00
4"	"	104.00	87.00	191.00
5"	"	221.00	121.00	342.00
6"	"	228.00	154.00	382.00
Locknuts				
1/2"	EA.	0.08	1.75	1.83
3/4"	"	0.09	1.75	1.84
1"	"	0.18	1.75	1.93
1-1/4"	"	0.23	1.75	1.98

BASIC MATERIALS	UNIT	MAT.	INST.	TOTAL
16110.24 GALVANIZED CONDUIT				
1-1/2"	EA.	0.33	2.05	2.38
2"	"	0.48	2.05	2.53
2-1/2"	"	0.13	2.80	2.93
3"	"	1.55	2.80	4.35
3-1/2"	"	2.65	2.80	5.45
4"	"	3.35	3.10	6.45
5"	"	7.15	3.10	10.25
6"	"	12.50	3.10	15.60
Plastic conduit bushings				
1/2"	EA.	0.08	4.25	4.33
3/4"	"	0.12	5.05	5.17
1"	"	0.18	6.60	6.78
1-1/4"	"	0.24	7.70	7.94
1-1/2"	"	0.32	8.70	9.02
2"	"	0.58	10.70	11.28
2-1/2"	"	1.45	17.35	18.80
3"	"	1.80	23.15	24.95
3-1/2"	"	1.85	27.80	29.65
4"	"	2.15	30.80	32.95
5"	"	4.90	39.70	44.60
6"	"	9.40	55.50	64.90
Conduit bushings, steel				
1/2"	EA.	0.12	4.25	4.37
3/4"	"	0.17	5.05	5.22
1"	"	0.25	6.60	6.85
1-1/4"	"	0.35	7.70	8.05
1-1/2"	"	0.53	8.70	9.23
2"	"	0.81	10.70	11.51
2-1/2"	"	1.85	17.35	19.20
3"	"	2.25	23.15	25.40
3-1/2"	"	4.70	27.80	32.50
4"	"	5.75	30.80	36.55
5"	"	12.15	39.70	51.85
6"	"	21.65	55.50	77.16
Pipe cap				
1/2"	EA.	0.14	1.75	1.89
3/4"	"	0.15	1.75	1.90
1"	"	0.26	1.75	2.01
1-1/4"	"	0.45	2.80	3.25
1-1/2"	"	0.69	2.80	3.49
2"	"	0.78	2.80	3.58
2-1/2"	"	1.30	3.10	4.40
3"	"	1.65	3.10	4.75
3-1/2"	"	2.25	3.10	5.35
4"	"	2.90	3.70	6.60
5"	"	3.80	5.05	8.85
6"	"	4.80	6.95	11.75
GRS elbows, 36" radius				
2"	EA.	64.00	23.15	87.15
2-1/2"	"	88.00	28.00	116.00
3"	"	116.00	36.50	152.50
3-1/2"	"	160.00	43.40	203.40

BASIC MATERIALS	UNIT	MAT.	INST.	TOTAL
16110.24 GALVANIZED CONDUIT				
4"	EA.	176.00	52.50	228.50
5"	"	284.00	87.00	371.00
6"	"	299.00	132.00	431.00
42" radius				
2"	EA.	71.00	28.00	99.00
2-1/2"	"	97.50	34.70	132.20
3"	"	128.00	43.40	171.40
3-1/2"	"	178.00	52.50	230.50
4"	"	216.00	60.50	276.50
5"	"	316.00	99.00	415.00
6"	"	333.00	139.00	472.00
48" radius				
2"	EA.	80.50	32.30	112.80
2-1/2"	"	110.00	39.10	149.10
3"	"	144.00	49.60	193.60
3-1/2"	"	201.00	60.50	261.50
4"	"	244.00	75.00	319.00
5"	"	358.00	107.00	465.00
6"	"	376.00	154.00	530.00
Threaded couplings				
1/2"	EA.	0.90	1.75	2.65
3/4"	"	1.10	2.05	3.15
1"	"	1.55	2.80	4.35
1-1/4"	"	1.95	3.10	5.05
1-1/2"	"	2.45	3.45	5.90
2"	"	3.25	3.70	6.95
2-1/2"	"	7.25	4.25	11.50
3"	"	9.95	5.05	15.00
3-1/2"	"	13.25	5.05	18.30
4"	"	13.95	5.55	19.50
5"	"	31.60	6.30	37.90
6"	"	42.00	6.60	48.60
Threadless couplings				
1/2"	EA.	1.55	3.45	5.00
3/4"	"	2.25	4.25	6.50
1"	"	3.85	5.05	8.90
1-1/4"	"	5.85	6.60	12.45
1-1/2"	"	7.40	8.70	16.10
2"	"	16.85	10.70	27.55
2-1/2"	"	48.90	17.35	66.25
3"	"	65.00	21.35	86.35
3-1/2"	"	87.00	28.00	115.00
4"	"	111.00	34.70	145.70
5"	"	245.00	43.40	288.40
6"	"	309.00	185.00	494.00
Threadless connectors				
1/2"	EA.	0.98	34.70	35.68
3/4"	"	1.55	4.25	5.80
1"	"	2.20	5.05	7.25
1-1/4"	"	4.00	6.60	10.60
1-1/2"	"	5.55	8.70	14.25
2"	"	10.90	10.70	21.60

BASIC MATERIALS

16110.24 GALVANIZED CONDUIT

	UNIT	MAT.	INST.	TOTAL
2-1/2"	EA.	34.70	17.35	52.05
3"	"	46.40	21.35	67.75
3-1/2"	"	60.50	28.00	88.50
4"	"	73.50	34.70	108.20
5"	"	188.00	43.40	231.40
6"	"	244.00	52.50	296.50
Setscrew connectors				
1/2"	EA.	0.89	2.80	3.69
3/4"	"	1.25	3.10	4.35
1"	"	1.95	3.45	5.40
1-1/4"	"	3.40	4.25	7.65
1-1/2"	"	4.90	5.05	9.95
2"	"	9.80	6.60	16.40
2-1/2"	"	29.20	8.70	37.90
3"	"	41.50	10.70	52.20
3-1/2"	"	54.00	13.20	67.20
4"	"	68.00	17.35	85.35
5"	"	119.00	21.35	140.35
6"	"	154.00	28.00	182.00
Clamp type entrance caps				
1/2"	EA.	2.55	10.70	13.25
3/4"	"	3.05	13.20	16.25
1"	"	3.70	15.40	19.10
1-1/4"	"	3.75	0.38	4.13
1-1/2"	"	7.75	21.35	29.10
2"	"	14.50	25.20	39.70
3-1/2"	"	38.20	32.70	70.90
3"	"	51.50	39.10	90.60
3-1/2"	"	56.00	47.90	103.90
4"	"	75.00	84.00	159.00
"LB" condulets				
1/2"	EA.	2.95	10.70	13.65
3/4"	"	3.55	13.20	16.75
1"	"	5.30	15.40	20.70
1-1/4"	"	9.20	17.35	26.55
1-1/2"	"	12.00	21.35	33.35
2"	"	19.75	25.20	44.95
2-1/2"	"	41.40	34.70	76.10
3"	"	55.00	47.90	102.90
3-1/2"	"	99.00	60.50	159.50
4"	"	112.00	73.00	185.00
"T" condulets				
1/2"	EA.	3.65	13.20	16.85
3/4"	"	4.40	15.40	19.80
1"	"	6.60	17.35	23.95
1-1/4"	"	9.70	19.85	29.55
1-1/2"	"	12.95	21.35	34.30
2"	"	20.05	25.20	45.25
2-1/2"	"	43.40	39.10	82.50
3"	"	57.00	52.50	109.50
3-1/2"	"	106.00	64.50	170.50
4"	"	119.00	77.00	196.00

BASIC MATERIALS

16110.24 GALVANIZED CONDUIT

	UNIT	MAT.	INST.	TOTAL
"X" condulets				
1/2"	EA.	5.45	15.40	20.85
3/4"	"	5.85	17.35	23.20
1"	"	9.65	19.85	29.50
1-1/4"	"	12.60	21.35	33.95
1-1/2"	"	16.25	23.15	39.40
2"	"	33.40	30.50	63.90
Blank steel condulet covers				
1/2"	EA.	0.85	3.45	4.30
3/4"	"	1.05	3.45	4.50
1"	"	1.45	3.45	4.90
1-1/4"	"	1.70	4.25	5.95
1-1/2"	"	1.80	4.25	6.05
2"	"	3.05	4.25	7.30
2-1/2"	"	4.75	5.05	9.80
3"	"	5.15	5.05	10.20
3-1/2"	"	5.65	5.05	10.70
4"	"	6.15	6.95	13.10
Solid condulet gaskets				
1/2"	EA.	1.05	1.75	2.80
3/4"	"	1.05	1.75	2.80
1"	"	1.30	1.75	3.05
1-1/4"	"	1.60	2.80	4.40
1-1/2"	"	1.70	2.80	4.50
2"	"	1.90	2.80	4.70
2-1/2"	"	3.15	3.45	6.60
3"	"	3.25	3.45	6.70
3-1/2"	"	3.95	3.45	7.40
4"	"	4.30	5.05	9.35
One-hole malleable straps				
1/2"	EA.	0.15	1.40	1.55
3/4"	"	0.20	1.40	1.60
1"	"	0.29	1.40	1.69
1-1/4"	"	0.51	1.75	2.26
1-1/2"	"	0.62	1.75	2.37
2"	"	1.20	1.75	2.95
2-1/2"	"	2.50	2.05	4.55
3"	"	3.55	2.05	5.60
3-1/2"	"	5.10	2.05	7.15
4"	"	11.30	2.80	14.10
5"	"	39.80	2.80	42.60
6"	"	42.70	2.80	45.50
One-hole steel straps				
1/2"	EA.	0.11	1.40	1.51
3/4"	"	0.12	1.40	1.52
1"	"	0.22	1.40	1.62
1-1/4"	"	0.30	1.75	2.05
1-1/2"	"	0.36	1.75	2.11
2"	"	0.69	1.75	2.44
2-1/2"	"	1.05	2.05	3.10
3"	"	1.15	2.05	3.20
3-1/2"	"	1.70	2.05	3.75

BASIC MATERIALS	UNIT	MAT.	INST.	TOTAL
16110.24 GALVANIZED CONDUIT				
4"	EA.	2.15	2.80	4.95
Bushed chase nipples				
1/2"	EA.	0.21	2.05	2.26
3/4"	"	0.28	2.40	2.68
1"	"	0.51	3.10	3.61
1-1/4"	"	0.81	0.35	1.16
1-1/2"	"	1.05	4.25	5.30
2"	"	1.60	5.05	6.65
2-1/2"	"	3.75	5.05	8.80
3"	"	5.45	6.30	11.75
3-1/2"	"	11.50	8.70	20.20
4"	"	18.90	10.30	29.20
Offset nipples				
1/2"	EA.	1.50	2.05	3.55
3/4"	"	1.55	2.40	3.95
1"	"	1.90	3.10	5.00
1-1/4"	"	4.20	3.70	7.90
1-1/2"	"	5.15	4.25	9.40
2"	"	8.15	5.05	13.20
3"	"	26.70	6.30	33.00
Short elbows				
1/2"	EA.	1.00	5.05	6.05
3/4"	"	1.40	6.95	8.35
1"	"	2.40	8.70	11.10
1-1/4"	"	6.35	10.30	16.65
1-1/2"	"	9.35	12.05	21.40
2"	"	15.60	13.90	29.50
Pulling elbows, female to female				
1/2"	EA.	2.55	8.70	11.25
3/4"	"	3.00	10.30	13.30
1"	"	5.05	13.90	18.95
1-1/4"	"	7.35	18.50	25.85
1-1/2"	"	13.70	25.20	38.90
2"	"	11.30	29.50	40.80
Grounding locknuts				
1/2"	EA.	0.67	2.80	3.47
3/4"	"	0.84	2.80	3.64
1"	"	1.20	2.80	4.00
1-1/4"	"	1.25	3.10	4.35
1-1/2"	"	1.35	3.10	4.45
2"	"	1.85	3.10	4.95
2-1/2"	"	3.40	3.45	6.85
3"	"	4.25	3.45	7.70
3-1/2"	"	6.95	3.45	10.40
4"	"	9.30	5.05	14.35
Insulated grounding metal bushings				
1/2"	EA.	1.95	6.60	8.55
3/4"	"	2.50	7.70	10.20
1"	"	3.40	8.70	12.10
1-1/4"	"	4.05	10.70	14.75
1-1/2"	"	4.65	13.20	17.85

16 ELECTRICAL

BASIC MATERIALS

16110.24 GALVANIZED CONDUIT	UNIT	MAT.	INST.	TOTAL
2"	EA.	5.50	15.40	20.90
2-1/2"	"	7.45	23.15	30.60
3"	"	8.60	28.00	36.60
3-1/2"	"	12.45	32.70	45.15
4"	"	15.50	36.50	52.00
5"	"	31.10	54.50	85.60
6"	"	46.80	60.50	107.30
Nipples				
1/2" x				
4"	EA.	1.10	5.05	6.15
6"	"	1.45	5.05	6.50
8"	"	2.50	5.05	7.55
10"	"	2.85	5.05	7.90
12"	"	3.30	5.05	8.35
3/4" x				
4"	EA.	1.25	5.05	6.30
6"	"	1.65	5.05	6.70
8"	"	2.80	5.05	7.85
10"	"	3.35	5.05	8.40
12"	"	3.75	5.05	8.80
1" x				
4"	EA.	1.80	5.05	6.85
6"	"	2.25	5.05	7.30
8"	"	3.65	5.05	8.70
10"	"	4.55	5.05	9.60
12"	"	5.20	5.05	10.25
1-1/4" x				
4"	EA.	2.20	8.70	10.90
6"	"	2.95	8.70	11.65
8"	"	4.75	8.70	13.45
10"	"	5.95	8.70	14.65
12"	"	6.95	8.70	15.65
1-1/2" x				
4"	EA.	2.95	8.70	11.65
6"	"	4.05	8.70	12.75
8"	"	5.95	8.70	14.65
10"	"	7.15	8.70	15.85
12"	"	7.75	8.70	16.45
2" x				
4"	EA.	3.85	8.70	12.55
6"	"	4.90	8.70	13.60
8"	"	7.40	8.70	16.10
10"	"	8.85	8.70	17.55
12"	"	10.10	8.70	18.80
2-1/2" x				
6"	EA.	10.70	10.40	21.10
8"	"	14.05	10.40	24.45
10"	"	16.30	10.40	26.70
12"	"	19.00	10.40	29.40
3" x				
6"	EA.	12.45	10.40	22.85
8"	"	16.70	10.40	27.10

16 ELECTRICAL

BASIC MATERIALS	UNIT	MAT.	INST.	TOTAL
16110.24 GALVANIZED CONDUIT				
10"	EA.	21.95	10.40	32.35
12"	"	23.95	10.40	34.35
3-1/2" x				
6"	EA.	15.20	10.40	25.60
8"	"	19.10	10.40	29.50
10"	"	23.05	10.40	33.45
12"	"	27.00	10.40	37.40
4" x				
8"	EA.	21.55	13.90	35.45
10"	"	26.60	13.90	40.50
10"	"	0.00	13.90	13.90
12"	"	31.70	13.90	45.60
5" x				
8"	EA.	35.90	13.90	49.80
12"	"	51.00	13.90	64.90
6" x				
8"	EA.	44.60	13.90	58.50
10"	"	55.50	13.90	69.40
12"	"	62.00	13.90	75.90
16110.25 PLASTIC CONDUIT				
PVC conduit, schedule 40				
1/2"	L.F.	0.20	1.05	1.25
3/4"	"	0.25	1.05	1.30
1"	"	0.37	1.40	1.77
1-1/4"	"	0.51	1.40	1.91
1-1/2"	"	0.61	1.75	2.36
2"	"	0.79	1.75	2.54
2-1/2"	"	1.30	2.05	3.35
3"	"	1.50	2.05	3.55
3-1/2"	"	1.95	2.80	4.75
4"	"	2.15	2.80	4.95
5"	"	3.10	3.10	6.20
6"	"	4.10	3.45	7.55
Couplings				
1/2"	EA.	0.18	1.75	1.93
3/4"	"	0.19	1.75	1.94
1"	"	0.30	1.75	2.05
1-1/4"	"	0.38	2.05	2.43
1-1/2"	"	0.52	2.05	2.57
2"	"	0.70	2.05	2.75
2-1/2"	"	1.25	2.05	3.30
3"	"	2.00	2.80	4.80
3-1/2"	"	2.30	2.80	5.10
4"	"	3.10	3.45	6.55
5"	"	7.75	3.45	11.20
6"	"	9.95	3.45	13.40
90 degree elbows				
1/2"	EA.	0.68	3.45	4.13
3/4"	"	0.74	4.25	4.99
1"	"	0.90	4.25	5.15

BASIC MATERIALS

16110.25 PLASTIC CONDUIT

	UNIT	MAT.	INST.	TOTAL
1-1/4"	EA.	1.30	5.05	6.35
1-1/2"	"	1.80	6.60	8.40
2"	"	2.55	7.70	10.25
2-1/2"	"	4.55	8.70	13.25
3"	"	7.95	10.70	18.65
3-1/2"	"	11.05	13.20	24.25
4"	"	16.45	17.35	33.80
5"	"	24.05	21.35	45.40
6"	"	43.70	25.20	68.90
Terminal adapters				
1/2"	EA.	0.23	3.45	3.68
3/4"	"	0.40	3.45	3.85
1"	"	0.47	3.45	3.92
1-1/4"	"	0.62	5.55	6.17
1-1/2"	"	0.76	5.55	6.31
2"	"	1.05	5.55	6.60
2-1/2"	"	1.80	7.70	9.50
3"	"	2.55	7.70	10.25
3-1/2"	"	3.30	7.70	11.00
4"	"	4.25	13.20	17.45
5"	"	8.50	13.20	21.70
6"	"	10.20	13.20	23.40
End bells				
1"	EA.	1.30	34.70	36.00
1-1/4"	"	1.40	5.55	6.95
1-1/2"	"	1.45	5.55	7.00
2"	"	2.15	5.55	7.70
2-1/2"	"	2.30	7.70	10.00
3"	"	2.60	7.70	10.30
3-1/2"	"	2.65	7.70	10.35
4"	"	3.10	13.20	16.30
5"	"	4.70	13.20	17.90
6"	"	5.55	13.20	18.75
Lb conduit body				
1/2"	EA.	1.65	6.60	8.25
3/4"	"	2.20	6.60	8.80
1	"	2.45	6.60	9.05
1-1/4"	"	3.65	10.70	14.35
1-1/2"	"	4.65	10.70	15.35
2"	"	6.75	10.70	17.45
2-1/2"	"	23.95	15.40	39.35
3"	"	24.65	18.50	43.15
3-1/2"	"	25.70	21.35	47.05
4"	"	27.00	25.20	52.20
Direct burial, conduit				
2"	L.F.	0.40	1.75	2.15
3"	"	0.73	2.05	2.78
4"	"	1.40	2.80	4.20
5"	"	2.00	3.10	5.10
6"	"	2.85	3.45	6.30
Encased burial conduit				
2"	L.F.	0.41	1.75	2.16

BASIC MATERIALS	UNIT	MAT.	INST.	TOTAL
16110.25 PLASTIC CONDUIT				
3"	L.F.	0.66	2.05	2.71
4"	"	1.00	2.80	3.80
5"	"	1.45	3.10	4.55
6"	"	1.90	3.45	5.35
"EB" and "DB" duct, 90 degree elbows				
1-1/2"	EA.	4.95	5.05	10.00
2"	"	5.60	7.95	13.55
3"	"	7.10	13.20	20.30
3-1/2"	"	12.15	15.40	27.55
4"	"	13.00	18.50	31.50
5"	"	15.70	23.15	38.85
6"	"	25.30	31.20	56.50
45 degree elbows				
1-1/2"	EA.	3.95	7.95	11.90
2"	"	4.05	7.95	12.00
3"	"	5.05	13.20	18.25
3-1/2"	"	7.20	15.40	22.60
4"	"	7.30	18.50	25.80
5"	"	10.85	23.15	34.00
6"	"	37.10	31.20	68.30
Couplings				
1-1/2"	EA.	0.64	2.05	2.69
2"	"	0.84	2.05	2.89
3"	"	0.98	2.80	3.78
3-1/2"	"	1.10	2.80	3.90
4"	"	2.55	3.45	6.00
5"	"	3.70	3.45	7.15
6"	"	5.10	5.55	10.65
Bell ends				
1-1/2"	EA.	2.45	5.55	8.00
2"	"	3.10	5.55	8.65
3"	"	3.50	7.70	11.20
3-1/2"	"	3.65	7.70	11.35
4"	"	3.90	13.20	17.10
5"	"	5.30	13.20	18.50
6"	"	5.70	13.20	18.90
Female adapters, 1-1/2"	"	0.75	6.95	7.70
5 degree couplings				
1-1/2"	EA.	2.60	2.40	5.00
2"	"	2.95	2.40	5.35
3"	"	3.80	3.45	7.25
4"	"	2.00	5.05	7.05
5"	"	2.60	5.05	7.65
6"	"	3.95	5.05	9.00
45 degree elbows				
1/2"	EA.	0.51	4.25	4.76
3/4"	"	0.62	5.05	5.67
1"	"	0.85	5.05	5.90
1-1/4"	"	1.25	6.30	7.55
1-1/2"	"	1.70	7.95	9.65
2"	"	2.50	9.25	11.75
2-1/2"	"	4.65	10.30	14.95

BASIC MATERIALS

16110.25 PLASTIC CONDUIT

	UNIT	MAT.	INST.	TOTAL
3"	EA.	8.00	13.20	21.20
3-1/2"	"	9.00	15.40	24.40
4"	"	12.65	21.35	34.00
5"	"	20.15	25.20	45.35
6"	"	29.30	31.20	60.50
Female adapters				
1/2"	EA.	0.24	4.25	4.49
3/4"	"	0.41	4.25	4.66
1"	"	0.52	4.25	4.77
1-1/4"	"	0.64	6.95	7.59
1-1/2"	"	0.73	6.95	7.68
2"	"	1.00	6.95	7.95
2-1/2"	"	1.75	9.25	11.00
3"	"	2.90	9.25	12.15
3-1/2"	"	3.60	9.25	12.85
4"	"	4.60	15.40	20.00
5"	"	10.25	15.40	25.65
6"	"	11.65	15.40	27.05
Expansion couplings				
1/2"	EA.	8.25	4.25	12.50
3/4"	"	8.45	4.25	12.70
1"	"	8.60	5.05	13.65
1-1/4"	"	8.70	6.95	15.65
1-1/2"	"	8.85	6.95	15.80
2"	"	9.35	6.95	16.30
2-1/2"	"	13.40	10.30	23.70
3"	"	17.10	10.30	27.40
3-1/2"	"	19.70	10.30	30.00
4"	"	24.10	15.40	39.50
5"	"	35.20	15.40	50.60
6"	"	47.00	15.40	62.40
Plugs				
2"	EA.	0.52	6.95	7.47
3"	"	0.66	10.30	10.96
3-1/2"	"	0.81	10.30	11.11
4"	"	0.98	15.40	16.38
5"	"	1.30	15.40	16.70
6"	"	2.55	17.35	19.90
PVC cement				
1 pint	EA.			4.95
1 quart	"			7.15
1 gallon	"			21.85
Type "T" condulets				
1/2"	EA.	2.80	10.30	13.10
3/4"	"	3.20	10.30	13.50
1"	"	3.60	10.30	13.90
1-1/4"	"	4.50	17.35	21.85
1-1/2"	"	5.65	17.35	23.00
2"	"	7.60	17.35	24.95
EB & DB female adapters				
2"	EA.	1.00	8.70	9.70
3"	"	3.75	13.20	16.95

BASIC MATERIALS	UNIT	MAT.	INST.	TOTAL
16110.25 PLASTIC CONDUIT				
3-1/2"	EA.	4.55	21.35	25.90
4"	"	4.95	25.20	30.15
5"	"	10.90	34.70	45.60
6"	"	12.30	0.54	12.84
16110.27 PLASTIC COATED CONDUIT				
Rigid steel conduit, plastic coated				
1/2"	L.F.	2.15	1.75	3.90
3/4"	"	2.50	2.05	4.55
1"	"	3.25	2.80	6.05
1-1/4"	"	4.10	3.45	7.55
1-1/2"	"	4.95	4.25	9.20
2"	"	6.50	5.05	11.55
2-1/2"	"	9.90	6.60	16.50
3"	"	12.45	7.70	20.15
3-1/2"	"	15.15	8.70	23.85
4"	"	18.55	10.70	29.25
5"	"	31.80	13.20	45.00
90 degree elbows				
1/2"	EA.	8.45	10.70	19.15
3/4"	"	8.60	13.20	21.80
1"	"	10.00	15.40	25.40
1-1/4"	"	12.55	17.35	29.90
1-1/2"	"	15.15	21.35	36.50
2"	"	21.10	27.80	48.90
2-1/2"	"	40.80	39.70	80.50
3"	"	41.20	46.30	87.50
3-1/2"	"	88.50	56.50	145.00
4"	"	93.50	69.50	163.00
5"	"	216.00	87.00	303.00
Couplings				
1/2"	EA.	2.45	2.05	4.50
3/4"	"	2.60	2.80	5.40
1"	"	3.50	3.10	6.60
1-1/4"	"	4.10	3.70	7.80
1-1/2"	"	5.30	4.25	9.55
2"	"	7.10	5.05	12.15
2-1/2"	"	17.80	6.30	24.10
3"	"	20.75	6.60	27.35
3-1/2"	"	28.60	6.95	35.55
4"	"	35.00	7.70	42.70
5"	"	99.00	8.70	107.70
1 hole conduit straps				
3/4"	EA.	2.85	1.75	4.60
1"	"	4.10	1.75	5.85
1-1/4"	"	5.15	2.05	7.20
1-1/2"	"	6.25	2.05	8.30
2"	"	7.15	2.05	9.20
3"	"	11.15	2.80	13.95
3-1/2"	"	12.45	2.80	15.25
4"	"	16.95	3.45	20.40
"L.B." condulets with covers				

16 ELECTRICAL

BASIC MATERIALS	UNIT	MAT.	INST.	TOTAL
16110.27 PLASTIC COATED CONDUIT				
1/2"	EA.	21.85	17.35	39.20
3/4"	"	24.20	17.35	41.55
1"	"	32.50	21.35	53.85
1-1/4"	"	47.10	25.20	72.30
1-1/2"	"	57.50	30.50	88.00
2"	"	85.00	34.70	119.70
2-1/2"	"	150.00	47.90	197.90
3"	"	189.00	60.50	249.50
3-1/2"	"	274.00	75.00	349.00
4"	"	308.00	87.00	395.00
"T" condulets with covers				
1/2"	EA.	25.30	19.85	45.15
3/4"	"	28.20	21.35	49.55
1"	"	37.70	23.15	60.85
1-1/4"	"	53.00	28.00	81.00
1-1/2"	"	66.50	32.70	99.20
2"	"	96.50	36.50	133.00
2-1/2"	"	157.00	52.50	209.50
3-1/2"	"	301.00	77.00	378.00
4"	"	327.00	92.50	419.50
5"	"	210.00	66.00	276.00
16110.28 STEEL CONDUIT				
Intermediate metal conduit (IMC)				
1/2"	L.F.	0.71	1.05	1.76
3/4"	"	0.85	1.40	2.25
1"	"	1.15	1.75	2.90
1-1/4"	"	1.50	2.05	3.55
1-1/2"	"	1.75	2.80	4.55
2"	"	2.45	3.10	5.55
2-1/2"	"	3.80	4.15	7.95
3"	"	5.15	5.05	10.20
3-1/2"	"	6.80	6.30	13.10
4"	"	8.10	6.60	14.70
90 degree ell				
1/2"	EA.	2.20	8.70	10.90
3/4"	"	2.90	10.70	13.60
1"	"	4.20	13.20	17.40
1-1/4"	"	5.85	15.40	21.25
1-1/2"	"	7.40	17.35	24.75
2"	"	10.70	19.85	30.55
2-1/2"	"	18.50	23.15	41.65
3"	"	28.30	30.80	59.10
3-1/2"	"	50.00	39.70	89.70
4"	"	58.00	46.30	104.30
Couplings				
1/2"	EA.	0.88	1.75	2.63
3/4"	"	1.10	2.05	3.15
1"	"	1.55	2.80	4.35
1-1/4"	"	1.95	3.10	5.05
1-1/2"	"	2.45	3.45	5.90

BASIC MATERIALS

BASIC MATERIALS	UNIT	MAT.	INST.	TOTAL
16110.28 STEEL CONDUIT				
2"	EA.	3.25	3.70	6.95
2-1/2"	"	7.20	4.25	11.45
3"	"	9.85	5.05	14.90
3-1/2"	"	13.15	5.05	18.20
4"	"	13.85	5.55	19.40
16110.32 FLEXIBLE WIRING SYSTEMS				
Single circuit cables				
5'	EA.	9.10	2.05	11.15
10'	"	12.95	3.45	16.40
15'	"	16.05	5.05	21.10
20'	"	22.50	6.95	29.45
25'	"	29.00	9.25	38.25
30'	"	35.50	10.30	45.80
40'	"	48.40	13.90	62.30
Two circuit cables				
5'	EA.	12.35	2.05	14.40
10'	"	14.50	3.45	17.95
15'	"	18.15	5.05	23.20
20'	"	25.70	6.95	32.65
25'	"	33.00	9.25	42.25
30'	"	40.30	10.30	50.60
40'	"	55.00	13.90	68.90
Two wire switch and receptacle cables				
5'	EA.	5.10	2.05	7.15
10'	"	10.35	3.45	13.80
15'	"	13.25	5.05	18.30
20'	"	22.80	6.95	29.75
25'	"	30.30	9.25	39.55
30'	"	40.90	10.30	51.20
40'	"	52.50	12.05	64.55
Three wire switch				
5'	EA.	5.85	2.05	7.90
10'	"	11.80	3.45	15.25
15'	"	15.45	5.05	20.50
20'	"	22.80	6.95	29.75
25'	"	30.30	9.25	39.55
30'	"	40.90	10.30	51.20
40'	"	52.50	12.05	64.55
Distribution boxes				
2 circuit	EA.	9.55	18.50	28.05
3 circuit	"	11.00	23.15	34.15
4 circuit	"	16.55	27.80	44.35
6 circuit	"	22.10	38.00	60.10
12 circuit	"	34.10	77.00	111.10
18 circuit	"	53.50	103.00	156.50
Tap boxes				
1 single pole switch	EA.	19.25	13.90	33.15
2 single pole switches	"	21.20	18.50	39.70
1 3 way switch	"	22.45	15.40	37.85
1 4 way switch	"	27.90	18.50	46.40

16 ELECTRICAL

BASIC MATERIALS	UNIT	MAT.	INST.	TOTAL
16110.32 FLEXIBLE WIRING SYSTEMS				
1 receptacle	EA.	17.35	13.90	31.25
2 receptacles	"	21.70	18.50	40.20
4 receptacles	"	27.10	27.80	54.90
1 clock	"	21.70	13.90	35.60
2 clocks	"	25.10	18.50	43.60
4 clocks	"	28.30	27.80	56.10
Dust cap	"	2.85	3.45	6.30
Cable coupler	"	6.70	6.95	13.65
Reversing connector	"	9.55	10.30	19.85
16110.35 WIREMOLD				
Wiremold raceway with fittings, surface mounted				
#200	L.F.	0.46	1.05	1.51
#500	"	0.55	1.05	1.60
#700	"	0.61	1.40	2.01
#800	"	0.99	1.40	2.39
Fittings, #200, 90 degree flat elbow	EA.	0.88	1.75	2.63
Internal elbow	"	1.75	1.75	3.50
Extension adapter	"	3.10	2.05	5.15
#200, #500, #700				
Single pole switch and box	EA.	5.90	13.90	19.80
Duplex receptacle with box	"	6.80	10.70	17.50
#500, #700				
90 deg. flat elbow	EA.	0.76	2.80	3.56
Internal elbow	"	0.99	2.80	3.79
Junction box	"	5.00	4.65	9.65
Fixture box	"	5.05	4.65	9.70
Shallow switch and receptacle box	"	4.10	2.80	6.90
#800				
90 deg. flat elbow	EA.	0.94	2.80	3.74
Internal elbow	"	0.94	2.80	3.74
Junction box	"	2.95	4.25	7.20
#1500 series				
Raceway	L.F.	0.96	1.75	2.71
Wire clip	EA.	0.23	1.40	1.63
2 hole clip	"	0.29	2.05	2.34
Flat elbow	"	3.35	6.95	10.30
Internal elbow	"	3.40	6.95	10.35
External elbow	"	3.70	6.95	10.65
Adapter fitting	"	3.75	1.75	5.50
Duplex receptacle box	"	7.00	5.05	12.05
#2000 series				
Raceway	L.F.	1.05	1.75	2.80
Wire clip	EA.	0.28	1.40	1.68
Coupling	"	0.33	1.75	2.08
Supporting clip	"	0.62	1.40	2.02
Entrance end	"	3.15	5.05	8.20
Flat elbow	"	3.00	5.05	8.05
Tee	"	5.95	6.95	12.90
Single pole switch	"	7.80	6.95	14.75
#2100 series				

BASIC MATERIALS	UNIT	MAT.	INST.	TOTAL
16110.35 WIREMOLD				
Raceway	L.F.	0.66	1.75	2.41
Cover	"	0.52	0.35	0.87
Wireclip	EA.	0.29	1.40	1.69
Coupling	"	0.38	1.75	2.13
Entrance end	"	3.10	6.95	10.05
Flat elbow	"	3.40	6.95	10.35
Tee	"	4.20	6.95	11.15
Device box	"	4.85	5.05	9.90
#2220 series				
Raceway	L.F.	2.05	3.45	5.50
Cover	"	0.41	0.69	1.10
Wire cup	EA.	0.48	1.40	1.88
Coupling	"	1.10	1.75	2.85
Entrance end	"	7.45	6.95	14.40
Flat elbow	"	13.70	6.95	20.65
Wall box connection	"	5.50	6.95	12.45
External elbow	"	7.25	5.05	12.30
#2600 series				
Raceway	L.F.	2.30	3.45	5.75
Wireclip	EA.	0.35	1.40	1.75
Bushing	"	0.55	1.40	1.95
Flat elbow	"	4.75	6.95	11.70
Junction box	"	6.45	6.95	13.40
#3000 series				
Raceway	L.F.	2.45	3.45	5.90
Wire clip	EA.	0.66	1.40	2.06
Coupling	"	1.25	1.40	2.65
Supporting clip	"	1.35	1.75	3.10
Device bracket	"	2.90	2.05	4.95
Entrance end	"	4.05	6.95	11.00
Flat elbow	"	9.05	8.70	17.75
Tee	"	13.20	12.05	25.25
Device box	"	5.55	6.95	12.50
Circuit breaker housing	"	13.25	8.70	21.95
Panel connection	"	5.00	17.35	22.35
Conduit connector	"	17.70	8.70	26.40
#4000 series				
Raceway	L.F.	2.95	3.45	6.40
Cover	"	1.75	1.75	3.50
Divider	"	0.52	1.75	2.27
Wire clip	EA.	1.10	1.75	2.85
Coupling	"	3.55	2.05	5.60
Device bracket	"	3.35	2.05	5.40
Entrance end	"	10.65	10.70	21.35
Flat elbow	"	12.15	8.70	20.85
Tee	"	21.50	10.70	32.20
Internal elbow	"	6.85	10.70	17.55
External elbow	"	15.55	10.70	26.25
Panel connector	"	7.70	25.20	32.90
#6000 series				
Raceway	L.F.	4.60	5.15	9.75
Cover	"	1.90	1.75	3.65

BASIC MATERIALS	UNIT	MAT.	INST.	TOTAL
16110.35 WIREMOLD				
Wire cup	EA.	1.25	1.75	3.00
Coupling	"	4.05	3.45	7.50
Device bracket	"	4.50	3.45	7.95
Entrance end	"	12.05	15.40	27.45
Flat elbow	"	16.10	13.90	30.00
Tee	"	24.55	17.35	41.90
Internal elbow	"	15.20	13.90	29.10
External elbow	"	15.40	15.40	30.80
Panel connector	"	7.15	31.20	38.35
Telepower poles				
# 21TP2	EA.	78.00	25.20	103.20
# 21TP4	"	86.00	34.70	120.70
# 30TP4	"	151.00	34.70	185.70
Fittings				
Telephone	EA.	5.60	6.95	12.55
T-bar clamp	"	4.00	6.95	10.95
C-hanger clamp	"	6.50	6.95	13.45
Tap-off fitting	"	3.90	6.95	10.85
Take-off connector	"	26.70	25.20	51.90
16110.40 UNDERFLOOR DUCT				
Underfloor blank duct, insert duct				
7/8"	L.F.	4.85	1.75	6.60
1-3/8"	"	5.30	1.75	7.05
1-7/8"	"	5.40	1.75	7.15
Box opening plugs	EA.	1.85	5.05	6.90
Duct end plugs	"	2.15	5.05	7.20
Sleeve couplings	"	5.00	12.05	17.05
Expansion couplings	"	28.50	12.05	40.55
Vertical elbow	"	24.00	12.05	36.05
Offset elbow	"	19.20	12.05	31.25
Horizontal elbow	"	46.20	12.05	58.25
Adjustable elbow	"	8.55	12.05	20.60
Cabinet connector	"	7.10	41.40	48.50
Y-take off	"	21.35	21.35	42.70
Underfloor duct leveling legs	"	2.15	5.05	7.20
Conduit adapters				
1/2"	EA.	8.50	8.70	17.20
3/4"	"	8.65	8.70	17.35
1"	"	8.70	8.70	17.40
1-1/4"	"	9.70	10.30	20.00
2"	"	13.50	12.05	25.55
Reducer bushings				
1-1/4" x 3/4"	EA.	4.30	6.95	11.25
1-1/4" x 1"	"	5.00	6.95	11.95
2" x 1-1/2"	"	5.85	8.70	14.55
Support couplers				
1 standard	EA.	9.95	8.70	18.65
2 standard	"	12.10	10.30	22.40
3 standard	"	14.95	12.05	27.00
Supports				

BASIC MATERIALS

16110.40 UNDERFLOOR DUCT

BASIC MATERIALS	UNIT	MAT.	INST.	TOTAL
1 duct	EA.	10.60	5.05	15.65
2 duct	"	12.10	5.90	18.00
3 duct	"	12.75	6.60	19.35
4 duct	"	16.40	8.70	25.10
5 duct	"	17.90	10.30	28.20
Single level junction box				
1 standard	EA.	118.00	27.80	145.80
2 standard	"	177.00	52.50	229.50
3 standard	"	340.00	103.00	443.00
4 standard	"	438.00	139.00	577.00
Two level junction boxes				
1 standard	EA.	133.00	34.70	167.70
2 standard	"	261.00	69.50	330.50
Sealing compound	"			2.65
Insert adapters	"	6.40	5.05	11.45
Ellipsoids	"	7.10	12.05	19.15
Insert closing cap	"	0.74	5.05	5.79
Marker Screws	"	2.00	5.05	7.05
Access boxes	"	142.00	34.70	176.70
Closing caps	"	0.74	5.05	5.79
Afterset markers	"	1.95	5.05	7.00
Cell markers	"	4.30	5.05	9.35
Tie down straps	"	4.00	5.05	9.05
Plastic grommets	"	2.00	13.90	15.90
Metal grommets	"	13.70	13.90	27.60
Receptacle				
Duplex, 20a	EA.	29.90	17.35	47.25
Single				
30a	EA.	39.70	17.35	57.05
50a	"	43.60	21.35	64.95
Double duplex	"	31.20	21.35	52.55
Single, 20a	"	27.40	15.40	42.80
Double single	"	32.40	17.35	49.75
Twist lock	"	29.90	15.40	45.30
1 conduit opening	"	28.70	15.40	44.10
2 conduit openings	"	32.40	17.35	49.75
1 bushed opening	"	28.70	15.40	44.10
2 bushed openings	"	32.40	17.35	49.75
Amphenol connector				
1"	EA.	47.40	17.35	64.75
2"	"	53.50	18.50	72.00
5"	"	63.50	21.35	84.85
Standpipes				
Aluminum	EA.	35.80	8.70	44.50
Brass	"	37.70	8.70	46.40
Abandonment plates				
Aluminum	EA.	10.10	8.70	18.80
Brass	"	16.80	8.70	25.50
Split bell caps				
Aluminum	EA.	57.50	8.70	66.20
Brass	"	71.50	8.70	80.20
Flush floor receptacles				

BASIC MATERIALS	UNIT	MAT.	INST.	TOTAL
16110.40 UNDERFLOOR DUCT				
Aluminum	EA.	48.70	25.20	73.90
Brass	"	53.00	25.20	78.20
Flush floor telephone				
Aluminum	EA.	34.40	17.35	51.75
Brass	"	42.90	17.35	60.25
Super underfloor duct blank duct				
1/2"	L.F.	8.50	2.05	10.55
7/8"	"	8.70	2.05	10.75
1-3/8"	"	9.35	2.05	11.40
1-7/8"	"	9.95	2.05	12.00
Box opening plugs	EA.	2.05	5.05	7.10
End plugs	"	2.15	5.05	7.20
Conduit adapters	"	8.55	25.20	33.75
Sleeve coupling	"	6.40	15.40	21.80
Expansion coupling	"	42.70	15.40	58.10
Reducing coupling	"	29.40	15.40	44.80
Vertical elbow	"	30.20	15.40	45.60
Offset elbow	"	31.10	15.40	46.50
Horizontal elbow	"	60.50	15.40	75.90
Adjustable elbow	"	17.80	15.40	33.20
Cabinet connector	"	13.50	52.50	66.00
Super underfloor duct Y-take off	"	60.00	25.20	85.20
Leveling legs	"	2.50	5.05	7.55
Support couplers				
1 super	EA.	12.10	8.70	20.80
2 super	"	16.40	10.30	26.70
1 super, 1 standard	"	14.95	10.30	25.25
2 super, 2 standard	"	17.75	12.05	29.80
Single level junction boxes				
1 super	EA.	163.00	52.50	215.50
2 super	"	288.00	69.50	357.50
4 super	"	444.00	121.00	565.00
Double level junction boxes				
1 super	EA.	183.00	41.40	224.40
2 super	"	261.00	52.50	313.50
16110.50 WALL DUCT				
Lay-in wall duct, 10"	L.F.	19.90	2.05	21.95
Horizontal elbow	EA.	94.50	25.20	119.70
Edgewise elbow	"	50.50	25.20	75.70
Tee	"	94.50	31.20	125.70
Cross	"	88.00	38.00	126.00
Cabinet connector	"	56.50	55.50	112.00
Reverse elbow	"	50.50	18.50	69.00
Sweep elbow	"	92.00	25.20	117.20
Partition	"	17.15	2.05	19.20
Straight tunnel	"	17.60	9.55	27.15
Elbow tunnel	"	26.40	12.05	38.45
Tee kit	"	27.70	13.90	41.60
Ceiling dropout	"	103.00	34.70	137.70
Coupling device	"	5.80	5.05	10.85

16 ELECTRICAL

BASIC MATERIALS

	UNIT	MAT.	INST.	TOTAL
16110.50 WALL DUCT				
End cap	EA.	12.60	9.55	22.15
Lay-in wall duct, 18"	L.F.	26.60	2.80	29.40
Horizontal elbow	EA.	116.00	27.80	143.80
Edgeware elbow	"	66.50	34.70	101.20
Tee	"	113.00	38.00	151.00
Cross	"	116.00	43.40	159.40
Reverse elbow	"	68.00	25.20	93.20
Sweep elbow	"	113.00	27.80	140.80
Partition	"	17.60	3.45	21.05
Straight tunnel	"	21.40	13.90	35.30
Elbow tunnel	"	31.50	17.35	48.85
Tee kit	"	30.20	18.50	48.70
Ceiling dropout	"	104.00	38.00	142.00
Coupling device	"	8.70	6.95	15.65
Reducer coupling	"	26.90	13.90	40.80
Cabinet connector	"	32.70	69.50	102.20
End cap	"	17.60	13.90	31.50
16110.60 TRENCH DUCT				
Trench duct, with cover				
9"	L.F.	57.00	5.90	62.90
12"	"	63.50	6.95	70.45
18"	"	85.00	9.25	94.25
24"	"	108.00	12.05	120.05
30"	"	136.00	13.90	149.90
36"	"	161.00	19.85	180.85
Tees				
9"	EA.	216.00	60.50	276.50
12"	"	240.00	69.50	309.50
18"	"	308.00	77.00	385.00
24"	"	431.00	87.00	518.00
30"	"	584.00	103.00	687.00
36"	"	764.00	117.00	881.00
Vertical elbows				
9"	EA.	73.50	27.80	101.30
12"	"	77.50	38.00	115.50
18"	"	99.00	47.10	146.10
24"	"	107.00	58.00	165.00
30"	"	123.00	69.50	192.50
36"	"	136.00	87.00	223.00
Cabinet connectors				
9"	EA.	95.50	69.50	165.00
12"	"	107.00	73.00	180.00
18"	"	126.00	84.00	210.00
24"	"	154.00	87.00	241.00
30"	"	182.00	95.50	277.50
36"	"	201.00	103.00	304.00
End closers				
9"	EA.	22.85	21.35	44.20
12"	"	23.80	23.15	46.95
18"	"	35.60	27.80	63.40

16 ELECTRICAL

BASIC MATERIALS	UNIT	MAT.	INST.	TOTAL
16110.60 TRENCH DUCT				
24"	EA.	49.50	38.00	87.50
30"	"	57.00	44.80	101.80
36"	"	71.00	50.50	121.50
Horizontal elbows				
9"	EA.	216.00	52.50	268.50
12"	"	239.00	60.50	299.50
18"	"	309.00	73.00	382.00
24"	"	431.00	87.00	518.00
30"	"	584.00	99.00	683.00
36"	"	766.00	111.00	877.00
Crosses				
9"	EA.	344.00	69.50	413.50
12"	"	358.00	77.00	435.00
18"	"	431.00	87.00	518.00
24"	"	554.00	95.50	649.50
30"	"	648.00	111.00	759.00
36"	"	771.00	121.00	892.00
16110.80 WIREWAYS				
Wireway, hinge cover type				
2-1/2" x 2-1/2"				
1' section	EA.	7.70	5.35	13.05
2'	"	10.95	6.60	17.55
3'	"	14.90	8.70	23.60
5'	"	24.60	13.20	37.80
10'	"	51.00	23.15	74.15
4" x 4"				
1'	EA.	7.45	8.70	16.15
2'	"	12.40	8.70	21.10
3'	"	18.35	10.70	29.05
4'	"	24.95	10.70	35.65
10'	"	56.00	27.80	83.80
6" x 6"				
1'	EA.	16.20	13.20	29.40
2'	"	20.25	13.20	33.45
3'	"	28.70	15.40	44.10
4'	"	37.40	15.40	52.80
5'	"	39.90	19.85	59.75
10'	"	96.00	30.80	126.80
8" x 8"				
1'	EA.	26.20	15.40	41.60
2'	"	39.90	15.40	55.30
3'	"	60.00	17.35	77.35
4'	"	72.50	17.35	89.85
5'	"	81.00	21.35	102.35
12" x 12"				
1'	EA.	36.10	21.35	57.45
2'	"	70.00	21.35	91.35
3'	"	103.00	25.20	128.20
4'	"	124.00	25.20	149.20
5'	"	143.00	30.80	173.80

16 ELECTRICAL

BASIC MATERIALS

16110.80 WIREWAYS

	UNIT	MAT.	INST.	TOTAL
Fittings				
2-1/2" x 2-1/2"				
Drop hanger	EA.	4.50	4.25	8.75
Bracket hanger	"	6.45	4.25	10.70
Panel adapter	"	7.70	17.35	25.05
End plate	"	2.55	4.25	6.80
U-connector	"	2.55	4.25	6.80
Tee	"	21.20	6.95	28.15
Cross	"	22.90	8.70	31.60
90 degree elbow	"	18.05	6.95	25.00
Sweep elbow	"	31.20	6.95	38.15
45 degree elbow	"	18.00	6.95	24.95
Lay-in adapter	"	8.20	5.05	13.25
4" x 4"				
Drop hanger	EA.	5.75	5.05	10.80
Bracket hanger	"	7.70	5.05	12.75
Panel adapter	"	9.05	21.35	30.40
End plate	"	3.20	5.05	8.25
U-connector	"	3.40	5.05	8.45
Tee	"	24.60	8.70	33.30
Cross	"	28.00	12.05	40.05
90 degree elbow	"	20.60	8.70	29.30
Sweep elbow	"	35.60	8.70	44.30
45 degree elbow	"	22.00	8.70	30.70
Lay-in adapter	"	11.95	6.95	18.90
6" x 6"				
Drop hanger	EA.	10.35	5.05	15.40
Bracket hanger	"	11.00	5.05	16.05
Reducing bushing	"	22.90	5.05	27.95
Panel adapter	"	11.85	25.20	37.05
End plate	"	3.90	5.05	8.95
U-connector	"	4.35	5.05	9.40
Tee	"	28.00	8.70	36.70
Cross	"	34.70	12.05	46.75
90 degree elbow	"	22.90	8.70	31.60
Sweep elbow	"	51.50	8.70	60.20
45 degree elbow	"	22.90	8.70	31.60
Lay-in adapter	"	16.10	8.70	24.80
8" x 8"				
Drop hanger	EA.	13.55	5.05	18.60
Bracket hanger	"	11.65	5.05	16.70
Reducing bushing	"	27.10	5.05	32.15
Panel adapter	"	16.10	31.20	47.30
End plate	"	5.15	5.05	10.20
U-connector	"	6.45	5.05	11.50
Tee	"	51.50	13.90	65.40
Cross	"	55.00	15.40	70.40
90 degree elbow	"	37.30	13.90	51.20
Sweep elbow	"	33.10	13.90	47.00
45 degree elbow	"	37.30	13.90	51.20
Lay-in adapter	"	20.05	8.70	28.75
10" x 10"				

16 ELECTRICAL

BASIC MATERIALS	UNIT	MAT.	INST.	TOTAL
16110.80 WIREWAYS				
Drop hanger	EA.	13.65	5.05	18.70
Bracket hanger	"	13.90	5.05	18.95
Reducing bushing	"	38.60	5.05	43.65
Panel adapter	"	16.25	34.70	50.95
End plate	"	6.30	5.05	11.35
U-connector	"	6.55	5.05	11.60
Tee	"	50.00	15.40	65.40
Cross	"	74.00	17.35	91.35
90 degree elbow	"	42.10	15.40	57.50
Sweep elbow	"	40.90	15.40	56.30
45 degree elbow	"	42.20	15.40	57.60
Lay-in adapter	"	22.85	10.30	33.15
12" x 12"				
Drop hanger	EA.	38.10	5.05	43.15
Bracket hanger	"	26.10	5.05	31.15
Reducing bushing	"	39.00	5.05	44.05
Panel adapter	"	27.10	43.40	70.50
End plate	"	11.60	5.05	16.65
U-connector	"	15.40	21.35	36.75
Tee	"	102.00	185.00	287.00
Cross	"	103.00	21.35	124.35
90 degree elbow	"	68.00	21.35	89.35
Sweep elbow	"	95.50	21.35	116.85
45 degree elbow	"	100.00	21.35	121.35
Lay-in adapter	"	25.10	18.50	43.60
Raintight wireway, 4" x 4"				
1' section	EA.	19.05	13.90	32.95
5'	"	45.00	13.90	58.90
10'	"	81.50	34.70	116.20
Fittings				
90 degree elbow	EA.	32.80	13.90	46.70
Tee	"	36.00	15.40	51.40
Cross	"	61.50	17.35	78.85
Panel adapter	"	15.35	25.20	40.55
End plate	"	7.95	5.05	13.00
Gusset bracket	"	5.80	5.05	10.85
6" x 6"				
1' section	EA.	27.50	17.35	44.85
5'	"	65.00	25.20	90.20
10'	"	116.00	52.50	168.50
Fittings				
90 degree elbow	EA.	41.30	17.35	58.65
Tee	"	50.50	17.35	67.85
Cross	"	71.50	25.20	96.70
Panel adapter	"	20.30	34.70	55.00
End plate	"	9.00	5.05	14.05
Gusset bracket	"	7.40	5.05	12.45
16120.41 ALUMINUM CONDUCTORS				
Type XHHW, stranded aluminum, 600v				
#8	L.F.	0.12	0.17	0.29

BASIC MATERIALS

16120.41 ALUMINUM CONDUCTORS	UNIT	MAT.	INST.	TOTAL
#6	L.F.	0.18	0.21	0.39
#4	"	0.23	0.28	0.51
#2	"	0.29	0.31	0.60
1/0	"	0.46	0.38	0.84
2/0	"	0.55	0.42	0.97
3/0	"	0.66	0.49	1.15
4/0	"	0.80	0.52	1.32
300 MCM	"	1.15	0.69	1.84
350 MCM	"	1.25	0.79	2.04
400 MCM	"	1.45	0.97	2.42
500 MCM	"	1.75	1.15	2.90
600 MCM	"	2.00	1.40	3.40
700 MCM	"	2.15	1.65	3.80
750 MCM	"	2.90	1.80	4.70
THW, stranded				
#8	L.F.	0.10	0.17	0.27
#6	"	0.11	0.21	0.32
#4	"	0.14	0.28	0.42
#3	"	0.20	0.31	0.51
#1	"	0.29	0.35	0.64
1/0	"	0.32	0.38	0.70
2/0	"	0.37	0.41	0.78
3/0	"	0.46	0.41	0.87
4/0	"	0.53	0.52	1.05
250 MCM	"	0.63	0.62	1.25
300 MCM	"	0.83	0.69	1.52
350 MCM	"	0.89	0.79	1.68
400 MCM	"	0.98	0.97	1.95
500 MCM	"	1.15	1.15	2.30
600 MCM	"	1.40	1.40	2.80
700 MCM	"	1.60	1.65	3.25
750 MCM	"	1.70	1.80	3.50
XLP, stranded				
#6	L.F.	0.18	0.17	0.35
#4	"	0.23	0.28	0.51
#2	"	0.29	0.31	0.60
#1	"	0.40	0.35	0.75
1/0	"	0.46	0.38	0.84
2/0	"	0.55	0.42	0.97
3/0	"	0.66	0.49	1.15
4/0	"	0.80	0.52	1.32
250 MCM	"	0.98	0.56	1.54
300 MCM	"	1.15	0.69	1.84
350 MCM	"	1.25	0.79	2.04
400 MCM	"	1.45	0.97	2.42
500 MCM	"	1.75	1.15	2.90
600 MCM	"	2.00	1.40	3.40
700 MCM	"	2.15	1.65	3.80
750 MCM	"	2.90	1.80	4.70
1000 MCM	"	3.75	2.00	5.75
Bare stranded aluminum wire				
#4	L.F.	0.22	0.28	0.50

BASIC MATERIALS	UNIT	MAT.	INST.	TOTAL
16120.41 ALUMINUM CONDUCTORS				
#2	L.F.	0.23	0.31	0.54
1/0	"	0.29	0.38	0.67
2/0	"	0.33	0.42	0.75
3/0	"	0.36	0.49	0.85
4/0	"	0.40	0.52	0.92
Triplex XLP cable				
#4	L.F.	0.48	0.52	1.00
#2	"	0.62	0.69	1.31
1/0	"	0.98	1.05	2.03
4/0	"	1.80	1.70	3.50
Aluminum quadruplex XLP cable				
#4	L.F.	0.62	0.62	1.24
#2	"	0.81	0.79	1.60
1/0	"	1.35	2.65	4.00
2/0	"	1.60	1.45	3.05
4/0	"	2.35	2.20	4.55
Triplexed URD-XLP cable				
#6	L.F.	0.42	0.38	0.80
#4	"	0.57	0.49	1.06
#2	"	0.75	0.62	1.37
1/0	"	1.20	0.97	2.17
2/0	"	1.35	1.15	2.50
3/0	"	1.65	1.40	3.05
4/0	"	1.90	1.65	3.55
250 MCM	"	2.20	1.90	4.10
350 MCM	"	2.85	2.00	4.85
Type S.E.U. cable				
#8/3	L.F.	0.62	0.87	1.49
#6/3	"	0.62	0.97	1.59
#4/3	"	0.83	1.20	2.03
#2/3	"	1.10	1.30	2.40
#1/3	"	1.45	1.40	2.85
1/0-3	"	1.70	1.45	3.15
2/0-3	"	1.90	1.55	3.45
3/0-3	"	2.45	1.80	4.25
4/0-3	"	2.65	2.00	4.65
Type S.E.R. cable with ground				
#8/3	L.F.	0.75	0.97	1.72
#6/3	"	0.88	1.20	2.08
#4/3	"	1.00	1.30	2.30
#2/3	"	1.50	1.40	2.90
#1/3	"	1.90	1.55	3.45
1/0-3	"	2.25	1.75	4.00
2/0-3	"	2.60	1.90	4.50
3/0-3	"	3.10	2.05	5.15
4/0-3	"	3.70	2.30	6.00
#6/4	"	1.45	1.30	2.75
#4/4	"	1.70	1.55	3.25
#2/4	"	2.55	1.55	4.10
#1/4	"	3.15	1.75	4.90
1/0-4	"	3.80	1.80	5.60
2/0-4	"	4.40	2.00	6.40

BASIC MATERIALS

BASIC MATERIALS	UNIT	MAT.	INST.	TOTAL
16120.41 ALUMINUM CONDUCTORS				
3/0-4	L.F.	5.25	2.20	7.45
4/0-4	"	6.40	2.65	9.05
16120.43 COPPER CONDUCTORS				
Copper conductors, type THW, solid				
#14	L.F.	0.04	0.14	0.18
#12	"	0.07	0.17	0.24
#10	"	0.09	0.21	0.30
Stranded				
#14	L.F.	0.07	0.14	0.21
#12	"	0.09	0.17	0.26
#10	"	0.11	0.21	0.32
#8	"	0.18	0.28	0.46
#6	"	0.23	0.31	0.54
#4	"	0.35	0.35	0.70
#3	"	0.50	0.35	0.85
#2	"	0.61	0.42	1.03
#1	"	0.77	0.49	1.26
1/0	"	0.94	0.56	1.50
2/0	"	1.10	0.69	1.79
3/0	"	1.35	0.87	2.22
4/0	"	1.70	0.97	2.67
250 MCM	"	2.10	1.05	3.15
300 MCM	"	2.50	1.15	3.65
350 MCM	"	3.05	1.40	4.45
400 MCM	"	3.40	1.55	4.95
500 MCM	"	4.10	1.80	5.90
600 MCM	"	5.80	2.05	7.85
750 MCM	"	7.45	2.30	9.75
1000 MCM	"	8.25	2.65	10.90
THHN-THWN, solid				
#14	L.F.	0.04	0.14	0.18
#12	"	0.07	0.17	0.24
#10	"	0.09	0.21	0.30
Stranded				
#14	L.F.	0.07	0.14	0.21
#12	"	0.09	0.17	0.26
#10	"	0.11	0.21	0.32
#8	"	0.18	0.28	0.46
#6	"	0.23	0.31	0.54
#4	"	0.35	0.35	0.70
#2	"	0.61	0.42	1.03
#1	"	0.77	0.49	1.26
1/0	"	0.94	0.56	1.50
2/0	"	1.10	0.69	1.79
3/0	"	1.35	0.87	2.22
4/0	"	1.70	0.97	2.67
250 MCM	"	2.10	1.05	3.15
350 MCM	"	3.05	1.40	4.45
XHHW				
#14	L.F.	0.10	0.14	0.24
#10	"	0.17	0.21	0.38

BASIC MATERIALS

16120.43 COPPER CONDUCTORS

BASIC MATERIALS	UNIT	MAT.	INST.	TOTAL
#8	L.F.	0.23	0.28	0.51
#6	"	0.25	0.31	0.56
#4	"	0.38	0.31	0.69
#2	"	0.61	0.38	0.99
#1	"	0.79	0.49	1.28
1/0	"	0.94	0.56	1.50
2/0	"	1.20	0.66	1.86
3/0	"	1.45	0.87	2.32
XLP, 600v				
#12	L.F.	0.11	0.17	0.28
#10	"	0.17	0.21	0.38
#8	"	0.23	0.28	0.51
#6	"	0.28	0.31	0.59
#4	"	0.41	0.35	0.76
#3	"	0.50	0.38	0.88
#2	"	0.63	0.42	1.05
#1	"	0.83	0.49	1.32
1/0	"	0.99	0.56	1.55
2/0	"	1.20	0.69	1.89
3/0	"	1.50	0.90	2.40
4/0	"	1.75	0.97	2.72
250 MCM	"	2.20	1.05	3.25
300 MCM	"	2.60	1.15	3.75
350 MCM	"	2.95	1.35	4.30
400 MCM	"	3.50	1.55	5.05
500 MCM	"	4.10	1.80	5.90
600 MCM	"	5.95	2.05	8.00
750 MCM	"	7.15	2.30	9.45
1000 MCM	"	10.00	2.65	12.65
Bare solid wire				
#14	L.F.	0.06	0.14	0.20
#12	"	0.08	0.17	0.25
#10	"	0.10	0.21	0.31
#8	"	0.18	0.28	0.46
#6	"	0.23	0.31	0.54
#4	"	0.35	0.35	0.70
#2	"	0.57	0.42	0.99
Bare stranded wire				
#8	L.F.	0.18	0.28	0.46
#6	"	0.24	0.35	0.59
#4	"	0.35	0.35	0.70
#2	"	0.59	0.38	0.97
#1	"	0.76	0.49	1.25
1/0	"	0.92	0.62	1.54
2/0	"	1.10	0.69	1.79
3/0	"	1.35	0.87	2.22
4/0	"	1.70	0.97	2.67
250 MCM	"	2.05	1.05	3.10
300 MCM	"	2.40	1.15	3.55
350 MCM	"	2.80	1.40	4.20
400 MCM	"	3.35	1.55	4.90
500 MCM	"	4.00	1.80	5.80

BASIC MATERIALS

	UNIT	MAT.	INST.	TOTAL
16120.43 COPPER CONDUCTORS				
Type "BX" solid armored cable				
#14/2	L.F.	0.32	0.87	1.19
#14/3	"	0.40	0.97	1.37
#14/4	"	0.54	1.05	1.59
#12/2	"	0.34	0.97	1.31
#12/3	"	0.50	1.05	1.55
#12/4	"	0.70	1.20	1.90
#10/2	"	0.58	1.05	1.63
#10/3	"	0.75	1.20	1.95
#10/4	"	1.15	1.40	2.55
#8/2	"	0.94	1.20	2.14
#8/3	"	1.20	1.40	2.60
Steel type, metal clad cable, solid, with ground				
#14/2	L.F.	0.34	0.62	0.96
#14/3	"	0.46	0.69	1.15
#14/4	"	0.65	0.79	1.44
#12/2	"	0.41	0.69	1.10
#12/3	"	0.57	0.87	1.44
#12/4	"	0.76	1.05	1.81
#10/2	"	0.67	0.79	1.46
#10/3	"	1.00	0.97	1.97
#10/4	"	1.35	1.15	2.50
Metal clad cable, stranded, with ground				
#8/2	L.F.	1.10	0.97	2.07
#8/3	"	1.60	1.20	2.80
#8/4	"	2.05	1.45	3.50
#6/2	"	1.65	1.05	2.70
#6/3	"	2.00	1.30	3.30
#6/4	"	2.40	1.55	3.95
#4/2	"	2.20	1.40	3.60
#4/3	"	2.80	1.55	4.35
#4/4	"	3.30	1.90	5.20
#3/3	"	3.45	1.75	5.20
#3/4	"	4.10	2.05	6.15
#2/3	"	4.10	2.00	6.10
#2/4	"	5.05	2.30	7.35
#1/3	"	4.95	2.65	7.60
#4	"	1.60	1.20	2.80
16120.45 FLAT CONDUCTOR CABLE				
Flat conductor cable, with shield, 3 conductor				
#12 awg	L.F.	3.55	2.05	5.60
#10 awg	"	4.15	2.05	6.20
4 conductor				
#12 awg	L.F.	4.80	2.80	7.60
#10 awg	"	5.40	2.80	8.20
Transition boxes				
#12 awg	L.F.	5.95	3.10	9.05
#10 awg	"	6.70	3.10	9.80
Flat conductor cable communication, with shield				
10 conductor	L.F.	2.35	2.05	4.40
16 conductor	"	2.70	2.40	5.10

BASIC MATERIALS

BASIC MATERIALS	UNIT	MAT.	INST.	TOTAL
16120.45 FLAT CONDUCTOR CABLE				
24 conductor	L.F.	3.00	3.45	6.45
Power and communication heads, duplex receptacle	EA.	36.00	27.80	63.80
Double duplex receptacle	"	42.00	33.00	75.00
Telephone	"	24.00	27.80	51.80
Receptacle and telephone	"	60.00	33.00	93.00
Blank cover	"	6.00	5.05	11.05
Transition boxes				
Surface	EA.	108.00	25.20	133.20
Flush	"	66.00	34.70	100.70
Flat conductor cable fittings				
End caps	EA.	1.15	5.05	6.20
Insulators	"	12.55	10.30	22.85
Splice connectors	"	0.89	15.40	16.29
Tap connectors	"	0.96	15.40	16.36
Cable connectors	"	1.05	15.40	16.45
Terminal blocks	"	7.30	21.35	28.65
Tape	"			11.20
16120.47 SHEATHED CABLE				
Non-metallic sheathed cable				
Type NM cable with ground				
#14/2	L.F.	0.12	0.52	0.64
#12/2	"	0.18	0.56	0.74
#10/2	"	0.29	0.62	0.91
#8/2	"	0.59	0.69	1.28
#6/2	"	0.88	0.87	1.75
#14/3	"	0.20	0.90	1.10
#12/3	"	0.29	0.93	1.22
#10/3	"	0.43	0.94	1.37
#8/3	"	0.91	0.96	1.87
#6/3	"	1.25	0.97	2.22
#4/3	"	1.85	1.10	2.95
#2/3	"	2.75	1.20	3.95
Type U.F. cable with ground				
#14/2	L.F.	0.15	0.56	0.71
#12/2	"	0.21	0.66	0.87
#10/2	"	0.31	0.69	1.00
#8/2	"	0.91	0.79	1.70
#6/2	"	1.10	0.94	2.04
#14/3	"	0.22	0.69	0.91
#12/3	"	0.31	0.76	1.07
#10/3	"	0.44	0.87	1.31
#8/3	"	1.20	0.97	2.17
#6/3	"	1.70	1.10	2.80
Type S.F.U. cable, 3 conductor				
#8	L.F.	0.80	0.97	1.77
#6	"	1.10	1.05	2.15
#3	"	2.05	1.40	3.45
#2	"	2.55	1.55	4.10
#1	"	3.90	1.75	5.65
#1/0	"	4.40	1.90	6.30

16 ELECTRICAL

BASIC MATERIALS	UNIT	MAT.	INST.	TOTAL
16120.47 SHEATHED CABLE				
#2/0	L.F.	5.20	2.20	7.40
#3/0	"	6.40	2.40	8.80
#4/0	"	7.15	2.65	9.80
Type SER cable, 4 conductor				
#6	L.F.	1.45	1.25	2.70
#4	"	2.25	1.35	3.60
#3	"	2.60	1.55	4.15
#2	"	3.45	1.70	5.15
#1	"	4.75	1.90	6.65
#1/0	"	6.25	2.20	8.45
#2/0	"	7.15	2.30	9.45
#3/0	"	8.25	2.65	10.90
#4/0	"	0.62	2.90	3.52
Flexible cord, type STO cord				
#18/2	L.F.	0.63	0.14	0.77
#18/3	"	0.74	0.17	0.91
#18/4	"	1.00	0.21	1.21
#16/2	"	0.71	0.14	0.85
#16/3	"	0.86	0.15	1.01
#16/4	"	1.15	0.17	1.32
#14/2	"	1.15	0.17	1.32
#14/3	"	1.45	0.21	1.66
#14/4	"	1.70	0.24	1.94
#12/2	"	1.55	0.21	1.76
#12/3	"	1.85	0.23	2.08
#12/4	"	2.10	0.28	2.38
#10/2	"	1.90	0.24	2.14
#10/3	"	2.40	0.28	2.68
#10/4	"	2.65	0.31	2.96
#8/2	"	3.20	0.28	3.48
#8/3	"	3.55	0.31	3.86
#8/4	"	4.95	0.35	5.30
Flexible cord, copper, type STO cord				
16130.10 FLOOR BOXES				
Adjustable floor boxes, steel	EA.	22.45	18.50	40.95
Cast bronze round	"	33.00	25.20	58.20
1 gang	"	38.30	27.80	66.10
2 gang	"	72.50	33.00	105.50
3 gang	"	103.00	34.70	137.70
Aluminum round	"	51.50	25.20	76.70
1 gang	"	43.60	27.80	71.40
2 gang	"	78.00	33.00	111.00
3 gang	"	104.00	34.70	138.70
Steel plate single recept	"	14.50	5.05	19.55
Duplex receptacle	"	13.85	6.30	20.15
Twist lock receptacle	"	14.70	6.30	21.00
Plug, 3/4"	"	18.50	5.05	23.55
1" plug	"	17.15	5.05	22.20
Carpet flange	"	22.45	5.05	27.50
Adjustable bronze plates for round cast boxes				

BASIC MATERIALS	UNIT	MAT.	INST.	TOTAL
16130.10 FLOOR BOXES				
1/2" plug	EA.	19.80	5.05	24.85
3/4" plug	"	20.10	5.05	25.15
1" plug	"	20.40	5.05	25.45
1-1/4" plug	"	23.75	6.30	30.05
2" plug	"	24.50	6.95	31.45
Combination plug	"	20.40	6.95	27.35
Duplex receptacle plug	"	26.40	6.95	33.35
Adjustable aluminum plates for round cast boxes				
1/2" plug	EA.	22.45	5.05	27.50
3/4" plug	"	22.80	5.05	27.85
1" plug	"	23.40	5.05	28.45
1-1/4" plug	"	25.10	6.30	31.40
2" plug	"	25.30	6.95	32.25
Combination plug	"	22.75	6.95	29.70
Duplex receptacle plug	"	38.30	6.95	45.25
Adjustable bronze plates for gang type boxes				
1/2" plug	EA.	23.60	5.05	28.65
3/4" plug	"	23.95	5.05	29.00
1" plug	"	24.20	5.05	29.25
1-1/4" plug	"	25.30	6.30	31.60
2" plug	"	26.40	6.95	33.35
Carpet plate				
1 gang	EA.	20.85	5.05	25.90
2 gang	"	31.70	5.05	36.75
3 gang	"	42.30	6.95	49.25
Adjustable aluminum plates for gang type boxes				
1/2" plug	EA.	21.10	5.05	26.15
3/4" plug	"	21.60	5.05	26.65
1" plug	"	21.90	5.05	26.95
1-1/4" plug	"	25.10	6.30	31.40
2" plug	"	25.80	6.95	32.75
Duplex recept	"	21.25	6.95	28.20
Carpet plate				
1 gang	EA.	35.70	5.05	40.75
2 gang	"	57.00	5.05	62.05
3 gang	"	83.00	6.95	89.95
4 gang carpet plate	"	34.30	19.85	54.15
Telephone	"	31.20	17.35	48.55
Floor box nozzles, horizontal				
Duplex recept	EA.	95.00	18.50	113.50
Single recept	"	94.00	18.50	112.50
Double duplex recept	"	102.00	25.20	127.20
Vertical with duplex recept	"	95.00	21.35	116.35
Double duplex recept	"	99.50	25.20	124.70
Floor box bell nozzles split bell	"	17.15	8.70	25.85
One piece bell	"	54.00	8.70	62.70
Floor box standpipe				
1/2" x 3"	EA.	14.90	5.05	19.95
1/2" x 1"	"	14.70	5.05	19.75
Poke thru floor outlets				
2" floor	EA.	46.20	34.70	80.90
3" floor	"	50.00	41.40	91.40

BASIC MATERIALS	UNIT	MAT.	INST.	TOTAL
16130.10 FLOOR BOXES				
4" floor	EA.	51.00	44.80	95.80
7" floor	"	54.00	52.50	106.50
9" floor	"	76.50	55.50	132.00
11" floor	"	79.50	63.00	142.50
13" floor	"	81.50	69.50	151.00
16130.40 BOXES				
Round cast box, type SEH				
1/2"	EA.	13.20	12.05	25.25
3/4"	"	14.40	14.60	29.00
SEHC				
1/2"	EA.	16.60	12.05	28.65
3/4"	"	17.40	14.60	32.00
SEHL				
1/2"	EA.	20.45	12.05	32.50
3/4"	"	21.60	15.40	37.00
SEHT				
1/2"	EA.	23.30	14.60	37.90
3/4"	"	24.00	17.35	41.35
SEHX				
1/2"	EA.	25.20	17.35	42.55
3/4"	"	26.20	21.35	47.55
Blank cover	"	3.30	5.05	8.35
1/2", hub cover	"	2.50	5.05	7.55
Cover with gasket	"	2.75	6.15	8.90
Rectangle, type FS boxes				
1/2"	EA.	6.60	12.05	18.65
3/4"	"	7.00	13.90	20.90
1"	"	7.55	17.35	24.90
FSA				
1/2"	EA.	6.70	12.05	18.75
3/4"	"	7.60	13.90	21.50
FSC				
1/2"	EA.	7.35	12.05	19.40
3/4"	"	8.00	14.60	22.60
1"	"	10.00	17.35	27.35
FSL				
1/2"	EA.	8.10	12.05	20.15
3/4"	"	8.35	13.90	22.25
FSR				
1/2"	EA.	7.95	12.05	20.00
3/4"	"	10.10	13.90	24.00
FSS				
1/2"	EA.	7.05	12.05	19.10
3/4"	"	7.75	13.90	21.65
FSLA				
1/2"	EA.	8.10	12.05	20.15
3/4"	"	9.15	13.90	23.05
FSCA				
1/2"	EA.	10.75	12.05	22.80
3/4"	"	12.30	13.90	26.20

BASIC MATERIALS

16130.40 BOXES

BASIC MATERIALS	UNIT	MAT.	INST.	TOTAL
FSCC				
1/2"	EA.	8.60	13.90	22.50
3/4"	"	11.45	17.35	28.80
FSCT				
1/2"	EA.	9.00	13.90	22.90
3/4"	"	11.20	17.35	28.55
1"	"	14.15	19.85	34.00
FST				
1/2"	EA.	11.65	17.35	29.00
3/4"	"	13.10	19.85	32.95
FSX				
1/2"	EA.	11.80	21.35	33.15
3/4"	"	13.75	25.20	38.95
FSCD boxes				
1/2"	EA.	13.05	21.35	34.40
3/4"	"	13.80	25.20	39.00
Rectangle, type FS, 2 gang boxes				
1/2"	EA.	12.00	12.05	24.05
3/4"	"	12.90	13.90	26.80
1"	"	14.15	17.35	31.50
FSC, 2 gang boxes				
1/2"	EA.	13.10	12.05	25.15
3/4"	"	13.30	13.90	27.20
1"	"	16.60	17.35	33.95
FSS, 2 gang boxes				
3/4"	EA.	13.75	13.90	27.65
FS, tandem boxes				
1/2"	EA.	15.30	13.90	29.20
3/4"	"	17.10	15.40	32.50
FSC, tandem boxes				
1/2"	EA.	17.00	13.90	30.90
3/4"	"	18.00	15.40	33.40
FS, three gang boxes				
3/4"	EA.	18.45	15.40	33.85
1"	"	20.35	17.35	37.70
FSS, three gang boxes, 3/4"	"	23.80	17.35	41.15
Weatherproof cast aluminum boxes, 1 gang, 3 outlets				
1/2"	EA.	3.40	13.90	17.30
3/4"	"	3.70	17.35	21.05
2 gang, 3 outlets				
1/2"	EA.	6.55	17.35	23.90
3/4"	"	6.85	18.50	25.35
1 gang, 4 outlets				
1/2"	EA.	4.20	21.35	25.55
3/4"	"	4.50	25.20	29.70
2 gang, 4 outlets				
1/2"	EA.	7.10	21.35	28.45
3/4"	"	7.45	25.20	32.65
1 gang, 5 outlets				
1/2"	EA.	4.85	25.20	30.05
3/4"	"	5.85	27.80	33.65
2 gang, 5 outlets				

BASIC MATERIALS

16130.40 BOXES	UNIT	MAT.	INST.	TOTAL
1/2"	EA.	8.75	25.20	33.95
3/4"	"	10.80	27.80	38.60
2 gang, 6 outlets				
1/2"	EA.	10.10	29.50	39.60
3/4"	"	10.80	31.20	42.00
2 gang, 7 outlets				
1/2"	EA.	10.35	34.70	45.05
3/4"	"	11.00	38.00	49.00
Weatherproof and type FS box covers, blank, 1 gang	"	1.50	5.05	6.55
Tumbler switch, 1 gang	"	1.80	5.05	6.85
1 gang, single recept	"	1.90	5.05	6.95
Duplex recept	"	1.80	5.05	6.85
Despard	"	1.75	5.05	6.80
Red pilot light	"	11.55	5.05	16.60
SW and				
Single recept	EA.	5.40	6.95	12.35
Duplex recept	"	4.45	6.95	11.40
2 gang				
Blank	EA.	2.00	6.30	8.30
Tumbler switch	"	2.75	6.30	9.05
Single recept	"	3.05	6.30	9.35
Duplex recept	"	3.30	6.30	9.60
3 gang				
Blank	EA.	3.70	6.95	10.65
Tumbler switch	"	4.60	6.95	11.55
4 gang				
Tumbler switch	EA.	5.85	8.70	14.55
Explosion proof boxes type E				
1/2"	EA.	18.35	12.05	30.40
3/4"	"	17.65	13.90	31.55
1"	"	18.25	17.35	35.60
1-1/4"	"	31.60	19.85	51.45
1-1/2"	"	72.00	21.35	93.35
Type L.B.				
1/2"	EA.	17.50	13.90	31.40
3/4"	"	18.60	17.35	35.95
1"	"	19.40	19.85	39.25
1-1/4"	"	34.10	23.15	57.25
1-1/2"	"	68.00	25.20	93.20
2"	"	76.50	27.80	104.30
Type C				
1/2"	EA.	18.35	13.90	32.25
3/4"	"	19.65	17.35	37.00
1"	"	20.25	19.85	40.10
1-1/4"	"	34.10	23.15	57.25
1-1/2"	"	71.00	25.20	96.20
2"	"	80.00	27.80	107.80
Type CA				
1/2"	EA.	18.30	19.85	38.15
3/4"	"	20.80	25.20	46.00
Type L				
1/2"	EA.	19.10	13.90	33.00

BASIC MATERIALS	UNIT	MAT.	INST.	TOTAL
16130.40 BOXES				
3/4"	EA.	20.25	17.35	37.60
1"	"	21.20	19.85	41.05
1-1/4"	"	35.70	23.15	58.85
1-1/2"	"	70.00	25.20	95.20
2"	"	73.00	27.80	100.80
Type N				
1/2"	EA.	19.40	13.90	33.30
3/4"	"	20.20	17.35	37.55
1"	"	20.80	21.35	42.15
1-1/4"	"	36.30	23.15	59.45
Type T				
1/2"	EA.	18.60	18.50	37.10
3/4"	"	19.75	25.20	44.95
1"	"	20.45	29.50	49.95
1-1/4"	"	34.80	34.70	69.50
1-1/2"	"	71.50	40.20	111.70
2"	"	74.00	44.80	118.80
Type TA				
1/2"	EA.	21.40	25.20	46.60
3/4"	"	22.60	27.80	50.40
Type X				
1/2"	EA.	19.40	25.20	44.60
3/4"	"	20.75	29.50	50.25
1"	"	22.70	34.70	57.40
1-1/4"	"	40.90	40.20	81.10
1-1/2"	"	77.00	44.80	121.80
2"	"	86.00	50.50	136.50
With union hubs				
1/2"	EA.	40.00	25.20	65.20
3/4"	"	41.60	27.80	69.40
Box covers				
Surface	EA.	8.55	6.95	15.50
Sealing	"	9.35	6.95	16.30
Dome	"	12.90	6.95	19.85
1/2" nipple	"	16.45	6.95	23.40
3/4" nipple	"	16.95	6.95	23.90
16130.45 EXPLOSION PROOF FITTINGS				
Flexible couplings with female unions				
1/2" x 18"	EA.	74.00	6.95	80.95
3/4" x 18"	"	93.50	9.55	103.05
1" x 18"	"	156.00	12.05	168.05
1-1/4" x 18"	"	273.00	14.60	287.60
1-1/2" x 18"	"	338.00	17.35	355.35
2" x 18"	"	446.00	19.85	465.85
1/2" x 24"	"	87.00	8.70	95.70
3/4" x 24"	"	111.00	10.30	121.30
1" x 24"	"	200.00	13.90	213.90
1-1/4" x 24"	"	314.00	15.40	329.40
1-1/2" x 24"	"	383.00	19.85	402.85
2" x 24"	"	547.00	21.35	568.35
Female seal-offs				

BASIC MATERIALS

16130.45 EXPLOSION PROOF FITTINGS

	UNIT	MAT.	INST.	TOTAL
1/2"	EA.	5.65	19.85	25.50
3/4"	"	6.65	23.15	29.80
1"	"	8.60	25.20	33.80
1-1/4"	"	10.40	29.50	39.90
1-1/2"	"	15.75	34.70	50.45
2"	"	20.45	40.20	60.65
2-1/2"	"	31.40	60.50	91.90
3"	"	41.60	75.00	116.60
4"	"	160.00	92.50	252.50
Conduit plugs				
1/2"	EA.	0.83	5.05	5.88
3/4"	"	0.95	5.05	6.00
1"	"	1.15	5.05	6.20
1-1/4"	"	1.45	8.70	10.15
1-1/2"	"	1.80	8.70	10.50
2"	"	3.15	10.30	13.45
2-1/2"	"	5.35	10.30	15.65
3"	"	7.60	12.05	19.65
4"	"	9.60	12.05	21.65
Sealing cement				
1 pound	EA.			5.10
5 pound	"			14.60
Fibre				
1 ounce	EA.			3.75
8 ounce	"			12.65
Male unions				
1/2"	EA.	4.40	6.95	11.35
3/4"	"	6.25	8.40	14.65
1"	"	10.60	9.55	20.15
1-1/4"	"	19.45	10.30	29.75
1-1/2"	"	21.10	12.05	33.15
2"	"	27.20	14.60	41.80
2-1/2"	"	46.00	17.35	63.35
3"	"	60.00	25.20	85.20
4"	"	95.50	31.20	126.70
Female unions				
1/2"	EA.	4.05	6.95	11.00
3/4"	"	5.75	8.40	14.15
1"	"	10.05	9.55	19.60
1-1/4"	"	15.10	10.30	25.40
1-1/2"	"	19.40	12.05	31.45
2"	"	24.85	14.60	39.45
2-1/2"	"	37.50	17.35	54.85
3"	"	51.50	25.20	76.70
4"	"	81.00	31.20	112.20
Male elbows				
1/2"	EA.	5.40	8.70	14.10
3/4"	"	6.10	10.30	16.40
1"	"	8.90	12.05	20.95
1-1/4"	"	11.65	15.40	27.05
Female elbows				
1/2"	EA.	3.65	8.70	12.35

BASIC MATERIALS

BASIC MATERIALS	UNIT	MAT.	INST.	TOTAL
16130.45 EXPLOSION PROOF FITTINGS				
3/4"	EA.	4.05	10.30	14.35
1"	"	5.45	12.05	17.50
1-1/4"	"	8.90	15.40	24.30
Pulling elbows				
1/2"	EA.	25.70	12.05	37.75
3/4"	"	27.40	15.40	42.80
1"	"	69.00	17.35	86.35
1-1/4"	"	76.00	21.35	97.35
1-1/2"	"	106.00	25.20	131.20
2"	"	110.00	66.00	176.00
2-1/2"	"	243.00	87.00	330.00
3"	"	233.00	103.00	336.00
3-1/2"	"	437.00	121.00	558.00
4"	"	428.00	146.00	574.00
Male expansion couplings				
1/2"	EA.	10.25	8.70	18.95
3/4"	"	13.40	10.30	23.70
1"	"	25.00	15.40	40.40
Female expansion couplings				
1/2"	EA.	9.20	8.70	17.90
3/4"	"	14.45	10.30	24.75
1"	"	25.00	15.40	40.40
16130.60 PULL AND JUNCTION BOXES				
4"				
Octagon box	EA.	1.05	3.95	5.00
Box extension	"	1.55	2.05	3.60
Plaster ring	"	0.83	2.05	2.88
Cover blank	"	0.35	2.05	2.40
Square box	"	1.30	3.95	5.25
Box extension	"	1.40	2.05	3.45
Plaster ring	"	0.65	2.05	2.70
Cover blank	"	0.41	2.05	2.46
4-11/16"				
Square box	EA.	2.30	3.95	6.25
Box extension	"	3.10	2.05	5.15
Plaster ring	"	1.80	2.05	3.85
Cover blank	"	0.70	2.05	2.75
Switch and device boxes				
2 gang	EA.	6.65	3.95	10.60
3 gang	"	8.20	3.95	12.15
4 gang	"	11.60	5.55	17.15
Device covers				
2 gang	EA.	3.25	2.05	5.30
3 gang	"	3.85	2.05	5.90
4 gang	"	5.60	2.05	7.65
Handy box	"	1.00	3.95	4.95
Extension	"	1.35	2.05	3.40
Switch cover	"	0.41	2.05	2.46
Switch box with knockout	"	1.10	5.05	6.15
Weatherproof cover, spring type	"	3.85	2.80	6.65
Cover plate, dryer receptacle 1 gang plastic	"	1.80	3.45	5.25

16 ELECTRICAL

BASIC MATERIALS

16130.60 PULL AND JUNCTION BOXES

	UNIT	MAT.	INST.	TOTAL
For 4" receptacle, 2 gang	EA.	3.10	3.45	6.55
Duplex receptacle cover plate, plastic	"	0.35	2.05	2.40
4", vertical bracket box, 1-1/2" with				
RMX clamps	EA.	1.60	5.05	6.65
BX clamps	"	2.45	5.05	7.50
4", octagon device cover				
1 switch	EA.	1.10	2.05	3.15
1 duplex recept	"	1.05	2.05	3.10
4", octagon swivel hanger box, 1/2" hub	"	3.80	2.05	5.85
3/4" hub	"	4.15	2.05	6.20
4" octagon adjustable bar hangers				
18-1/2"	EA.	1.30	1.75	3.05
26-1/2"	"	1.55	1.75	3.30
With clip				
18-1/2"	EA.	1.40	1.75	3.15
26-1/2"	"	1.35	1.75	3.10
4", square face bracket boxes, 1-1/2"				
RMX	EA.	2.65	5.05	7.70
BX	"	2.85	5.05	7.90
4" square to round plaster rings	"	1.05	2.05	3.10
2 gang device plaster rings	"	1.30	2.05	3.35
Surface covers				
1 gang switch	EA.	1.25	2.05	3.30
2 gang switch	"	1.30	2.05	3.35
1 single recept	"	1.25	2.05	3.30
1 20a twist lock recept	"	1.55	2.05	3.60
1 30a twist lock recept	"	1.60	2.05	3.65
1 duplex recept	"	0.95	2.05	3.00
2 duplex recept	"	1.10	2.05	3.15
Switch and duplex recept	"	1.35	2.05	3.40
4-11/16" square to round plaster rings	"	2.85	2.05	4.90
2 gang device plaster rings	"	2.25	2.05	4.30
Surface covers				
1 gang switch	EA.	3.45	2.05	5.50
2 gang switch	"	3.65	2.05	5.70
1 single recept	"	3.45	2.05	5.50
1 20a twist lock recept	"	3.65	2.05	5.70
1 30a twist lock recept	"	3.85	2.05	5.90
1 duplex recept	"	3.45	2.05	5.50
2 duplex recept	"	3.50	2.05	5.55
Switch and duplex recept	"	3.55	2.05	5.60
4" plastic round boxes, ground straps				
box only	EA.	0.79	5.05	5.84
Box w/clamps	"	0.98	6.95	7.93
Box w/16" bar	"	1.40	7.95	9.35
Box w/24" bar	"	1.50	8.70	10.20
4" plastic round box covers				
blank cover	EA.	0.46	2.05	2.51
Plaster ring	"	0.52	2.05	2.57
4" plastic square boxes				
box only	EA.	0.59	5.05	5.64
Box w/clamps	"	0.73	6.95	7.68

16 ELECTRICAL

BASIC MATERIALS	UNIT	MAT.	INST.	TOTAL
16130.60 PULL AND JUNCTION BOXES				
Box w/hanger	EA.	0.89	8.70	9.59
Box w/nails and clamp	"	1.30	8.70	10.00
4" plastic square box covers				
Blank cover	EA.	0.52	2.05	2.57
1 gang ring	"	0.48	2.05	2.53
2 gang ring	"	0.52	2.05	2.57
Round ring	"	0.59	2.05	2.64
16130.65 PULL BOXES AND CABINETS				
Galvanized pull boxes, screw cover				
4x4x4	EA.	3.60	6.60	10.20
4x6x4	"	4.50	6.60	11.10
6x6x4	"	5.40	6.60	12.00
6x8x4	"	7.80	6.60	14.40
8x8x4	"	9.10	8.70	17.80
8x10x4	"	10.55	8.40	18.95
8x12x4	"	11.80	8.70	20.50
Screw cover				
10x10x4	EA.	11.80	10.70	22.50
12x12x6	"	18.00	15.40	33.40
12x15x6	"	20.40	15.40	35.80
12x18x6	"	23.85	17.35	41.20
15x18x6	"	27.00	19.85	46.85
18x24x6	"	48.90	21.35	70.25
18x30x6	"	54.00	25.20	79.20
24x36x6	"	84.00	25.20	109.20
Cast iron junction box, unflanged				
6x6x4				
3/4" tap	EA.	58.50	17.35	75.85
1" tap	"	60.00	17.35	77.35
Two 1/2" taps	"	62.50	17.35	79.85
3/4" taps	"	63.00	17.35	80.35
6" adapter plate	"	26.40	12.05	38.45
6" exterior collar	"	43.00	12.05	55.05
Screw cover cabinet				
12x12x4	EA.	50.50	21.35	71.85
12x16x4	"	66.50	21.35	87.85
12x16x6	"	74.50	21.35	95.85
12x18x4	"	72.50	23.15	95.65
12x18x6	"	85.00	23.15	108.15
18x18x4	"	86.50	34.70	121.20
18x18x6	"	98.50	34.70	133.20
18x24x6	"	140.00	39.70	179.70
24x24x6	"	161.00	46.30	207.30
24x36x6	"	233.00	58.00	291.00
36x48x6	"	468.00	87.00	555.00
NEMA 3R, rain tight screw cover enclosures				
6x6x4	EA.	10.70	7.30	18.00
8x6x4	"	12.25	10.30	22.55
8x8x4	"	14.10	10.30	24.40
10x8x4	"	15.90	13.90	29.80

BASIC MATERIALS

BASIC MATERIALS	UNIT	MAT.	INST.	TOTAL
16130.65 PULL BOXES AND CABINETS				
10x10x4	EA.	18.05	13.90	31.95
12x8x4	"	17.80	15.40	33.20
12x12x4	"	22.75	15.40	38.15
15x12x4	"	26.60	18.50	45.10
8x8x6	"	16.55	13.90	30.45
10x8x6	"	18.75	15.40	34.15
10x10x6	"	21.25	15.40	36.65
12x8x6	"	20.90	18.50	39.40
12x10x6	"	23.65	19.00	42.65
12x12x6	"	26.40	19.00	45.40
18x12x6	"	34.80	24.35	59.15
16130.80 RECEPTACLES				
Contractor grade duplex receptacles, 15 a 120v				
Duplex	EA.	1.25	6.95	8.20
125 volt, 20a, duplex, grounding type, standard grade	"	4.75	6.95	11.70
Ground fault interrupter type	"	37.10	10.30	47.40
250 volt, 20a, 2 pole, single receptacle, ground type	"	5.55	6.95	12.50
120/208v, 4 pole, single receptacle, twist lock				
20a	EA.	11.95	12.05	24.00
50a	"	22.75	12.05	34.80
125/250v, 3 pole, flush receptacle				
30a	EA.	7.60	10.30	17.90
50a	"	8.75	10.30	19.05
60a	"	35.30	12.05	47.35
277 v, 20a, 2 pole, grounding type, twist lock	"	8.75	6.95	15.70
Dryer receptacle, 250v, 30a/50a, 3 wire	"	14.65	10.30	24.95
Clock receptacle, 2 pole, grounding type	"	6.10	6.95	13.05
125v, 20a single recept. grounding type				
Standard grade	EA.	4.45	6.95	11.40
Specification	"	6.90	6.95	13.85
Hospital	"	7.10	6.95	14.05
Isolated ground orange	"	12.40	8.70	21.10
Duplex				
Specification grade	EA.	7.15	6.95	14.10
Hospital	"	9.25	6.95	16.20
Isolated ground orange	"	13.95	8.70	22.65
250v, 20a, duplex, 2 pole, grounding, spec. grade	"	10.60	6.95	17.55
Combination recepts, 20a, 125v and 250v, duplex	"	9.50	6.95	16.45
GFI hospital grade recepts, 20a, 125v, duplex	"	42.90	10.30	53.20
125/250v, 3 pole, 3 wire surface recepts				
30a	EA.	14.65	10.30	24.95
50a	"	16.25	10.30	26.55
60a	"	35.50	12.05	47.55
Cord set, 3 wire, 6' cord				
30a	EA.	15.85	10.30	26.15
50a	"	22.05	10.30	32.35
125/250v, 3 pole, 3 wire cap				
30a	EA.	12.95	13.90	26.85
50a	"	23.50	13.90	37.40
60a	"	30.30	15.40	45.70

BASIC MATERIALS

16198.10 ELECTRIC MANHOLES

BASIC MATERIALS	UNIT	MAT.	INST.	TOTAL
Precast, handhole, 4' deep				
2'x2'	EA.	283.00	121.00	404.00
3'x3'	"	385.00	193.00	578.00
4'x4'	"	826.00	356.00	1,182
Power manhole, complete, precast, 8' deep				
4'x4'	EA.	1,250	487.00	1,737
6'x6'	"	1,640	694.00	2,334
8'x8'	"	1,900	731.00	2,631
6' deep, 9' x 12'	"	2,210	868.00	3,078
Cast in place, power manhole, 8' deep				
4'x4'	EA.	1,470	487.00	1,957
6'x6'	"	1,810	694.00	2,504
8'x8'	"	2,040	731.00	2,771

16199.10 UTILITY POLES & FITTINGS

BASIC MATERIALS	UNIT	MAT.	INST.	TOTAL
Wood pole, creosoted				
25'	EA.	124.00	81.50	205.50
30'	"	146.00	103.00	249.00
35'	"	317.00	121.00	438.00
40'	"	566.00	132.00	698.00
45'	"	638.00	241.00	879.00
50'	"	710.00	250.00	960.00
55'	"	780.00	262.00	1,042
Treated, wood preservative, 6"x6"				
8'	EA.	11.35	17.35	28.70
10'	"	28.30	27.80	56.10
12'	"	31.70	30.80	62.50
14'	"	39.60	46.30	85.90
16'	"	56.50	55.50	112.00
18'	"	85.00	69.50	154.50
20'	"	113.00	69.50	182.50
Aluminum, brushed, no base				
8'	EA.	243.00	69.50	312.50
10'	"	283.00	92.50	375.50
15'	"	425.00	95.50	520.50
20'	"	657.00	111.00	768.00
25'	"	894.00	132.00	1,026
30'	"	1,590	153.00	1,743
35'	"	1,360	174.00	1,534
40'	"	2,150	217.00	2,367
Steel, no base				
10'	EA.	294.00	87.00	381.00
15'	"	340.00	103.00	443.00
20'	"	430.00	132.00	562.00
25'	"	566.00	157.00	723.00
30'	"	821.00	177.00	998.00
35'	"	1,250	217.00	1,467
Concrete, no base				
13'	EA.	453.00	191.00	644.00
16'	"	680.00	252.00	932.00
18'	"	821.00	305.00	1,126

BASIC MATERIALS

16199.10 UTILITY POLES & FITTINGS

	UNIT	MAT.	INST.	TOTAL
25'	EA.	986.00	347.00	1,333
30'	"	1,290	421.00	1,711
35'	"	1,670	487.00	2,157
40'	"	1,930	555.00	2,485
45'	"	2,340	591.00	2,931
50'	"	2,950	631.00	3,581
55'	"	3,260	661.00	3,921
60'	"	3,740	694.00	4,434
Pole line hardware				
Wood crossarm				
4'	EA.	24.00	46.30	70.30
8'	"	48.00	58.00	106.00
10'	"	96.00	71.00	167.00
Angle steel brace				
1 piece	EA.	5.75	8.70	14.45
2 piece	"	12.25	12.05	24.30
Eye nut, 5/8"	"	2.85	1.75	4.60
Bolt (14-16"), 5/8"	"	7.50	6.95	14.45
Transformer, ground connection	"	3.35	8.70	12.05
Stirrup	"	10.20	10.70	20.90
Secondary lead support	"	11.40	13.90	25.30
Spool insulator	"	8.40	6.95	15.35
Guy grip, preformed				
7/16"	EA.	1.80	5.05	6.85
1/2"	"	2.70	5.05	7.75
Hook	"	1.80	8.70	10.50
Strain insulator	"	14.65	12.60	27.25
Wire				
5/16"	L.F.	0.30	0.17	0.47
7/16"	"	0.59	0.21	0.80
1/2"	"	1.00	0.28	1.28
Soft drawn ground, copper, #8	"	0.24	0.28	0.52
Ground clamp	EA.	3.90	10.70	14.60
Perforated strapping for conduit, 1-1/2"	L.F.	1.50	5.05	6.55
Hot line clamp	EA.	11.40	27.80	39.20
Lightning arrester				
3kv	EA.	60.00	34.70	94.70
10kv	"	96.00	55.50	151.50
30kv	"	179.00	69.50	248.50
36kv	"	360.00	87.00	447.00
Fittings				
Plastic molding	L.F.	1.75	5.05	6.80
Molding staples	EA.	0.42	1.75	2.17
Ground wires staples	"	0.17	1.05	1.22
Copper butt plate	"	0.48	10.30	10.78
Anchor bond clamp	"	2.15	5.05	7.20
Guy wire				
1/4"	L.F.	0.23	1.05	1.28
3/8"	"	0.38	1.75	2.13
Guy grip				
1/4"	EA.	1.20	1.75	2.95
3/8"	"	2.10	1.75	3.85

POWER GENERATION

16210.10 GENERATORS

	UNIT	MAT.	INST.	TOTAL
Diesel generator, with auto transfer switch				
30kw	EA.	15,010	1,070	16,080
50kw	"	18,420	1,070	19,490
75kw	"	24,200	1,460	25,660
100kw	"	26,790	1,630	28,420
125kw	"	28,600	1,740	30,340
150kw	"	33,000	1,980	34,980
175kw	"	34,320	2,310	36,630
200kw	"	35,640	2,780	38,420
250kw	"	39,000	3,080	42,080
300kw	"	46,800	3,470	50,270
350kw	"	50,160	3,970	54,130
400kw	"	61,600	4,630	66,230
450kw	"	66,000	5,050	71,050
500kw	"	71,500	5,550	77,050
600kw	"	97,900	6,940	104,840
750kw	"	137,500	6,940	144,440

16230.10 CAPACITORS

	UNIT	MAT.	INST.	TOTAL
Three phase capacitors				
240v				
1.5 kvar	EA.	184.00	87.00	271.00
2.5 kvar	"	214.00	111.00	325.00
3.0 kvar	"	240.00	139.00	379.00
4 kvar	"	253.00	174.00	427.00
5 kvar	"	282.00	185.00	467.00
6 kvar	"	311.00	198.00	509.00
7.5 kvar	"	382.00	214.00	596.00
10 kvar	"	460.00	278.00	738.00
15 kvar	"	635.00	330.00	965.00
20 kvar	"	757.00	416.00	1,173
25 kvar	"	887.00	451.00	1,338
40 kvar	"	1,810	625.00	2,435
50 kvar	"	2,170	729.00	2,899
60 kvar	"	2,790	746.00	3,536
75 kvar	"	3,090	868.00	3,958
100 kvar	"	3,790	1,040	4,830
480v				
1.5 kvar	EA.	154.00	87.00	241.00
2.5 kvar	"	161.00	111.00	272.00
3 kvar	"	177.00	139.00	316.00
4 kvar	"	198.00	174.00	372.00
5 kvar	"	216.00	185.00	401.00
6 kvar	"	235.00	198.00	433.00
7.5 kvar	"	262.00	214.00	476.00
10 kvar	"	323.00	278.00	601.00
12.5 kvar	"	364.00	330.00	694.00
15 kvar	"	397.00	416.00	813.00
18 kvar	"	427.00	434.00	861.00
20 kvar	"	444.00	451.00	895.00
22.5 kvar	"	459.00	468.00	927.00

POWER GENERATION

16230.10 CAPACITORS

	UNIT	MAT.	INST.	TOTAL
25 kvar	EA.	495.00	515.00	1,010
30 kvar	"	554.00	515.00	1,069
35 kvar	"	654.00	555.00	1,209
40 kvar	"	695.00	625.00	1,320
45 kvar	"	764.00	694.00	1,458
50 kvar	"	804.00	729.00	1,533
60 kvar	"	1,020	763.00	1,783
70 kvar	"	1,130	834.00	1,964
75 kvar	"	1,230	868.00	2,098
80 kvar	"	1,430	938.00	2,368
90 kvar	"	1,520	1,010	2,530
100 kvar	"	1,620	1,040	2,660
125 kvar	"	2,180	1,150	3,330
150 kvar	"	2,510	1,290	3,800

16320.10 TRANSFORMERS

	UNIT	MAT.	INST.	TOTAL
Floor mounted, single phase, int. dry, 480v-120/240v				
3 kva	EA.	217.00	63.00	280.00
5 kva	"	294.00	107.00	401.00
7.5 kva	"	441.00	121.00	562.00
10 kva	"	499.00	132.00	631.00
15 kva	"	648.00	149.00	797.00
25 kva	"	824.00	262.00	1,086
37.5 kva	"	1,150	327.00	1,477
50 kva	"	1,300	356.00	1,656
75 kva	"	1,770	370.00	2,140
100 kva	"	2,360	402.00	2,762
Three phase, 480v-120/280v				
15 kva	EA.	824.00	209.00	1,033
30 kva	"	1,060	327.00	1,387
45 kva	"	1,410	375.00	1,785
75 kva	"	2,000	380.00	2,380
112.5 kva	"	2,710	441.00	3,151
150 kva	"	3,420	471.00	3,891
225 kva	"	4,650	534.00	5,184
Single phase, dry type, 2400v				
167 kva	EA.	6,420	780.00	7,200
250 kva	"	8,360	1,040	9,400
333 kva	"	10,270	780.00	11,050
5000v				
167 kva	EA.	6,880	780.00	7,660
250 kva	"	8,930	1,040	9,970
333 kva	"	10,930	1,300	12,230
8660v				
167 kva	EA.	7,230	954.00	8,184
250 kva	"	9,780	1,210	10,990
333 kva	"	11,720	2,350	14,070
1500v				
167 kva	EA.	8,150	954.00	9,104
250 kva	"	10,560	1,210	11,770
333 kva	"	12,610	1,480	14,090
Three phase, dry type transformer, 2400v				

POWER GENERATION

	UNIT	MAT.	INST.	TOTAL
16320.10 TRANSFORMERS				
225 kva	EA.	9,340	868.00	10,208
300 kva	"	11,100	954.00	12,054
500 kva	"	14,650	1,480	16,130
750 kva	"	19,240	1,830	21,070
5000v				
225.0 kva	EA.	9,420	868.00	10,288
300 kva	"	11,310	954.00	12,264
500 kva	"	14,730	1,480	16,210
750 kva	"	19,320	1,830	21,150
8660v				
225.0 kva	EA.	10,270	1,040	11,310
300 kva	"	12,890	1,130	14,020
500 kva	"	16,640	1,650	18,290
750 kva	"	20,880	2,000	22,880
1500v				
225 kva	EA.	11,160	1,040	12,200
300 kva	"	14,020	1,130	15,150
500 kva	"	18,050	1,650	19,700
750 kva	"	22,650	2,000	24,650
Buck boost transformers				
.25 kva	EA.	58.50	34.70	93.20
.50 kva	"	76.00	43.40	119.40
.75 kva	"	99.00	52.50	151.50
1.00 kva	"	129.00	60.50	189.50
1.50 kva	"	152.00	69.50	221.50
2.00 kva	"	211.00	87.00	298.00
3.00 kva	"	294.00	103.00	397.00
16350.10 CIRCUIT BREAKERS				
Molded case, 240v, 15-60a, bolt-on				
1 pole	EA.	7.55	8.70	16.25
2 pole	"	16.80	12.05	28.85
70-100a, 2 pole	"	44.40	18.50	62.90
15-60a, 3 pole	"	54.50	13.90	68.40
70-100a, 3 pole	"	79.50	21.35	100.85
480v, 2 pole				
15-60a	EA.	121.00	10.30	131.30
70-100a	"	153.00	13.90	166.90
3 pole				
15-60a	EA.	153.00	13.90	166.90
70-100a	"	183.00	15.40	198.40
70-225a	"	403.00	21.35	424.35
Draw out air circuit breakers				
600a	EA.	6,970	555.00	7,525
800a	"	9,010	631.00	9,641
1600a	"	14,410	841.00	15,251
2000a	"	19,220	957.00	20,177
3000a	"	33,630	1,110	34,740
4000a	"	51,650	1,320	52,970
Load center circuit breakers, 240v				
1 pole, 10-60a	EA.	7.35	8.70	16.05

POWER GENERATION

16350.10 CIRCUIT BREAKERS

Description	UNIT	MAT.	INST.	TOTAL
2 pole				
10-60a	EA.	16.80	13.90	30.70
70-100a	"	16.15	23.15	39.30
110-150a	"	102.00	25.20	127.20
3 pole				
10-60a	EA.	57.50	17.35	74.85
70-100a	"	86.50	25.20	111.70
Load center, G.F.I. breakers, 240v				
1 pole, 15-30a	EA.	57.50	10.30	67.80
2 pole, 15-30a	"	102.00	13.90	115.90
Key operated breakers, 240v, 1 pole, 10-30a	"	9.60	10.30	19.90
Tandem breakers, 240v				
1 pole, 15-30a	EA.	15.60	13.90	29.50
2 pole, 15-30a	"	30.00	18.50	48.50
Bolt-on, G.F.I. breakers, 240v, 1 pole, 15-30a	"	71.50	12.05	83.55
Enclosed breaker, 120v, 1 pole, 15-50a, NEMA 1	"	83.00	27.80	110.80
240v, 2 pole				
15-60a, NEMA 1	EA.	112.00	43.40	155.40
70-100a, NEMA 1	"	150.00	60.50	210.50
3 pole				
15-60a, NEMA 1	EA.	150.00	52.50	202.50
70-100a, NEMA 1	"	197.00	77.00	274.00
Enclosed circuit breakers				
120v, 1 pole, NEMA 3R, 15-50a	EA.	155.00	31.20	186.20
240v, 2 pole, NEMA 3R				
15-60a	EA.	196.00	43.40	239.40
70-100a	"	227.00	60.50	287.50
3 pole, NEMA 3R				
15-60a	EA.	216.00	52.50	268.50
70-100a	"	265.00	77.00	342.00
480v, NEMA 1				
1 pole, 15-50a	EA.	95.50	27.80	123.30
2 pole, 15-60a	"	173.00	43.40	216.40
70-100a	"	213.00	52.50	265.50
3 pole, NEMA 1				
15-60a	EA.	211.00	52.50	263.50
70-100a	"	244.00	77.00	321.00
480v, 1 pole, 15-50a, NEMA 3R	"	167.00	34.70	201.70
2 pole				
2 pole, 15-60a, NEMA 3R	EA.	244.00	43.40	287.40
70-100a, NEMA 3R	"	285.00	60.50	345.50
3 pole				
15-60a, NEMA 3R	EA.	282.00	52.50	334.50
70-100a, NEMA 3R	"	314.00	77.00	391.00
70-100a, NEMA 3R	"	194.00	43.40	237.40
70-100a, NEMA 1	"	236.00	60.50	296.50
3 pole				
15-60a, NEMA 1	EA.	240.00	60.50	300.50
70-100a, NEMA 1	"	285.00	77.00	362.00
Enclosed breakers, 600v, 2 phase, NEMA 3R				
15-60a	EA.	253.00	43.40	296.40
70-100a	"	314.00	60.50	374.50

POWER GENERATION

16350.10 CIRCUIT BREAKERS

POWER GENERATION	UNIT	MAT.	INST.	TOTAL
3 phase, NEMA 3R				
15-60a	EA.	295.00	52.50	347.50
70-100a	"	338.00	77.00	415.00
600v, 3 phase, NEMA 1				
125a	EA.	594.00	77.00	671.00
150a	"	594.00	103.00	697.00
175a	"	594.00	103.00	697.00
200a	"	594.00	103.00	697.00
225a	"	594.00	103.00	697.00
250a	"	1,040	214.00	1,254
300a	"	1,040	214.00	1,254
350a	"	1,040	214.00	1,254
400a	"	1,040	214.00	1,254
500a	"	1,720	338.00	2,058
600a	"	1,720	338.00	2,058
700a	"	2,210	373.00	2,583
800a	"	2,210	373.00	2,583
900a	"	2,800	521.00	3,321
100a	"	2,800	521.00	3,321
1200a	"	6,290	643.00	6,933
1400a	"	6,290	643.00	6,933
1600a	"	6,410	834.00	7,244
1800a	"	6,580	1,040	7,620
2000a	"	6,580	1,040	7,620
600v, 3 phase, NEMA 3R				
125-225a	EA.	708.00	77.00	785.00
250-400a	"	1,330	198.00	1,528
500-600a	"	2,050	338.00	2,388
700-800a	"	2,630	382.00	3,012
900-1000a	"	3,150	521.00	3,671
1000-1200a	"	7,380	659.00	8,039
1400-1600a	"	7,490	834.00	8,324
1800-2000a	"	7,620	1,040	8,660

16360.10 SAFETY SWITCHES

POWER GENERATION	UNIT	MAT.	INST.	TOTAL
Fused, 3 phase, 30 amp, 600v, heavy duty				
NEMA 1	EA.	107.00	39.70	146.70
NEMA 3r	"	182.00	39.70	221.70
NEMA 4	"	459.00	55.50	514.50
NEMA 12	"	177.00	60.50	237.50
60a				
NEMA 1	EA.	129.00	39.70	168.70
NEMA 3r	"	212.00	39.70	251.70
NEMA 4	"	516.00	55.50	571.50
NEMA 12	"	180.00	60.50	240.50
100a				
NEMA 1	EA.	240.00	60.50	300.50
NEMA 3r	"	333.00	60.50	393.50
NEMA 4	"	1,010	69.50	1,080
NEMA 12	"	282.00	87.00	369.00
200a				

16 ELECTRICAL

POWER GENERATION

16360.10 SAFETY SWITCHES

POWER GENERATION	UNIT	MAT.	INST.	TOTAL
NEMA 1	EA.	323.00	87.00	410.00
NEMA 3r	"	459.00	87.00	546.00
NEMA 4	"	1,410	95.50	1,506
NEMA 12	"	450.00	121.00	571.00
400a				
NEMA 1	EA.	904.00	191.00	1,095
NEMA 3r	"	1,080	191.00	1,271
NEMA 4	"	1,810	200.00	2,010
NEMA 12	"	994.00	244.00	1,238
600a				
NEMA 1	EA.	1,520	278.00	1,798
NEMA 3r	"	2,120	278.00	2,398
NEMA 4	"	3,570	312.00	3,882
NEMA 12	"	1,670	427.00	2,097
Non-fused, 240-600v, heavy duty, 3 phase, 30 amp				
NEMA 1	EA.	55.50	39.70	95.20
NEMA 3r	"	100.00	39.70	139.70
NEMA 4	"	405.00	60.50	465.50
NEMA 12	"	123.00	60.50	183.50
60a				
NEMA1	EA.	100.00	39.70	139.70
NEMA 3r	"	177.00	39.70	216.70
NEMA 4	"	482.00	60.50	542.50
NEMA 12	"	153.00	60.50	213.50
100a				
NEMA 1	EA.	156.00	60.50	216.50
NEMA 3r	"	252.00	60.50	312.50
NEMA 4	"	983.00	87.00	1,070
NEMA 12	"	219.00	87.00	306.00
200a, NEMA 1	"	268.00	87.00	355.00
600a, NEMA 12	"	1,300	427.00	1,727
Bolt-on hubs				
3/4" - 1-1/2"	EA.	8.15	8.70	16.85
2"	"	14.75	10.30	25.05
2-1/2"	"	23.35	10.30	33.65
3"	"	43.20	12.05	55.25
3-1/2"	"	63.50	13.90	77.40
4"	"	81.50	13.90	95.40
Watertight hubs				
1/2"	EA.	6.60	8.70	15.30
3/4"	"	9.50	10.30	19.80
1"	"	10.00	13.90	23.90
1-1/4"	"	11.50	15.40	26.90
1-1/2"	"	16.90	16.35	33.25
2"	"	25.10	17.35	42.45
2-1/2"	"	29.90	18.50	48.40
3"	"	38.10	21.35	59.45
3-1/2"	"	60.00	27.80	87.80
4"	"	80.50	29.50	110.00
Non-fused, 600v, 3 pole, NEMA 7				
600a	EA.	482.00	77.00	559.00
100a	"	582.00	111.00	693.00

16 ELECTRICAL

POWER GENERATION	UNIT	MAT.	INST.	TOTAL
16360.10 SAFETY SWITCHES				
225a	EA.	1,300	139.00	1,439
NEMA 9				
60a	EA.	500.00	87.00	587.00
100a	"	588.00	116.00	704.00
225a	"	1,410	146.00	1,556
Fusible bolted pressure switches, 600v/3 pole, NEMA 1				
800a	EA.	2,950	555.00	3,505
1200a	"	3,570	763.00	4,333
1600a	"	3,850	868.00	4,718
2000a	"	3,970	1,040	5,010
2500a	"	4,530	1,210	5,740
3000a	"	6,120	1,560	7,680
4000a	"	8,160	1,800	9,960
Non-fusible				
800a	EA.	2,830	503.00	3,333
1200a	"	3,170	694.00	3,864
1600a	"	3,400	798.00	4,198
2000a	"	3,600	971.00	4,571
2500a	"	4,130	1,210	5,340
3000a	"	5,890	1,560	7,450
4000a	"	7,930	1,800	9,730
Fusible load interrupter switches, 4.16 kv, NEMA 1				
200a	EA.	2,610	1,040	3,650
600a	"	4,090	2,440	6,530
Fusible load interrupter switch, 13.8 kv				
NEMA 1, 600a	EA.	4,530	3,470	8,000
NEMA 3R, 600a	"	5,440	3,470	8,910
4.16 kv, NEMA 3R				
200a	EA.	2,830	1,040	3,870
600a	"	5,100	2,440	7,540
Non-fused load interrupter switch, 4.16 kv, NEMA 1				
200a	EA.	2,270	1,040	3,310
600a	"	3,850	2,440	6,290
13.8 kv, NEMA 1, 600a	"	4,190	3,470	7,660
4.16 kv, NEMA 3R				
200a	EA.	2,380	1,040	3,420
600a	"	4,760	2,440	7,200
13.8 kv, NEMA 3R, 600a	"	5,100	3,470	8,570
Interrupter switch accessories, strip heater	"			311.00
Cable lugs	"			56.50
Key interlock	"			396.00
Auxiliary switch	"			220.00
5 kv lightning arrester				
5 kva	EA.			934.00
15 kv	"			1,020
16365.10 FUSES				
Fuse, one-time, 250v				
30a	EA.	0.88	1.75	2.63
60a	"	1.10	1.75	2.85
100a	"	4.75	1.75	6.50

POWER GENERATION	UNIT	MAT.	INST.	TOTAL
16365.10 FUSES				
200a	EA.	11.90	1.75	13.65
400a	"	21.40	1.75	23.15
600a	"	30.90	1.75	32.65
600v				
30a	EA.	3.55	1.75	5.30
60a	"	4.75	1.75	6.50
100a	"	10.70	1.75	12.45
200a	"	20.20	1.75	21.95
400a	"	39.20	1.75	40.95
Fusetron, 600v				
200a	EA.	19.60	1.75	21.35
400a	"	39.20	1.75	40.95
Fuse, amp-trap, K1, 250v				
30a	EA.	5.80	1.75	7.55
60a	"	12.75	1.75	14.50
100a	"	29.50	1.75	31.25
200a	"	59.50	1.75	61.25
400a	"	119.00	1.75	120.75
600a	"	148.00	1.75	149.75
600v				
30a	EA.	12.55	1.75	14.30
60a	"	23.20	1.75	24.95
100a	"	47.60	1.75	49.35
200a	"	77.50	1.75	79.25
400a	"	150.00	1.75	151.75
K5, 250v				
30a	EA.	2.65	1.75	4.40
60a	"	4.85	1.75	6.60
100a	"	11.30	1.75	13.05
200a	"	22.00	1.75	23.75
400a	"	41.60	1.75	43.35
600a	"	62.00	1.75	63.75
600v				
30a	EA.	5.20	1.75	6.95
60a	"	10.05	1.75	11.80
100a	"	19.80	1.75	21.55
200a	"	38.70	1.75	40.45
400a	"	75.50	1.75	77.25
600a	"	108.00	1.75	109.75
J, 600v				
30a	EA.	9.60	1.75	11.35
60a	"	16.80	1.75	18.55
100a	"	38.40	1.75	40.15
200a	"	75.00	1.75	76.75
400a	"	142.00	1.75	143.75
L, 600v				
1200a	EA.	249.00	13.90	262.90
1600a	"	381.00	13.90	394.90
2000a	"	428.00	13.90	441.90
2500a	"	558.00	13.90	571.90
3000a	"	642.00	13.90	655.90
4000a	"	867.00	13.90	880.90

POWER GENERATION

16365.10 FUSES

	UNIT	MAT.	INST.	TOTAL
5000a	EA.	1,310	13.90	1,324
Fuse cl-ay 250v				
600a	EA.	235.00	10.30	245.30
1200a	"	235.00	10.30	245.30
1600a	"	295.00	10.30	305.30
2000a	"	390.00	10.30	400.30
600v				
1200a	EA.	227.00	10.30	237.30
1600a	"	295.00	10.30	305.30
2000a	"	390.00	10.30	400.30
Reducers, 600v				
60a-30a	EA.	4.75	5.05	9.80
100a-30a	"	5.95	5.05	11.00
100a-60a	"	16.90	5.05	21.95
200a-60a	"	31.10	8.70	39.80
200a-100a	"	11.90	8.70	20.60
400a-100a	"	45.90	12.05	57.95
400a-200a	"	48.50	12.05	60.55
600a-100a	"	81.50	1,390	1,472
600a-200a	"	79.00	1,390	1,469
600a-400a	"	81.50	1,390	1,472

16395.10 GROUNDING

	UNIT	MAT.	INST.	TOTAL
Ground rods, copper clad, 1/2" x				
6'	EA.	6.55	23.15	29.70
8'	"	8.50	25.20	33.70
10'	"	10.45	34.70	45.15
5/8" x				
5'	EA.	6.65	21.35	28.00
6'	"	7.90	25.20	33.10
8'	"	9.80	34.70	44.50
10'	"	12.75	43.40	56.15
3/4" x				
8'	EA.	15.85	25.20	41.05
10'	"	18.80	27.80	46.60
Ground rod clamp				
5/8"	EA.	2.10	4.25	6.35
3/4"	"	2.70	4.25	6.95
Coupling, on threaded rods, 3/4"	"	6.65	1.75	8.40
Ground receptacles	"	13.15	8.70	21.85
Bus bar, copper, 2" x 1/4"	L.F.	3.05	5.05	8.10
Copper braid, 1" x 1/8", for door ground	EA.	2.55	3.45	6.00
Brazed connection for				
#6 wire	EA.	8.50	17.35	25.85
#2 wire	"	10.80	27.80	38.60
#2/0 wire	"	12.75	34.70	47.45
#4/0 wire	"	19.10	39.70	58.80
Ground rod couplings				
1/2"	EA.	3.80	3.45	7.25
5/8"	"	4.90	3.45	8.35
Ground rod, driving stud				

16 ELECTRICAL

POWER GENERATION

16395.10 GROUNDING

POWER GENERATION	UNIT	MAT.	INST.	TOTAL
1/2"	EA.	2.60	3.45	6.05
5/8"	"	2.70	3.45	6.15
3/4"	"	2.75	3.45	6.20
Ground rod clamps, #8-2 to				
1" pipe	EA.	2.80	6.95	9.75
2" pipe	"	3.85	8.70	12.55
3" pipe	"	18.45	10.30	28.75
5" pipe	"	27.90	12.05	39.95
6" piep	"	42.40	15.40	57.80
#4-4/0 to				
1" pipe	EA.	7.90	6.95	14.85
2" pipe	"	12.85	8.70	21.55
3" pipe	"	20.85	10.30	31.15
3" pipe	"	31.30	12.05	43.35
8 pipe	"	47.90	15.40	63.30
8 pipe	"	52.00	23.15	75.15
10 pipe	"	65.50	33.00	98.50
12 pipe	"	75.00	44.80	119.80

SERVICE AND DISTRIBUTION

16425.10 SWITCHBOARDS

SERVICE AND DISTRIBUTION	UNIT	MAT.	INST.	TOTAL
Switchboard, 90" high, no main disconnect, 208/120v				
400a	EA.	1,340	275.00	1,615
600a	"	2,070	278.00	2,348
1000a	"	2,610	278.00	2,888
1200a	"	2,790	347.00	3,137
1600a	"	3,060	414.00	3,474
2000a	"	3,290	487.00	3,777
2500a	"	3,630	555.00	4,185
240/480v				
600a	EA.	2,380	283.00	2,663
800a	"	2,610	283.00	2,893
1600a	"	3,290	414.00	3,704
2000a	"	3,510	487.00	3,997
2500a	"	3,740	555.00	4,295
3000a	"	4,310	957.00	5,267
4000a	"	5,210	1,030	6,240
Main breaker sections, 600v				
1200a, GFI	EA.	9,400	589.00	9,989
1600a, GFI	"	11,100	677.00	11,777
2000a, GFI	"	11,780	694.00	12,474
2500a, GFI	"	13,030	868.00	13,898
3000a, GFI	"	14,620	1,040	15,660

SERVICE AND DISTRIBUTION

16425.10 SWITCHBOARDS

SERVICE AND DISTRIBUTION	UNIT	MAT.	INST.	TOTAL
4000a, GFI	EA.	17,220	1,210	18,430
Switchboard meter sections, 600v				
400a	EA.	1,700	278.00	1,978
600a	"	2,610	347.00	2,957
800a	"	3,170	382.00	3,552
1000a	"	3,850	468.00	4,318
2000a	"	4,760	555.00	5,315
2500a	"	5,330	694.00	6,024
3000a	"	5,550	868.00	6,418
4000a	"	6,230	1,040	7,270
Insulated case, draw out compartment, 208/120v				
800a	EA.	1,130	87.00	1,217
1600a	"	1,360	103.00	1,463
2000a	"	1,700	121.00	1,821
2500a	"	2,040	121.00	2,161
3000a	"	3,400	139.00	3,539
4000a	"	4,760	166.00	4,926
Accessories for power trip breakers				
Shunt trip	EA.	543.00	17.35	560.35
Key interlock	"	263.00	77.00	340.00
Lifting and transport truck	"	1,700	156.00	1,856
Lifting device	"	175.00	46.30	221.30
Bus duct connection, 3 phase, 4 wire				
225a	EA.	238.00	116.00	354.00
400a	"	296.00	103.00	399.00
600a	"	381.00	116.00	497.00
800a	"	416.00	139.00	555.00
2500a	"	772.00	209.00	981.00
3000a	"	1,160	259.00	1,419
4000a	"	2,080	305.00	2,385
Provision for mounting current transformers				
800a & below primary	EA.	872.00	103.00	975.00
1000 to 1500a primary	"	1,090	103.00	1,193
2000 to 6000a primary	"	1,310	103.00	1,413
Provision for mounting potential transformers				
2000a max	EA.	3,120	132.00	3,252
Switchboard instruments				
Voltmeter	EA.	1,120	34.70	1,155
Ammeter, incoming line	"	1,120	34.70	1,155
Wattmeter	"	1,780	34.70	1,815
Varmeter	"	1,840	34.70	1,875
Power factor meter	"	2,120	34.70	2,155
Frequency meter	"	2,380	34.70	2,415
Recording voltmeter	"	5,050	69.50	5,120
Wattmeter	"	5,050	69.50	5,120
Power factor meter	"	5,890	69.50	5,960
Frequency meter	"	5,890	69.50	5,960
Instrument phase select switch	"	227.00	17.35	244.35
Enclosure, 90" high, 3 phase, 4 wire				
1000a	EA.	1,930	237.00	2,167
1200a	"	2,040	244.00	2,284
1600a	"	2,270	298.00	2,568

16 ELECTRICAL

SERVICE AND DISTRIBUTION

	UNIT	MAT.	INST.	TOTAL
16425.10 SWITCHBOARDS				
2000a	EA.	2,380	463.00	2,843
5500a	"	2,950	544.00	3,494
3000a	"	3,290	631.00	3,921
4000a	"	4,310	816.00	5,126
Circuit breakers, 600v, 100a, frame				
15-30a, 1 pole	EA.	62.50	10.30	72.80
15-60a, 2 pole	"	164.00	12.05	176.05
70-100a, 2 pole	"	199.00	13.90	212.90
15-60a, 3 pole	"	206.00	15.40	221.40
70-100a, 3 pole	"	241.00	17.35	258.35
Bolt on breakers, 600v, 225a frame, 110-225a				
2 pole	EA.	471.00	21.35	492.35
3 pole	"	588.00	38.00	626.00
400a frame, 250-400a, 2 pole	"	859.00	43.40	902.40
800a frame				
450-600a, 2 pole	EA.	1,410	66.00	1,476
700-800a, 2 pole	"	1,770	87.00	1,857
450-600a, 3 pole	"	1,730	146.00	1,876
700-800a, 3 pole	"	2,240	154.00	2,394
Bolt on branch breakers, 600v				
1000-2000a, 2 pole	EA.	3,770	185.00	3,955
2500a, 2 pole	"	6,950	373.00	7,323
1000-2000a, 3 pole	"	4,830	278.00	5,108
2500a, 3 pole	"	8,480	382.00	8,862
3000a, 3 pole	"	16,140	694.00	16,834
Metal clad substation switch board, selector switch				
600a, 5kv	EA.	12,460	1,460	13,920
600a, 15kv	"	13,820	1,660	15,480
Fused switch, 600a				
5kv	EA.	9,060	1,210	10,270
15kv	"	12,580	1,210	13,790
1200a				
5kv	EA.	10,080	1,390	11,470
15kv	"	13,480	1,390	14,870
Oil cutout switch				
5 kv	EA.	3,510	521.00	4,031
15 kv	"	4,870	625.00	5,495
Liquid air terminal section	"	453.00	278.00	731.00
Dry air terminal section	"	906.00	295.00	1,201
Auxillary compartment	"	6,460	1,040	7,500
16430.20 METERING				
Outdoor wp meter sockets, 1 gang, 240v, 1 phase		0.00	52.50	52.50
Includes sealing ring 100a	EA.	23.10	52.50	75.60
150a	"	30.80	61.50	92.30
200a	"	38.50	69.50	108.00
Die cast hubs, 1-1/4"	"	3.60	11.10	14.70
1-1/2"	"	4.10	11.10	15.20
2"	"	4.95	11.10	16.05
Indoor meter center, main switch single phase, 240v				
400a	EA.	559.00	278.00	837.00
600a	"	1,090	382.00	1,472

SERVICE AND DISTRIBUTION

16430.20 METERING

	UNIT	MAT.	INST.	TOTAL
800a	EA.	1,800	406.00	2,206
Main breaker				
400a	EA.	999.00	278.00	1,277
600a	"	1,510	382.00	1,892
800a	"	1,930	406.00	2,336
1000a	"	2,800	555.00	3,355
Terminal box	"	4,060	572.00	4,632
1600a	"	6,100	625.00	6,725
Terminal box				
800a	EA.	189.00	347.00	536.00
1600a	"	771.00	625.00	1,396
Main switch, three phase, 208v				
400a	EA.	683.00	295.00	978.00
600a	"	1,230	416.00	1,646
800a	"	2,550	468.00	3,018
Main breaker				
400a	EA.	1,190	295.00	1,485
600a	"	1,870	416.00	2,286
800a	"	2,480	468.00	2,948
1000a	"	3,180	589.00	3,769
1200a	"	4,430	625.00	5,055
1600a	"	6,530	729.00	7,259
Terminal box				
800a	EA.	212.00	451.00	663.00
1600a	"	811.00	729.00	1,540
Indoor meter center				
2 meters	EA.	198.00	174.00	372.00
3 meters	"	249.00	214.00	463.00
4 meters	"	320.00	252.00	572.00
5 meters	"	406.00	278.00	684.00
6 meters	"	513.00	312.00	825.00
Plug on breakers, single phase, 208v				
60a	EA.	15.70	8.70	24.40
70a	"	31.40	8.70	40.10
80a	"	43.20	8.70	51.90
90a	"	45.20	8.70	53.90
100a	"	47.60	12.05	59.65
Indoor meter center, single phase, 125a breakers				
3 meters	EA.	261.00	214.00	475.00
4 meters	"	346.00	252.00	598.00
5 meters	"	432.00	278.00	710.00
6 meters	"	520.00	295.00	815.00
7 meters	"	607.00	347.00	954.00
8 meters	"	695.00	382.00	1,077
10 meters	"	868.00	416.00	1,284
150a breakers				
3 meters	EA.	1,260	214.00	1,474
4 meters	"	1,680	252.00	1,932
6 meters	"	2,520	278.00	2,798
7 meters	"	2,940	347.00	3,287
8 meters	"	3,360	382.00	3,742
200a breakers				

SERVICE AND DISTRIBUTION

16430.20 METERING

	UNIT	MAT.	INST.	TOTAL
3 meters	EA.	1,320	214.00	1,534
4 meters	"	1,760	252.00	2,012
6 meters	"	2,630	278.00	2,908
7 meters	"	3,070	347.00	3,417
8 meters	"	3,510	382.00	3,892
Indoor meter center, three phase, 125a breakers				
3 meters	EA.	280.00	214.00	494.00
4 meters	"	374.00	252.00	626.00
5 meters	"	466.00	278.00	744.00
6 meters	"	560.00	312.00	872.00
7 meters	"	653.00	347.00	1,000
8 meters	"	747.00	382.00	1,129
10 meters	"	934.00	416.00	1,350
150a breakers				
3 meters	EA.	1,290	214.00	1,504
4 meters	"	1,720	252.00	1,972
6 meters	"	2,600	295.00	2,895
7 meters	"	3,000	382.00	3,382
8 meters	"	3,430	416.00	3,846
200a breakers				
3 meters	EA.	1,350	231.00	1,581
4 meters	"	1,800	252.00	2,052
6 meters	"	2,690	312.00	3,002
7 meters	"	3,140	382.00	3,522
8 meters	"	3,590	416.00	4,006
NEMA 3R, meter center, main switch, 1 phase, 240v				
400a	EA.	729.00	278.00	1,007
600a	"	1,380	347.00	1,727
800a	"	2,110	382.00	2,492
Main breaker				
400a	EA.	1,760	278.00	2,038
600a	"	2,250	347.00	2,597
800a	"	2,770	252.00	3,022
1000a	"	3,260	521.00	3,781
1200a	"	4,600	555.00	5,155
Terminal box				
225a	EA.	172.00	252.00	424.00
800a	"	255.00	399.00	654.00
1600a	"	839.00	625.00	1,464
NEMA 3R, three phase, 280v				
400a	EA.	824.00	295.00	1,119
600a	"	1,660	416.00	2,076
800a	"	3,080	451.00	3,531
Main breaker				
400a	EA.	2,040	295.00	2,335
600a	"	2,620	416.00	3,036
800a	"	3,320	451.00	3,771
1000a	"	3,720	589.00	4,309
1200a	"	5,100	625.00	5,725
Terminal box				
225a	EA.	194.00	278.00	472.00
800a	"	284.00	451.00	735.00

SERVICE AND DISTRIBUTION

16430.20 METERING	UNIT	MAT.	INST.	TOTAL
1600a	EA.	879.00	729.00	1,608
NEMA 3R meter center, single phase, 208v, 100a				
2 meters	EA.	211.00	174.00	385.00
3 meters	"	255.00	214.00	469.00
4 meters	"	374.00	252.00	626.00
5 meters	"	448.00	278.00	726.00
6 meters	"	713.00	312.00	1,025
4 meters	"	374.00	252.00	626.00
6 meters	"	560.00	278.00	838.00
7 meters	"	653.00	347.00	1,000
8 meters	"	747.00	382.00	1,129
125a, 3 meters	"	1,280	214.00	1,494
4 meters	"	1,710	252.00	1,962
6 meters	"	2,560	286.00	2,846
7 meters	"	2,990	347.00	3,337
8 meters	"	3,420	382.00	3,802
150a, 3 meters	"	1,340	214.00	1,554
4 meters	"	1,780	252.00	2,032
6 meters	"	2,680	295.00	2,975
7 meters	"	3,130	347.00	3,477
8 meters	"	3,570	382.00	3,952
NEMA 3R center, 3 phase, 208v, 125a breakers				
3 meters	EA.	3,010	214.00	3,224
4 meters	"	402.00	252.00	654.00
6 meters	"	604.00	295.00	899.00
7 meters	"	703.00	347.00	1,050
8 meters	"	805.00	382.00	1,187
150a				
3 meters	EA.	1,330	231.00	1,561
4 meters	"	1,740	252.00	1,992
6 meters	"	2,610	312.00	2,922
7 meters	"	3,050	364.00	3,414
8 meters	"	3,480	399.00	3,879
200a				
3 meters	EA.	1,370	231.00	1,601
4 meters	"	1,820	252.00	2,072
6 meters	"	2,730	312.00	3,042
7 meters	"	3,190	382.00	3,572
8 meters	"	3,640	399.00	4,039
NEMA 3R, center plug-on breakers, 208v, 1 phase				
60a	EA.	15.70	87.00	102.70
70a	"	31.40	8.70	40.10
90a	"	45.20	8.70	53.90
100a	"	47.60	12.05	59.65
125a	"	91.50	13.90	105.40

16460.10 TRANSFORMERS	UNIT	MAT.	INST.	TOTAL
Pad mounted, single phase, dry type, 480v-120/240v				
15 kva	EA.	923.00	278.00	1,201
25 kva	"	1,370	312.00	1,682
37.5 kva	"	2,030	347.00	2,377

16 ELECTRICAL

SERVICE AND DISTRIBUTION

16460.10 TRANSFORMERS

SERVICE AND DISTRIBUTION	UNIT	MAT.	INST.	TOTAL
50 kva	EA.	2,240	380.00	2,620
3 phase				
225 kva	EA.	4,760	868.00	5,628
300 kva	"	6,120	1,070	7,190
500 kva	"	9,630	1,320	10,950
750 kva	"	15,520	1,630	17,150
1000 kva	"	18,810	1,740	20,550
1500 kva	"	22,180	1,980	24,160
Substation transformers, outdoor, 5 kv - 208v				
112.5 kva	EA.	7,820	763.00	8,583
150 kva	"	8,500	834.00	9,334
225 kva	"	9,630	971.00	10,601
300 kva	"	11,100	1,040	12,140
500 kva	"	14,730	1,560	16,290
750 kva	"	20,390	1,910	22,300
1000 kva	"	25,150	2,260	27,410
15 kv, 208v				
112 kva	EA.	9,790	971.00	10,761
150 kva	"	10,010	1,040	11,050
225 kva	"	10,340	1,210	11,550
300 kva	"	11,660	1,390	13,050
500 kva	"	16,390	1,740	18,130
750 kva	"	20,790	2,090	22,880
1000 kva	"	24,640	2,440	27,080
5kv, 480v				
112kva	EA.	7,480	763.00	8,243
150 kva	"	7,920	834.00	8,754
225 kva	"	8,910	971.00	9,881
300 kva	"	10,010	1,040	11,050
500 kva	"	13,420	1,560	14,980
750 kva	"	18,260	1,910	20,170
1000 kva	"	21,450	2,260	23,710
1500 kva	"	28,050	2,590	30,640
2000 kva	"	34,540	3,120	37,660
2500 kva	"	41,030	3,800	44,830
15 kv, 480v				
112.5 kva	EA.	9,570	971.00	10,541
150 kva	"	9,790	1,040	10,830
225 kva	"	9,900	1,210	11,110
300 kva	"	10,780	1,390	12,170
500 kva	"	15,290	1,740	17,030
750 kva	"	18,590	2,090	20,680
1000 kva	"	21,450	2,440	23,890
1500 kva	"	28,160	2,780	30,940
2000 kva	"	34,650	3,120	37,770
2500 kva	"	41,140	4,140	45,280
Pad mounted 3 phase, 15 kv outdoor				
50 kva	EA.	2,310	356.00	2,666
75 kva	"	3,250	408.00	3,658
112 kva	"	4,180	448.00	4,628
150 kva	"	5,940	505.00	6,445
225 kva	"	7,150	534.00	7,684

SERVICE AND DISTRIBUTION

SERVICE AND DISTRIBUTION	UNIT	MAT.	INST.	TOTAL
16460.10 TRANSFORMERS				
300 kva	EA.	8,800	591.00	9,391
500 kva	"	10,780	957.00	11,737
750 kva	"	13,750	1,260	15,010
1000 kva	"	16,280	1,540	17,820
1500 kva	"	20,790	1,850	22,640
Dry type, for power gear, 5 kv indoor				
75 kva	EA.	6,380	555.00	6,935
112.5 kva	"	7,150	646.00	7,796
150 kva	"	8,250	731.00	8,981
225 kva	"	10,230	816.00	11,046
300 kva	"	12,100	868.00	12,968
500 kva	"	15,980	957.00	16,937
750 kva	"	21,010	1,260	22,270
16470.10 PANELBOARDS				
Indoor load center, 1 phase 240v main lug only				
30a - 2 spaces	EA.	9.35	69.50	78.85
100a - 8 spaces	"	31.60	84.00	115.60
150a - 16 spaces	"	77.00	103.00	180.00
200a - 24 spaces	"	100.00	121.00	221.00
200a - 42 spaces	"	185.00	139.00	324.00
Main circuit breaker				
100a - 8 spaces	EA.	78.00	84.00	162.00
100a - 16 spaces	"	100.00	95.50	195.50
150a - 16 spaces	"	162.00	103.00	265.00
150a - 24 spaces	"	186.00	111.00	297.00
200a - 24 spaces	"	189.00	121.00	310.00
200a - 42 spaces	"	261.00	126.00	387.00
3 phase, 480/277v, main lugs only, 120a, 30 circuits	"	515.00	121.00	636.00
277/480v, 4 wire, flush surface				
225a, 30 circuits	EA.	549.00	139.00	688.00
400a, 30 circuits	"	629.00	174.00	803.00
600a, 42 circuits	"	783.00	209.00	992.00
208/120v, main circuit breaker, 3 phase, 4 wire				
100a				
12 circuits	EA.	451.00	177.00	628.00
20 circuits	"	554.00	219.00	773.00
30 circuits	"	817.00	244.00	1,061
225a				
30 circuits	EA.	697.00	270.00	967.00
42 circuits	"	1,260	330.00	1,590
400a				
30 circuits	EA.	1,720	514.00	2,234
42 circuits	"	2,060	555.00	2,615
600a, 42 circuits	"	2,290	631.00	2,921
120/208v, flush, 3 ph., 4 wire, main only				
100a				
12 circuits	EA.	315.00	177.00	492.00
20 circuits	"	434.00	219.00	653.00
30 circuits	"	629.00	244.00	873.00
225a				

SERVICE AND DISTRIBUTION

16470.10 PANELBOARDS

	UNIT	MAT.	INST.	TOTAL
30 circuits	EA.	658.00	270.00	928.00
42 circuits	"	835.00	330.00	1,165
400a				
30 circuits	EA.	1,260	514.00	1,774
42 circuits	"	1,830	555.00	2,385
600a, 42 circuits	"	2,860	631.00	3,491
Panelboard accessories				
Grounding bus	EA.	12.60	12.05	24.65
Handle lock device	"	1.70	5.05	6.75
Factory assembled panel				
1 pole space	EA.	9.15	12.05	21.20
2 pole space	"	19.45	5.05	24.50
3 pole space	"	28.60	4.65	33.25
Panelboards 1 phase, 240/120v main circuit breaker				
Single phase, 3 wire, 120/240v flush				
100a, 20 circuits	EA.	372.00	121.00	493.00
225a, 30 circuits	"	744.00	139.00	883.00
240/120v, main lugs only	"			
100a				
8 circuits	EA.	222.00	103.00	325.00
12 circuits	"	240.00	103.00	343.00
20 circuits	"	252.00	103.00	355.00
225a				
24 circuits	EA.	268.00	121.00	389.00
30 circuits	"	306.00	132.00	438.00
42 circuits	"	338.00	132.00	470.00
Distribution panelboards, 3 ph, main breaker				
225a	EA.	846.00	555.00	1,401
400a	"	1,540	625.00	2,165
600a	"	2,400	763.00	3,163
800a	"	3,090	834.00	3,924
1000a	"	4,000	971.00	4,971
1200a	"	4,920	1,040	5,960
Single phase				
225a	EA.	744.00	486.00	1,230
400a	"	1,370	555.00	1,925
600a	"	1,940	694.00	2,634
800a	"	2,630	834.00	3,464
1000a	"	3,430	971.00	4,401
1200a	"	4,350	1,040	5,390
Fusible distribution panelboards, 3 phase, 600v				
100a	EA.	801.00	486.00	1,287
200a	"	915.00	555.00	1,470
400a	"	1,600	694.00	2,294
600a	"	2,060	834.00	2,894
800a	"	3,090	971.00	4,061
Single phase				
100a	EA.	686.00	416.00	1,102
200a	"	801.00	486.00	1,287
400a	"	1,720	625.00	2,345
600a	"	1,830	763.00	2,593
800a	"	2,520	901.00	3,421

SERVICE AND DISTRIBUTION

	UNIT	MAT.	INST.	TOTAL
16470.10 PANELBOARDS				
Hospital panels, operating room				
3kv - 208v	EA.	2,520	214.00	2,734
3kv - 277v	"	2,630	214.00	2,844
5kv - 208v	"	2,630	214.00	2,844
5kv - 277v	"	2,860	214.00	3,074
Coronary care				
3kv - 208v	EA.	2,970	252.00	3,222
3kv - 277v	"	3,200	252.00	3,452
5kv - 208v	"	3,200	252.00	3,452
5kv - 277v	"	3,430	252.00	3,682
Intensive care				
3kv - 208v	EA.	3,320	278.00	3,598
3kv - 277v	"	3,430	278.00	3,708
5kv - 208v	"	3,430	278.00	3,708
5kv - 277v	"	3,660	278.00	3,938
15kv - 208v	"	6,520	416.00	6,936
15kv - 277v	"	6,750	416.00	7,166
25kv - 208v	"	6,980	555.00	7,535
25kv - 277v	"	7,210	555.00	7,765
Explosion proof, 240v, m.l.b. 20a, single phase				
6 breakers	EA.	1,830	382.00	2,212
8 breakers	"	2,060	408.00	2,468
10 breakers	"	2,400	434.00	2,834
12 breakers	"	2,750	460.00	3,210
14 breakers	"	3,320	486.00	3,806
16 breakers	"	4,000	486.00	4,486
18 breakers	"	4,350	538.00	4,888
20 breakers	"	4,580	564.00	5,144
22 breakers	"	4,690	589.00	5,279
24 breakers	"	4,920	616.00	5,536
16480.10 MOTOR CONTROLS				
Motor generator set, 3 phase, 480/277v, w/controls				
10kw	EA.	6,920	957.00	7,877
15kw	"	9,020	1,070	10,090
20kw	"	10,010	1,110	11,120
25kw	"	11,550	1,210	12,760
30kw	"	12,920	1,260	14,180
40kw	"	14,120	1,320	15,440
50kw	"	15,730	1,390	17,120
60kw	"	17,820	1,540	19,360
75kw	"	20,020	1,740	21,760
100kw	"	23,100	2,140	25,240
125kw	"	41,800	2,310	44,110
150kw	"	46,200	2,310	48,510
200kw	"	52,800	2,520	55,320
250kw	"	57,200	2,520	59,720
300kw	"	66,000	2,780	68,780
2 pole, 230 volt starter, w/NEMA-1				
1 hp, 9 amp, size 00	EA.	100.00	34.70	134.70
2 hp, 18amp, size 0	"	128.00	34.70	162.70
3 hp, 27amp, size 1	"	139.00	34.70	173.70

SERVICE AND DISTRIBUTION

16480.10 MOTOR CONTROLS

	UNIT	MAT.	INST.	TOTAL
5 hp, 45amp, size 1p	EA.	179.00	34.70	213.70
7-1/2 hp, 45a, size 2	"	302.00	34.70	336.70
15 hp, 90a, size 3	"	460.00	34.70	494.70
2 pole, w/NEMA-4 enclosure				
2 hp, 18a, size 1	EA.	261.00	55.50	316.50
5 hp, 45amp, size 1p	"	331.00	55.50	386.50
7-1/2 hp, 45a, size 2	"	524.00	55.50	579.50
3 pole, 2 hp, 9a, 200-575v starter				
W/NEMA-1, size 00	EA.	128.00	46.30	174.30
W/NEMA-4 enclosure, size 00	"	261.00	60.50	321.50
5hp, 18a				
W/NEMA-1 enclosure, size 0	EA.	170.00	46.30	216.30
W/NEMA-4 enclosure, size 0	"	290.00	60.50	350.50
7.5-10hp, 27a				
7.5-10hp 27a, w/NEMA-1 enclosure, size 1	EA.	174.00	46.30	220.30
W/NEMA-4 enclosure size 1	"	290.00	60.50	350.50
10-25hp, 45a				
W/NEMA-1 enclosure, size 2	EA.	302.00	46.30	348.30
W/NEMA-4 enclosure, size 2	"	536.00	60.50	596.50
25-50hp, 90a				
W/NEMA-1 enclosure, size 3	EA.	524.00	60.50	584.50
W/NEMA-4 enclosure, size 3	"	515.00	87.00	602.00
40-100hp, 135a				
W/NEMA-1 enclosure, size 4	EA.	1,130	87.00	1,217
W/NEMA-4 enclosure, size 4	"	1,750	121.00	1,871
75-200hp, 270a				
W/NEMA-1 enclosure, size 5	EA.	2,670	191.00	2,861
W/NEMA-4 enclosure, size 5	"	3,380	244.00	3,624
Magnetic starter accessories				
On-off-auto selector switch kit	EA.	34.50	11.10	45.60
With pilot light	"	63.00	12.05	75.05
Control center main lug only, 208v, 3 phase				
600a	EA.	1,070	416.00	1,486
1200a	"	2,380	555.00	2,935
Main circuit breakers, 208v, 3 phase				
400a	EA.	2,140	347.00	2,487
600a	"	2,380	486.00	2,866
800a	"	2,740	555.00	3,295
1000a	"	2,970	625.00	3,595
1200a	"	5,470	694.00	6,164
Non-reversing starters				
Size 1	EA.	416.00	25.20	441.20
Size 2	"	486.00	43.40	529.40
Size 3	"	547.00	52.50	599.50
Size 4	"	1,020	60.50	1,081
Reversing starters				
Size 1	EA.	547.00	25.20	572.20
Size 2	"	748.00	38.00	786.00
Fusible switch, non-revolving starters				
Size 1	EA.	367.00	25.20	392.20
Size 2	"	463.00	43.40	506.40
Size 3	"	594.00	52.50	646.50

SERVICE AND DISTRIBUTION

16480.10 MOTOR CONTROLS

SERVICE AND DISTRIBUTION	UNIT	MAT.	INST.	TOTAL
Size 4	EA.	952.00	60.50	1,013
Reversing starters				
Size 1	EA.	392.00	25.20	417.20
Size 2	"	571.00	38.00	609.00
Two speed, non-reversing starter				
Size 1	EA.	547.00	25.20	572.20
Size 2	"	802.00	38.00	840.00
Magnetic starter, 600v, 2 pole, NEMA 3R				
Size 0, 2 hp	EA.	180.00	34.70	214.70
Size 1, 5hp	"	238.00	38.00	276.00
NEMA 3R				
Size 2, 7.5 hp	EA.	385.00	39.70	424.70
Size 3, 15 hp	"	844.00	41.40	885.40
NEMA 7				
Size 0, 2 hp	EA.	578.00	60.50	638.50
Size 1, 5 hp	"	651.00	69.50	720.50
Size 2, 7.5 hp	"	962.00	77.00	1,039
NEMA 12				
Size 0, 2 hp	EA.	180.00	52.50	232.50
Size 1, 5 hp	"	238.00	60.50	298.50
Size 2, 7.5 hp	"	385.00	69.50	454.50
Size 3, 15 hp	"	578.00	77.00	655.00
3 pole, NEMA 1				
Size 00	EA.			
Size 0	"			
Size 1	"			
Size 2	"			
Size 3	"			
Size 4	"			
Size 5	"			
Size 6	"	7,820	347.00	8,167
Size 7	"	10,540	416.00	10,956
Size 8	"	14,620	555.00	15,175
NEMA 4				
Size 00	EA.			
Size 0	"			
Size 1	"			
Size 2	"			
Size 3	"			
Size 4	"			
Size 5	"			
Size 6	"	9,860	486.00	10,346
Size 7	"	12,580	555.00	13,135
Size 8	"	16,660	694.00	17,354
NEMA 3R				
Size 0	EA.	192.00	43.40	235.40
Size 1	"	214.00	47.10	261.10
Size 2	"	407.00	63.00	470.00
Size 3	"	6,350	66.00	6,416
Size 4	"	1,590	95.50	1,686
NEMA 7				
Size 0	EA.	588.00	69.50	657.50

SERVICE AND DISTRIBUTION

16480.10 MOTOR CONTROLS

	UNIT	MAT.	INST.	TOTAL
Size 1	EA.	623.00	75.00	698.00
Size 2	"	997.00	77.00	1,074
Size 3	"	1,470	95.50	1,566
Size 4	"	2,490	154.00	2,644
Size 5	"	5,780	338.00	6,118
Size 6	"	14,960	555.00	15,515
NEMA 12				
Size 00	EA.	101.00	60.50	161.50
Size 0	"	135.00	64.50	199.50
Size 1	"	153.00	66.00	219.00
Size 2	"	283.00	69.50	352.50
Size 3	"	453.00	87.00	540.00
Size 4	"	1,030	121.00	1,151
Size 5	"	2,490	252.00	2,742
Size 6	"	6,000	434.00	6,434
Size 7	"	8,610	486.00	9,096
Size 8	"	12,690	694.00	13,384
Reversing magnetic starters, 600v, 3 pole, NEMA 1				
Size 00	EA.	272.00	43.40	315.40
Size 0	"	317.00	44.80	361.80
Size 1	"	362.00	47.10	409.10
Size 2	"	713.00	52.50	765.50
Size 3	"	1,250	60.50	1,311
Size 4	"	2,830	69.50	2,900
Size 5	"	6,120	185.00	6,305
Size 6	"	13,710	330.00	14,040
Size 7	"	18,690	382.00	19,072
Size 8	"	26,630	625.00	27,255
NEMA 4				
Size 0	EA.	509.00	60.50	569.50
Size 4	"	623.00	63.00	686.00
Size 2	"	1,110	66.00	1,176
Size 3	"	1,700	69.50	1,770
Size 4	"	3,850	87.00	3,937
Size 5	"	7,020	252.00	7,272
Size 6	"	15,750	434.00	16,184
Size 7	"	20,620	521.00	21,141
Size 8	"	28,550	694.00	29,244
NEMA 7				
Size 0	EA.	1,040	69.50	1,110
Size 1	"	1,090	77.00	1,167
Size 2	"	1,810	87.00	1,897
Size 3	"	2,950	103.00	3,053
NEMA 12				
Size 0	EA.	396.00	60.50	456.50
Size 1	"	441.00	66.00	507.00
Size 2	"	826.00	69.50	895.50
Size 3	"	1,470	77.00	1,547
Size 4	"	3,170	87.00	3,257
Size 5	"	7,020	252.00	7,272
Size 6	"	14,840	486.00	15,326
Size 7	"	19,830	555.00	20,385

SERVICE AND DISTRIBUTION

16480.10 MOTOR CONTROLS

	UNIT	MAT.	INST.	TOTAL
Size 8	EA.	27,760	694.00	28,454
Electrically held lighting contactors, NEMA 1, 20a				
2 pole	EA.	120.00	34.70	154.70
3 pole	"	132.00	43.40	175.40
4 pole	"	169.00	52.50	221.50
6 pole	"	243.00	69.50	312.50
8 pole	"	317.00	87.00	404.00
10 pole	"	362.00	103.00	465.00
12 pole	"	419.00	121.00	540.00
30a				
2 pole	EA.	130.00	34.70	164.70
3 pole	"	141.00	43.40	184.40
4 pole	"	169.00	52.50	221.50
5 pole	"	227.00	60.50	287.50
60a				
2 pole	EA.	266.00	34.70	300.70
3 pole	"	283.00	43.40	326.40
4 pole	"	351.00	52.50	403.50
5 pole	"	509.00	60.50	569.50
100a				
2 pole	EA.	441.00	43.40	484.40
3 pole	"	470.00	60.50	530.50
4 pole	"	578.00	77.00	655.00
5 pole	"	826.00	95.50	921.50
200a				
2 pole	EA.	1,030	95.50	1,126
3 pole	"	1,100	87.00	1,187
4 pole	"	1,470	111.00	1,581
300a				
2 pole	EA.	2,180	146.00	2,326
3 pole	"	2,380	185.00	2,565
400a				
2 pole	EA.	5,710	146.00	5,856
3 pole	"	6,460	185.00	6,645
600a				
2 pole	EA.	7,020	231.00	7,251
3 pole	"	7,820	321.00	8,141
800a				
2 pole	EA.	8,380	278.00	8,658
3 pole	"	9,290	382.00	9,672
Mechanically held lighting contactors, NEMA 1, 20a				
2 pole	EA.	192.00	34.70	226.70
3 pole	"	204.00	43.40	247.40
4 pole	"	214.00	52.50	266.50
6 pole	"	351.00	69.50	420.50
8 pole	"	385.00	87.00	472.00
10 pole	"	430.00	103.00	533.00
30a				
2 pole	EA.	204.00	34.70	238.70
3 pole	"	214.00	43.40	257.40
4 pole	"	227.00	52.50	279.50
5 pole	"	283.00	60.50	343.50

16 ELECTRICAL

SERVICE AND DISTRIBUTION

	UNIT	MAT.	INST.	TOTAL
16480.10 MOTOR CONTROLS				
60a				
2 pole	EA.	407.00	34.70	441.70
3 pole	"	425.00	43.40	468.40
4 pole	"	509.00	52.50	561.50
5 pole	"	657.00	60.50	717.50
100a				
2 pole	EA.	566.00	43.40	609.40
3 pole	"	600.00	60.50	660.50
4 pole	"	719.00	69.50	788.50
5 pole	"	968.00	87.00	1,055
200a				
2 pole	EA.	1,460	60.50	1,521
3 pole	"	1,650	87.00	1,737
4 pole	"	2,020	111.00	2,131
300a				
2 pole	EA.	2,610	146.00	2,756
3 pole	"	2,830	185.00	3,015
400a				
2 pole	EA.	6,230	146.00	6,376
3 pole	"	7,020	185.00	7,205
600a				
2 pole	EA.	7,480	231.00	7,711
3 pole	"	8,380	308.00	8,688
800a				
2 pole	EA.	8,840	278.00	9,118
3 pole	"	9,860	397.00	10,257
AC relays, control type open, 15a, 600v				
2 pole	EA.	68.00	34.70	102.70
3 pole	"	81.50	43.40	124.90
4 pole	"	91.50	52.50	144.00
6 pole	"	136.00	69.50	205.50
8 pole	"	160.00	87.00	247.00
10 pole	"	229.00	103.00	332.00
12 pole	"	245.00	121.00	366.00
16490.10 SWITCHES				
Oil switches, medium voltage, bus components				
Switches, 277/120v, toggle device only	EA.	283.00	55.50	338.50
With oil 35kv, g&w gram 44, 4 way switch	"	9,350	278.00	9,628
Weatherproof enclosure				
3 way switch	EA.	11,550	347.00	11,897
4 way switch	"	12,100	380.00	12,480
Fused interrupter load, 35kv				
20A				
1 pole	EA.	13,200	555.00	13,755
2 pole	"	14,300	591.00	14,891
3 way	"	15,400	591.00	15,991
4 way	"	16,500	631.00	17,131
30a, 1 pole	"	13,200	555.00	13,755
3 way	"	15,400	591.00	15,991
4 way	"	16,500	631.00	17,131
Weatherproof switch, including box & cover, 20a				

SERVICE AND DISTRIBUTION

16490.10 SWITCHES

	UNIT	MAT.	INST.	TOTAL
1 pole	EA.	14,300	555.00	14,855
2 pole	"	15,400	591.00	15,991
3 way	"	16,500	631.00	17,131
4 way	"	17,600	631.00	18,231
3 way, oil switch, 15kv enclosure	"	12,100	414.00	12,514
Pedestal for 35kv double breaker switch	"	462.00	174.00	636.00
Bus terminal connector, 2	"	404.00	87.00	491.00
2 to 3	"	491.00	87.00	578.00
Support connector, 3	"	260.00	55.50	315.50
Tee connector, 2 to 3	"	375.00	69.50	444.50
Flexible bus stud connector	"	317.00	60.50	377.50
End cap 3	"	346.00	46.30	392.30
Weldment connection, 3	"	144.00	34.70	178.70
Plate switch, 1 gang	"	0.34	1.75	2.09
Start stop stations, manual motor starters	"	34.70	25.20	59.90
Lockout switch	"	5.80	8.70	14.50
Forward-reverse switch	"	40.40	25.20	65.60
On-off switch	"	40.40	25.20	65.60
Open-close switch	"	40.40	25.20	65.60
Forward-reverse-stop switch	"	46.20	34.70	80.90
Standard 3 button switch any standard legend	"	46.20	34.70	80.90
Standard 3 button with lockout	"	51.50	34.70	86.20
Manual motor starters, tog, 115/230v				
Size 1 gp	EA.	69.50	34.70	104.20
Size 2	"	92.50	34.70	127.20
Button				
Size 0	EA.	52.00	34.70	86.70
Size 1	"	75.00	34.70	109.70
Size 2	"	98.00	34.70	132.70
3-phase				
Size 0	EA.	86.00	46.30	132.30
Size 1	"	110.00	46.30	156.30
Time & float switches	"	173.00	55.50	228.50
Astronomical time switch, 40a, 240v	"	271.00	34.70	305.70
Timer switch 0-5 minute, with box	"	17.35	17.35	34.70
Single pole/single throw time, 277v, NEMA-1	"	58.00	25.20	83.20
Single toggle switch, 20a, 120v, with pilot	"	11.80	8.70	20.50
3-way toggle	"	32.40	10.30	42.70
Photo electric switches				
1000 watt				
105-135v	EA.	18.00	25.20	43.20
208-277v	"	24.00	25.20	49.20
3000 watt, 105-130v				
Double throw	EA.	72.00	34.70	106.70
Single throw	"	66.00	34.70	100.70
Double pole/single throw, 210-250v	"	83.50	46.30	129.80
Dimmer switch and switch plate				
600 w	EA.	18.00	10.70	28.70
1000 w	"	42.00	12.05	54.05
Dimmer switch incandescent				
1500w	EA.	94.50	24.35	118.85
2000w	"	136.00	25.90	161.90

16 ELECTRICAL

SERVICE AND DISTRIBUTION	UNIT	MAT.	INST.	TOTAL
16490.10 SWITCHES				
Fluorescent				
12 lamps	EA.	42.00	17.35	59.35
20 lamps	"	72.00	19.15	91.15
30 lamps	"	120.00	20.85	140.85
40 lamps	"	144.00	24.35	168.35
Time clocks with skip, 40a, 120v				
SPST	EA.	48.00	25.90	73.90
SPDT	"	54.00	25.90	79.90
DPST	"	54.00	25.90	79.90
DPDT	"	83.50	34.70	118.20
SPST	"	89.00	34.70	123.70
Astronomic time clocks with skip, 40a, 120v				
DPST	EA.	83.50	25.90	109.40
SPST	"	131.00	34.70	165.70
SPDT	"	95.50	25.90	121.40
Raintight time clocks, 40a, 120v				
SPDT	EA.	60.00	34.70	94.70
DPST	"	72.00	34.70	106.70
Contractor grade wall switch 15a, 120v				
Single pole	EA.	1.45	5.55	7.00
Three way	"	2.40	6.95	9.35
Four way	"	12.00	9.25	21.25
Specification grade toggle switches, 20a, 120-277v				
Single pole	EA.	9.00	6.95	15.95
Double pole	"	10.35	10.30	20.65
3 way	"	9.55	8.70	18.25
4 way	"	24.25	10.30	34.55
30a, 120-277v				
Single pole	EA.	13.45	6.95	20.40
Double pole	"	17.25	10.30	27.55
3 way	"	16.65	8.70	25.35
Specification grade key switches, 20a, 120-277v				
Single pole	EA.	13.80	6.95	20.75
Double pole	"	15.60	10.30	25.90
3 way	"	15.10	8.70	23.80
4 way	"	31.20	10.30	41.50
Red pilot light handle switches, 20a, 120-277v				
Single pole	EA.			
Double pole	"	18.70	10.30	29.00
3 way	"	24.50	8.70	33.20
30a, 120-277v				
Single pole	EA.	17.40	6.95	24.35
Double pole	"	21.25	10.30	31.55
3 way	"	32.40	8.70	41.10
Momentary contact switches, 20a				
SPDT, ivory	EA.	16.35	8.70	25.05
SPDT, locking	"	21.60	10.30	31.90
Maintained contact switches				
SPDT ivory	EA.	33.00	8.70	41.70
DPDT ivory	"	33.40	8.70	42.10
SPDT locking	"	38.50	10.30	48.80
DPDT locking	"	39.00	12.05	51.05

SERVICE AND DISTRIBUTION

	UNIT	MAT.	INST.	TOTAL
16490.10 SWITCHES				
Mercury switch, 3 way	EA.	8.15	8.70	16.85
Door switches, open on or off	"	14.40	17.35	31.75
Combination switch and pilot light, single pole	"	6.60	10.30	16.90
3 way	"	8.10	12.05	20.15
Combination switch and receptacle, single pole	"	9.40	10.30	19.70
3 way	"	9.50	10.30	19.80
Combination two switches, single pole/single pole	"	7.80	8.70	16.50
3 way	"	9.60	13.90	23.50
Switch plates, plastic ivory				
1 gang	EA.	0.34	2.80	3.14
2 gang	"	0.69	3.45	4.14
3 gang	"	1.05	4.15	5.20
4 gang	"	1.75	5.05	6.80
5 gang	"	3.95	5.55	9.50
6 gang	"	4.60	6.30	10.90
Stainless steel				
1 gang	EA.	1.15	2.80	3.95
2 gang	"	2.25	3.45	5.70
3 gang	"	3.95	4.25	8.20
4 gang	"	5.40	5.05	10.45
5 gang	"	7.45	5.55	13.00
6 gang	"	9.25	5.05	14.30
Brass				
1 gang	EA.	2.50	2.80	5.30
2 gang	"	4.95	3.45	8.40
3 gang	"	9.00	4.25	13.25
4 gang	"	12.10	5.05	17.15
5 gang	"	15.00	5.05	20.05
6 gang	"	17.90	6.30	24.20
16490.20 TRANSFER SWITCHES				
Automatic transfer switch 600v, 3 pole				
30a	EA.	1,320	121.00	1,441
60a	"	1,870	121.00	1,991
100a	"	2,750	165.00	2,915
150a	"	3,470	209.00	3,679
225a	"	4,180	278.00	4,458
260a	"	4,510	278.00	4,788
400a	"	6,050	347.00	6,397
600a	"	8,250	524.00	8,774
800a	"	10,120	631.00	10,751
1000a	"	14,520	731.00	15,251
1200a	"	16,500	793.00	17,293
1600a	"	20,350	868.00	21,218
2000a	"	21,780	1,030	22,810
2600a	"	38,500	1,460	39,960
Automatic transfer switches, 600v, 3 phase, 3000a	"	60,500	1,740	62,240
16490.80 SAFETY SWITCHES				
Safety switch, 600v, 3 pole, heavy duty, NEMA-1				
30a	EA.	108.00	34.70	142.70

SERVICE AND DISTRIBUTION

16490.80 SAFETY SWITCHES

	UNIT	MAT.	INST.	TOTAL
60a	EA.	129.00	39.70	168.70
100a	"	252.00	55.50	307.50
200a	"	323.00	87.00	410.00
400a	"	930.00	191.00	1,121
600a	"	1,520	278.00	1,798
800a	"	2,710	365.00	3,075
1200a	"	3,360	496.00	3,856

LIGHTING

16510.05 INTERIOR LIGHTING

	UNIT	MAT.	INST.	TOTAL
Recessed fluorescent fixtures, 2'x2'				
2 lamp	EA.	45.30	25.20	70.50
4 lamp	"	61.00	25.20	86.20
2 lamp w/flange	"	57.00	34.70	91.70
4 lamp w/flange	"	72.00	34.70	106.70
1'x4'				
2 lamp	EA.	46.10	23.15	69.25
3 lamp	"	63.50	23.15	86.65
2 lamp w/flange	"	57.00	25.20	82.20
3 lamp w/flange	"	74.50	25.20	99.70
2'x4'				
2 lamp	EA.	57.00	25.20	82.20
3 lamp	"	66.50	25.20	91.70
4 lamp	"	65.00	25.20	90.20
2 lamp w/flange	"	67.50	34.70	102.20
3 lamp w/flange	"	77.00	34.70	111.70
4 lamp w/flange	"	76.00	34.70	110.70
4'x4'				
4 lamp	EA.	136.00	34.70	170.70
6 lamp	"	160.00	34.70	194.70
8 lamp	"	168.00	34.70	202.70
4 lamp w/flange	"	164.00	52.50	216.50
6 lamp w/flange	"	207.00	52.50	259.50
8 lamp, w/flange	"	195.00	52.50	247.50
Surface mounted incandescent fixtures				
40w	EA.	41.20	23.15	64.35
75w	"	44.80	23.15	67.95
100w	"	53.00	23.15	76.15
150w	"	62.50	23.15	85.65
Pendant				
40w	EA.	44.80	27.80	72.60
75w	"	49.50	27.80	77.30

16 ELECTRICAL

LIGHTING	UNIT	MAT.	INST.	TOTAL
16510.05 INTERIOR LIGHTING				
100w	EA.	56.50	27.80	84.30
150w	"	64.50	27.80	92.30
Contractor grade recessed down lights				
100 watt housing only	EA.	16.50	34.70	51.20
150 watt housing only	"	19.80	34.70	54.50
100 watt trim	"	22.00	17.35	39.35
150 watt trim	"	27.50	17.35	44.85
Recessed incandescent fixtures				
40w	EA.	76.00	52.50	128.50
75w	"	79.50	52.50	132.00
100w	"	81.50	52.50	134.00
150w	"	93.50	52.50	146.00
Exit lights, 120v				
Recessed	EA.	59.00	43.40	102.40
Back mount	"	41.20	25.20	66.40
Universal mount	"	47.10	25.20	72.30
Emergency battery units, 6v-120v, 50 unit	"	88.00	52.50	140.50
With 1 head	"	102.00	52.50	154.50
With 2 heads	"	118.00	52.50	170.50
Mounting bucket	"	17.65	25.20	42.85
Light track single circuit				
2'	EA.	20.30	17.35	37.65
4'	"	37.90	17.35	55.25
8'	"	67.50	34.70	102.20
12'	"	107.00	52.50	159.50
Fittings and accessories				
Dead end	EA.	3.65	5.05	8.70
Starter kit	"	18.95	8.70	27.65
Conduit feed	"	23.00	5.05	28.05
Straight connector	"	12.80	5.05	17.85
Center feed	"	26.10	5.05	31.15
L-connector	"	20.30	5.05	25.35
T-connector	"	31.20	5.05	36.25
X-connector	"	40.60	6.95	47.55
Cord and plug	"	29.80	3.45	33.25
Rigid corner	"	39.30	5.05	44.35
Flex connector	"	61.00	5.05	66.05
2 way connector	"	88.00	6.95	94.95
Spacer clip	"	1.35	1.75	3.10
Grid box	"	7.45	5.05	12.50
T-bar clip	"	2.55	1.75	4.30
Utility hook	"	8.15	5.05	13.20
Fixtures, square				
R-20	EA.	54.00	5.05	59.05
R-30	"	56.50	5.05	61.55
40w flood	"	90.00	5.05	95.05
40w spot	"	93.50	5.05	98.55
100w flood	"	104.00	5.05	109.05
100w spot	"	107.00	5.05	112.05
Mini spot	"	75.00	5.05	80.05
Mini flood	"	77.00	5.05	82.05
Quartz, 500w	"	206.00	5.05	211.05

LIGHTING

16510.05 INTERIOR LIGHTING

LIGHTING	UNIT	MAT.	INST.	TOTAL
R-20 sphere	EA.	77.00	5.05	82.05
R-30 sphere	"	90.00	5.05	95.05
R-20 cylinder	"	42.00	5.05	47.05
R-30 cylinder	"	47.40	5.05	52.45
R-40 cylinder	"	67.50	5.05	72.55
R-30 wall wash	"	81.50	5.05	86.55
R-40 wall wash	"	102.00	5.05	107.05
Explosion proof, incan., surface mounted				
100w - 200w	EA.	191.00	60.50	251.50
300w	"	254.00	60.50	314.50
500w	"	359.00	60.50	419.50
With guard				
100w-200w	EA.	199.00	77.00	276.00
300w	"	266.00	77.00	343.00
500w	"	378.00	77.00	455.00
Reflectors for incan. light fixtures, dome	"	29.40	8.70	38.10
Angle	"	33.00	8.70	41.70
Highbay	"	70.50	10.30	80.80
Explosion proof fluor. fixtures, 800 ms.				
1 lamp	EA.	790.00	77.00	867.00
2 lamp	"	984.00	92.50	1,077
3 lamp	"	1,470	103.00	1,573
4 lamp	"	1,900	111.00	2,011
Explosion proof hp sodium fixtures				
50w-70w	EA.	535.00	77.00	612.00
100w	"	549.00	77.00	626.00
150w	"	572.00	87.00	659.00
200w	"	582.00	87.00	669.00
250w	"	606.00	87.00	693.00
310w	"	629.00	87.00	716.00
400w	"	942.00	87.00	1,029
With guard				
50w-70w	EA.	544.00	87.00	631.00
100w	"	559.00	87.00	646.00
150w	"	594.00	95.50	689.50
200w	"	606.00	95.50	701.50
250w	"	629.00	95.50	724.50
310w	"	656.00	95.50	751.50
400w	"	984.00	95.50	1,080
Explosion proof metal halide fixtures				
175w	EA.	464.00	87.00	551.00
250w	"	505.00	87.00	592.00
400w	"	747.00	87.00	834.00
With guard, 175w	"	476.00	95.50	571.50
250w	"	517.00	95.50	612.50
400w	"	779.00	95.50	874.50
Energy saving rapid start fluor. lamps				
F30cw	EA.	6.20	3.45	9.65
F40 cw	"	3.20	3.45	6.65
F40 cwx	"	5.50	3.45	8.95
F30 ww	"	7.45	3.45	10.90
F40 ww	"	4.05	3.45	7.50

LIGHTING

16510.05 INTERIOR LIGHTING

	UNIT	MAT.	INST.	TOTAL
F40 wwx	EA.	5.50	3.45	8.95
Slimline				
F48 cw	EA.	7.70	5.05	12.75
F96 cwx	"	11.90	5.05	16.95
F48 ww	"	9.10	5.05	14.15
F96 ww	"	8.70	5.05	13.75
F96 wwx	"	11.90	5.05	16.95
High output	"	7.75	5.05	12.80
F96 cwx	"	11.50	5.05	16.55
F96 cw	"	10.20	5.05	15.25
Power groove, F48 cw	"	18.55	5.05	23.60
Circle				
Fc6 cw	EA.	7.05	3.45	10.50
Fc8 cw	"	5.95	3.45	9.40
Fc12 cw	"	6.80	3.45	10.25
Fc16 cw	"	9.35	3.45	12.80
Fc6 ww	"	7.65	3.45	11.10
Fc8 ww	"	6.45	3.45	9.90
Fc12 ww	"	7.80	3.45	11.25
Fc16 ww	"	11.45	3.45	14.90
Incandescent lamps				
200w	EA.	3.20	3.45	6.65
300w	"	3.50	3.45	6.95
500w	"	7.85	3.45	11.30
750w	"	19.15	5.05	24.20
1000w	"	20.90	6.95	27.85
1500w	"	30.60	6.95	37.55
Energy saving reflector floodlight lamps				
25w	EA.	7.40	3.45	10.85
30w	"	5.95	3.45	9.40
50w	"	6.60	3.45	10.05
75w	"	5.80	3.45	9.25
120w	"	6.95	3.45	10.40
150w	"	4.85	3.45	8.30
200w	"	10.05	3.45	13.50
300w	"	9.95	3.45	13.40
500w	"	23.10	5.05	28.15
750w	"	29.20	5.05	34.25
Reflector spotlight				
75w	EA.	5.25	3.45	8.70
100w	"	10.95	3.45	14.40
125w	"	14.40	3.45	17.85
150w	"	18.90	3.45	22.35
250w	"	13.90	3.45	17.35
300w	"	14.45	3.45	17.90
400w	"	14.75	5.05	19.80
500w	"	23.90	5.05	28.95
1000w	"	53.00	6.95	59.95
Medium par flood lamps				
75w	EA.	5.20	3.45	8.65
100w	"	5.80	3.45	9.25
150w	"	6.25	3.45	9.70

16 ELECTRICAL

LIGHTING	UNIT	MAT.	INST.	TOTAL
16510.05 INTERIOR LIGHTING				
200w	EA.	23.90	3.45	27.35
300w	"	26.60	3.45	30.05
500w	"	52.00	5.05	57.05
Medium par spot lamps				
75w	EA.	6.45	3.45	9.90
120w	"	6.65	3.45	10.10
150w	"	6.95	3.45	10.40
Tubular quartz lamps				
100w	EA.	28.40	5.05	33.45
150w	"	17.00	5.05	22.05
200w	"	33.30	5.05	38.35
400w	"	30.80	5.05	35.85
500w	"	25.40	6.95	32.35
750w	"	31.40	6.95	38.35
1000w	"	44.50	8.70	53.20
1250w	"	47.00	8.70	55.70
1500w	"	38.10	8.70	46.80
Ballast replacements rapid start fluor				
1f-40-120v	EA.	14.00	25.20	39.20
1f-40-277v	"	23.55	25.20	48.75
1f-96-120v	"	72.50	25.20	97.70
1f-96-277v	"	88.00	25.20	113.20
2f-40-120v	"	20.00	25.20	45.20
2f-40-277v	"	23.55	25.20	48.75
2f-96-120v	"	62.00	25.20	87.20
2f-96-277v	"	65.00	25.20	90.20
Circline, 1fc6-1fc16	"	18.45	25.20	43.65
Very high output, 1500ma				
1f48-120v	EA.	106.00	25.20	131.20
1f48-277v	"	108.00	25.20	133.20
1f96-120v	"	93.50	25.20	118.70
1f96-277v	"	108.00	25.20	133.20
2f48-120v	"	93.50	25.20	118.70
2f48-277v	"	108.00	25.20	133.20
2f96-120v	"	107.00	25.20	132.20
2f96-277v	"	119.00	25.20	144.20
Mercury, multi tap				
475w	EA.	70.50	34.70	105.20
100w	"	67.00	34.70	101.70
175w	"	80.00	34.70	114.70
250w	"	95.50	34.70	130.20
400w	"	108.00	34.70	142.70
1000W	"	162.00	34.70	196.70
Metal halide, multi tap				
175w	EA.	96.50	34.70	131.20
250w	"	123.00	34.70	157.70
400w	"	153.00	34.70	187.70
1000w	"	229.00	34.70	263.70
1500w	"	294.00	34.70	328.70
High pressure sodium				
70w	EA.	134.00	34.70	168.70
100w	"	146.00	34.70	180.70

16 ELECTRICAL

LIGHTING

16510.05 INTERIOR LIGHTING

	UNIT	MAT.	INST.	TOTAL
150w	EA.	158.00	34.70	192.70
250w	"	229.00	34.70	263.70
400w	"	252.00	34.70	286.70
1000w	"	346.00	34.70	380.70
Surface mounted fluorescent, wrap around lens				
1 lamp	EA.	41.60	27.80	69.40
2 lamps	"	45.20	30.80	76.00
4 lamps	"	71.50	34.70	106.20
Wall mounted fluorescent				
2-20w lamps	EA.	33.30	17.35	50.65
2-30w lamps	"	35.70	17.35	53.05
2-40w lamps	"	41.60	23.15	64.75
Indirect, with wood shielding, 2049w lamps				
4'	EA.	71.50	34.70	106.20
8'	"	101.00	55.50	156.50
Industrial fluorescent, 2 lamp				
4'	EA.	71.50	25.20	96.70
8'	"	113.00	46.30	159.30
Strip fluorescent				
4'				
1 lamp	EA.	23.80	23.15	46.95
2 lamps	"	29.70	23.15	52.85
8'				
1 lamp	EA.	47.60	25.20	72.80
2 lamps	"	53.50	30.80	84.30
Wire guard for strip fixture, 4' long	"	11.90	12.05	23.95
Strip fluorescent, 8' long, two 4' lamps	"	71.50	46.30	117.80
With four 4' lamps	"	95.00	55.50	150.50
Wet location fluorescent, plastic housing				
4' long				
1 lamp	EA.	83.50	34.70	118.20
2 lamps	"	89.00	46.30	135.30
8' long				
2 lamps	EA.	147.00	55.50	202.50
4 lamps	"	201.00	60.50	261.50
Parabolic troffer, 2'x2'				
With 2 "U" lamps	EA.	101.00	34.70	135.70
With 3 "U" lamps	"	119.00	39.70	158.70
2'x4'				
With 2 40w lamps	EA.	135.00	39.70	174.70
With 3 40w lamps	"	154.00	46.30	200.30
With 4 40w lamps	"	166.00	46.30	212.30
1'x4'				
With 1 t12 lamp, 9 cell	EA.	101.00	25.20	126.20
With 2 t12 lamps	"	113.00	30.80	143.80
With 1 t12 lamp, 20 cell	"	113.00	25.20	138.20
With 2 t12 lamps	"	124.00	30.80	154.80
Steel sided surface fluorescent, 2'x4'				
3 lamps	EA.	89.00	46.30	135.30
4 lamps	"	95.00	46.30	141.30
Outdoor sign fluor., 1 lamp, remote ballast				
4' long	EA.	1,700	209.00	1,909

LIGHTING	UNIT	MAT.	INST.	TOTAL
16510.05 INTERIOR LIGHTING				
6' long	EA.	2,040	278.00	2,318
Recess mounted, commercial, 2'x2', 13" high				
100w	EA.	566.00	139.00	705.00
250w	"	623.00	156.00	779.00
High pressure sodium, hi-bay open				
400w	EA.	311.00	60.50	371.50
1000w	"	538.00	84.00	622.00
Enclosed				
400w	EA.	484.00	84.00	568.00
1000w	"	713.00	103.00	816.00
Metal halide hi-bay, open				
400w	EA.	283.00	60.50	343.50
1000w	"	481.00	84.00	565.00
Enclosed				
400w	EA.	453.00	84.00	537.00
1000w	"	680.00	103.00	783.00
High pressure sodium, low bay, surface mounted				
100w	EA.	198.00	34.70	232.70
150w	"	214.00	39.70	253.70
250w	"	243.00	46.30	289.30
400w	"	311.00	55.50	366.50
Metal halide, low bay, pendant mounted				
175w	EA.	214.00	46.30	260.30
250w	"	283.00	55.50	338.50
400w	"	351.00	77.00	428.00
Indirect luminare, square, metal halide, freestanding				
175w	EA.	566.00	34.70	600.70
250w	"	578.00	34.70	612.70
400w	"	623.00	34.70	657.70
High pressure sodium				
150w	EA.	578.00	34.70	612.70
250w	"	606.00	34.70	640.70
400w	"	657.00	34.70	691.70
Round, metal halide				
175w	EA.	554.00	34.70	588.70
250w	"	566.00	34.70	600.70
400w	"	600.00	34.70	634.70
High pressure sodium				
150w	EA.	543.00	34.70	577.70
250w	"	588.00	34.70	622.70
400w	"	612.00	34.70	646.70
Wall mounted, metal halide				
175w	EA.	441.00	87.00	528.00
250w	"	475.00	87.00	562.00
400w	"	781.00	111.00	892.00
High pressure sodium				
150w	EA.	509.00	87.00	596.00
250w	"	623.00	87.00	710.00
400w	"	815.00	111.00	926.00
Wall pack lithonia, high pressure sodium				
35w	EA.	135.00	30.80	165.80
55w	"	153.00	34.70	187.70

LIGHTING

16510.05 INTERIOR LIGHTING

LIGHTING	UNIT	MAT.	INST.	TOTAL
150w	EA.	169.00	55.50	224.50
250w	"	214.00	60.50	274.50
Low pressure sodium				
35w	EA.	192.00	60.50	252.50
55w	"	294.00	69.50	363.50
Wall pack hubbell, high pressure sodium				
35w	EA.	146.00	30.80	176.80
150w	"	186.00	55.50	241.50
250w	"	238.00	60.50	298.50
Compact fluorescent				
2-7w	EA.	90.00	34.70	124.70
2-13w	"	101.00	46.30	147.30
1-18w	"	124.00	46.30	170.30
Handball & racquet ball court, 2'x2', metal halide				
250w	EA.	317.00	87.00	404.00
400w	"	396.00	95.50	491.50
High pressure sodium				
250w	EA.	351.00	87.00	438.00
400w	"	385.00	95.50	480.50
Bollard light, 42" w/found., high pressure sodium				
70w	EA.	538.00	89.50	627.50
100w	"	550.00	89.50	639.50
150w	"	554.00	89.50	643.50
Light fixture lamps				
Lamp				
20w med. bipin base, cool white, 24"	EA.	4.10	5.05	9.15
30w cool white, rapid start, 36"	"	5.35	5.05	10.40
40w cool white "U", 3"	"	10.25	5.05	15.30
40w cool white, rapid start, 48"	"	2.45	5.05	7.50
70w high pressure sodium, mogul base	"	53.00	6.95	59.95
75w slimline, 96"	"	6.85	6.95	13.80
100w				
Incandescent, 100a, inside frost	EA.	2.10	3.45	5.55
Mercury vapor, clear, mogul base	"	25.90	6.95	32.85
High pressure sodium, mogul base	"	56.50	6.95	63.45
150w				
Par 38 flood or spot, incandescent	EA.	6.35	3.45	9.80
High pressure sodium, 1/2 mogul base	"	59.00	6.95	65.95
175w				
Mercury vapor, clear, mogul base	EA.	44.80	6.95	51.75
Mercury halide, clear, mogul base	"	20.00	6.95	26.95
250w				
High pressure sodium, mogul base	EA.	62.50	6.95	69.45
Mercury vapor, clear, mogul base	"	35.30	6.95	42.25
Metal halide, clear, mogul base	"	56.50	6.95	63.45
High pressure sodium, mogul base	"	62.50	6.95	69.45
400w				
Mercury vapor, clear, mogul base	EA.	28.30	6.95	35.25
Metal halide, clear, mogul base	"	53.00	6.95	59.95
High pressure sodium, mogul base	"	67.00	6.95	73.95
1000w				
Mercury vapor, clear, mogul base	EA.	61.50	8.70	70.20

16 ELECTRICAL

LIGHTING	UNIT	MAT.	INST.	TOTAL
16510.05 INTERIOR LIGHTING				
High pressure sodium, mogul base	EA.	157.00	8.70	165.70
16510.30 EXTERIOR LIGHTING				
Exterior light fixtures				
Rectangle, high pressure sodium				
70w	EA.	356.00	87.00	443.00
100w	"	367.00	89.50	456.50
150w	"	396.00	89.50	485.50
250w	"	532.00	95.50	627.50
400w	"	600.00	121.00	721.00
Flood, rectangular, high pressure sodium				
70w	EA.	367.00	87.00	454.00
100w	"	385.00	89.50	474.50
150w	"	425.00	89.50	514.50
400w	"	634.00	121.00	755.00
1000w	"	962.00	156.00	1,118
Round				
400w	EA.	657.00	121.00	778.00
1000w	"	1,030	156.00	1,186
Round, metal halide				
400w	EA.	725.00	121.00	846.00
1000w	"	1,080	156.00	1,236
Light fixture arms, cobra head, 6', high press. sodium				
100w	EA.	396.00	69.50	465.50
150w	"	623.00	87.00	710.00
250w	"	651.00	87.00	738.00
400w	"	673.00	103.00	776.00
Flood, metal halide				
400w	EA.	691.00	121.00	812.00
1000w	"	1,050	156.00	1,206
1500w	"	1,420	209.00	1,629
Mercury vapor				
250w	EA.	407.00	95.50	502.50
400w	"	704.00	121.00	825.00
Incandescent				
300w	EA.	73.50	60.50	134.00
500w	"	124.00	69.50	193.50
1000w	"	135.00	111.00	246.00
16510.90 POWER LINE FILTERS				
Heavy duty power line filter, 240v				
100a	EA.	2,750	347.00	3,097
300a	"	9,130	555.00	9,685
600a	"	12,650	841.00	13,491
16600.20 CENTRAL INVERTER SYSTEMS				
Central inverter systems				
500va	EA.	6,060	103.00	6,163
1000va	"	6,620	139.00	6,759

16 ELECTRICAL

LIGHTING

LIGHTING	UNIT	MAT.	INST.	TOTAL
16600.20 CENTRAL INVERTER SYSTEMS				
1500va	EA.	7,740	185.00	7,925
2400va	"	9,990	231.00	10,221
3000va	"	10,550	295.00	10,845
4500va	"	16,040	347.00	16,387
6000va	"	20,200	382.00	20,582
7500va	"	23,340	486.00	23,826
10,000va	"	26,930	555.00	27,485
16,600va	"	46,680	798.00	47,478
25,000va	"	55,650	1,210	56,860
16610.30 UNINTERRUPTIBLE POWER				
Uninterruptible power systems, (U.P.S.)	EA.	10,940	278.00	11,218
5 kva	"	13,530	382.00	13,912
7.5 kva	"	16,500	555.00	17,055
10 kva	"	23,100	763.00	23,863
15 kva	"	26,330	925.00	27,255
20 kva	"	35,200	834.00	36,034
25 kva	"	43,890	868.00	44,758
30 kva	"	46,640	901.00	47,541
35 kva	"	49,430	938.00	50,368
40 kva	"	52,250	971.00	53,221
45 kva	"	54,980	1,010	55,990
50 kva	"	57,750	1,040	58,790
62.5 kva	"	66,440	1,110	67,550
75 kva	"	75,130	1,210	76,340
100 kva	"	85,470	1,250	86,720
150 kva	"	106,260	1,740	108,000
200 kva	"	110,000	1,910	111,910
300 kva	"	110,000	2,590	112,590
400 kva	"	110,000	3,120	113,120
500 kva	"	110,000	3,800	113,800
16670.10 LIGHTNING PROTECTION				
Lightning protection				
Copper point, nickel plated, 12'				
1/2" dia.	EA.	26.40	34.70	61.10
5/8" dia.	"	31.20	34.70	65.90

COMMUNICATIONS

COMMUNICATIONS	UNIT	MAT.	INST.	TOTAL
16720.10 FIRE ALARM SYSTEMS				
Master fire alarm box, pedestal mounted	EA.	3,970	555.00	4,525
Master fire alarm box	"	2,040	209.00	2,249

COMMUNICATIONS

16720.10 FIRE ALARM SYSTEMS

	UNIT	MAT.	INST.	TOTAL
Box light	EA.	113.00	17.35	130.35
Ground assembly for box	"	56.50	23.15	79.65
Bracket for pole type box	"	73.50	25.20	98.70
Pull station				
Waterproof	EA.	56.50	17.35	73.85
Manual	"	34.00	13.90	47.90
Horn, waterproof	"	51.00	34.70	85.70
Interior alarm	"	34.00	25.20	59.20
Coded transmitter, automatic	"	600.00	69.50	669.50
Control panel, 8 zone	"	1,020	278.00	1,298
Battery charger and cabinet	"	363.00	69.50	432.50
Batteries, nickel cadmium or lead calcium	"	141.00	174.00	315.00
CO2 pressure switch connection	"	56.50	25.20	81.70
Annunciator panels				
Fire detection annunciator, remote type, 8-zone	EA.	204.00	63.00	267.00
12-zone	"	261.00	69.50	330.50
16-zone	"	328.00	87.00	415.00
Fire alarm systems				
Bell	EA.	62.50	21.35	83.85
Weatherproof bell	"	77.00	23.15	100.15
Horn	"	36.30	25.20	61.50
Siren	"	368.00	69.50	437.50
Chime	"	45.30	21.35	66.65
Audio/visual	"	65.50	25.20	90.70
Strobe light	"	60.50	25.20	85.70
Smoke detector	"	96.50	23.15	119.65
Heat detection	"	17.00	17.35	34.35
Thermal detector	"	15.85	17.35	33.20
Ionization detector	"	79.50	18.50	98.00
Duct detector	"	266.00	95.50	361.50
Test switch	"	45.30	17.35	62.65
Remote indicator	"	22.90	19.85	42.75
Door holder	"	98.50	25.20	123.70
Telephone jack	"	17.00	10.30	27.30
Fireman phone	"	243.00	34.70	277.70
Speaker	"	48.50	27.80	76.30
Remote fire alarm annunciator panel				
24 zone	EA.	1,250	231.00	1,481
48 zone	"	2,490	451.00	2,941
Control panel				
12 zone	EA.	849.00	103.00	952.00
16 zone	"	1,110	154.00	1,264
24 zone	"	1,700	231.00	1,931
48 zone	"	3,170	555.00	3,725
Power supply	"	198.00	52.50	250.50
Status command	"	5,380	174.00	5,554
Printer	"	2,100	52.50	2,153
Transponder	"	173.00	31.20	204.20
Transformer	"	124.00	23.15	147.15
Transceiver	"	169.00	25.20	194.20
Relays	"	68.00	17.35	85.35
Flow switch	"	220.00	69.50	289.50

16 ELECTRICAL

COMMUNICATIONS

COMMUNICATIONS	UNIT	MAT.	INST.	TOTAL
16720.10 FIRE ALARM SYSTEMS				
Tamper switch	EA.	133.00	103.00	236.00
End of line resistor	"	9.60	12.05	21.65
Printed ckt. card	"	85.00	17.35	102.35
Central processing unit	"	11,100	214.00	11,314
UPS backup to c.p.u.	"	10,760	312.00	11,072
Smoke detector, fixed temp. & rate of rise comb.	"	169.00	55.50	224.50
16720.50 SECURITY SYSTEMS				
Sensors				
Balanced magnetic door switch, surface mounted	EA.	81.00	17.35	98.35
With remote test	"	109.00	34.70	143.70
Flush mounted	"	77.50	64.50	142.00
Mounted bracket	"	5.95	12.05	18.00
Mounted bracket spacer	"	5.35	12.05	17.40
Photoelectric sensor, for fence				
6 beam	EA.	8,710	95.50	8,806
9 beam	"	10,640	148.00	10,788
Photoelectric sensor, 12 volt dc				
500' range	EA.	261.00	55.50	316.50
800' range	"	285.00	69.50	354.50
Capacitance wire grid kit				
Surface	EA.	71.50	34.70	106.20
Duct	"	53.50	55.50	109.00
Tube grid kit	"	89.00	17.35	106.35
Vibration sensor, 30 max per zone	"	107.00	17.35	124.35
Audio sensor, 30 max per zone	"	119.00	17.35	136.35
Inertia sensor				
Outdoor	EA.	83.00	25.20	108.20
Indoor	"	53.50	17.35	70.85
Ultrasonic transmitter, 20 max per zone				
Omni-directional	EA.	59.50	55.50	115.00
Directional	"	65.50	46.30	111.80
Transceiver				
Omni-directional	EA.	65.50	34.70	100.20
Directional	"	71.50	34.70	106.20
Passive infra-red sensor, 20 max per zone	"	463.00	55.50	518.50
Access/secure control unit, balanced magnetic switch	"	285.00	55.50	340.50
Photoelectric sensor	"	475.00	55.50	530.50
Photoelectric fence sensor	"	486.00	55.50	541.50
Capacitance sensor	"	564.00	60.50	624.50
Audio and vibration sensor	"	498.00	55.50	553.50
Inertia sensor	"	664.00	55.50	719.50
Ultrasonic sensor	"	760.00	60.50	820.50
Infra-red sensor	"	475.00	69.50	544.50
Monitor panel, with access/secure tone, standard	"	320.00	60.50	380.50
High security	"	475.00	69.50	544.50
Emergency power indicator	"	195.00	17.35	212.35
Monitor rack with 115v power supply				
1 zone	EA.	267.00	34.70	301.70
10 zone	"	1,370	87.00	1,457
Monitor cabinet, wall mounted				
1 zone	EA.	416.00	34.70	450.70

COMMUNICATIONS

	UNIT	MAT.	INST.	TOTAL
16720.50 SECURITY SYSTEMS				
5 zone	EA.	1,490	55.50	1,546
10 zone	"	653.00	60.50	713.50
20 zone	"	2,070	69.50	2,140
Floor mounted, 50 zone	"	2,150	139.00	2,289
Security system accessories				
Tamper assembly for monitor cabinet	EA.	53.50	15.40	68.90
Monitor panel blank	"	7.15	12.05	19.20
Audible alarm	"	60.50	17.35	77.85
Audible alarm control	"	249.00	12.05	261.05
Termination screw, terminal cabinet				
25 pair	EA.	178.00	55.50	233.50
50 pair	"	285.00	87.00	372.00
150 pair	"	469.00	174.00	643.00
Universal termination, cabinets & panel				
Remote test	EA.	41.60	60.50	102.10
No remote test	"	29.70	25.20	54.90
High security line supervision termination	"	212.00	34.70	246.70
Door cord for capacitance sensor, 12"	"	7.15	17.35	24.50
Insulation block kit for capacitance sensor	"	33.30	12.05	45.35
Termination block for capacitance sensor	"	7.25	12.05	19.30
Guard alert display	"	772.00	21.35	793.35
Uninterrupted power supply				
Plug-in 40kva transformer				
12 volt	EA.	29.70	12.05	41.75
18 volt	"	20.20	12.05	32.25
24 volt	"	14.25	12.05	26.30
Test relay	"	45.80	12.05	57.85
Coaxial cable, 50 ohm	L.F.	0.19	0.21	0.40
Door openers	EA.	53.50	17.35	70.85
Push buttons				
Standard	EA.	11.90	12.05	23.95
Weatherproof	"	17.85	15.40	33.25
Bells	"	45.20	25.20	70.40
Horns				
Standard	EA.	47.60	34.70	82.30
Weatherproof	"	89.00	43.40	132.40
Chimes	"	69.00	23.15	92.15
Flasher	"	50.00	21.35	71.35
Motion detectors	"	207.00	52.50	259.50
Intercom units	"	47.60	25.20	72.80
Remote annunciator	"	2,140	174.00	2,314
16730.20 CLOCK SYSTEMS				
Clock systems				
Single face	EA.	114.00	27.80	141.80
Double face	"	315.00	27.80	342.80
Skeleton	"	240.00	95.50	335.50
Master	"	2,520	174.00	2,694
Signal generator	"	2,170	139.00	2,309
Elapsed time indicator				
Controller	"	434.00	27.80	461.80
	"	91.50	18.50	110.00

16 ELECTRICAL

COMMUNICATIONS

16730.20 CLOCK SYSTEMS

	UNIT	MAT.	INST.	TOTAL
Clock and speaker	EA.	189.00	38.00	227.00
Bell				
Standard	EA.	63.00	18.50	81.50
Weatherproof	"	80.00	27.80	107.80
Horn				
Standard	EA.	40.00	25.20	65.20
Weatherproof	"	51.50	33.00	84.50
Chime	"	45.80	18.50	64.30
Buzzer	"	17.15	18.50	35.65
Flasher	"	63.00	21.35	84.35
Control Board	"	286.00	121.00	407.00
Program unit	"	286.00	174.00	460.00
Block back box	"	17.15	17.35	34.50
Double clock back box	"	34.30	23.15	57.45
Wire guard	"	11.45	6.95	18.40

16740.10 TELEPHONE SYSTEMS

	UNIT	MAT.	INST.	TOTAL
Communication cable				
25 pair	L.F.	0.35	0.90	1.25
100 pair	"	1.80	0.99	2.79
150 pair	"	2.40	1.15	3.55
200 pair	"	3.25	1.40	4.65
300 pair	"	4.65	1.45	6.10
400 pair	"	5.95	1.55	7.50
Cable tap in manhole or junction box				
25 pair cable	EA.	3.05	132.00	135.05
50 pair cable	"	6.05	262.00	268.05
75 pair cable	"	9.10	391.00	400.10
100 pair cable	"	12.10	524.00	536.10
150 pair cable	"	18.20	771.00	789.20
200 pair cable	"	24.25	1,030	1,054
300 pair cable	"	36.40	1,540	1,576
400 pair cable	"	48.50	2,140	2,189
Cable terminations, manhole or junction box				
25 pair cable	EA.	3.05	130.00	133.05
50 pair cable	"	6.05	259.00	265.05
100 pair cable	"	12.10	524.00	536.10
150 pair cable	"	18.20	771.00	789.20
200 pair cable	"	24.25	1,030	1,054
300 pair cable	"	36.40	1,540	1,576
400 pair cable	"	36.40	2,140	2,176
Telephones, standard				
1 button	EA.	63.00	103.00	166.00
2 button	"	79.00	121.00	200.00
6 button	"	118.00	185.00	303.00
12 button	"	239.00	264.00	503.00
18 button	"	275.00	308.00	583.00
Hazardous area				
Desk	EA.	1,100	252.00	1,352
Wall	"	512.00	174.00	686.00
Accessories				

COMMUNICATIONS	UNIT	MAT.	INST.	TOTAL
16740.10 TELEPHONE SYSTEMS				
Standard ground	EA.	18.20	55.50	73.70
Push button	"	18.05	55.50	73.55
Buzzer	"	19.40	55.50	74.90
Interface device	"	10.30	27.80	38.10
Long cord	"	10.90	27.80	38.70
Interior jack	"	6.65	13.90	20.55
Exterior jack	"	13.35	21.35	34.70
Hazardous area				
Selector switch	EA.	113.00	111.00	224.00
Bell	"	162.00	111.00	273.00
Horn	"	251.00	146.00	397.00
Horn relay	"	209.00	107.00	316.00
16740.30 CALL SYSTEMS				
Call systems, single bed station	EA.	60.50	18.50	79.00
Double bed station	"	97.00	25.20	122.20
Call-in cord	"	36.40	6.95	43.35
Pull cord	"	54.50	6.95	61.45
Pillow speaker	"	36.40	9.55	45.95
Dome light	"	24.25	18.50	42.75
Zone light	"	26.70	18.50	45.20
Stake station	"	82.50	21.35	103.85
Duty station	"	60.50	17.35	77.85
Utility station	"	79.00	21.35	100.35
Nurses station	"	82.50	18.50	101.00
Surgical station	"	198.00	25.20	223.20
Master station	"	2,060	87.00	2,147
Control station	"	727.00	278.00	1,005
Annunciator	"	272.00	69.50	341.50
Power supply	"	206.00	53.50	259.50
Speakers	"	30.30	27.80	58.10
Foot switch	"	36.40	10.30	46.70
Code blue systems				
Bed station	EA.	199.00	25.20	224.20
Dome light	"	30.30	23.15	53.45
Zone light	"	54.50	25.20	79.70
Pull cord	"	60.50	8.70	69.20
Nurses station	"	85.00	18.50	103.50
Annunciator	"	272.00	69.50	341.50
Power supply	"	198.00	50.50	248.50
Nurse station indicator, alarm annunciators, flush				
4 circuit	EA.	1,580	139.00	1,719
6 circuit	"	2,060	278.00	2,338
12 circuit	"	3,520	417.00	3,937
Desktop				
4 circuit	EA.	1,940	121.00	2,061
6 circuit	"	2,420	121.00	2,541
12 circuit	"	3,760	174.00	3,934
16750.20 SIGNALING SYSTEMS				
Signaling systems				
4" bell	EA.	64.50	20.85	85.35

16 ELECTRICAL

COMMUNICATIONS

COMMUNICATIONS	UNIT	MAT.	INST.	TOTAL
16750.20 SIGNALING SYSTEMS				
6" bell	EA.	78.00	22.55	100.55
10" bell	"	91.50	25.90	117.40
Buzzer				
Size 0	EA.	17.15	15.40	32.55
Size 1	"	18.30	15.40	33.70
Size 2	"	19.45	17.35	36.80
Size 3	"	20.60	18.50	39.10
Horn	"	57.00	21.35	78.35
Chime	"	91.50	18.50	110.00
Push button				
Standard	EA.	19.45	13.90	33.35
Weatherproof	"	29.50	17.35	46.85
Door opener				
Mortise	EA.	20.60	17.35	37.95
Rim	"	28.90	13.90	42.80
Transformer	"	10.40	15.40	25.80
Contractor grade doorbell chime kit				
Chime	EA.	19.80	34.70	54.50
Doorbutton	"	2.75	11.10	13.85
Transformer	"	8.80	17.35	26.15
16770.30 SOUND SYSTEMS				
Power amplifiers	EA.	552.00	121.00	673.00
Pre-amplifiers	"	440.00	95.50	535.50
Tuner	"	284.00	50.50	334.50
Horn				
Equilizer	EA.	712.00	55.50	767.50
Mixer	"	297.00	77.00	374.00
Tape recorder	"	932.00	64.50	996.50
Microphone				
Cassette Player	EA.	472.00	75.00	547.00
Record player	"	41.20	66.00	107.20
Equipment rack	"	53.00	44.80	97.80
Speaker				
Ceiling	EA.	101.00	47.10	148.10
Wall	"	294.00	139.00	433.00
Paging	"	106.00	27.80	133.80
Column	"	158.00	18.50	176.50
Single	"	35.30	21.35	56.65
Double	"	111.00	154.00	265.00
Volume control	"	35.30	18.50	53.80
Plug-in	"	111.00	27.80	138.80
Desk	"	88.00	13.90	101.90
Outlet	"	17.60	13.90	31.50
Stand	"	35.30	10.30	45.60
Console	"	8,470	278.00	8,748
Power supply	"	158.00	44.80	202.80
16780.10 ANTENNAS AND TOWERS				
Guy cable, alumaweld				
1x3, 7/32"	L.F.	0.24	1.75	1.99

COMMUNICATIONS

	UNIT	MAT.	INST.	TOTAL
16780.10 ANTENNAS AND TOWERS				
1x3, 1/4"	L.F.	0.30	1.75	2.05
1x3, 25/64"	"	0.42	2.05	2.47
1x19, 1/2"	"	1.05	2.40	3.45
1x7, 35/64"	"	1.20	2.80	4.00
1x19, 13/16"	"	1.30	3.45	4.75
Preformed alumaweld end grip				
1/4" cable	EA.	1.85	3.45	5.30
3/8" cable	"	2.45	3.45	5.90
1/2" cable	"	3.05	5.05	8.10
9/16" cable	"	3.80	6.95	10.75
5/8" cable	"	4.65	8.70	13.35
Fiberglass guy rod, white epoxy coated				
1/4" dia.	L.F.	1.20	5.05	6.25
3/8" dia	"	1.85	5.05	6.90
1/2" dia	"	2.45	6.95	9.40
5/8" dia	"	3.05	8.70	11.75
Preformed glass grip end grip, guy rod				
1/4" dia.	EA.	6.75	5.05	11.80
3/8" dia.	"	7.75	6.95	14.70
1/2" dia.	"	9.55	8.70	18.25
5/8" dia.	"	10.70	8.70	19.40
Spelter socket end grip, 1/4" dia. guy rod				19.90
Standard strength	EA.	19.90	17.35	37.25
High performance	"	24.50	17.35	41.85
3/8" dia. guy rod				
Standard strength	EA.	19.35	12.05	31.40
High performance	"	24.50	17.35	41.85
Timber pole, Douglas Fir				
80-85 ft	EA.	1,380	677.00	2,057
90-95 ft	"	1,550	771.00	2,321
Southern yellow pine				
35-45 ft	EA.	795.00	380.00	1,175
50-55 ft	"	990.00	487.00	1,477
16780.50 TELEVISION SYSTEMS				
TV outlet, self terminating, w/cover plate	EA.	6.55	10.70	17.25
Thru splitter	"	14.25	55.50	69.75
End of line	"	11.90	46.30	58.20
In line splitter multitap				
4 way	EA.	23.80	63.00	86.80
2 way	"	17.85	59.00	76.85
Equipment cabinet	"	59.50	55.50	115.00
Antenna				
Broad band uhf	EA.	119.00	121.00	240.00
Lightning arrester	"	32.10	25.20	57.30
TV cable	L.F.	0.26	0.17	0.43
Coaxial cable rg	"	0.19	0.17	0.36
Cable drill, with replacement tip	EA.	5.95	17.35	23.30
Cable blocks for in-line taps	"	11.90	25.20	37.10
In-line taps ptu-series 36 tv system	"	14.25	39.70	53.95
Control receptacles	"	9.35	15.60	24.95

16 ELECTRICAL

COMMUNICATIONS	UNIT	MAT.	INST.	TOTAL
16780.50 TELEVISION SYSTEMS				
Coupler	EA.	17.85	84.00	101.85
Head end equipment	"	2,250	231.00	2,481
TV camera	"	1,220	58.00	1,278
TV power bracket	"	112.00	27.80	139.80
TV monitor	"	935.00	50.50	985.50
Video recorder	"	1,810	73.00	1,883
Console	"	3,750	295.00	4,045
Selector switch	"	586.00	47.90	633.90
TV controller	"	280.00	48.70	328.70

RESISTANCE HEATING	UNIT	MAT.	INST.	TOTAL
16850.10 ELECTRIC HEATING				
Baseboard heater				
2', 375w	EA.	34.30	34.70	69.00
3', 500w	"	42.30	34.70	77.00
4', 750w	"	52.50	39.70	92.20
5', 935w	"	70.00	46.30	116.30
6', 1125w	"	78.00	55.50	133.50
7', 1310w	"	97.00	63.00	160.00
8', 1500w	"	109.00	69.50	178.50
9', 1680w	"	120.00	77.00	197.00
10', 1875w	"	125.00	79.50	204.50
Unit heater wall mounted				
1500w	EA.	160.00	55.50	215.50
2500w	"	172.00	63.00	235.00
4000w	"	229.00	79.50	308.50
Thermostat				
Integral	EA.	26.30	17.35	43.65
Line voltage	"	26.30	17.35	43.65
Electric heater connection	"	1.15	8.70	9.85
Fittings				
Inside corner	EA.	17.15	13.90	31.05
Outside corner	"	18.30	13.90	32.20
Receptacle section	"	19.45	13.90	33.35
Blank section	"	24.00	13.90	37.90
Unit heater, wall mounted				
750w	EA.	97.00	55.50	152.50
1500w	"	120.00	58.00	178.00
2000w	"	165.00	60.50	225.50
3000w	"	200.00	69.50	269.50
Infrared heaters				
600w	EA.	125.00	34.70	159.70

RESISTANCE HEATING

16850.10 ELECTRIC HEATING

	UNIT	MAT.	INST.	TOTAL
2000w	EA.	148.00	41.40	189.40
3000w	"	194.00	69.50	263.50
4000w	"	240.00	87.00	327.00
1000w	"	91.50	34.70	126.20
2000w	"	109.00	34.70	143.70
3000w	"	125.00	43.40	168.40
4000w	"	131.00	44.80	175.80
Controller	"	4.60	23.15	27.75
Wall bracket	"	9.25	25.20	34.45
Radiant ceiling heater panels				
500w	EA.	116.00	34.70	150.70
750w	"	139.00	34.70	173.70
Unit heaters, suspended, single phase				
3.0 kw	EA.	189.00	95.50	284.50
5.0 kw	"	213.00	95.50	308.50
7.5 kw	"	326.00	111.00	437.00
10.0 kw	"	360.00	132.00	492.00
Three phase				
5 kw	EA.	302.00	95.50	397.50
7.5 kw	"	327.00	111.00	438.00
10 kw	"	404.00	132.00	536.00
15 kw	"	650.00	146.00	796.00
20 kw	"	826.00	185.00	1,011
25 kw	"	976.00	52.50	1,029
30 kw	"	1,140	278.00	1,418
35 kw	"	1,440	278.00	1,718
Unit heater thermostat	"	30.70	18.50	49.20
Mounting bracket	"	31.60	25.20	56.80
Relay	"	40.00	21.35	61.35
Duct heaters, three phase				
10 kw	EA.	360.00	132.00	492.00
15 kw	"	422.00	132.00	554.00
17.5 kw	"	446.00	139.00	585.00
20 kw	"	474.00	214.00	688.00

CONTROLS

16910.40 CONTROL CABLE

	UNIT	MAT.	INST.	TOTAL
Control cable, 600v, #14 THWN, PVC jacket				
2 wire	L.F.	0.18	0.28	0.46
4 wire	"	0.25	0.35	0.60
6 wire	"	0.42	4.55	4.97
8 wire	"	0.48	5.05	5.53
10 wire	"	0.57	5.55	6.12

CONTROLS	UNIT	MAT.	INST.	TOTAL
16910.40 CONTROL CABLE				
12 wire	L.F.	0.95	6.30	7.25
14 wire	"	0.97	7.30	8.27
16 wire	"	1.10	7.70	8.80
18 wire	"	1.20	8.40	9.60
20 wire	"	1.35	8.70	10.05
22 wire	"	1.40	9.90	11.30
Audio cables, shielded, #24 gauge				
3 conductor	L.F.	0.13	0.14	0.27
4 conductor	"	0.17	0.21	0.38
5 conductor	"	0.19	0.24	0.43
6 conductor	"	0.22	0.31	0.53
7 conductor	"	0.24	0.38	0.62
8 conductor	"	0.28	0.42	0.70
9 conductor	"	0.30	0.49	0.79
10 conductor	"	0.32	0.52	0.84
15 conductor	"	0.56	0.62	1.18
20 conductor	"	0.76	0.79	1.55
25 conductor	"	0.95	0.94	1.89
30 conductor	"	1.15	1.05	2.20
40 conductor	"	1.50	1.25	2.75
50 conductor	"	1.85	1.45	3.30
#22 gauge				
3 conductor	L.F.	0.12	0.14	0.26
4 conductor	"	0.28	0.21	0.49
#20 gauge				
3 conductor	L.F.	0.15	0.14	0.29
10 conductor	"	0.51	0.52	1.03
15 conductor	"	0.65	0.62	1.27
#18 gauge				
3 conductor	L.F.	0.19	0.14	0.33
4 conductor	"	0.24	0.21	0.45
Microphone cables, #24 gauge				
2 conductor	L.F.	0.19	0.14	0.33
3 conductor	"	0.22	0.17	0.39
#20 gauge				
1 conductor	L.F.	0.21	0.14	0.35
2 conductor	"	0.33	0.14	0.47
2 conductor	"	0.40	0.14	0.54
3 conductor	"	0.46	0.21	0.67
4 conductor	"	0.66	0.24	0.90
5 conductor	"	0.81	0.31	1.12
7 conductor	"	0.90	0.38	1.28
8 conductor	"	1.00	0.42	1.42
Computer cables shielded, #24 gauge				
1 pair	L.F.	0.13	0.14	0.27
2 pair	"	0.19	0.14	0.33
3 pair	"	0.22	0.21	0.43
4 pair	"	0.26	0.24	0.50
5 pair	"	0.32	0.31	0.63
6 pair	"	0.38	0.38	0.76
7 pair	"	0.41	0.42	0.83
8 pair	"	0.46	0.49	0.95

CONTROLS

16910.40 CONTROL CABLE

	UNIT	MAT.	INST.	TOTAL
50 pair	L.F.	2.80	1.35	4.15
Coaxial cables				
RG 6/u	L.F.	0.13	0.21	0.34
RG 6a/u	"	0.30	0.21	0.51
RG 8/u	"	0.33	0.21	0.54
RG 8a/u	"	0.42	0.21	0.63
RG 9/u	"	1.85	0.21	2.06
RG 11/u	"	0.34	0.21	0.55
RG 58/u	"	0.13	0.21	0.34
RG 59/u	"	0.19	0.21	0.40
RG 62/u	"	0.15	0.21	0.36
RG 174/u	"	0.13	0.21	0.34
RG 213/u	"	0.42	0.21	0.63
MATV and CCTV camera cables				
1 conductor	L.F.	0.19	0.14	0.33
2 conductor	"	0.24	0.17	0.41
4 conductor	"	0.57	0.21	0.78
7 conductor	"	0.78	0.03	0.81
12 conductor	"	1.20	0.52	1.72
13 conductor	"	1.30	0.56	1.86
14 conductor	"	1.35	0.62	1.97
28 conductor	"	3.15	0.94	4.09
Fire alarm cables, #22 gauge				
6 conductor	L.F.	0.19	0.35	0.54
9 conductor	"	0.24	0.52	0.76
12 conductor	"	0.28	0.56	0.84
#18 gauge				
2 conductor	L.F.	0.19	0.17	0.36
4 conductor	"	0.24	0.24	0.48
#16 gauge				
2 conductor	L.F.	0.19	0.24	0.43
4 conductor	"	0.26	0.28	0.54
#14 gauge				
2 conductor	L.F.	0.28	0.28	0.56
#12 gauge				
2 conductor	L.F.	0.34	0.35	0.69
Plastic jacketed thermostat cable				
2 conductor	L.F.	0.13	0.14	0.27
3 conductor	"	0.20	0.17	0.37
4 conductor	"	0.24	0.21	0.45
5 conductor	"	0.28	0.28	0.56
6 conductor	"	0.33	0.31	0.64
7 conductor	"	0.36	0.42	0.78
8 conductor	"	0.41	0.45	0.86

Man-Hour Tables

The man-hour productivities used to develop the labor costs are listed in the following section of this book. These productivities represent typical installation labor for thousands of construction items. The data takes into account all activities involved in normal construction under commonly experienced working conditions. As with the Costbook pages, these items are listed according to the CSI MASTERFORMAT. In order to best use the information in this book, please review this sample page and read the "Features in this Book" section.

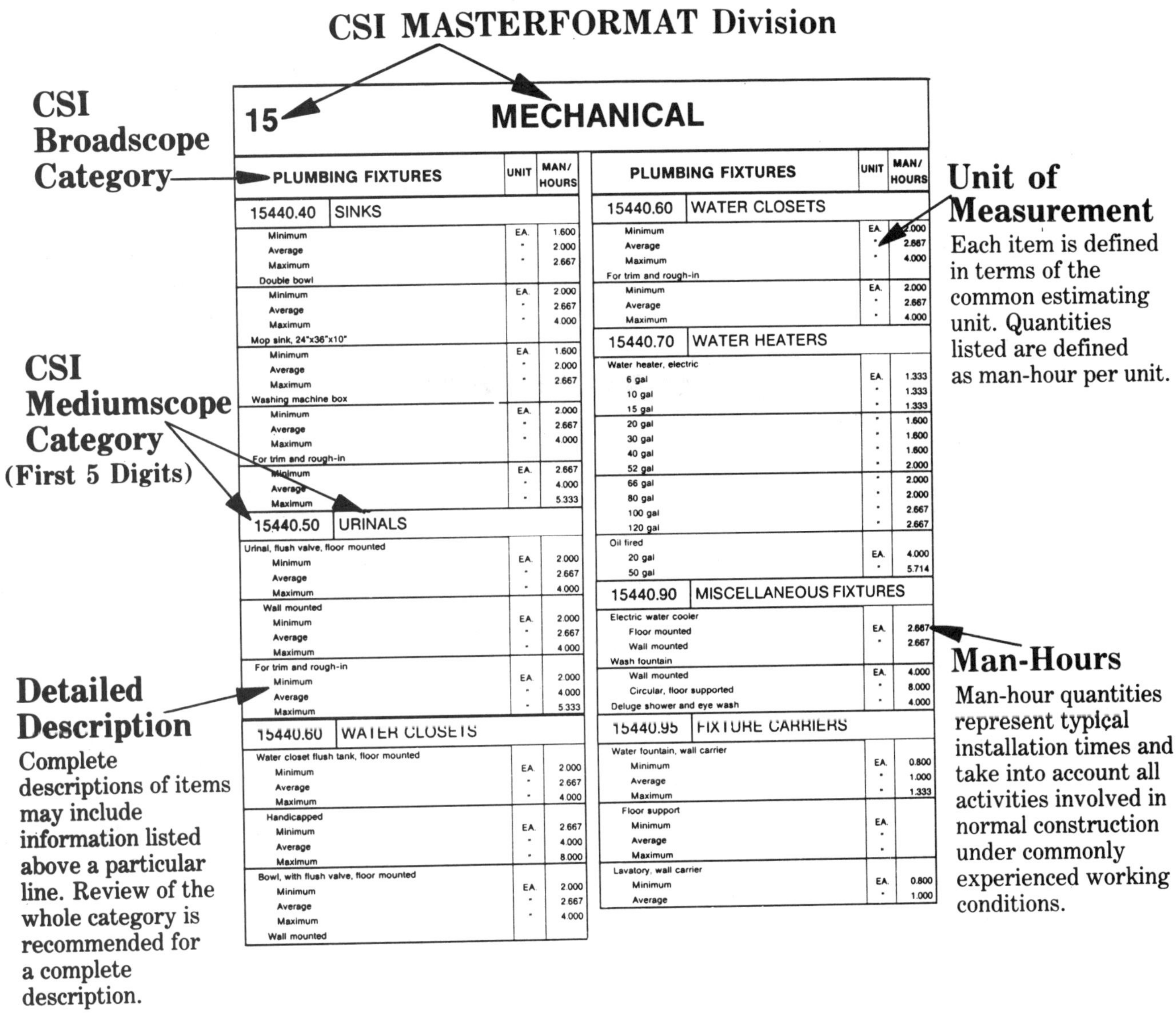

PLUMBING FIXTURES	UNIT	MAN/ HOURS
15440.40 SINKS		
Minimum	EA.	1.600
Average	"	2.000
Maximum	"	2.667
Double bowl		
Minimum	EA.	2.000
Average	"	2.667
Maximum	"	4.000
Mop sink, 24"x36"x10"		
Minimum	EA.	1.600
Average	"	2.000
Maximum	"	2.667
Washing machine box		
Minimum	EA.	2.000
Average	"	2.667
Maximum	"	4.000
For trim and rough-in		
Minimum	EA.	2.667
Average	"	4.000
Maximum	"	5.333
15440.50 URINALS		
Urinal, flush valve, floor mounted		
Minimum	EA.	2.000
Average	"	2.667
Maximum	"	4.000
Wall mounted		
Minimum	EA.	2.000
Average	"	2.667
Maximum	"	4.000
For trim and rough-in		
Minimum	EA.	2.000
Average	"	4.000
Maximum	"	5.333
15440.60 WATER CLOSETS		
Water closet flush tank, floor mounted		
Minimum	EA.	2.000
Average	"	2.667
Maximum	"	4.000
Handicapped		
Minimum	EA.	2.667
Average	"	4.000
Maximum	"	8.000
Bowl, with flush valve, floor mounted		
Minimum	EA.	2.000
Average	"	2.667
Maximum	"	4.000
Wall mounted		

PLUMBING FIXTURES	UNIT	MAN/ HOURS
15440.60 WATER CLOSETS		
Minimum	EA.	2.000
Average	"	2.667
Maximum	"	4.000
For trim and rough-in		
Minimum	EA.	2.000
Average	"	2.667
Maximum	"	4.000
15440.70 WATER HEATERS		
Water heater, electric		
6 gal	EA.	1.333
10 gal	"	1.333
15 gal	"	1.333
20 gal	"	1.600
30 gal	"	1.600
40 gal	"	1.600
52 gal	"	2.000
66 gal	"	2.000
80 gal	"	2.000
100 gal	"	2.667
120 gal	"	2.667
Oil fired		
20 gal	EA.	4.000
50 gal	"	5.714
15440.90 MISCELLANEOUS FIXTURES		
Electric water cooler		
Floor mounted	EA.	2.667
Wall mounted	"	2.667
Wash fountain		
Wall mounted	EA.	4.000
Circular, floor supported	"	8.000
Deluge shower and eye wash	"	4.000
15440.95 FIXTURE CARRIERS		
Water fountain, wall carrier		
Minimum	EA.	0.800
Average	"	1.000
Maximum	"	1.333
Floor support		
Minimum	EA.	
Average	"	
Maximum	"	
Lavatory, wall carrier		
Minimum	EA.	0.800
Average	"	1.000

BNi Building News™

02 SITEWORK

DEMOLITION	UNIT	MAN/ HOURS
02060.10 BUILDING DEMOLITION		
Cut-outs		
Concrete, elevated slabs, mesh reinforcing		
Under 5 cf	C.F.	0.800
Over 5 cf	"	0.667
Bar reinforcing		
Under 5 cf	C.F.	1.333
Over 5 cf	"	1.000
Rubbish handling		
Load in dumpster or truck		
Minimum	C.F.	0.018
Maximum	"	0.027
For use of elevators, add		
Minimum	C.F.	0.004
Maximum	"	0.008
Rubbish hauling		
Hand loaded on trucks, 2 mile trip	C.Y.	0.320
Machine loaded on trucks, 2 mile trip	"	0.240
02075.80 CORE DRILLING		
Concrete		
6" thick		
3" dia.	EA.	0.571
4" dia.	"	0.667
6" dia.	"	0.800
8" dia.	"	1.333
8" thick		
3" dia.	EA.	0.800
4" dia.	"	1.000
6" dia.	"	1.143
8" dia.	"	1.600
10" thick		
3" dia.	EA.	1.000
4" dia.	"	1.143
6" dia.	"	1.333
8" dia.	"	2.000
12" thick		
3" dia.	EA.	1.333
4" dia.	"	1.600
6" dia.	"	2.000
8" dia.	"	2.667

HAZARDOUS WASTE	UNIT	MAN/ HOURS
02080.10 ASBESTOS REMOVAL		
Enclosure using wood studs & poly, install & remove	S.F.	0.016
02080.12 DUCT INSULATION REMOVAL		
Remove duct insulation, duct size		
6" x 12"	L.F.	0.040
x 18"	"	0.053
x 24"	"	0.067
8" x 12"	"	0.057
x 18"	"	0.062
x 24"	"	0.073
12" x 12"	"	0.057
x 18"	"	0.067
x 24"	"	0.080
02080.15 PIPE INSULATION REMOVAL		
Removal, asbestos insulation		
2" thick, pipe		
1" to 3" dia.	L.F.	0.067
4" to 6" dia.	"	0.076
3" thick		
7" to 8" dia.	L.F.	0.080
9" to 10" dia.	"	0.084
11" to 12" dia.	"	0.089
13" to 14" dia.	"	0.094
15" to 18" dia.	"	0.100

SITE DEMOLITION	UNIT	MAN/ HOURS
02105.43 GAS PIPING		
Remove welded steel pipe, not including excavation		
4" dia.	L.F.	0.150
5" dia.	"	0.240
6" dia.	"	0.300
8" dia.	"	0.480
10" dia.	EA.	0.600
02105.45 SANITARY PIPING		
Remove sewer pipe, not including excavation		
4" dia.	L.F.	0.096

SITE DEMOLITION

02105.45 SANITARY PIPING

SITE DEMOLITION	UNIT	MAN/HOURS
6" dia.	L.F.	0.109
8" dia.	"	0.120
10" dia.	"	0.126
12" dia.	"	0.133
15" dia.	"	0.141
18" dia.	"	0.160
24" dia.	"	0.200
30" dia.	"	0.240
36" dia.	"	0.300

02105.48 WATER PIPING

SITE DEMOLITION	UNIT	MAN/HOURS
Remove water pipe, not including excavation		
4" dia.	L.F.	0.109
6" dia.	"	0.114
8" dia.	"	0.126
10" dia.	"	0.133
12" dia.	"	0.141
14" dia.	"	0.150
16" dia.	"	0.160
18" dia.	"	0.171
20" dia.	"	0.185
Remove valves		
6"	EA.	1.200
10"	"	1.333
14"	"	1.500
18"	"	2.000

02105.60 UNDERGROUND TANKS

SITE DEMOLITION	UNIT	MAN/HOURS
Remove underground storage tank, and backfill		
50 to 250 gals	EA.	8.000
600 gals	"	8.000
1000 gals	"	12.000
4000 gals	"	19.200
5000 gals	"	19.200
10,000 gals	"	32.000
12,000 gals	"	40.000
15,000 gals	"	48.000
20,000 gals	"	60.000

02105.66 SEPTIC TANKS

SITE DEMOLITION	UNIT	MAN/HOURS
Remove septic tank		
1000 gals	EA.	2.000
2000 gals	"	2.400
5000 gals	"	3.000
15,000 gals	"	24.000
25,000 gals	"	32.000
40,000 gals	"	48.000

SITE DEMOLITION

02162.10 TRENCH SHEETING

SITE DEMOLITION	UNIT	MAN/HOURS
Closed timber, including pull and salvage, excavation		
8' deep	S.F.	0.064
20' deep	"	0.091

EARTHWORK

02210.10 HAULING MATERIAL

EARTHWORK	UNIT	MAN/HOURS
Haul material by 10 cy dump truck, round trip distance		
1 mile	C.Y.	0.044
2 mile	"	0.053
5 mile	"	0.073
10 mile	"	0.080
20 mile	"	0.089
30 mile	"	0.107
Site grading, cut & fill, sandy clay, 200' haul, 75 hp dozer	"	0.032
Spread topsoil by equipment on site	"	0.036
Site grading (cut and fill to 6") less than 1 acre		
75 hp dozer	C.Y.	0.053
1.5 cy backhoe/loader	"	0.080

02210.30 BULK EXCAVATION

EARTHWORK	UNIT	MAN/HOURS
Hydraulic excavator		
1 cy capacity		
Light material	C.Y.	0.040
Medium material	"	0.048
Wet material	"	0.060
Blasted rock	"	0.069
Wheel mounted front-end loader		
7/8 cy capacity		
Light material	C.Y.	0.020
Medium material	"	0.023
Wet material	"	0.027
Blasted rock	"	0.032
Track mounted front-end loader		
1-1/2 cy capacity		
Light material	C.Y.	0.013
Medium material	"	0.015
Wet material	"	0.016
Blasted rock	"	0.018

02 SITEWORK

EARTHWORK

EARTHWORK	UNIT	MAN/ HOURS
02220.40 BUILDING EXCAVATION		
Structural excavation, unclassified earth		
3/8 cy backhoe	C.Y.	0.107
3/4 cy backhoe	"	0.080
1 cy backhoe	"	0.067
Foundation backfill and compaction by machine	"	0.160
02220.50 UTILITY EXCAVATION		
Trencher, sandy clay, 8" wide trench		
18" deep	L.F.	0.018
24" deep	"	0.020
36" deep	"	0.023
Trench backfill, 95% compaction		
Tamp by hand	C.Y.	0.500
Vibratory compaction	"	0.400
Trench backfilling, with borrow sand, place & compact	"	0.400
02220.60 TRENCHING		
Trenching and continuous footing excavation		
By gradall		
1 cy capacity		
Light soil	C.Y.	0.023
Medium soil	"	0.025
Heavy/wet soil	"	0.027
Loose rock	"	0.029
Blasted rock	"	0.031
By hydraulic excavator		
1/2 cy capacity		
Light soil	C.Y.	0.027
Medium soil	"	0.029
Heavy/wet soil	"	0.032
Loose rock	"	0.036
Blasted rock	"	0.040
Hand excavation		
Bulk, wheeled 100'		
Normal soil	C.Y.	0.889
Sand or gravel	"	0.800
Medium clay	"	1.143
Heavy clay	"	1.600
Loose rock	"	2.000
Trenches, up to 2' deep		
Normal soil	C.Y.	1.000
Sand or gravel	"	0.889
Medium clay	"	1.333
Heavy clay	"	2.000
Loose rock	"	2.667
Trenches, to 6' deep		
Normal soil	C.Y.	1.143
Sand or gravel	"	1.000

EARTHWORK	UNIT	MAN/ HOURS
02220.60 TRENCHING		
Medium clay	C.Y.	1.600
Heavy clay	"	2.667
Loose rock	"	4.000
Backfill trenches		
With compaction		
By hand	C.Y.	0.667
By 60 hp tracked dozer	"	0.020
By 200 hp tracked dozer	"	0.009
By small front-end loader	"	0.023
Backfill trenches, sand bedding, no compaction		
By hand	C.Y.	0.667
By small front-end loader	"	0.023
02220.90 HAND EXCAVATION		
Excavation		
To 2' deep		
Normal soil	C.Y.	0.889
Sand and gravel	"	0.800
Medium clay	"	1.000
Heavy clay	"	1.143
Loose rock	"	1.333
To 6' deep		
Normal soil	C.Y.	1.143
Sand and gravel	"	1.000
Medium clay	"	1.333
Heavy clay	"	1.600
Loose rock	"	2.000
Compaction of backfill around structures or in trench		
By hand with air tamper	C.Y.	0.571
By hand with vibrating plate tamper	"	0.533
1 ton roller	"	0.400
Miscellaneous hand labor		
Trim slopes, sides of excavation	S.F.	0.001
Trim bottom of excavation	"	0.002
Excavation around obstructions and services	C.Y.	2.667
02240.05 SOIL STABILIZATION		
Straw bale secured with rebar	L.F.	0.027
Filter barrier, 18" high filter fabric	"	0.080
Sediment fence, 36" fabric with 6" mesh	"	0.100
02605.30 MANHOLES		
Precast sections, 48" dia.		2.000
Base section	EA.	2.000
1'0" riser	"	1.600
1'4" riser	"	1.714
2'8" riser	"	1.846

EARTHWORK

	UNIT	MAN/HOURS
02605.30 MANHOLES		
4'0" riser	EA.	2.000
2'8" cone top	"	2.400
Precast manholes, 48" dia.		
4' deep	EA.	4.800
6' deep	"	6.000
7' deep	"	6.857
8' deep	"	8.000
10' deep	"	9.600
Cast-in-place, 48" dia., with frame and cover		
5' deep	EA.	12.000
6' deep	"	13.714
8' deep	"	16.000
10' deep	"	19.200
Brick manholes, 48" dia. with cover, 8" thick		
4' deep	EA.	8.000
6' deep	"	8.889
8' deep	"	10.000
10' deep	"	11.429
12' deep	"	13.333
14' deep	"	16.000
Inverts for manholes		
Single channel	EA.	3.200
Triple channel	"	4.000
Frames and covers, 24" diameter		
300 lb	EA.	0.800
400 lb	"	0.889
500 lb	"	1.143
Watertight, 350 lb	"	2.667
For heavy equipment, 1200 lb	"	4.000
Steps for manholes		
7" x 9"	EA.	0.160
8" x 9"	"	0.178
Curb inlet, 4' throat, cast in place		
12"-30" pipe	EA.	12.000
36"-48" pipe	"	13.714
Raise exist frame and cover, repaving	"	4.800
02610.10 CAST IRON FLANGED PIPE		
Cast iron flanged sections		
4" pipe, with one bolt set		
3' section	EA.	0.218
4' section	"	0.240
5' section	"	0.267
6' section	"	0.300
8' section	"	0.343
10' section	"	0.480
12' section	"	0.800
15' section	"	1.200
18' section	"	1.600

EARTHWORK

	UNIT	MAN/HOURS
02610.10 CAST IRON FLANGED PIPE		
6" pipe, with one bolt set		
3' section	EA.	0.240
4' section	"	0.282
5' section	"	0.320
6' section	"	0.369
8' section	"	0.533
10' section	"	0.600
12' section	"	0.800
15' section	"	1.200
18' section	"	1.714
8" pipe, with one bolt set		
3' section	EA.	0.300
4' section	"	0.343
5' section	"	0.400
6' section	"	0.480
8' section	"	0.686
10' section	"	0.800
12' section	"	1.200
15' section	"	1.600
18' section	"	2.000
10" pipe, with one bolt set		
3' section	EA.	0.308
4' section	"	0.353
5' section	"	0.414
6' section	"	0.500
8' section	"	0.727
10' section	"	0.857
12' section	"	1.333
15' section	"	1.714
18' section	"	2.400
12" pipe, with one bolt set		
3' section	EA.	0.333
4' section	"	0.387
5' section	"	0.462
6' section	"	0.545
8' section	"	0.800
10' section	"	0.923
12' section	"	1.500
15' section	"	2.000
18' section	"	2.667
02610.11 CAST IRON FITTINGS		
Mechanical joint, with 2 bolt kits		
90 deg bend		
4"	EA.	0.533
6"	"	0.615
8"	"	0.800
10"	"	1.143
12"	"	1.600

02 SITEWORK

EARTHWORK	UNIT	MAN/HOURS
02610.11 CAST IRON FITTINGS		
14"	EA.	2.000
16"	"	2.667
45 deg bend		
4"	EA.	0.533
6"	"	0.615
8"	"	0.800
10"	"	1.143
12"	"	1.600
14"	"	2.000
16"	"	2.667
Tee, with 3 bolt kits		
4" x 4"	EA.	0.800
6" x 6"	"	1.000
8" x 8"	"	1.333
10" x 10"	"	2.000
12" x 12"	"	2.667
Wye, with 3 bolt kits		
6" x 6"	EA.	1.000
8" x 8"	"	1.333
10" x 10"	"	2.000
12" x 12"	"	2.667
Reducer, with 2 bolt kits		
6" x 4"	EA.	1.000
8" x 6"	"	1.333
10" x 8"	"	2.000
12" x 10"	"	2.667
Flanged, 90 deg bend, 125 lb.		
4"	EA.	0.667
6"	"	0.800
8"	"	1.000
10"	"	1.333
12"	"	2.000
14"	"	2.667
16"	"	2.667
Tee		
4"	EA.	1.000
6"	"	1.143
8"	"	1.333
10"	"	1.600
12"	"	2.000
14"	"	2.667
16"	"	4.000
02610.13 GATE VALVES		
Gate valve, (AWWA) mechanical joint, with adjustable box		
4" valve	EA.	0.800
6" valve	"	0.960
8" valve	"	1.200
10" valve	"	1.412

EARTHWORK	UNIT	MAN/HOURS
02610.13 GATE VALVES		
12" valve	EA.	1.714
14" valve	"	2.000
16" valve	"	2.182
18" valve	"	2.400
Flanged, with box, post indicator (AWWA)		
4" valve	EA.	0.960
6" valve	"	1.091
8" valve	"	1.333
10" valve	"	1.600
12" valve	"	2.000
14" valve	"	2.400
16" valve	"	3.000
02610.15 WATER METERS		
Water meter, displacement type		
1"	EA.	0.800
1-1/2"	"	0.889
2"	"	1.000
02610.17 CORPORATION STOPS		
Stop for flared copper service pipe		
3/4"	EA.	0.400
1"	"	0.444
1-1/4"	"	0.533
1-1/2"	"	0.667
2"	"	0.800
02610.40 DUCTILE IRON PIPE		
Ductile iron pipe, cement lined, slip-on joints		
4"	L.F.	0.067
6"	"	0.071
8"	"	0.075
10"	"	0.080
12"	"	0.096
14"	"	0.120
16"	"	0.133
18"	"	0.150
20"	"	0.171
Mechanical joint pipe		
4"	L.F.	0.092
6"	"	0.100
8"	"	0.109
10"	"	0.120
12"	"	0.160
14"	"	0.185
16"	"	0.218
18"	"	0.240
20"	"	0.267
Fittings, mechanical joint		

EARTHWORK	UNIT	MAN/ HOURS
02610.40 DUCTILE IRON PIPE		
90 degree elbow		
4"	EA.	0.533
6"	"	0.615
8"	"	0.800
10"	"	1.143
12"	"	1.600
14"	"	2.000
16"	"	2.667
18"	"	3.200
20"	"	4.000
45 degree elbow		
4"	EA.	0.533
6"	"	0.615
8"	"	0.800
10"	"	1.143
12"	"	1.600
14"	"	2.000
16"	"	2.667
18"	"	4.000
20"	"	4.000
Tee		
4"x3"	EA.	1.000
4"x4"	"	1.000
6"x3"	"	1.143
6"x4"	"	1.143
6"x6"	"	1.143
8"x4"	"	1.333
8"x6"	"	1.333
8"x8"	"	1.333
10"x4"	"	1.600
10"x6"	"	1.600
10"x8"	"	1.600
10"x10"	"	1.600
12"x4"	"	2.000
12"x6"	"	2.000
12"x8"	"	2.000
12"x10"	"	2.000
12"x12"	"	2.133
14"x4"	"	2.286
14"x6"	"	2.286
14"x8"	"	2.286
14"x10"	"	2.286
14"x12"	"	2.462
14"x14"	"	2.462
16"x4"	"	2.667
16"x6"	"	2.667
16"x8"	"	2.667
16"x10"	"	2.667
16"x12"	"	2.667

EARTHWORK	UNIT	MAN/ HOURS
02610.40 DUCTILE IRON PIPE		
16"x14"	EA.	2.667
16"x16"	"	2.667
18"x6"	"	2.909
18"x8"	"	2.909
18"x10"	"	2.909
18"x12"	"	2.909
18"x14"	"	2.909
18"x16"	"	2.909
18"x18"	"	2.909
20"x6"	"	3.200
20"x8"	"	3.200
20"x10"	"	3.200
20"x12"	"	3.200
20"x14"	"	3.200
20"x16"	"	3.200
20"x18"	"	3.200
20"x20"	"	3.200
Cross		
4"x3"	EA.	1.333
4"x4"	"	1.333
6"x3"	"	1.600
6"x4"	"	1.600
6"x6"	"	1.600
8"x4"	"	1.778
8"x6"	"	1.778
8"x8"	"	1.778
10"x4"	"	2.000
10"x6"	"	2.000
10"x8"	"	2.000
10"x10"	"	2.000
12"x4"	"	2.286
12"x6"	"	2.286
12"x8"	"	2.286
12"x10"	"	2.462
12"x12"	"	2.462
14"x4"	"	2.667
14"x6"	"	2.667
14"x8"	"	2.667
14"x10"	"	2.667
14"x12"	"	2.909
14"x14"	"	2.909
16"x4"	"	3.200
16"x6"	"	3.200
16"x8"	"	3.200
16"x10"	"	3.200
16"x12"	"	3.200
16"x14"	"	3.200
16"x16"	"	3.200
18"x6"	"	3.556

EARTHWORK	UNIT	MAN/ HOURS
02610.40 DUCTILE IRON PIPE		
18"x8"	EA.	3.556
18"x10"	"	3.556
18"x12"	"	3.556
18"x14"	"	3.556
18"x16"	"	3.556
18"x18"	"	3.556
20"x6"	"	3.810
20"x8"	"	3.810
20"x10"	"	3.810
20"x12"	"	3.810
20"x14"	"	3.810
20"x16"	"	3.810
20"x18"	"	4.000
20"x20"	"	4.000
02610.60 PLASTIC PIPE		
PVC, class 150 pipe		
4" dia.	L.F.	0.060
6" dia.	"	0.065
8" dia.	"	0.069
10" dia.	"	0.075
12" dia.	"	0.080
Schedule 40 pipe		
1-1/2" dia.	L.F.	0.047
2" dia.	"	0.050
2-1/2" dia.	"	0.053
3" dia.	"	0.057
4" dia.	"	0.067
6" dia.	"	0.080
90 degree elbows		
1"	EA.	0.133
1-1/2"	"	0.133
2"	"	0.145
2-1/2"	"	0.160
3"	"	0.178
4"	"	0.200
6"	"	0.267
45 degree elbows		
1"	EA.	0.133
1-1/2"	"	0.133
2"	"	0.145
2-1/2"	"	0.160
3"	"	0.178
4"	"	0.200
6"	"	0.267
Tees		
1"	EA.	0.160
1-1/2"	"	0.160
2"	"	0.178

EARTHWORK	UNIT	MAN/ HOURS
02610.60 PLASTIC PIPE		
2-1/2"	EA.	0.200
3"	"	0.229
4"	"	0.267
6"	"	0.320
Couplings		
1"	EA.	0.133
1-1/2"	"	0.133
2"	"	0.145
2-1/2"	"	0.160
3"	"	0.178
4"	"	0.200
6"	"	0.267
Drainage pipe		
PVC schedule 80		
1" dia.	L.F.	0.047
1-1/2" dia.	"	0.047
ABS, 2" dia.	"	0.050
2-1/2" dia.	"	0.053
3" dia.	"	0.057
4" dia.	"	0.067
6" dia.	"	0.080
8" dia.	"	0.063
10" dia.	"	0.075
12" dia.	"	0.080
90 degree elbows		
1"	EA.	0.133
1-1/2"	"	0.133
2"	"	0.145
2-1/2"	"	0.160
3"	"	0.178
4"	"	0.200
6"	"	0.267
45 degree elbows		
1"	EA.	0.133
1-1/2"	"	0.133
2"	"	0.145
2-1/2"	"	0.160
3"	"	0.178
4"	"	0.200
6"	"	0.267
Tees		
1"	EA.	0.160
1-1/2"	"	0.160
2"	"	0.178
2-1/2"	"	0.200
3"	"	0.229
4"	"	0.267
6"	"	0.320
Couplings		

EARTHWORK	UNIT	MAN/HOURS
02610.60 PLASTIC PIPE		
1"	EA.	0.133
1-1/2"	"	0.133
2"	"	0.145
2-1/2"	"	0.160
3"	"	0.178
4"	"	0.200
6"	"	0.267
Pressure pipe		
PVC, class 200 pipe		
3/4"	L.F.	0.040
1"	"	0.042
1-1/4"	"	0.044
1-1/2"	"	0.047
2"	"	0.050
2-1/2"	"	0.053
3"	"	0.057
4"	"	0.067
6"	"	0.080
8"	"	0.069
90 degree elbows		
3/4"	EA.	0.133
1"	"	0.133
1-1/4"	"	0.133
1-1/2"	"	0.133
2"	"	0.145
2-1/2"	"	0.160
3"	"	0.178
4"	"	0.200
6"	"	0.267
8"	"	0.400
45 degree elbows		
3/4"	EA.	0.133
1"	"	0.133
1-1/4"	"	0.133
1-1/2"	"	0.133
2"	"	0.145
2-1/2"	"	0.160
3"	"	0.178
4"	"	0.200
6"	"	0.267
8"	"	0.400
Tees		
3/4"	EA.	0.160
1"	"	0.160
1-1/4"	"	0.160
1-1/2"	"	0.160
2"	"	0.178
2-1/2"	"	0.200
3"	"	0.229

EARTHWORK	UNIT	MAN/HOURS
02610.60 PLASTIC PIPE		
4"	EA.	0.267
6"	"	0.320
8"	"	0.444
Couplings		
3/4"	EA.	0.133
1"	"	0.133
1-1/4"	"	0.133
1-1/2"	"	0.133
2"	"	0.145
2-1/2"	"	0.160
3"	"	0.178
4"	"	0.178
6"	"	0.200
8"	"	0.267
02610.90 VITRIFIED CLAY PIPE		
Vitrified clay pipe, extra strength		
6" dia.	L.F.	0.109
8" dia.	"	0.114
10" dia.	"	0.120
12" dia.	"	0.160
15" dia.	"	0.240
18" dia.	"	0.267
24" dia.	"	0.343
30" dia.	"	0.480
36" dia.	"	0.686
02630.10 TAPPING SADDLES & SLEEVES		
Tapping saddle, tap size to 2"		
4" saddle	EA.	0.400
6" saddle	"	0.500
8" saddle	"	0.667
10" saddle	"	0.800
12" saddle	"	1.143
14" saddle	"	1.600
Tapping sleeve		
4x4	EA.	0.533
6x4	"	0.615
6x6	"	0.615
8x4	"	0.800
8x6	"	0.800
10x4	"	0.960
10x6	"	0.960
10x8	"	0.960
10x10	"	1.000
12x4	"	1.000
12x6	"	1.091
12x8	"	1.200
12x10	"	1.333

02 SITEWORK

EARTHWORK	UNIT	MAN/ HOURS
02630.10 TAPPING SADDLES & SLEEVES		
12x12	EA.	1.500
Tapping valve, mechanical joint		
4" valve	EA.	3.000
6" valve	"	4.000
8" valve	"	6.000
10" valve	"	8.000
12" valve	"	12.000
Tap hole in pipe		
4" hole	EA.	1.000
6" hole	"	1.600
8" hole	"	2.667
10" hole	"	3.200
12" hole	"	4.000
02640.15 VALVE BOXES		
Valve box, adjustable, for valves up to 20"		
3' deep	EA.	0.267
4' deep	"	0.320
5' deep	"	0.400
02640.19 THRUST BLOCKS		
Thrust block, 3000# concrete		
1/4 c.y.	EA.	1.333
1/2 c.y.	"	1.600
3/4 c.y.	"	2.667
1 c.y.	"	5.333
02645.10 FIRE HYDRANTS		
Standard, 3 way post, 6" mechanical joint		
2' deep	EA.	8.000
4' deep	"	9.600
6' deep	"	12.000
8' deep	"	13.714
02665.10 CHILLED WATER SYSTEMS		
Chilled water pipe, 2" thick insulation, w/casing		
Align and tack weld on sleepers		
1-1/2" dia.	L.F.	0.022
3" dia.	"	0.034
4" dia.	"	0.048
6" dia.	"	0.060
8" dia.	"	0.069
10" dia.	"	0.080
12" dia.	"	0.096
14" dia.	"	0.104

EARTHWORK	UNIT	MAN/ HOURS
02665.10 CHILLED WATER SYSTEMS		
16" dia.	L.F.	0.120
Align and tack weld on trench bottom		
18" dia.	L.F.	0.133
20" dia.	"	0.150
Preinsulated fittings		
Align and tack weld on sleepers		
Elbows		
1-1/2"	EA.	0.500
3"	"	0.800
4"	"	1.000
6"	"	1.333
8"	"	1.600
Tees		
1-1/2"	EA.	0.533
3"	"	0.889
4"	"	1.143
6"	"	1.600
8"	"	2.000
Reducers		
3"	EA.	0.667
4"	"	0.800
6"	"	1.000
8"	"	1.333
Anchors, not including concrete		
4"	EA.	1.000
6"	"	1.000
Align and tack weld on trench bottom		
Elbows		
10"	EA.	1.500
12"	"	1.714
14"	"	1.846
16"	"	2.000
18"	"	2.182
20"	"	2.400
Tees		
10"	EA.	1.500
12"	"	1.714
14"	"	1.846
16"	"	2.000
18"	"	2.182
20"	"	2.400
Reducers		
10"	EA.	1.000
12"	"	1.091
14"	"	1.200
16"	"	1.333
18"	"	1.500
20"	"	1.714
Anchors, not including concrete		

02 SITEWORK

EARTHWORK	UNIT	MAN/ HOURS
02665.10 CHILLED WATER SYSTEMS		
10"	EA.	1.000
12"	"	1.091
14"	"	1.200
16"	"	1.333
18"	"	1.500
20"	"	1.714
02670.10 WELLS		
Domestic water, drilled and cased		
4" dia.	L.F.	1.600
6" dia.	"	1.846
8" dia.	"	2.400
02685.10 GAS DISTRIBUTION		
Gas distribution lines		
Polyethylene, 60 psi coils		
1-1/4" dia.	L.F.	0.053
1-1/2" dia.	"	0.057
2" dia.	"	0.067
3" dia.	"	0.080
30' pipe lengths		
3" dia.	L.F.	0.089
4" dia.	"	0.100
6" dia.	"	0.133
8" dia.	"	0.160
Steel, schedule 40, plain end		
1" dia.	L.F.	0.067
2" dia.	"	0.073
3" dia.	"	0.080
4" dia.	"	0.160
5" dia.	"	0.171
6" dia.	"	0.200
8" dia.	"	0.218
Natural gas meters, direct digital reading, threaded		
250 cfh @ 5 lbs	EA.	1.600
425 cfh @ 10 lbs	"	1.600
800 cfh @ 20 lbs	"	2.000
1000 cfh @ 25 lbs	"	2.000
1,400 cfh @ 100 lbs	"	2.667
2,300 cfh @ 100 lbs	"	4.000
5,000 cfh @ 100 lbs	"	8.000
Gas pressure regulators		
Threaded		
3/4"	EA.	1.000
1"	"	1.333
1-1/4"	"	1.333
1-1/2"	"	1.333
2"	"	1.600
Flanged		

EARTHWORK	UNIT	MAN/ HOURS
02685.10 GAS DISTRIBUTION		
3"	EA.	2.000
4"	"	2.667
02690.10 STORAGE TANKS		
Oil storage tank, underground		
Steel		
500 gals	EA.	3.000
1,000 gals	"	4.000
4,000 gals	"	8.000
5,000 gals	"	12.000
10,000 gals	"	24.000
Fiberglass, double wall		
550 gals	EA.	4.000
1,000 gals	"	4.000
2,000 gals	"	6.000
4,000 gals	"	12.000
6,000 gals	"	16.000
8,000 gals	"	24.000
10,000 gals	"	30.000
12,000 gals	"	40.000
15,000 gals	"	53.333
20,000 gals	"	60.000
Above ground		
Steel		
275 gals	EA.	2.400
500 gals	"	4.000
1,000 gals	"	4.800
1,500 gals	"	6.000
2,000 gals	"	8.000
5,000 gals	"	12.000
Fill cap	"	0.800
Vent cap	"	0.800
Level indicator	"	0.800
02695.80 STEAM METERS		
In-line turbine, direct reading, 300 lb, flanged		
2"	EA.	1.000
3"	"	1.333
4"	"	1.600
Threaded, 2"		
5" line	EA.	8.000
6" line	"	8.000
8" line	"	8.000
10" line	"	8.000
12" line	"	8.000
14" line	"	8.000
16" line	"	8.000

SEWERAGE AND DRAINAGE

02720.10 CATCH BASINS

SEWERAGE AND DRAINAGE	UNIT	MAN/HOURS
Standard concrete catch basin		
Cast in place, 3'8" x 3'8", 6" thick wall		
2' deep	EA.	6.000
3' deep	"	6.000
4' deep	"	8.000
5' deep	"	8.000
6' deep	"	9.600
4'x4', 8" thick wall, cast in place		
2' deep	EA.	6.000
3' deep	"	6.000
4' deep	"	8.000
5' deep	"	8.000
6' deep	"	9.600
Frames and covers, cast iron		
Round		
24" dia.	EA.	2.000
26" dia.	"	2.000
28" dia.	"	2.000
Rectangular		
23"x23"	EA.	2.000
27"x20"	"	2.000
24"x24"	"	2.000
26"x26"	"	2.000
Curb inlet frames and covers		
27"x27"	EA.	2.000
24"x36"	"	2.000
24"x25"	"	2.000
24"x22"	"	2.000
20"x22"	"	2.000
Airfield catch basin frame and grating, galvanized		
2'x4'	EA.	2.000
2'x2'	"	2.000

02720.40 STORM DRAINAGE

SEWERAGE AND DRAINAGE	UNIT	MAN/HOURS
Concrete pipe		
Plain, bell and spigot joint, class II		
6" pipe	L.F.	0.109
8" pipe	"	0.120
10" pipe	"	0.126
12" pipe	"	0.133
15" pipe	"	0.141
18" pipe	"	0.150
21" pipe	"	0.160
24" pipe	"	0.171
Reinforced, class III, tongue and groove joint		
12" pipe	L.F.	0.133
15" pipe	"	0.141
18" pipe	"	0.150
21" pipe	"	0.160
24" pipe	L.F.	0.171
27" pipe	"	0.185
30" pipe	"	0.200
36" pipe	"	0.218
42" pipe	"	0.240
48" pipe	"	0.267
54" pipe	"	0.300
60" pipe	"	0.343
66" pipe	"	0.400
72" pipe	"	0.480
Flared end-section, concrete		
12" pipe	L.F.	0.133
15" pipe	"	0.141
18" pipe	"	0.150
24" pipe	"	0.171
30" pipe	"	0.200
36" pipe	"	0.218
42" pipe	"	0.240
48" pipe	"	0.267
54" pipe	"	0.300
Corrugated metal pipe, coated, paved invert		
16 ga.		
8" pipe	L.F.	0.080
10" pipe	"	0.083
12" pipe	"	0.086
15" pipe	"	0.092
18" pipe	"	0.100
21" pipe	"	0.109
24" pipe	"	0.120
30" pipe	"	0.133
36" pipe	"	0.150
12 ga., 48" pipe	"	0.171
10 ga.		
60" pipe	L.F.	0.200
72" pipe	"	0.240
Galvanized or aluminum, plain		
16 ga.		
8" pipe	L.F.	0.080
10" pipe	"	0.083
12" pipe	"	0.086
15" pipe	"	0.092
18" pipe	"	0.100
24" pipe	"	0.120
30" pipe	"	0.133
36" pipe	"	0.150
12 ga., 48" pipe	"	0.171
10 ga., 60" pipe	"	0.200
Galvanized or aluminum, coated oval arch		
16 ga.		

SEWERAGE AND DRAINAGE

02720.40 STORM DRAINAGE

SEWERAGE AND DRAINAGE	UNIT	MAN/ HOURS
17" x 13"	L.F.	0.109
21" x 15"	"	0.120
14 ga.		
28" x 20"	L.F.	0.133
35" x 24"	"	0.171
12 ga.		
42" x 29"	L.F.	0.200
57" x 38"	"	0.240
64" x 43"	"	0.253
Oval arch culverts, plain		
16 ga.		
17" x 13"	L.F.	0.109
21" x 15"	"	0.120
14 ga.		
28" x 20"	L.F.	0.133
35" x 24"	"	0.171
12 ga.		
57" x 38"	L.F.	0.200
64" x 43"	"	0.240
71" x 47"	"	0.253
Nestable corrugated metal pipe		
16 ga.		
10" pipe	L.F.	0.083
12" pipe	"	0.086
15" pipe	"	0.092
18" pipe	"	0.100
24" pipe	"	0.120
30" pipe	"	0.133
14 ga., 36" pipe	"	0.150
Headwalls, cast in place, 30 deg wingwall		
12" pipe	EA.	2.000
15" pipe	"	2.000
18" pipe	"	2.286
24" pipe	"	2.286
30" pipe	"	2.667
36" pipe	"	4.000
42" pipe	"	4.000
48" pipe	"	5.333
54" pipe	"	6.667
60" pipe	"	8.000
4" cleanout for storm drain		
4" pipe	EA.	1.000
6" pipe	"	1.000
8" pipe	"	1.000
Connect new drain line		
To existing manhole	EA.	2.667
To new manhole	"	1.600

SEWERAGE AND DRAINAGE

02730.10 SANITARY SEWERS

SEWERAGE AND DRAINAGE	UNIT	MAN/ HOURS
Clay		
6" pipe	L.F.	0.080
8" pipe	"	0.086
10" pipe	"	0.092
12" pipe	"	0.100
PVC		
4" pipe	L.F.	0.060
6" pipe	"	0.063
8" pipe	"	0.067
10" pipe	"	0.071
12" pipe	"	0.075
Cleanout		
4" pipe	EA.	1.000
6" pipe	"	1.000
8" pipe	"	1.000
Connect new sewer line		
To existing manhole	EA.	2.667
To new manhole	"	1.600

02740.10 DRAINAGE FIELDS

SEWERAGE AND DRAINAGE	UNIT	MAN/ HOURS
Perforated PVC pipe, for drain field		
4" pipe	L.F.	0.053
6" pipe	"	0.057

02740.50 SEPTIC TANKS

SEWERAGE AND DRAINAGE	UNIT	MAN/ HOURS
Septic tank, precast concrete		
1000 gals	EA.	4.000
2000 gals	"	6.000
5000 gals	"	12.000
25,000 gals	"	48.000
40,000 gals	"	80.000
Leaching pit, precast concrete, 72" diameter		
3' deep	EA.	3.000
6' deep	"	3.429
8' deep	"	4.000

02760.10 PIPELINE RESTORATION

SEWERAGE AND DRAINAGE	UNIT	MAN/ HOURS
Relining existing water main		
6" dia.	L.F.	0.240
8" dia.	"	0.253
10" dia.	"	0.267
12" dia.	"	0.282
14" dia.	"	0.300
16" dia.	"	0.320
18" dia.	"	0.343
20" dia.	"	0.369
24" dia.	"	0.400
36" dia.	"	0.480

02 SITEWORK

SEWERAGE AND DRAINAGE

02760.10 PIPELINE RESTORATION

	UNIT	MAN/HOURS
48" dia.	L.F.	0.533
72" dia.	"	0.600
Replacing in line gate valves		
6" valve	EA.	3.200
8" valve	"	4.000
10" valve	"	4.800
12" valve	"	6.000
16" valve	"	6.857
18" valve	"	8.000
20" valve	"	9.600
24" valve	"	12.000
36" valve	"	16.000

POWER & COMMUNICATIONS

02780.20 HIGH VOLTAGE CABLE

	UNIT	MAN/HOURS
High voltage XLP copper cable, shielded, 5000v		
#6 awg	L.F.	0.013
#4 awg	"	0.016
#2 awg	"	0.019
#1 awg	"	0.021
#1/0 awg	"	0.024
#2/0 awg	"	0.029
#3/0 awg	"	0.034
#4/0 awg	"	0.036
#250 awg	"	0.043
#300 awg	"	0.048
#350 awg	"	0.053
#500 awg	"	0.073
#750 awg	"	0.080
Ungrounded, 15,000v		
#1 awg	L.F.	0.031
#1/0 awg	"	0.034
#2/0 awg	"	0.036
#3/0 awg	"	0.040
#4/0 awg	"	0.046
#250 awg	"	0.048
#300 awg	"	0.053
#350 awg	"	0.062
#500 awg	"	0.080
#750 awg	"	0.098

POWER & COMMUNICATIONS

02780.20 HIGH VOLTAGE CABLE

	UNIT	MAN/HOURS
#1000 awg	L.F.	0.123
Aluminum cable, shielded, 5000v		
#6 awg	L.F.	0.011
#4 awg	"	0.013
#2 awg	"	0.015
#1 awg	"	0.017
#1/0 awg	"	0.019
#2/0 awg	"	0.020
#3/0 awg	"	0.021
#4/0 awg	"	0.024
#250 awg	"	0.026
#300 awg	"	0.031
#350 awg	"	0.034
#500 awg	"	0.036
#750 awg	"	0.044
#1000 awg	"	0.050
Ungrounded, 15,000v		
#1 awg	L.F.	0.021
#1/0 awg	"	0.025
#2/0 awg	"	0.027
#3/0 awg	"	0.028
#4/0 awg	"	0.029
#250 awg	"	0.031
#300 awg	"	0.032
#350 awg	"	0.036
#500 awg	"	0.043
#750 awg	"	0.052
#1000 awg	"	0.064
Indoor terminations, 5000v		
#6 - #4	EA.	0.157
#2 - #2/0	"	0.157
#3/0 - #250	"	0.157
#300 - #750	"	2.759
#1000	"	3.810
In-line splice, 5000v		
#6 - #4/0	EA.	3.810
#250 - #500	"	10.000
#750 - #1000	"	13.008
T-splice, 5000v		
#2 - #4/0	EA.	11.994
#250 - #500	"	20.000
#750 - #1000	"	25.000
Indoor terminations, 15,000v		
#2 - #2/0	EA.	3.478
#3/0 - #500	"	5.333
#750 - #1000	"	6.154
In-line splice, 15,000v		
#2 - #4/0	EA.	8.999
#250 - #500	"	11.994

02 SITEWORK

POWER & COMMUNICATIONS

02780.20 HIGH VOLTAGE CABLE

POWER & COMMUNICATIONS	UNIT	MAN/HOURS
#750 - #1000	EA.	18.018
T-splice, 15,000v		
#4	EA.	18.018
#250 - #500	"	29.963
#750 - #1000	"	44.944
Compression lugs, 15,000v		
#4	EA.	0.400
#2	"	0.533
#1	"	0.533
#1/0	"	0.667
#2/0	"	0.667
#3/0	"	0.851
#4/0	"	0.851
#250	"	0.952
#300	"	0.952
#350	"	1.159
#500	"	1.250
#750	"	1.509
#1000	"	1.905
Compression splices, 15,000v		
#4	EA.	0.667
#2	"	0.727
#1	"	0.899
#1/0	"	1.000
#2/0	"	1.159
#3/0	"	1.250
#4/0	"	1.404
#250	"	1.509
#350	"	1.739
#500	"	2.000
#750	"	2.500

02780.40 SUPPORTS & CONNECTORS

POWER & COMMUNICATIONS	UNIT	MAN/HOURS
Cable supports for conduit		
1-1/2"	EA.	0.348
2"	"	0.348
2-1/2"	"	0.400
3"	"	0.400
3-1/2"	"	0.500
4"	"	0.500
5"	"	0.667
6"	"	0.727
Split bolt connectors		
#10	EA.	0.200
#8	"	0.200
#6	"	0.200
#4	"	0.400
#3	"	0.400

POWER & COMMUNICATIONS

02780.40 SUPPORTS & CONNECTORS

POWER & COMMUNICATIONS	UNIT	MAN/HOURS
#2	EA.	0.400
#1/0	"	0.667
#2/0	"	0.667
#3/0	"	0.667
#4/0	"	0.667
#250	"	1.000
#350	"	1.000
#500	"	1.000
#750	"	1.509
#1000	"	1.509
Single barrel lugs		
#6	EA.	0.250
#1/0	"	0.500
#250	"	0.667
#350	"	0.667
#500	"	0.667
#600	"	0.899
#800	"	0.899
#1000	"	0.899
Double barrel lugs		
#1/0	EA.	0.899
#250	"	1.290
#350	"	1.290
#600	"	1.905
#800	"	1.905
#1000	"	1.905
Three barrel lugs		
#2/0	EA.	1.290
#250	"	1.905
#350	"	1.905
#600	"	2.667
#800	"	2.667
#1000	"	2.667
Four barrel lugs		
#250	EA.	2.759
#350	"	2.759
#600	"	3.478
#800	"	3.478
Compression conductor adapters		
#6	EA.	0.296
#4	"	0.348
#2	"	0.444
#1	"	0.444
#1/0	"	0.533
#250	"	0.800
#350	"	0.851
#500	"	1.096
#750	"	1.143
Terminal blocks, 2 screw		

POWER & COMMUNICATIONS	UNIT	MAN/ HOURS
02780.40 SUPPORTS & CONNECTORS		
3 circuit	EA.	0.200
6 circuit	"	0.200
8 circuit	"	0.200
10 circuit	"	0.296
12 circuit	"	0.296
18 circuit	"	0.296
24 circuit	"	0.348
36 circuit	"	0.348
Compression splice		
#8 awg	EA.	0.381
#6 awg	"	0.276
#4 awg	"	0.276
#2 awg	"	0.533
#1 awg	"	0.533
#1/0 awg	"	0.533
#2/0 awg	"	0.851
#3/0 awg	"	0.851
#4/0 awg	"	0.851
#250 awg	"	1.356
#300 awg	"	1.356
#350 awg	"	1.404
#400 awg	"	1.404
#500 awg	"	1.509
#600 awg	"	1.509
#750 awg	"	1.739
#1000 awg	"	1.739

SITE IMPROVEMENTS	UNIT	MAN/ HOURS

03 CONCRETE

FORMWORK	UNIT	MAN/ HOURS
03110.20 ELEVATED SLAB FORMWORK		
Elevated slab formwork		
Slab, with drop panels		
1 use	S.F.	0.064
5 uses	"	0.055
Floor slab, hung from steel beams		
1 use	S.F.	0.062
5 uses	"	0.053
Floor slab, with pans or domes		
1 use	S.F.	0.073
5 uses	"	0.062
Equipment curbs, 12" high		
1 use	L.F.	0.080
5 uses	"	0.067
03110.25 EQUIPMENT PAD FORMWORK		
Equipment pad, job built		
1 use	S.F.	0.100
2 uses	"	0.094
3 uses	"	0.089
4 uses	"	0.084
5 uses	"	0.080
03110.35 FOOTING FORMWORK		
Wall footings, job built, continuous		
1 use	S.F.	0.080
5 uses	"	0.067
Column footings, spread		
1 use	S.F.	0.100
5 uses	"	0.080
03110.55 SLAB/MAT FORMWORK		
Edge forms		
6" high		
1 use	L.F.	0.073
5 uses	"	0.062
12" high		
1 use	L.F.	0.080
5 uses	"	0.067
Formwork for openings		
1 use	S.F.	0.160
5 uses	"	0.114
03110.65 WALL FORMWORK		
Wall forms, exterior, job built		
Up to 8' high wall		
1 use	S.F.	0.080
5 uses	"	0.067

FORMWORK	UNIT	MAN/ HOURS
03110.65 WALL FORMWORK		
Over 8' high wall		
1 use	S.F.	0.100
5 uses	"	0.080
Column pier and pilaster		
1 use	S.F.	0.160
5 uses	"	0.114
Interior wall forms		
Up to 8' high		
1 use	S.F.	0.073
5 uses	"	0.062
Over 8' high		
1 use	S.F.	0.089
5 uses	"	0.073

REINFORCEMENT	UNIT	MAN/ HOURS
03210.20 ELEVATED SLAB REINFORCING		
Elevated slab		
#3 - #4	TON	10.000
#5 - #6	"	8.889
Galvanized		
#3 - #4	TON	10.000
#5 - #6	"	8.889
03210.25 EQUIP. PAD REINFORCING		
Equipment pad		
#3 - #4	TON	16.000
#5 - #6	"	14.545
#7 - #8	"	13.333
#9 - #10	"	12.308
#11 - #12	"	11.429
03210.35 FOOTING REINFORCING		
Footings		
Grade 50		
#3 - #4	TON	13.333
#5 - #6	"	11.429
Grade 60		
#3 - #4	TON	13.333
#5 - #6	"	11.429

REINFORCEMENT

03210.55 SLAB/MAT REINFORCING

REINFORCEMENT	UNIT	MAN/HOURS
Bars, slabs		
#3 - #4	TON	13.333
#5 - #6	"	11.429
#7 - #8	"	10.000
Galvanized		
#3 - #4	TON	13.333
#5 - #6	"	11.429

03210.65 WALL REINFORCING

REINFORCEMENT	UNIT	MAN/HOURS
Walls		
#3 - #4	TON	11.429
#5 - #6	"	10.000
Galvanized		
#3 - #4	TON	11.429
#5 - #6	"	10.000
Masonry wall (horizontal)		
#3 - #4	TON	32.000
#5 - #6	"	26.667
Galvanized		
#3 - #4	TON	32.000
#5 - #6	"	26.667
Masonry wall (vertical)		
#3 - #4	TON	40.000
#5 - #6	"	32.000
Galvanized		
#3 - #4	TON	40.000
#5 - #6	"	32.000

PLACING CONCRETE

03380.20 ELEVATED SLAB CONCRETE

PLACING CONCRETE	UNIT	MAN/HOURS
Elevated slab		
2500# or 3000# concrete		
By crane	C.Y.	0.480
By pump	"	0.369
By hand buggy	"	0.800
5000# concrete		
By crane	C.Y.	0.480
By pump	"	0.369
By hand buggy	"	0.800

PLACING CONCRETE

03380.25 EQUIPMENT PAD CONCRETE

PLACING CONCRETE	UNIT	MAN/HOURS
Equipment pad		
2500# or 3000# concrete		
By chute	C.Y.	0.267
By pump	"	0.686
By crane	"	0.800
3500# or 4000# concrete		
By chute	C.Y.	0.267
By pump	"	0.686
By crane	"	0.800
5000# concrete		
By chute	C.Y.	0.267
By pump	"	0.686
By crane	"	0.800

03380.35 FOOTING CONCRETE

PLACING CONCRETE	UNIT	MAN/HOURS
Continuous footing		
2500# or 3000# concrete		
By chute	C.Y.	0.267
By pump	"	0.600
By crane	"	0.686
5000# concrete		
By chute	C.Y.	0.267
By pump	"	0.600
By crane	"	0.686
Spread footing		
2500# or 3000# concrete		
Under 5 cy		
By chute	C.Y.	0.267
By pump	"	0.640
By crane	"	0.738
5000# concrete		
Under 5 c.y.		
By chute	C.Y.	0.267
By pump	"	0.640
By crane	"	0.738

03380.55 SLAB/MAT CONCRETE

PLACING CONCRETE	UNIT	MAN/HOURS
Slab on grade		
2500# or 3000# concrete		
By chute	C.Y.	0.200
By crane	"	0.400
By pump	"	0.343
By hand buggy	"	0.533
5000# concrete		
By chute	C.Y.	0.200
By crane	"	0.400
By pump	"	0.343
By hand buggy	"	0.533

PLACING CONCRETE	UNIT	MAN/ HOURS
03380.65 WALL CONCRETE		
Walls		
2500# or 3000# concrete		
To 4'		
By chute	C.Y.	0.229
By crane	"	0.800
By pump	"	0.738
To 8'		
By crane	C.Y.	0.873
By pump	"	0.800
Filled block (CMU)		
3000# concrete, by pump		
4" wide	S.F.	0.034
6" wide	"	0.040
8" wide	"	0.048
10" wide	"	0.056
12" wide	"	0.069
Pilasters, 3000# concrete	C.F.	0.960
Wall cavity, 2" thick, 3000# concrete	S.F.	0.032

04 MASONRY

MORTAR AND GROUT

04100.10 MASONRY GROUT	UNIT	MAN/ HOURS
Grout, non shrink, non-metallic, trowelable	C.F.	0.016
Grout-filled individual cmu cells		
4" wide	L.F.	0.012
8" wide	"	0.012

04150.50 MASONRY FLASHING	UNIT	MAN/ HOURS
Through-wall flashing		
5 oz. coated copper	S.F.	0.067
0.030" elastomeric	"	0.053

UNIT MASONRY

04210.10 BRICK MASONRY	UNIT	MAN/ HOURS
Standard size brick, running bond		
Face brick, red (6.4/sf)		
Veneer	S.F.	0.133
Cavity wall	"	0.114
9" solid wall	"	0.229
Chimney, standard brick, including flue		
16" x 16"	L.F.	0.800
16" x 20"	"	0.800
16" x 24"	"	0.800
20" x 20"	"	1.000
20" x 24"	"	1.000
20" x 32"	"	1.143

04220.10 CONCRETE MASONRY UNITS	UNIT	MAN/ HOURS
Hollow, load bearing		
4"	S.F.	0.059
8"	"	0.067
Solid, load bearing		
4"	S.F.	0.059
8"	"	0.067
Back-up block, 8" x 16"		
4"	S.F.	0.047
8"	"	0.053
Steel angles and plates		
Minimum	Lb.	0.011
Maximum	"	0.020
Various size angle lintels		

UNIT MASONRY

04220.10 CONCRETE MASONRY UNITS	UNIT	MAN/ HOURS
1/4" stock		
3" x 3"	L.F.	0.050
3" x 3-1/2"	"	0.050
3/8" stock		
3" x 4"	L.F.	0.050
4" x 4"	"	0.050
1/2" stock		
6" x 4"	L.F.	0.050

04295.10 PARGING/MASONRY PLASTER	UNIT	MAN/ HOURS
Parging		
1/2" thick	S.F.	0.053
3/4" thick	"	0.067

MASONRY RESTORATION

04520.10 RESTORATION AND CLEANING	UNIT	MAN/ HOURS
Masonry cleaning		
Washing brick		
Smooth surface	S.F.	0.013
Rough surface	"	0.018
Pointing masonry		
Brick	S.F.	0.032
Concrete block	"	0.023
Brick removal and replacement		
Minimum	EA.	0.100
Average	"	0.133
Maximum	"	0.400

04550.10 REFRACTORIES	UNIT	MAN/ HOURS
Flue liners		
Rectangular		
8" x 12"	L.F.	0.133
12" x 12"	"	0.145
12" x 18"	"	0.160
16" x 16"	"	0.178
18" x 18"	"	0.190
20" x 20"	"	0.200
24" x 24"	"	0.229

MASONRY RESTORATION	UNIT	MAN/ HOURS
04550.10 REFRACTORIES		
Round		
18" dia.	L.F.	0.190
24" dia.	"	0.229

METAL FASTENING

METAL FASTENING	UNIT	MAN/ HOURS
05050.10 STRUCTURAL WELDING		
Welding		
Single pass		
1/8"	L.F.	0.040
3/16"	"	0.053
1/4"	"	0.067
Miscellaneous steel shapes		
Plain	Lb.	0.002
Galvanized	"	0.003
Plates		
Plain	Lb.	0.002
Galvanized	"	0.003
05050.95 METAL LINTELS		
Lintels, steel		
Plain	Lb.	0.020
Galvanized	"	0.020

COLD FORMED FRAMING

COLD FORMED FRAMING	UNIT	MAN/ HOURS
05410.10 METAL FRAMING		
Furring channel, galvanized		
Beams and columns, 3/4"		
12" o.c.	S.F.	0.080
16" o.c.	"	0.073
Walls, 3/4"		
12" o.c.	S.F.	0.040
16" o.c.	"	0.033
Stud, load bearing		
16" o.c.		
16 ga.		
2-1/2"	S.F.	0.036
3-5/8"	"	0.036
24" o.c.		
16 ga.		
2-1/2"	S.F.	0.031
3-5/8"	"	0.031

06 WOOD AND PLASTICS

FASTENERS AND ADHESIVES	UNIT	MAN/HOURS
06050.10 ACCESSORIES		
Anchors		
Bolts, threaded two ends, with nuts and washers		
1/2" dia.		
4" long	EA.	0.050
7-1/2" long	"	0.050
3/4" dia.		
7-1/2" long	EA.	0.050
15" long	"	0.050
Bolts, carriage		
1/4 x 4	EA.	0.080
5/16 x 6	"	0.084
3/8 x 6	"	0.084
1/2 x 6	"	0.084
Joist and beam hangers		
18 ga.		
2 x 4	EA.	0.080
2 x 6	"	0.080
2 x 8	"	0.080
2 x 10	"	0.089
2 x 12	"	0.100
Rafter anchors, 18 ga., 1-1/2" wide		
Strap ties, 14 ga., 1-3/8" wide		
12" long	EA.	0.067
18" long	"	0.073
24" long	"	0.080
36" long	"	0.089

ROUGH CARPENTRY	UNIT	MAN/HOURS
06110.10 BLOCKING		
Steel construction		
Walls		
2x4	L.F.	0.053
2x6	"	0.062
2x8	"	0.067
2x10	"	0.073
2x12	"	0.080
Ceilings		
2x4	L.F.	0.062
2x6	"	0.073
2x8	"	0.080

ROUGH CARPENTRY	UNIT	MAN/HOURS
06110.10 BLOCKING		
2x10	L.F.	0.089
2x12	"	0.100
Wood construction		
Walls		
2x4	L.F.	0.044
2x6	"	0.050
2x8	"	0.053
2x10	"	0.057
2x12	"	0.062
Ceilings		
2x4	L.F.	0.050
2x6	"	0.057
2x8	"	0.062
2x10	"	0.067
2x12	"	0.073
06110.20 CEILING FRAMING		
Ceiling joists		
12" o.c.		
2x4	S.F.	0.019
2x8	"	0.021
16" o.c.		
2x4	S.F.	0.015
2x8	"	0.017
Sister joists for ceilings		
2x4	L.F.	0.057
2x6	"	0.067
2x8	"	0.080
2x10	"	0.100
2x12	"	0.133
06110.30 FLOOR FRAMING		
Floor joists		
12" o.c.		
2x6	S.F.	0.016
2x8	"	0.016
2x10	"	0.017
2x12	"	0.017
16" o.c.		
2x6	S.F.	0.013
2x8	"	0.014
2x10	"	0.014
2x12	"	0.014
Sister joists for floors		
2x4	L.F.	0.050
2x6	"	0.057
2x8	"	0.067

06 WOOD AND PLASTICS

ROUGH CARPENTRY	UNIT	MAN/ HOURS
06110.30 FLOOR FRAMING		
2x10	L.F.	0.080
2x12	"	0.100
3x6	"	0.080
3x8	"	0.089
3x10	"	0.100
3x12	"	0.114
4x6	"	0.080
4x8	"	0.089
4x10	"	0.100
4x12	"	0.114
06110.40 FURRING		
Furring, wood strips		
Walls		
On masonry or concrete walls		
1x2 furring		
12" o.c.	S.F.	0.025
16" o.c.	"	0.023
1x3 furring		
12" o.c.	S.F.	0.025
16" o.c.	"	0.023
On wood walls		
1x2 furring		
12" o.c.	S.F.	0.018
16" o.c.	"	0.016
1x3 furring		
12" o.c.	S.F.	0.018
16" o.c.	"	0.016
Ceilings		
On masonry or concrete ceilings		
1x2 furring		
12" o.c.	S.F.	0.044
16" o.c.	"	0.040
1x3 furring		
12" o.c.	S.F.	0.044
16" o.c.	"	0.040
On wood ceilings		
1x2 furring		
12" o.c.	S.F.	0.030
16" o.c.	"	0.027
1x3		
12" o.c.	S.F.	0.030
16" o.c.	"	0.027
06110.50 ROOF FRAMING		
Roof framing		
Rafters, gable end		
0-2 pitch (flat to 2-in-12)		

ROUGH CARPENTRY	UNIT	MAN/ HOURS
06110.50 ROOF FRAMING		
12" o.c.		
2x4	S.F.	0.017
2x6	"	0.017
2x8	"	0.018
16" o.c.		
2x6	S.F.	0.014
2x8	"	0.015
Sister rafters		
2x4	L.F.	0.057
2x6	"	0.067
2x8	"	0.080
2x10	"	0.100
2x12	"	0.133
06110.60 SLEEPERS		
Sleepers, over concrete		
12" o.c.		
1x2	S.F.	0.018
1x3	"	0.019
2x4	"	0.022
2x6	"	0.024
16" o.c.		
1x2	S.F.	0.016
1x3	"	0.016
2x4	"	0.019
2x6	"	0.020
06110.65 SOFFITS		
Soffit framing		
2x3	L.F.	0.057
2x4	"	0.062
2x6	"	0.067
2x8	"	0.073
06110.70 WALL FRAMING		
Framing wall, studs		
12" o.c.		
2x3	S.F.	0.015
2x4	"	0.015
16" o.c.		
2x3	S.F.	0.013
2x4	"	0.013
24" o.c.		
2x3	S.F.	0.011
2x4	"	0.011
Plates, top or bottom		

06 WOOD AND PLASTICS

ROUGH CARPENTRY

06110.70 WALL FRAMING	UNIT	MAN/HOURS
2x3	L.F.	0.024
2x4	"	0.025
Headers, door or window		
2x6		
Single		
3' long	EA.	0.400
6' long	"	0.500
Double		
3' long	EA.	0.444
6' long	"	0.571
2x8		
Single		
4' long	EA.	0.500
8' long	"	0.615
Double		
4' long	EA.	0.571
8' long	"	0.727
2x10		
Single		
5' long	EA.	0.615
10' long	"	0.800
Double		
5' long	EA.	0.667
10' long	"	0.800

06115.10 FLOOR SHEATHING	UNIT	MAN/HOURS
Sub-flooring, plywood, CDX		
1/2" thick	S.F.	0.010
5/8" thick	"	0.011
3/4" thick	"	0.013
Structural plywood		
1/2" thick	S.F.	0.010
Board type subflooring		
1x6		
Minimum	S.F.	0.018
Maximum	"	0.020
1x8		
Minimum	S.F.	0.017
Maximum	"	0.019
1x10		
Minimum	S.F.	0.016
Maximum	"	0.018
Underlayment		
Hardboard, 1/4" tempered	S.F.	0.010
Plywood		
3/8" thick	S.F.	0.010
1/2" thick	"	0.011
5/8" thick	"	0.011
3/4" thick	"	0.012

ROUGH CARPENTRY

06115.20 ROOF SHEATHING	UNIT	MAN/HOURS
Sheathing		
3/8" thick	S.F.	0.010
1/2" thick	"	0.011
5/8" thick	"	0.011

06115.30 WALL SHEATHING	UNIT	MAN/HOURS
Sheathing		
Plywood, CDX		
3/8" thick	S.F.	0.012
1/2" thick	"	0.012
5/8" thick	"	0.013
3/4" thick	"	0.015

06220.10 MILLWORK	UNIT	MAN/HOURS
Countertop, laminated plastic		
25" x 7/8" thick		
Minimum	L.F.	0.200
Average	"	0.267
Maximum	"	0.320
25" x 1-1/4" thick		
Minimum	L.F.	0.267
Average	"	0.320
Maximum	"	0.400
Add for cutouts	EA.	0.500
Backsplash, 4" high, 7/8" thick	L.F.	0.160
Base cabinets, 34-1/2" high, 24" deep, hardwood, no tops		
Minimum	L.F.	0.320
Average	"	0.400
Maximum	"	0.533
Wall cabinets		
Minimum	L.F.	0.267
Average	"	0.320
Maximum	"	0.400

MOISTURE PROTECTION

07150.10 DAMPROOFING

MOISTURE PROTECTION	UNIT	MAN/ HOURS
Silicone dampproofing, sprayed on		
Concrete surface		
1 coat	S.F.	0.004
2 coats	"	0.006
Concrete block		
1 coat	S.F.	0.005
2 coats	"	0.007
Brick		
1 coat	S.F.	0.006
2 coats	"	0.008

07160.10 BITUMINOUS DAMPPROOFING

MOISTURE PROTECTION	UNIT	MAN/ HOURS
Building paper, asphalt felt		
15 lb	S.F.	0.032
30 lb	"	0.033

07190.10 VAPOR BARRIERS

MOISTURE PROTECTION	UNIT	MAN/ HOURS
Vapor barrier, polyethylene		
2 mil	S.F.	0.004
6 mil	"	0.004
8 mil	"	0.004
10 mil	"	0.004

INSULATION

07210.10 BATT INSULATION

INSULATION	UNIT	MAN/ HOURS
Ceiling, fiberglass, unfaced		
3-1/2" thick, R11	S.F.	0.009
6" thick, R19	"	0.011
9" thick, R30	"	0.012
Suspended ceiling, unfaced		
3-1/2" thick, R11	S.F.	0.009
6" thick, R19	"	0.010
9" thick, R30	"	0.011
Crawl space, unfaced		
3-1/2" thick, R11	S.F.	0.012
6" thick, R19	"	0.013
9" thick, R30	"	0.015
Wall, fiberglass		
Paper backed		
2" thick, R7	S.F.	0.008
3" thick, R8	"	0.009
4" thick, R11	"	0.009
6" thick, R19	"	0.010
Foil backed, 1 side		
2" thick, R7	S.F.	0.008
3" thick, R11	"	0.009
4" thick, R14	"	0.009
6" thick, R21	"	0.010
Foil backed, 2 sides		
2" thick, R7	S.F.	0.009
3" thick, R11	"	0.010
4" thick, R14	"	0.011
6" thick, R21	"	0.011
Unfaced		
2" thick, R7	S.F.	0.008
3" thick, R9	"	0.009
4" thick, R11	"	0.009
6" thick, R19	"	0.010
Mineral wool batts		
Paper backed		
2" thick, R6	S.F.	0.008
4" thick, R12	"	0.009
6" thick, R19	"	0.010
Fasteners, self adhering, attached to ceiling deck		
2-1/2" long	EA.	0.013
4-1/2" long	"	0.015
Capped, self-locking washers for fastening insulation	"	0.008

07210.20 BOARD INSULATION

INSULATION	UNIT	MAN/ HOURS
Insulation, rigid		
Fiberglass, roof		
0.75" thick, R2.78	S.F.	0.007
1.06" thick, R4.17	"	0.008
1.31" thick, R5.26	"	0.008
1.63" thick, R6.67	"	0.008
2.25" thick, R8.33	"	0.009
Perlite board, roof		
1.00" thick, R2.78	S.F.	0.007
1.50" thick, R4.17	"	0.007
2.00" thick, R5.92	"	0.007
2.50" thick, R6.67	"	0.008
3.00" thick, R8.33	"	0.008
4.00" thick, R10.00	"	0.008
5.25" thick, R14.29	"	0.009
1" thick, R3.0	"	0.007

07 THERMAL AND MOISTURE

INSULATION	UNIT	MAN/HOURS
07210.60 LOOSE FILL INSULATION		
Blown-in type		
Fiberglass		
5" thick, R11	S.F.	0.007
6" thick, R13	"	0.008
9" thick, R19	"	0.011
Rockwool, attic application		
6" thick, R13	S.F.	0.008
8" thick, R19	"	0.010
10" thick, R22	"	0.012
12" thick, R26	"	0.013
15" thick, R30	"	0.016
Poured type		
Fiberglass		
1" thick, R4	S.F.	0.005
2" thick, R8	"	0.006
3" thick, R12	"	0.007
4" thick, R16	"	0.008
Mineral wool		
1" thick, R3	S.F.	0.005
2" thick, R6	"	0.006
3" thick, R9	"	0.007
4" thick, R12	"	0.008
Vermiculite or perlite		
2" thick, R4.8	S.F.	0.006
3" thick, R7.2	"	0.007
4" thick, R9.6	"	0.008
Masonry, poured vermiculite or perlite		
4" block	S.F.	0.004
6" block	"	0.005
8" block	"	0.006
10" block	"	0.006
12" block	"	0.007
07210.70 SPRAYED INSULATION		
Foam, sprayed on		
Polystyrene		
1" thick, R4	S.F.	0.008
2" thick, R8	"	0.011
Urethane		
1" thick, R7.7	S.F.	0.008
2" thick, R15.4	"	0.011
07250.10 FIREPROOFING		
Sprayed on		
1" thick		
On beams	S.F.	0.018
On columns	"	0.016
On decks		

INSULATION	UNIT	MAN/HOURS
07250.10 FIREPROOFING		
Flat surface	S.F.	0.008
Fluted surface	"	0.010
1-1/2" thick		
On beams	S.F.	0.023
On columns	"	0.020
On decks		
Flat surface	S.F.	0.010
Fluted surface	"	0.013

SHINGLES AND TILES	UNIT	MAN/HOURS
07310.10 ASPHALT SHINGLES		
Standard asphalt shingles, strip shingles		
210 lb/square	SQ.	0.800
235 lb/square	"	0.889
240 lb/square	"	1.000
260 lb/square	"	1.143
300 lb/square	"	1.333
385 lb/square	"	1.600
Roll roofing, mineral surface		
90 lb	SQ.	0.571
110 lb	"	0.667
140 lb	"	0.800
07310.30 METAL SHINGLES		
Aluminum, .020" thick		
Plain	SQ.	1.600
Colors	"	1.600
Steel, galvanized		
26 ga.		
Plain	SQ.	1.600
Colors	"	1.600
Porcelain enamel, 22 ga.		
Minimum	SQ.	2.000
Average	"	2.000
Maximum	"	2.000
Replacement shingles		
Small jobs	EA.	0.267
Large jobs	S.F.	0.133

07 THERMAL AND MOISTURE

SHINGLES AND TILES

07310.70 WOOD SHINGLES	UNIT	MAN/ HOURS
Wood shingles, on roofs		
White cedar, #1 shingles		
4" exposure	SQ.	2.667
5" exposure	"	2.000
On walls		
White cedar, #1 shingles		
4" exposure	SQ.	4.000
5" exposure	"	3.200

07310.80 WOOD SHAKES	UNIT	MAN/ HOURS
Shakes, hand split, 24" red cedar, on roofs		
5" exposure	SQ.	4.000
9" exposure	"	2.667

MEMBRANE ROOFING

07510.10 BUILT-UP ASPHALT ROOFING	UNIT	MAN/ HOURS
Built-up roofing, asphalt felt, including gravel		
2 ply	SQ.	2.000
3 ply	"	2.667
4 ply	"	3.200

07530.10 SINGLE-PLY ROOFING	UNIT	MAN/ HOURS
Elastic sheet roofing		
Neoprene, 1/16" thick	S.F.	0.010
EPDM rubber		
45 mil	S.F.	0.010
60 mil	"	0.010
PVC		
45 mil	S.F.	0.010
60 mil	"	0.010
Flashing		
Pipe flashing, 90 mil thick		
1" pipe	EA.	0.200
2" pipe	"	0.200
3" pipe	"	0.211
4" pipe	"	0.211
5" pipe	"	0.222
6" pipe	"	0.222
8" pipe	"	0.235

MEMBRANE ROOFING

07530.10 SINGLE-PLY ROOFING	UNIT	MAN/ HOURS
10" pipe	EA.	0.267
12" pipe	"	0.267
Neoprene flashing, 60 mil thick strip		
6" wide	L.F.	0.067
12" wide	"	0.100
18" wide	"	0.133
24" wide	"	0.200
Adhesives		
1/4" bead	L.F.	0.004
Ballast, 3/4" through 1-1/2" dia. river gravel, 100lb/sf	S.F.	0.800
Walkway for membrane roofs, 1/2" thick	"	0.027

FLASHING AND SHEET METAL

07610.10 METAL ROOFING	UNIT	MAN/ HOURS
Sheet metal roofing, copper, 16 oz, batten seam	SQ.	5.333
Standing seam	"	5.000
Aluminum roofing, natural finish		
Corrugated, on steel frame		
.0175" thick	SQ.	2.286
.032" thick	"	2.286
Corrugated galvanized steel roofing, on steel frame		
28 ga.	SQ.	2.286
22 ga.	"	2.286

07620.10 FLASHING AND TRIM	UNIT	MAN/ HOURS
Counter flashing		
Aluminum, .032"	S.F.	0.080
Stainless steel, .015"	"	0.080
16 oz.	"	0.080
Base flashing		
Aluminum, .040"	S.F.	0.067
Stainless steel, .018"	"	0.067
16 oz.	"	0.067
24 oz.	"	0.067
Scupper outlets		
10" x 10" x 4"	EA.	0.200
22" x 4" x 4"	"	0.200
8" x 8" x 5"	"	0.200

FLASHING AND SHEET METAL

07620.10 FLASHING AND TRIM

FLASHING AND SHEET METAL	UNIT	MAN/ HOURS
Drainage boots, roof, cast iron		
2 x 3	L.F.	0.100
3 x 4	"	0.100
4 x 5	"	0.107
4 x 6	"	0.107
5 x 7	"	0.114
6 x 6	EA.	0.200
8 x 8	"	0.200
8 x 10	"	0.200
8 x 12	"	0.200
Reglets, copper 10 oz.	L.F.	0.053
Stainless steel, .020"	"	0.053

07620.20 GUTTERS AND DOWNSPOUTS

FLASHING AND SHEET METAL	UNIT	MAN/ HOURS
Copper gutter and downspout		
Downspouts, 16 oz. copper		
Round		
3" dia.	L.F.	0.053
4" dia.	"	0.053
Rectangular, corrugated		
2" x 3"	L.F.	0.050
3" x 4"	"	0.050
Rectangular, flat surface		
2" x 3"	L.F.	0.053
3" x 4"	"	0.053
Lead-coated copper downspouts		
Round		
3" dia.	L.F.	0.050
4" dia.	"	0.057
Rectangular, corrugated		
2" x 3"	L.F.	0.053
3" x 4"	"	0.053
Rectangular, plain		
2" x 3"	L.F.	0.053
3" x 4"	"	0.053
Gutters, 16 oz. copper		
Half round		
4" wide	L.F.	0.080
5" wide	"	0.089
Type K		
4" wide	L.F.	0.080
5" wide	"	0.089
Lead-coated copper gutters		
Half round		
4" wide	L.F.	0.080
6" wide	"	0.089
Type K		
4" wide	L.F.	0.080
5" wide	"	0.089

FLASHING AND SHEET METAL

07620.20 GUTTERS AND DOWNSPOUTS

FLASHING AND SHEET METAL	UNIT	MAN/ HOURS
Aluminum gutter and downspout		
Downspouts		
2" x 3"	L.F.	0.053
3" x 4"	"	0.057
4" x 5"	"	0.062
Round		
3" dia.	L.F.	0.053
4" dia.	"	0.057
Gutters, stock units		
4" wide	L.F.	0.084
5" wide	"	0.089
Galvanized steel gutter and downspout		
Downspouts, round corrugated		
3" dia.	L.F.	0.053
4" dia.	"	0.053
5" dia.	"	0.057
6" dia.	"	0.057
Rectangular		
2" x 3"	L.F.	0.053
3" x 4"	"	0.050
4" x 4"	"	0.050
Gutters, stock units		
5" wide		
Plain	L.F.	0.089
Painted	"	0.089
6" wide		
Plain	L.F.	0.094
Painted	"	0.094

07700.10 MANUFACTURED SPECIALTIES

FLASHING AND SHEET METAL	UNIT	MAN/ HOURS
Ceiling access doors		
Swing up model, metal frame		
Steel door		
2'6" x 2'6"	EA.	0.800
2'6" x 3'0"	"	0.800
Aluminum door		
2'6" x 2'6"	EA.	0.800
2'6" x 3'0"	"	0.800
Swing down model, metal frame		
Steel door		
2'6" x 2'6"	EA.	0.800
2'6" x 3'0"	"	0.800
Aluminum door		
2'6" x 2'6"	EA.	0.800
2'6" x 3'0"	"	0.800
Gravity ventilators, with curb, base, damper and screen		
Stationary siphon		
6" dia.	EA.	0.533
12" dia.	"	0.533

07 THERMAL AND MOISTURE

FLASHING AND SHEET METAL

07700.10 MANUFACTURED SPECIALTIES

	UNIT	MAN/ HOURS
24" dia.	EA.	0.800
36" dia.	"	0.800
Wind driven spinner		
6" dia.	EA.	0.533
12" dia.	"	0.533
24" dia.	"	0.800
36" dia.	"	0.800
Stationary mushroom		
16" dia.	EA.	0.800
30" dia.	"	1.000
36" dia.	"	1.333
42" dia.	"	1.600

JOINT SEALERS

07920.10 CAULKING

	UNIT	MAN/ HOURS
Caulk exterior, two component		
1/4 x 1/2	L.F.	0.040
3/8 x 1/2	"	0.044
1/2 x 1/2	"	0.050
Caulk interior, single component		
1/4 x 1/2	L.F.	0.038
3/8 x 1/2	"	0.042
1/2 x 1/2	"	0.047

SPECIALTIES

10165.10 TOILET PARTITIONS

	UNIT	MAN/ HOURS
Toilet partition, plastic laminate		
Ceiling mounted	EA.	2.667
Floor mounted	"	2.000
Metal		
Ceiling mounted	EA.	2.667
Floor mounted	"	2.000
Wheel chair partition, plastic laminate		
Ceiling mounted	EA.	2.667
Floor mounted	"	2.000
Painted metal		
Ceiling mounted	EA.	2.667
Floor mounted	"	2.000
Urinal screen, plastic laminate		
Wall hung	EA.	1.000
Floor mounted	"	1.000
Porcelain enameled steel, floor mounted	"	1.000
Painted metal, floor mounted	"	1.000
Stainless steel, floor mounted	"	1.000
Metal toilet partitions		
Toilet partitions, front door and side divider, floor mounted		
Porcelain enameled steel	EA.	2.000
Painted steel	"	2.000
Stainless steel	"	2.000

10185.10 SHOWER STALLS

	UNIT	MAN/ HOURS
Shower receptors		
Precast, terrazzo		
32" x 32"	EA.	0.667
32" x 48"	"	0.800
Concrete		
32" x 32"	EA.	0.667
48" x 48"	"	0.889
Shower door, trim and hardware		
Economy, 24" wide, chrome frame, tempered glass	EA.	0.800
Porcelain enameled steel, flush	"	0.800
Baked enameled steel, flush	"	0.800
Aluminum frame, tempered glass, 48" wide, sliding	"	1.000
Folding	"	1.000
Aluminum frame and tempered glass, molded plastic		
Complete with receptor and door		
32" x 32"	EA.	2.000
36" x 36"	"	2.000
40" x 40"	"	2.286
Shower compartment, precast concrete receptor		
Single entry type		
Porcelain enameled steel	EA.	8.000
Baked enameled steel	"	8.000
Stainless steel	"	8.000
Double entry type		
Porcelain enameled steel	EA.	10.000
Baked enameled steel	"	10.000
Stainless steel	"	10.000

10 SPECIALTIES

SPECIALTIES	UNIT	MAN/ HOURS
10270.40 ACCESS & PEDESTAL FLOOR		
Panels, no covering, 2'x2'		
Plain	S.F.	0.010
Perforated	"	0.400
Pedestals		
For 6" to 12" clearance	EA.	0.080
Stringers		
2'	L.F.	0.038
6'	"	0.027
Accessories		
Ramp assembly	S.F.	0.032
Elevated floor assembly	"	0.030
Handrail	L.F.	0.400
Fascia plate	"	0.200
RF shielding components, floor liner		
Hot rolled steel sheet		
14 ga.	S.F.	0.020
11 ga.	"	0.062
10290.10 PEST CONTROL		
Termite control		
Under slab spraying		
Minimum	S.F.	0.002
Average	"	0.004
Maximum	"	0.008
10500.10 LOCKERS		
Locker bench, floor mounted, laminated maple		
4'	EA.	0.667
6'	"	0.667
Wardrobe locker, single tier type		
12" x 15" x 72"	EA.	0.800
18" x 15" x 72"	"	0.842
12" x 18" x 72"	"	0.889
18" x 18" x 72"	"	0.941
10520.10 FIRE PROTECTION		
Portable fire extinguishers		
Water pump tank type		
2.5 gal.		
Red enameled galvanized	EA.	0.533
Red enameled copper	"	0.533
Polished copper	"	0.533
Carbon dioxide type, red enamel steel		
Squeeze grip with hose and horn		
2.5 lb	EA.	0.533
5 lb	"	0.615
10 lb	"	0.800

SPECIALTIES	UNIT	MAN/ HOURS
10520.10 FIRE PROTECTION		
15 lb	EA.	1.000
20 lb	"	1.000
Wheeled type		
125 lb	EA.	1.600
250 lb	"	1.600
500 lb	"	1.600
Dry chemical, pressurized type		
Red enameled steel		
2.5 lb	EA.	0.533
5 lb	"	0.615
10 lb	"	0.800
20 lb	"	1.000
30 lb	"	1.000
Chrome plated steel, 2.5 lb	"	0.533
Other type extinguishers		
2.5 gal, stainless steel, pressurized water tanks	EA.	0.533
Soda and acid type	"	0.533
Cartridge operated, water type	"	0.533
Loaded stream, water type	"	0.533
Foam type	"	0.533
40 gal, wheeled foam type	"	1.600
Fire extinguisher cabinets		
Enameled steel		
8" x 12" x 27"	EA.	1.600
8" x 16" x 38"	"	1.600
Aluminum		
8" x 12" x 27"	EA.	1.600
8" x 16" x 38"	"	1.600
8" x 12" x 27"	"	1.600
Stainless steel		
8" x 16" x 38"	EA.	1.600
10670.10 SHELVING		
Metal storage shelving, baked enamel		
7 shelf unit, 72" or 84" high		
10" shelf	L.F.	0.800
18" shelf	"	0.941
4 shelf unit, 40" high		
10" shelf	L.F.	0.667
18" shelf	"	0.842
3 shelf unit, 32" high		
10" shelf	L.F.	0.400
18" shelf	"	0.471
Single shelf unit, attached to masonry		
10" shelf	L.F.	0.133
18" shelf	"	0.163

SPECIALTIES	UNIT	MAN/ HOURS
10750.10 TELEPHONE ENCLOSURES		
Telephone enclosure, wall mounted, shelf, 28" x 30" x 15"	EA.	2.000
Directory shelf, stainless steel, 3 binders	"	1.333
10800.10 BATH ACCESSORIES		
Ash receiver, wall mounted, aluminum	EA.	0.400
Grab bar, 1-1/2" dia., stainless steel, wall mounted		
24" long	EA.	0.400
36" long	"	0.421
42" long	"	0.444
48" long	"	0.471
52" long	"	0.500
1" dia., stainless steel		
12" long	EA.	0.348
18" long	"	0.364
24" long	"	0.400
30" long	"	0.421
36" long	"	0.444
48" long	"	0.471
Hand dryer, surface mounted, 110 volt	"	1.000
Medicine cabinet, 16 x 22, baked enamel, steel, lighted	"	0.320
With mirror, lighted	"	0.533
Mirror, 1/4" plate glass, up to 10 sf	S.F.	0.080
Mirror, stainless steel frame		
18"x24"	EA.	0.267
18"x32"	"	0.320
18"x36"	"	0.400
24"x30"	"	0.400
24"x36"	"	0.444
24"x48"	"	0.667
24"x60"	"	0.800
30"x30"	"	0.800
30"x72"	"	1.000
48"x72"	"	1.333
With shelf, 18"x24"	"	0.320
Sanitary napkin dispenser, stainless steel, wall mounted	"	0.533
Shower rod, 1" diameter		
Chrome finish over brass	EA.	0.400
Stainless steel	"	0.400
Soap dish, stainless steel, wall mounted	"	0.533
Toilet tissue dispenser, stainless, wall mounted		
Single roll	EA.	0.200
Double roll	"	0.229
Towel dispenser, stainless steel		
Flush mounted	EA.	0.444
Surface mounted	"	0.400
Combination towel dispenser and waste receptacle	"	0.533
Towel bar, stainless steel		
18" long	EA.	0.320
24" long	"	0.364

SPECIALTIES	UNIT	MAN/ HOURS
10800.10 BATH ACCESSORIES		
30" long	EA.	0.400
36" long	"	0.444
Toothbrush and tumbler holder	"	0.267
Waste receptacle, stainless steel, wall mounted	"	0.667

11 EQUIPMENT

ARCHITECTURAL EQUIPMENT	UNIT	MAN/ HOURS
11010.10 MAINTENANCE EQUIPMENT		
Vacuum cleaning system		
3 valves		
1.5 hp	EA.	8.889
2.5 hp	"	11.429
5 valves	"	16.000
7 valves	"	20.000
11060.10 THEATER EQUIPMENT		
Roll out stage, steel frame, wood floor		
Manual	S.F.	0.050
Electric	"	0.080
Portable stages		
8" high	S.F.	0.040
18" high	"	0.044
36" high	"	0.047
48" high	"	0.050
Band risers		
Minimum	S.F.	0.040
Maximum	"	0.040
Chairs for risers		
Minimum	EA.	0.036
Maximum	"	0.036
11080.10 POLICE EQUIPMENT		
Firing range equipment, rifle		
3 position	EA.	26.667
4 position	"	40.000
5 position	"	44.444
6 position	"	47.059
11090.10 CHECKROOM EQUIPMENT		
Motorized checkroom equipment		
No shelf system, 6'4" height		
7'6" length	EA.	8.000
14'6" length	"	8.000
28' length	"	8.000
One shelf, 6'8" height		
7'6" length	EA.	8.000
14'6" length	"	8.000
28' length	"	8.000
Two shelves, 7'5" height		
7'6" length	EA.	8.000
14'6" length	"	8.000
28' length	"	8.000
Three shelves, 8' height		
7'6" length	EA.	16.000
14'6" length	"	16.000
28' length	"	16.000

ARCHITECTURAL EQUIPMENT	UNIT	MAN/ HOURS
11090.10 CHECKROOM EQUIPMENT		
Four shelves, 8'7" height		
7'6" length	EA.	16.000
14'6" length	"	16.000
28' length	"	16.000
200 lb		
11110.10 LAUNDRY EQUIPMENT		
High capacity, heavy duty		
Washer extractors		
135 lb		
Standard	EA.	6.667
Pass through	"	6.667
200 lb		
Standard	EA.	6.667
Pass through	"	6.667
110 lb dryer	"	6.667
Hand operated presser	"	8.889
Mushroom press	"	8.889
Spreader feeders		
2 station	EA.	8.889
4 station	"	16.000
Delivery carts		
12 bushel	EA.	0.100
Low capacity		
Pressers		
Air operated	EA.	3.200
Hand operated	"	3.200
Extractor, low capacity	"	3.200
Ironer, 48"	"	1.600
Coin washers		
10 lb capacity	EA.	1.600
20 lb capacity	"	1.600
Coin dryer	"	1.000
Coin dry cleaner, 20 lb	"	3.200
11161.10 LOADING DOCK EQUIPMENT		
Dock leveler, 10 ton capacity		
6' x 8'	EA.	8.000
7' x 8'	"	8.000
11170.10 WASTE HANDLING		
Incinerator, electric		
100 lb/hr		
Minimum	EA.	8.000
Maximum	"	8.000
400 lb/hr		

11 EQUIPMENT

ARCHITECTURAL EQUIPMENT	UNIT	MAN/ HOURS
11170.10 WASTE HANDLING		
Minimum	EA.	16.000
Maximum	"	16.000
1000 lb/hr		
Minimum	EA.	24.242
Maximum	"	24.242
Incinerator, medical-waste		
25 lb/hr, 2-7 x 4-0	EA.	16.000
50 lb/hr, 2-11 x 4-11	"	16.000
75 lb/hr, 3-8 x 5-0	"	32.000
100 lb/hr, 3-8 x 6-0	"	32.000
Industrial compactor		
1 cy	EA.	8.889
3 cy	"	11.429
5 cy	"	16.000
11400.10 FOOD SERVICE EQUIPMENT		
Unit kitchens		
30" compact kitchen		
Refrigerator, with range, sink	EA.	4.000
Sink only	"	2.667
Range only	"	2.000
Cabinet for upper wall section	"	1.143
Stainless shield, for rear wall	"	0.320
Side wall	"	0.320
42" compact kitchen		
Refrigerator with range, sink	EA.	4.444
Sink only	"	4.000
Cabinet for upper wall section	"	1.333
Stainless shield, for rear wall	"	0.333
Side wall	"	0.333
54" compact kitchen		
Refrigerator, oven, range, sink	EA.	5.714
Cabinet for upper wall section	"	1.600
Stainless shield, for		
Rear wall	EA.	0.364
Side wall	"	0.364
60" compact kitchen		
Refrigerator, oven, range, sink	EA.	5.714
Cabinet for upper wall section	"	1.600
Stainless shield, for		
Rear wall	EA.	0.364
Side wall	"	0.364
72" compact kitchen		
Refrigerator, oven, range, sink	EA.	6.667
Cabinet for upper wall section	"	1.600
Stainless shield for		
Rear wall	EA.	0.400
Side wall	"	0.400
Bake oven		

ARCHITECTURAL EQUIPMENT	UNIT	MAN/ HOURS
11400.10 FOOD SERVICE EQUIPMENT		
Single deck		
Minimum	EA.	1.000
Maximum	"	2.000
Double deck		
Minimum	EA.	1.333
Maximum	"	2.000
Triple deck		
Minimum	EA.	1.333
Maximum	"	2.667
Convection type oven, electric, 40" x 45" x 57"		
Minimum	EA.	1.000
Maximum	"	2.000
Broiler, without oven, 69" x 26" x 39"		
Minimum	EA.	1.000
Maximum	"	1.333
Coffee urns, 10 gallons		
Minimum	EA.	2.667
Maximum	"	4.000
Fryer, with submerger		
Single		
Minimum	EA.	1.600
Maximum	"	2.667
Double		
Minimum	EA.	2.000
Maximum	"	2.667
Griddle, counter		
3' long		
Minimum	EA.	1.333
Maximum	"	1.600
5' long		
Minimum	EA.	2.000
Maximum	"	2.667
Kettles, steam, jacketed		
20 gallons		
Minimum	EA.	2.000
Maximum	"	4.000
40 gallons		
Minimum	EA.	2.000
Maximum	"	4.000
60 gallons		
Minimum	EA.	2.000
Maximum	"	4.000
Range		
Heavy duty, single oven, open top		
Minimum	EA.	1.000
Maximum	"	2.667
Fry top		
Minimum	EA.	1.000
Maximum	"	2.667

11 EQUIPMENT

ARCHITECTURAL EQUIPMENT

11400.10 FOOD SERVICE EQUIPMENT

ARCHITECTURAL EQUIPMENT	UNIT	MAN/ HOURS
Steamers, electric		
27 kw		
Minimum	EA.	2.000
Maximum	"	2.667
18 kw		
Minimum	EA.	2.000
Maximum	"	2.667
Dishwasher, rack type		
Single tank, 190 racks/hr	EA.	4.000
Double tank		
234 racks/hr	EA.	4.444
265 racks/hr	"	5.333
Dishwasher, automatic 100 meals/hr	"	2.667
Disposals		
100 gal/hr	EA.	2.667
120 gal/hr	"	2.759
250 gal/hr	"	2.857
Exhaust hood for dishwasher, gutter 4 sides, s-steel		
4'x4'x2'	EA.	2.963
4'x7'x2'	"	3.200
Food preparation machines		
Vertical cutter mixers		
25 quart	EA.	2.667
40 quart	"	2.667
80 quart	"	4.000
130 quart	"	6.667
Choppers		
5 lb	EA.	2.000
16 lb	"	2.667
40 lb	"	4.000
Mixers, floor models		
20 quart	EA.	1.000
60 quart	"	1.000
80 quart	"	1.143
140 quart	"	1.600
Ice cube maker		
50 lb per day		
Minimum	EA.	8.000
Maximum	"	8.000
500 lb per day		
Minimum	EA.	13.333
Maximum	"	13.333
Ice flakers		
300 lb per day	EA.	8.000
600 lb per day	"	13.333
1000 lb per day	"	17.778
2000 lb per day	"	20.000
Refrigerated cases		
Dairy products		

ARCHITECTURAL EQUIPMENT

11400.10 FOOD SERVICE EQUIPMENT

ARCHITECTURAL EQUIPMENT	UNIT	MAN/ HOURS
Multi deck type	L.F.	0.533
Delicatessen case, service deli		
Single deck	L.F.	4.000
Multi deck	"	5.000
Meat case		
Single deck	L.F.	4.706
Multi deck	"	5.000
Produce case		
Single deck	L.F.	4.706
Multi deck	"	5.000
Bottle coolers		
6' long		
Minimum	EA.	16.000
Maximum	"	16.000
10' long		
Minimum	EA.	26.667
Maximum	"	26.667
Frozen food cases		
Chest type	L.F.	4.706
Reach-in, glass door	"	5.000
Island case, single	"	4.706
Multi deck	"	5.000
Ice storage bins		
500 lb capacity	EA.	11.429
1000 lb capacity	"	22.857

11450.10 RESIDENTIAL EQUIPMENT

ARCHITECTURAL EQUIPMENT	UNIT	MAN/ HOURS
Compactor, 4 to 1 compaction	EA.	2.000
Dishwasher, built-in		
2 cycles	EA.	4.000
4 or more cycles	"	4.000
Disposal		
Garbage disposer	EA.	2.667
Heaters, electric, built-in		
Ceiling type	EA.	2.667
Wall type		2.667
Minimum	EA.	2.000
Maximum	"	2.667
Hood for range, 2-speed, vented		
30" wide	EA.	2.667
42" wide	"	2.667
Ice maker, automatic		
30 lb per day	EA.	1.143
50 lb per day	"	4.000
Ranges electric		
Built-in, 30", 1 oven	EA.	2.667
2 oven	"	2.667
Counter top, 4 burner, standard	"	2.000

11 EQUIPMENT

ARCHITECTURAL EQUIPMENT	UNIT	MAN/ HOURS
11450.10 RESIDENTIAL EQUIPMENT		
With grill	EA.	2.000
Free standing, 21", 1 oven	"	2.667
30", 1 oven	"	1.600
2 oven	"	1.600
Water softener		
30 grains per gallon	EA.	2.667
70 grains per gallon	"	4.000
11470.10 DARKROOM EQUIPMENT		
Dryers		
36" x 25" x 68"	EA.	4.000
48" x 25" x 68"	"	4.000
Processors, film		
Black and white	EA.	4.000
Color negatives	"	4.000
Prints	"	4.000
Transparencies	"	4.000
Sinks with cabinet and/or stand		
5" sink with stand		
24" x 48"	EA.	2.000
32" x 64"	"	2.667
38" x 52"	"	2.667
42" x 132"	"	4.000
48" x 52"	"	4.000
5" sink with cabinet		
24" x 48"	EA.	2.000
32" x 64"	"	2.667
38" x 52"	"	2.667
42" x 132"	"	4.000
48" x 52"	"	4.000
10" sink with stand		
24" x 48"	EA.	2.000
32" x 64"	"	2.667
38" x 52"	"	2.667
10" sink with cabinet		
24" x 48"	EA.	2.000
38" x 52"	"	2.667
11500.10 INDUSTRIAL EQUIPMENT		
Vehicular paint spray booth, solid back, 14'4" x 9'6"		
24' deep	EA.	8.000
26'6" deep	"	8.000
28'6" deep	"	8.000
Drive through, 14'9" x 9'6"		
24' deep	EA.	8.000
26'6" deep	"	8.000
28'6" deep	"	8.000
Water wash, paint spray booth		

ARCHITECTURAL EQUIPMENT	UNIT	MAN/ HOURS
11500.10 INDUSTRIAL EQUIPMENT		
5' x 11'2" x 10'8"	EA.	8.000
6' x 11'2" x 10'8"	"	8.000
8' x 11'2" x 10'8"	"	8.000
10' x 11'2" x 11'2"	"	8.000
12' x 12'2" x 11'2"	"	8.000
14' x 12'2" x 11'2"	"	8.000
16' x 12'2" x 11'2"	"	8.000
20' x 12'2" x 11'2"	"	8.000
Dry type spray booth, with paint arrestors		
5'4" x 7'2" x 6'8"	EA.	8.000
6'4" x 7'2" x 6'8"	"	8.000
8'4" x 7'2" x 9'2"	"	8.000
10'4" x 7'2" x 9'2"	"	8.000
12'4" x 7'6" x 9'2"	"	8.000
14'4" x 7'6" x 9'8"	"	8.000
16'4" x 7'7" x 9'8"	"	8.000
20'4" x 7'7" x 10'8"	"	8.000
Air compressor, electric		
1 hp		
115 volt	EA.	5.333
5 hp		
115 volt	EA.	8.000
230 volt	"	8.000
Hydraulic lifts		
8,000 lb capacity	EA.	20.000
11,000 lb capacity	"	32.000
24,000 lb capacity	"	53.333
Power tools		
Band saws		
10"	EA.	0.667
14"	"	0.800
Motorized shaper	"	0.615
Motorized lathe	"	0.667
Bench saws		
9" saw	EA.	0.533
10" saw	"	0.571
12" saw	"	0.667
Electric grinders		
1/3 hp	EA.	0.320
1/2 hp	"	0.348
3/4 hp	"	0.348
11600.10 LABORATORY EQUIPMENT		
Cabinets, base		
Minimum	L.F.	0.667
Maximum	"	0.667
Full storage, 7' high		
Minimum	L.F.	0.667

11 EQUIPMENT

ARCHITECTURAL EQUIPMENT	UNIT	MAN/ HOURS
11600.10 LABORATORY EQUIPMENT		
Maximum	L.F.	0.667
Wall		
Minimum	L.F.	0.800
Maximum	"	0.800
Counter tops		
Minimum	S.F.	0.100
Average	"	0.114
Maximum	"	0.133
Tables		
Open underneath	S.F.	0.400
Doors underneath	"	0.500
Medical laboratory equipment		
Analyzer		
Chloride	EA.	0.400
Blood	"	0.667
Bath, water, utility, countertop unit	"	0.800
Hot plate, lab, countertop	"	0.727
Stirrer	"	0.727
Incubator, anaerobic, 23x23x36"	"	4.000
Dry heat bath	"	1.333
Incinerator, for sterilizing	"	0.080
Meter, serum protein	"	0.100
Ph analog, general purpose	"	0.114
Refrigerator, blood bank, undercounter type 153 litres	"	1.333
5.4 cf, undercounter type	"	1.333
Refrigerator/freezer, 4.4 cf, undercounter type	"	1.333
Sealer, impulse, free standing, 20x12x4"	"	0.267
Timer, electric, 1-60 minutes, bench or wall mounted	"	0.444
Glassware washer - dryer, undercounter	"	10.000
Balance, torsion suspension, tabletop, 4.5 lb capacity	"	0.444
Binocular microscope, with in base illuminator	"	0.308
Centrifuge, table model, 19x16x13"	"	0.320
Clinical model, with four place head	"	0.178
11700.10 MEDICAL EQUIPMENT		
Hospital equipment, lights		
Examination, portable	EA.	0.667
Meters		
Air flow meter	EA.	0.444
Oxygen flow meters	"	0.333
Physical therapy		
Chair, hydrotherapy	EA.	0.133
Diathermy, shortwave, portable, on casters	"	0.320
Exercise bicycle, floor standing, 35" x 15"	"	0.267
Hydrocollator, 4 pack, portable, 129 x 90 x 160"	"	0.114
Lamp, infrared, mobile with variable heat control	"	0.615
Ultra violet, base mounted	"	0.615
Stimulator, galvanic-faradic, hand held	"	0.053

ARCHITECTURAL EQUIPMENT	UNIT	MAN/ HOURS
11700.10 MEDICAL EQUIPMENT		
Ultrasound, muscle stimulator, portable, 13x13x8"	EA.	0.067
Whirlpool, 85 gallon	"	4.000
65 gallon capacity	"	4.000
Radiology		
Radiographic table, motor driven tilting table	EA.	80.000
Fluoroscope image/tv system	"	160
Processor for washing and drying radiographs		
Water filter unit, 30" x 48-1/2" x 37-1/2"	EA.	13.333
Steam sterilizers		
For heat and moisture stable materials	EA.	0.800
For fast drying after sterilization	"	1.000
Compact unit	"	1.000
Semi-automatic	"	4.000
Floor loading		
Single door	EA.	6.667
Double door	"	8.000
Utensil washer, sanitizer	"	6.154
Automatic washer/sterilizer	"	16.000
16 x 16 x 26", including generator & accessories	"	26.667
Steam generator, elec., 10 kw to 180 kw	"	16.000
Surgical scrub		
Minimum	EA.	2.667
Maximum	"	2.667
Gas sterilizers		
Automatic, free standing, 21x19x29"	EA.	8.000
Surgical lights, ceiling mounted		
Minimum	EA.	13.333
Maximum	"	16.000
Water stills		
4 liters/hr	EA.	2.667
8 liters/hr	"	2.667
19 liters/hr	"	6.667
X-ray equipment		
Mobile unit		
Minimum	EA.	4.000
Maximum	"	8.000
Minimum	"	8.000
Maximum	"	8.000
Incubators		
15 cf	EA.	4.000
29 cf	"	6.667
Infant transport, portable	"	4.211
Headwall		
Aluminum, with back frame and console	EA.	4.000
Hospital ground detection system		
Power ground module	EA.	2.286
Ground slave module	"	1.739
Master ground module	"	1.509
Remote indicator	"	1.600

11 EQUIPMENT

ARCHITECTURAL EQUIPMENT	UNIT	MAN/ HOURS
11700.10 MEDICAL EQUIPMENT		
X-ray indicator	EA.	1.739
Micro ammeter	"	2.000
Supervisory module	"	1.739
Ground cords	"	0.296
Hospital isolation monitors, 5 ma		
120v	EA.	3.478
208v	"	3.478
240v	"	3.478
Digital clock-timers separate display	"	1.600
One display	"	1.600
Remote control	"	1.250
Battery pack	"	1.250
Surgical chronometer clock and 3 timers	"	2.500
Auxilary control	"	1.159
11700.20 DENTAL EQUIPMENT		
Dental care equipment		
Drill console with accessories	EA.	13.333
Amalgamator	"	0.400
Lathe	"	0.267
Finish polisher	"	0.533
Model trimer	"	0.364
Motor, wall mounted	"	0.364
Cleaner, ultrasonic	"	0.800
Curing unit, bench mounted	"	1.333
Oral evacuation system, dual pump	"	1.000
Sterilizer, table top, self contained	"	0.444
Dental lights		
Light, floor or ceiling mounted	EA.	4.000
X-ray unit		
Portable	EA.	2.000
Wall mounted with remote control	"	6.667
Illuminator, single panel	"	11.429
X-ray film processor	"	6.667
Shield, portable x-ray, lead lined	"	0.533

INTERIOR	UNIT	MAN/ HOURS
12302.10 CASEWORK		
Kitchen base cabinet, prefinished, 24" deep, 35" high		
12"wide	EA.	0.800
18" wide	"	0.800
24" wide	"	0.889
27" wide	"	0.889
36" wide	"	1.000
48" wide	"	1.000
Corner cabinet, 36" wide	"	1.000
Wall cabinet, 12" deep, 12" high		
30" wide	EA.	0.800
36" wide	"	0.800
15" high		
30" wide	EA.	0.889
36" wide	"	0.889
24" high		
30" wide	EA.	0.889
36" wide	"	0.889
30" high		
12" wide	EA.	1.000
18" wide	"	1.000
24" wide	"	1.000
27" wide	"	1.000
30" wide	"	1.143
36" wide	"	1.143
Corner cabinet, 30" high		
24" wide	EA.	1.333
30" wide	"	1.333
36" wide	"	1.333
Wardrobe	"	2.000
Vanity with top, laminated plastic		
24" wide	EA.	2.000
30" wide	"	2.000
36" wide	"	2.667
48" wide	"	3.200
12390.10 COUNTER TOPS		
Stainless steel, counter top, with backsplash	S.F.	0.200
Acid-proof, kemrock surface	"	0.133

13 SPECIAL

CONSTRUCTION	UNIT	MAN/ HOURS
13152.10 SWIMMING POOL EQUIPMENT		
Diving boards		
14' long		
Aluminum	EA.	4.444
Fiberglass	"	4.444
Ladders, heavy duty		
2 steps		
Minimum	EA.	1.600
Maximum	"	1.600
4 steps		
Minimum	EA.	2.000
Maximum	"	2.000
Lifeguard chair		
Minimum	EA.	8.000
Maximum	"	8.000
Lights, underwater		
12 volt, with transformer	EA.	2.000
110 volt		
Minimum	EA.	2.000
Maximum	"	2.000
Ground fault interrupter for 110 volt, each light	"	0.667
Pool covers		
Reinforced polyethylene	S.F.	0.062
Vinyl water tube		
Minimum	S.F.	0.062
Maximum	"	0.062
Slides with water tube		
Minimum	EA.	6.667
Maximum	"	6.667

14 CONVEYING

ELEVATORS

14210.10 ELEVATORS	UNIT	MAN/ HOURS
Passenger elevators, electric, geared		
Based on a shaft of 6 stops and 6 openings		
50 fpm, 2000 lb	EA.	24.000
100 fpm, 2000 lb	"	26.667
150 fpm		
2000 lb	EA.	30.000
3000 lb	"	34.286
4000 lb	"	40.000
Based on a shaft of 8 stops and 8 openings		
300 fpm		
3000 lb	EA.	48.000
3500 lb	"	48.000
4000 lb	"	53.333
5000 lb	"	57.143
4000 lb	"	53.333
Freight elevators, electric		
Based on a shaft of 6 stops and 6 openings		
50 fpm		
3500 lb	EA.	26.667
4000 lb	"	26.667
5000 lb	"	30.000
100 fpm		
3500 lb	EA.	30.000
4000 lb	"	30.000
5000 lb	"	34.286

14300.10 ESCALATORS	UNIT	MAN/ HOURS
Escalators		
32" wide, floor to floor		
12' high	EA.	40.000
15' high	"	48.000
18' high	"	60.000
22' high	"	80.000
25' high	"	96.000

LIFTS

14300.10 ESCALATORS

MATERIAL HANDLING

14580.10 PNEUMATIC SYSTEMS	UNIT	MAN/ HOURS
Pneumatic message tube system		
Average, 20 station job		
3" round system	E.A.	7.273
4" round system	"	8.000
6" round system	"	8.889
4" x 7" oval system	"	16.000
Trash and linen tube system		
10 stations	EA.	120
15 stations	"	160
20 stations	"	185
30 stations	"	218

HOISTS AND CRANES

14600.10 INDUSTRIAL HOISTS	UNIT	MAN/ HOURS
Industrial hoists, electric, light to medium duty		
500 lb	EA.	4.000
1000 lb	"	4.211
2000 lb	"	4.444
3000 lb	"	4.706
4000 lb	"	5.000
5000 lb	"	5.333
6000 lb	"	5.517
7500 lb	"	5.714
10,000 lb	"	5.926
15,000 lb	"	6.154
20,000 lb	"	6.667
25,000 lb	"	7.273
30,000 lb	"	8.000

HOISTS AND CRANES	UNIT	MAN/ HOURS
14600.10 INDUSTRIAL HOISTS		
Heavy duty		
500 lb	EA.	4.000
1000 lb	"	4.211
2000 lb	"	4.444
3000 lb	"	4.706
4000 lb	"	5.000
5000 lb	"	5.333
6000 lb	"	5.517
7500 lb	"	5.714
10,000 lb	"	5.926
15,000 lb	"	6.154
20,000 lb	"	6.667
25,000 lb	"	7.273
30,000 lb	"	8.000

15 MECHANICAL

BASIC MATERIALS	UNIT	MAN/ HOURS
15100.10 SPECIALTIES		
Wall penetration		
Concrete wall, 6" thick		
2" dia.	EA.	0.267
4" dia.	"	0.400
8" dia.	"	0.571
12" thick		
2" dia.	EA.	0.364
4" dia.	"	0.571
8" dia.	"	0.889
Non-destructive testing, piping systems		
X-ray of welds		
3" dia. pipe	EA.	0.800
4" dia. pipe	"	0.800
6" dia. pipe	"	0.800
8" dia. pipe	"	1.000
10" dia. pipe	"	1.000
Liquid penetration of welds		
2" dia. pipe	EA.	0.500
3" dia. pipe	"	0.500
4" dia. pipe	"	0.500
6" dia. pipe	"	0.500
8" dia. pipe	"	0.500
10" dia. pipe	"	0.500
15120.10 BACKFLOW PREVENTERS		
Backflow preventer, flanged, cast iron, with valves		
3" pipe	EA.	4.000
4" pipe	"	4.444
6" pipe	"	6.667
8" pipe	"	8.000
Threaded		
3/4" pipe	EA.	0.500
2" pipe	"	0.800
Reduced pressure assembly, bronze, threaded		
3/4"	EA.	0.500
1"	"	0.571
1-1/4"	"	0.667
1-1/2"	"	0.800
15140.10 PIPE HANGERS, HEAVY		
Hangers		
1/2" pipe, clevis pipe hanger		
Black steel	EA.	0.267
Galvanized	"	0.267
U bolt	"	0.080
3/4" pipe, clevis pipe hanger		
Black steel	EA.	0.267

BASIC MATERIALS	UNIT	MAN/ HOURS
15140.10 PIPE HANGERS, HEAVY		
Galvanized	EA.	0.267
U bolt	"	0.080
1" pipe, clevis pipe hanger		
Black steel	EA.	0.267
Galvanized	"	0.267
U bolt	"	0.080
1-1/4" pipe, clevis pipe hanger		
Black steel	EA.	0.267
Galvanized	"	0.267
U bolt	"	0.080
1-1/2" pipe, clevis pipe hanger		
Black steel	EA.	0.267
Galvanized	"	0.267
U bolt	"	0.080
2" pipe, clevis pipe hanger		
Black steel	EA.	0.267
Galvanized	"	0.267
Adjustable pipe roll stand	"	1.000
U bolt	"	0.080
Adjustable roller hanger	"	0.267
2-1/2" pipe, clevis pipe hanger		
Black steel	EA.	0.267
Galvanized	"	0.267
Adjustable pipe roll stand	"	0.267
Adjustable roller hanger	"	0.267
3" pipe, clevis pipe hanger		
Black steel	EA.	0.267
Galvanized	"	0.267
Adjustable pipe roll stand	"	0.267
U bolt	"	0.080
Adjustable roller hanger	"	0.267
3-1/2" pipe, clevis pipe hanger		
Black steel	EA.	0.267
Galvanized	"	0.267
Adjustable pipe roll stand	"	0.267
U bolt	"	0.080
Adjustable roller hanger	"	0.267
4" pipe, clevis pipe hanger		
Black steel	EA.	0.267
Galvanized	"	0.267
Adjustable pipe roll stand	"	0.267
U bolt	"	0.080
Adjustable roller hanger	"	0.267
5" pipe, clevis pipe hanger		
Black steel	EA.	0.320
Galvanized	"	0.320
Adjustable pipe roll stand	"	0.320
U bolt	"	0.080
Adjustable roller hanger	"	0.320

15 MECHANICAL

BASIC MATERIALS	UNIT	MAN/ HOURS
15140.10 PIPE HANGERS, HEAVY		
6" pipe, clevis pipe hanger		
Black steel	EA.	0.320
Galvanized	"	0.320
Adjustable pipe roll stand	"	0.320
U bolt	"	0.100
Adjustable roller hanger	"	0.320
8" pipe, clevis pipe hanger		
Black steel	EA.	0.320
Galvanized	"	0.320
Adjustable pipe roll stand	"	0.320
U bolt	"	0.100
Adjustable roller hanger	"	0.320
10" clevis pipe hanger		
Black steel	EA.	0.320
Galvanized	"	0.320
Adjustable pipe roll stand	"	0.320
Adjustable roller hanger	"	0.320
12" pipe, clevis pipe hanger		
Black steel	EA.	0.320
Galvanized	"	0.320
Adjustable pipe roll stand	"	0.320
Adjustable roller hanger	"	0.320
14" pipe, clevis pipe hanger		
Black steel	EA.	0.364
Galvanized	"	0.364
Threaded rod, galvanized		
C-clamp, steel, with lock nut		
3/8"	EA.	0.100
1/2"	"	0.100
5/8"	"	0.100
3/4"	"	0.100
7/8"	"	0.100
Angle support, medium, welded steel		
12"x18"	EA.	0.800
18"x24"	"	0.800
24"x30"	"	0.800
Heavy, welded steel		
12"x18"	EA.	0.800
18"x24"	"	0.800
24"x30"	"	0.800
15140.11 PIPE HANGERS, LIGHT		
A band, black iron		
1/2"	EA.	0.057
1"	"	0.059
1-1/4"	"	0.062
1-1/2"	"	0.067
2"	"	0.073
2-1/2"	"	0.080

BASIC MATERIALS	UNIT	MAN/ HOURS
15140.11 PIPE HANGERS, LIGHT		
3"	EA.	0.089
4"	"	0.100
5"	"	0.107
6"	"	0.114
8"	"	0.133
Copper		
1/2"	EA.	0.057
3/4"	"	0.059
1"	"	0.059
1-1/4"	"	0.062
1-1/2"	"	0.067
2"	"	0.073
2-1/2"	"	0.080
3"	"	0.089
4"	"	0.100
Black riser friction hangers		
3/4"	EA.	0.067
1"	"	0.070
1-1/4"	"	0.073
1-1/2"	"	0.076
2"	"	0.080
2-1/2"	"	0.089
3"	"	0.100
4"	"	0.114
5"	"	0.123
6"	"	0.133
8"	"	0.145
10"	"	0.160
Short pattern black riser clamps		
1-1/2"	EA.	0.073
2"	"	0.076
3"	"	0.080
4"	"	0.089
Copper riser friction hanger		
1/2"	EA.	0.062
3/4"	"	0.064
1"	"	0.067
1-1/4"	"	0.070
1-1/2"	"	0.073
2"	"	0.076
2-1/2"	"	0.080
3"	"	0.080
4"	"	0.089
Auto grip hangers, galvanized		
1/2"	EA.	0.057
3/4"	"	0.062
1"	"	0.064
1-1/4"	"	0.067
1-1/2"	"	0.070

BASIC MATERIALS	UNIT	MAN/ HOURS
15140.11 PIPE HANGERS, LIGHT		
2"	EA.	0.073
2-1/2"	"	0.076
3"	"	0.080
4"	"	0.089
Copper		
1/2"	EA.	0.057
3/4"	"	0.062
1"	"	0.064
1-1/4"	"	0.067
1-1/2"	"	0.070
2"	"	0.073
2-1/2"	"	0.076
3"	"	0.080
4"	"	0.089
Split rings (F&M), galvanized		
3/8"	EA.	0.057
1/2"	"	0.062
3/4"	"	0.064
1"	"	0.067
1-1/4"	"	0.070
1-1/2"	"	0.073
2"	"	0.076
2-1/2"	"	0.080
3"	"	0.084
4"	"	0.089
Copper		
1/4"	EA.	0.057
3/8"	"	0.057
1/2"	"	0.062
1"	"	0.067
1-1/4"	"	0.070
1-1/2"	"	0.073
2"	"	0.076
2-1/2"	"	0.080
3"	"	0.084
4"	"	0.089
F&M plates, galvanized		
3/8"	EA.	0.057
1/2"	"	0.059
Copper		
3/8"	EA.	0.057
1/2"	"	0.059
2 hole clips, galvanized		
3/4"	EA.	0.053
1"	"	0.055
1-1/4"	"	0.057
1-1/2"	"	0.059
2"	"	0.062
2-1/2"	"	0.064

BASIC MATERIALS	UNIT	MAN/ HOURS
15140.11 PIPE HANGERS, LIGHT		
3"	EA.	0.067
4"	"	0.073
Perforated strap		
3/4"		
Galvanized, 20 ga.	L.F.	0.040
Copper, 22 ga.	"	0.040
Threaded rod-couplings		
1/4"	EA.	0.050
3/4"	"	0.053
1/2"	"	0.057
5/8"	"	0.062
Reducing rod coupling, 1/2" x 3/8"	"	0.057
C-clamps		
3/4"	EA.	0.080
Top beam clamp		
3/8"	EA.	0.067
1/2"	"	0.073
Side beam connector		
3/8"	EA.	0.067
1/2"	"	0.073
J-Hooks		
1/2"	EA.	0.036
3/4"	"	0.036
1"	"	0.038
1-1/4"	"	0.039
1-1/2"	"	0.040
2"	"	0.040
3"	"	0.042
4"	"	0.042
PVC coated hangers, galvanized, 28 ga.		
1-1/2" x 12"	EA.	0.053
2" x 12"	"	0.057
3" x 12"	"	0.062
4" x 12"	"	0.067
Copper, 30 ga.		
1-1/2" x 12"	EA.	0.053
2" x 12"	"	0.057
3" x 12"	"	0.062
4" x 12"	"	0.067
2" x 24"	"	0.062
3" x 24"	"	0.067
4" x 24"	"	0.073
Milford hangers		
1/2" x 6"	EA.	0.057
1/2" x 12"	"	0.062
3/4" x 6"	"	0.059
3/4" x 12"	"	0.064
1" x 6"	"	0.062
1" x 12"	"	0.064

BASIC MATERIALS	UNIT	MAN/ HOURS
15140.11 PIPE HANGERS, LIGHT		
1-1/4" x 6"	EA.	0.064
1-1/4" x 12"	"	0.067
1-1/2" x 6"	"	0.067
1-1/2" x 12"	"	0.070
2" x 6"	"	0.070
2" x 12"	"	0.073
Wire hook hangers		
Black wire, 1/2" x		
4"	EA.	0.040
6"	"	0.042
8"	"	0.044
10"	"	0.044
12"	"	0.047
3/4" x		
4"	EA.	0.042
6"	"	0.044
8"	"	0.047
10"	"	0.050
12"	"	0.053
1" x		
4"	EA.	0.044
6"	"	0.047
8"	"	0.050
10"	"	0.053
12"	"	0.057
1-1/4" x		
4"	EA.	0.047
6"	"	0.050
8"	"	0.053
10"	"	0.057
12"	"	0.062
1-1/2" x		
6"	EA.	0.053
8"	"	0.057
10"	"	0.062
12"	"	0.067
2" x		
6"	EA.	0.057
8"	"	0.062
10"	"	0.067
12"	"	0.073
Copper wire hooks		
1/2" x		
4"	EA.	0.040
6"	"	0.042
8"	"	0.044
10"	"	0.047
12"	"	0.050
3/4" x		

BASIC MATERIALS	UNIT	MAN/ HOURS
15140.11 PIPE HANGERS, LIGHT		
4"	EA.	0.042
6"	"	0.044
8"	"	0.047
10"	"	0.050
12"	"	0.053
1" x		
4"	EA.	0.044
6"	"	0.047
8"	"	0.050
10"	"	0.053
12"	"	0.057
1-1/4" x		
6"	EA.	0.047
8"	"	0.050
10"	"	0.053
12"	"	0.057
1-1/2" x		
6"	EA.	0.053
8"	"	0.057
10"	"	0.062
12"	"	0.067
2" x		
6"	EA.	0.057
8"	"	0.062
10"	"	0.067
12"	"	0.073
15175.60 EXPANSION TANKS		
Expansion tank, 125 psi, galvanized steel		
20 gallon	EA.	1.000
30 gallon	"	1.333
65 gallon	"	2.000
80 gallon	"	2.286
15240.10 VIBRATION CONTROL		
Vibration isolator, in-line, stainless connector, screwed		
1/2"	EA.	0.444
3/4"	"	0.471
1"	"	0.500
1-1/4"	"	0.533
1-1/2"	"	0.571
2"	"	0.615
2-1/2"	"	0.667
3"	"	0.727
4"	"	0.800
6"	"	0.889
Flanged		
8"	EA.	1.000

15 MECHANICAL

BASIC MATERIALS	UNIT	MAN/ HOURS
15240.10 VIBRATION CONTROL		
10"	EA.	1.143
12"	"	1.333

INSULATION	UNIT	MAN/ HOURS
15260.10 FIBERGLASS PIPE INSULATION		
Fiberglass insulation on 1/2" pipe		
1" thick	L.F.	0.027
1-1/2" thick	"	0.033
3/4" pipe		
1" thick	L.F.	0.027
1-1/2" thick	"	0.033
1" pipe		
1" thick	L.F.	0.027
1-1/2" thick	"	0.033
2" thick	"	0.040
1-1/4" dia. pipe		
1" thick	L.F.	0.033
1-1/2" thick	"	0.036
2" thick	"	0.040
1-1/2" pipe		
1" thick	L.F.	0.033
1-1/2" thick	"	0.036
2" thick	"	0.038
2-1/2" thick	"	0.040
3-1/2" thick	"	0.042
2" pipe		
1" thick	L.F.	0.033
1-1/2" thick	"	0.036
2" thick	"	0.040
2-1/2" thick	"	0.044
3-1/2" thick	"	0.050
2-1/2" pipe		
1" thick	L.F.	0.033
1-1/2" thick	"	0.036
2" thick	"	0.040
2-1/2" thick	"	0.044
3" thick	"	0.050
3-1/2" thick	"	0.057
3" pipe		
1" thick	L.F.	0.038

INSULATION	UNIT	MAN/ HOURS
15260.10 FIBERGLASS PIPE INSULATION		
1-1/2" thick	L.F.	0.040
2" thick	"	0.044
2-1/2" thick	"	0.050
3" thick	"	0.057
3-1/2" thick	"	0.067
4" pipe		
1" thick	L.F.	0.038
1-1/2" thick	"	0.040
2" thick	"	0.044
2-1/2" thick	"	0.050
3" thick	"	0.057
3-1/2" thick	"	0.067
5" pipe		
1" thick	L.F.	0.038
2" thick	"	0.040
3" thick	"	0.047
4" thick	"	0.062
6" pipe		
1" thick	L.F.	0.042
2" thick	"	0.044
4" thick	"	0.053
6" thick	"	0.073
8" pipe		
2" thick	L.F.	0.042
3" thick	"	0.044
4" thick	"	0.053
6" thick	"	0.073
10" pipe		
2" thick	L.F.	0.042
3" thick	"	0.044
4" thick	"	0.053
6" thick	"	0.073
12" pipe		
2" thick	L.F.	0.042
3" thick	"	0.044
4" thick	"	0.053
6" thick	"	0.073
15260.20 CALCIUM SILICATE		
Calcium silicate insulation, 6" pipe		
2" thick	L.F.	0.057
2-1/2" thick	"	0.062
3" thick	"	0.067
4" thick	"	0.073
6" thick	"	0.080
8" pipe		
2" thick	L.F.	0.062
2-1/2" thick	"	0.067

INSULATION	UNIT	MAN/ HOURS
15260.20 CALCIUM SILICATE		
3" thick	L.F.	0.073
4" thick	"	0.080
6" thick	"	0.089
10" pipe		
2" thick	L.F.	0.062
2-1/2" thick	"	0.067
3" thick	"	0.073
4" thick	"	0.080
6" thick	"	0.089
12" pipe		
2" thick	L.F.	0.062
2-1/2" thick	"	0.067
3" thick	"	0.073
4" thick	"	0.080
6" thick	"	0.089
15260.60 EXTERIOR PIPE INSULATION		
Fiberglass insulation, aluminum jacket		
1/2" pipe		
1" thick	L.F.	0.062
1-1/2" thick	"	0.067
3/4" pipe		
1" thick	L.F.	0.062
1-1/2" thick	"	0.067
1" pipe		
1" thick	L.F.	0.062
1-1/2" thick	"	0.067
3-1/2" thick	"	0.089
1-1/4" pipe		
1" thick	L.F.	0.073
1-1/2" thick	"	0.076
3-1/2" thick	"	0.089
1-1/2" pipe		
1" thick	L.F.	0.073
1-1/2" thick	"	0.076
2" thick	"	0.080
2-1/2" thick	"	0.084
3-1/2" thick	"	0.089
2" pipe		
1" thick	L.F.	0.073
1-1/2" thick	"	0.076
2" thick	"	0.080
2-1/2" thick	"	0.084
3-1/2" thick	"	0.089
2-1/2" pipe		
1" thick	L.F.	0.073
1-1/2" thick	"	0.076
2" thick	"	0.080

INSULATION	UNIT	MAN/ HOURS
15260.60 EXTERIOR PIPE INSULATION		
2-1/2" thick	L.F.	0.084
3" thick	"	0.089
3-1/2" thick	"	0.094
3" pipe		
1" thick	L.F.	0.080
1-1/2" thick	"	0.084
2" thick	"	0.089
2-1/2" thick	"	0.094
3" thick	"	0.100
3-1/2" thick	"	0.107
4" pipe		
1" thick	L.F.	0.080
1-1/2" thick	"	0.084
2" thick	"	0.089
2-1/2" thick	"	0.094
3" thick	"	0.100
3-1/2" thick	"	0.107
5" pipe		
1" thick	L.F.	0.080
2" thick	"	0.084
2-1/2" thick	"	0.089
3" thick	"	0.094
3-1/2" thick	"	0.100
6" pipe		
1" thick	L.F.	0.089
2" thick	"	0.094
2-1/2" thick	"	0.100
3" thick	"	0.107
3-1/2" thick	"	0.114
4" thick	"	0.123
6" thick	"	0.133
8" pipe		
2" thick	L.F.	0.089
3" thick	"	0.094
3-1/2" thick	"	0.100
4" thick	"	0.107
6" thick	"	0.114
10" pipe		
2" thick	L.F.	0.089
3" thick	"	0.094
3-1/2" thick	"	0.100
4" thick	"	0.107
6" thick	"	0.114
12" pipe		
2" thick	L.F.	0.089
3" thick	"	0.094
3-1/2" thick	"	0.100
4" thick	"	0.107
6" thick	"	0.114

15 MECHANICAL

INSULATION	UNIT	MAN/ HOURS
15260.60 EXTERIOR PIPE INSULATION		
Calcium silicate with aluminum jacket, 6" pipe		
2" thick	L.F.	0.100
2-1/2" thick	"	0.107
3" thick	"	0.114
4" thick	"	0.133
8" pipe		
2" thick	L.F.	0.100
2-1/2" thick	"	0.107
3" thick	"	0.114
4" thick	"	0.123
6" thick	"	0.133
10" pipe		
2" thick	L.F.	0.114
2-1/2" thick	"	0.123
3" thick	"	0.133
4" thick	"	0.145
6" thick	"	0.160
12" pipe		
2" thick	L.F.	0.114
2-1/2" thick	"	0.123
3" thick	"	0.133
4" thick	"	0.145
6" thick	"	0.160
15260.90 PIPE INSULATION FITTINGS		
Insulation protection saddle		
1" thick covering		
1/2" pipe	EA.	0.320
3/4" pipe	"	0.320
1" pipe	"	0.320
1-1/4" pipe	"	0.320
1-1/2" pipe	"	0.320
2" pipe	"	0.320
2-1/2" pipe	"	0.320
3" pipe	"	0.364
4" pipe	"	0.400
6" pipe	"	0.500
1-1/2" thick covering		
3/4" pipe	EA.	0.320
1" pipe	"	0.320
1-1/4" pipe	"	0.320
1-1/2" pipe	"	0.320
2" pipe	"	0.320
2-1/2" pipe	"	0.320
3" pipe	"	0.320
4" pipe	"	0.400
6" pipe	"	0.500
8" pipe	"	0.667

INSULATION	UNIT	MAN/ HOURS
15260.90 PIPE INSULATION FITTINGS		
10" pipe	EA.	0.667
12" pipe	"	0.667
2" thick covering		
3/4" pipe	EA.	0.320
1" pipe	"	0.320
1-1/4" pipe	"	0.320
1-1/2" pipe	"	0.320
2" pipe	"	0.320
2-1/2" pipe	"	0.320
3" pipe	"	0.320
4" pipe	"	0.400
6" pipe	"	0.500
8" pipe	"	0.667
10" pipe	"	0.667
12" pipe	"	0.667
3" thick covering		
2" pipe	EA.	0.320
2-1/2" pipe	"	0.320
3" pipe	"	0.320
4" pipe	"	0.400
6" pipe	"	0.500
8" pipe	"	0.667
10" pipe	"	0.667
12" pipe	"	0.667
15280.10 EQUIPMENT INSULATION		
Equipment insulation, 2" thick, cellular glass	S.F.	0.050
Urethane, rigid, field applied jacket, plastered finish	"	0.100
Fiberglass, rigid, with vapor barrier	"	0.044
15290.10 DUCTWORK INSULATION		
Fiberglass duct insulation, plain blanket		
1-1/2" thick	S.F.	0.010
2" thick	"	0.013
With vapor barrier		
1-1/2" thick	S.F.	0.010
2" thick	"	0.013
Rigid with vapor barrier		
2" thick	S.F.	0.027
3" thick	"	0.032
4" thick	"	0.040
6" thick	"	0.053
Weatherproof, polystyrene, 3" thick, w/vapor barrier	"	0.080
Urethane board with vapor barrier	"	0.100

15 MECHANICAL

FIRE PROTECTION

15330.10 WET SPRINKLER SYSTEM

Item	UNIT	MAN/HOURS
Sprinkler head, 212 deg, brass, exposed piping	EA.	0.320
Chrome, concealed piping	"	0.444
Water motor alarm	"	1.333
Fire department inlet connection	"	1.600
Wall plate for fire dept connection	"	0.667
Swing check valve flanged iron body, 4"	"	2.667
Check valve, 6"	"	4.000
Wet pipe valve, flange to groove, 4"	"	0.889
Flange to flange		
6"	EA.	1.333
8"	"	2.667
Alarm valve, flange to flange, (wet valve)		
4"	EA.	0.889
8"	"	6.667
Inspector's test connection	"	0.667
Wall hydrant, polished brass, 2-1/2" x 2-1/2", single	"	0.571
2-way	"	0.571
3-way	"	0.571
Wet valve trim, includes retard chamber & guages, 4"-6"	"	0.667
Retard pressure switch for wet systems	"	1.600
Air maintenance device	"	0.667
Wall hydrant non-freeze, 8" thick wall, vacuum breaker	"	0.400
12" thick wall	"	0.400

15330.50 DRY SPRINKLER SYSTEM

Item	UNIT	MAN/HOURS
Dry pipe valve, flange to flange		
4"	EA.	1.600
6"	"	2.000
Trim, 4" and 6", includes gauges	"	0.667
Field testing and flushing	"	6.667
Disinfection	"	6.667
Pressure switch double circuit, open/close contacts	"	2.000
Low air		
Supervisory unit	EA.	1.333
Pressure switch	"	0.667

15330.70 CO2 SYSTEM

Item	UNIT	MAN/HOURS
CO2 system, high pressure, 75# cylinder with		
Valve assemblies	EA.	1.600
Storage rack	"	1.143
Manifold	"	5.714
Flexible loops	"	0.100
Beam scale for cylinders	"	1.333
Mechanically control head	"	0.533
Electrically control head	"	0.533
Stop valves	"	0.800
Check valves	"	1.000
Activation station	"	0.800

FIRE PROTECTION

15330.70 CO2 SYSTEM

Item	UNIT	MAN/HOURS
Nozzles	EA.	0.667
Hose reel with 75' of 3/4" hose	"	4.000
Main/reserve transfer switch	"	1.333
Pressure switch	"	0.800
Heat responsive device	"	1.333
Battery and charger	"	4.000
Low pressure		
Battery and charger	EA.	4.000
Pressure switch	"	0.889
Nozzles	"	0.727
Master selector valve	"	1.333
Selector valve	"	1.333
Low pressure hose reel with 75' of 3/4" hose	"	4.000
Tank fill lines	"	1.000
Activation stations	"	0.667
Electro manual pilot panels	"	1.000
Halon fire protection system, per computer room		

PLUMBING

15410.05 C.I. PIPE, ABOVE GROUND

Item	UNIT	MAN/HOURS
No hub pipe		
1-1/2" pipe	L.F.	0.057
2" pipe	"	0.067
3" pipe	"	0.080
4" pipe	"	0.133
6" pipe	"	0.160
8" pipe	"	0.267
10" pipe	"	0.320
No hub fittings, 1-1/2" pipe		
1/4 bend	EA.	0.267
1/8 bend	"	0.267
Sanitary tee	"	0.400
Sanitary cross	"	0.400
Wye	"	0.400
Tapped tee	"	0.267
P-trap	"	0.267
Tapped cross	"	0.267
2" pipe		
1/4 bend	EA.	0.320

15 MECHANICAL

PLUMBING

15410.05 C.I. PIPE, ABOVE GROUND	UNIT	MAN/HOURS
1/8 bend	EA.	0.320
Sanitary tee	"	0.533
Sanitary cross	"	0.533
Wye	"	0.667
Double wye	"	0.667
2x1-1/2" wye & 1/8 bend	"	0.500
Double wye & 1/8 bend	"	0.667
Test tee less 2" plug	"	0.320
Tapped tee		
2"x2"	EA.	0.320
2"x1-1/2"	"	0.320
P-trap		0.320
2"x2"	EA.	0.320
Tapped cross		
3" pipe		
1/4 bend	EA.	0.400
1/8 bend	"	0.400
Sanitary tee	"	0.500
3"x2" sanitary tee	"	0.500
3"x1-1/2" sanitary tee	"	0.500
Sanitary cross	"	0.667
3x2" sanitary cross	"	0.667
Wye	"	0.667
3x2" wye	"	0.667
Double wye	"	0.667
3x2" double wye	"	0.667
3x2" wye & 1/8 bend	"	0.571
3x1-1/2" wye & 1/8 bend	"	0.571
Double wye & 1/8 bend	"	0.667
3x2" double wye & 1/8 bend	"	0.667
3x2" reducer	"	0.364
Test tee, less 3" plug	"	0.400
3x3" tapped tee	"	0.400
3x2" tapped tee	"	0.400
3x1-1/2" tapped tee	"	0.400
P-trap	"	0.400
3x2" tapped cross	"	0.400
3x1-1/2" tapped cross	"	0.400
Closet flange, 3-1/2" deep	"	0.200
4" pipe		
1/4 bend	EA.	0.400
1/8 bend	"	0.400
Sanitary tee	"	0.667
4x3" sanitary tee	"	0.667
4x2" sanitary tee	"	0.667
Sanitary cross	"	0.800
4x3" sanitary cross	"	0.800
4x2" sanitary cross	"	0.800
Wye	"	0.667
4x3" wye	EA.	0.667
4x2" wye	"	0.667
Double wye	"	0.800
4x3" double wye	"	0.800
4x2" double wye	"	0.800
Wye & 1/8 bend	"	0.667
4x3" wye & 1/8 bend	"	0.667
4x2" wye & 1/8 bend	"	0.667
Double wye & 1/8 bend	"	0.800
4x3" double wye & 1/8 bend	"	0.800
4x2" double wye & 1/8 bend	"	0.800
4x3" reducer	"	0.400
4x2" reducer	"	0.400
Test tee, less 4" plug	"	0.400
4x2" tapped tee	"	0.400
4x1-1/2" tapped tee	"	0.400
P-trap	"	0.400
4x2" tapped cross	"	0.400
4x1-1/2" tapped cross	"	0.400
Closet flange		
3" deep	EA.	0.400
8" deep	"	0.400
6" pipe		
1/4 bend	EA.	0.667
1/8 bend	"	0.667
Sanitary tee	"	0.800
6x4" sanitary tee	"	0.800
Wye	"	0.800
6x4" wye	"	0.800
6x3" wye	"	0.800
6x2" wye	"	0.800
Double wye	"	1.000
6x4" double wye	"	1.000
Wye & 1/8 bend	"	0.800
6x4" wye & 1/8 bend	"	0.800
6x3" wye & 1/8 bend	"	0.800
6x2" wye & 1/8 bend	"	0.800
6x4" reducer	"	0.444
6x4" reducer	"	0.400
6x3" reducer	"	0.444
Test tee		
Less 6" plug	EA.	0.500
Plug	"	0.267
P-trap	"	0.500
8" pipe		
1/4 bend	EA.	0.667
1/8 bend	"	0.667
Sanitary tee	"	1.000
8x6" sanitary tee	"	1.000

PLUMBING	UNIT	MAN/ HOURS
15410.05 C.I. PIPE, ABOVE GROUND		
Wye	EA.	0.800
8x6" wye	"	0.667
8x4" wye	"	0.667
Double wye	"	0.667
Wye & 1/8 bend	"	0.667
8x6" wye & 1/8 bend	"	0.667
8x4" wye & 1/8 bend	"	0.667
8x6" reducer	"	0.667
8x4" reducer	"	0.667
8x3" reducer	"	0.667
8x2" reducer	"	0.444
Test tee		
Less 8" plug	EA.	0.667
Plug	"	
10" pipe		
1/4 bend	EA.	0.667
1/8 bend	"	0.667
Wye	"	1.333
10x8" wye	"	1.333
10x6" wye	"	1.333
10x4" wye	"	1.333
10x8" reducer	"	0.667
10x6" reducer	"	0.667
10x4" reducer	"	0.667
15410.06 C.I. PIPE, BELOW GROUND		
No hub pipe		
1-1/2" pipe	L.F.	0.040
2" pipe	"	0.044
3" pipe	"	0.050
4" pipe	"	0.067
6" pipe	"	0.073
8" pipe	"	0.089
10" pipe	"	0.100
Fittings, 1-1/2"		
1/4 bend	EA.	0.229
1/8 bend	"	0.229
Wye	"	0.320
Wye & 1/8 bend	"	0.229
P-trap	"	0.229
2"		
1/4 bend	EA.	0.267
1/8 bend	"	0.267
Double wye	"	0.500
Wye & 1/8 bend	"	0.400
Double wye & 1/8 bend	"	0.500
P-trap	"	0.267
3"		
1/4 bend	EA.	0.320

PLUMBING	UNIT	MAN/ HOURS
15410.06 C.I. PIPE, BELOW GROUND		
1/8 bend	EA.	0.320
Wye	"	0.500
3x2" wye	"	0.500
Wye & 1/8 bend	"	0.500
Double wye & 1/8 bend	"	0.500
3x2" double wye & 1/8 bend	"	0.500
3x2" reducer	"	0.320
P-trap	"	0.320
4"		
1/4 bend	EA.	0.320
1/8 bend	"	0.320
Wye	"	0.500
4x3" wye	"	0.500
4x2" wye	"	0.500
Double wye	"	0.667
4x3" double wye	"	0.667
4x2" double wye	"	0.667
Wye & 1/8 bend	"	0.500
4x3" wye & 1/8 bend	"	0.500
4x2" wye & 1/8 bend	"	0.500
Double wye & 1/8 bend	"	0.667
4x3" double wye & 1/8 bend	"	0.667
4x2" double wye & 1/8 bend	"	0.667
4x3" reducer	"	0.320
4x2" reducer	"	0.320
6"		
1/4 bend	EA.	0.500
1/8 bend	"	0.500
Wye & 1/8 bend	"	0.667
6x4" wye & 1/8 bend	"	0.667
6x3" wye & 1/8 bend	"	0.667
6x2" wye & 1/8 bend	"	0.667
6x3" reducer	"	0.364
P-trap	"	0.400
8"		
1/4 bend	EA.	0.500
1/8 bend	"	0.500
Wye	"	0.667
8x6" wye	"	0.500
8x4" wye	"	0.500
8x6" wye & 1/8 bend	"	0.500
8x4" reducer	"	0.400
8x3" reducer	"	0.400
8x2" reducer	"	0.400
10"		
1/4 bend	EA.	0.500
1/8 bend	"	0.500
Wye	"	1.000
10x8" wye	"	1.000

15 MECHANICAL

PLUMBING	UNIT	MAN/HOURS
15410.06 C.I. PIPE, BELOW GROUND		
10x6" wye	EA.	1.000
10x4" wye	"	1.000
10x8" reducer	"	0.500
10x6" reducer	"	0.500
15410.08 EXTRA HEAVY SOIL PIPE		
Extra heavy soil pipe, single hub		
2" x 5'	EA.	0.160
3" x 5'	"	0.170
4" x 5'	"	0.186
6" x 5'	"	0.200
Double hub		
2" x 5'	EA.	0.200
4" x 5'	"	0.211
5" x 5'	"	0.222
6" x 5'	"	0.229
Single hub		
3" x 10'	EA.	0.133
4" x 10'	"	0.138
6" x 10'	"	0.145
8" x 10'	"	0.154
"Mini", single hub		
2" x 42"	EA.	0.160
3" x 42"	"	0.170
4" x 42"	"	0.186
6" x 42"	"	0.200
8" x 42"	"	0.211
Fittings, 1/4" bend		
2"	EA.	0.267
3"	"	0.320
4"	"	0.320
1/8 bend		
2"	EA.	0.267
3"	"	0.320
4"	"	0.320
5"	"	0.400
6"	"	0.500
8"	"	0.500
Long sweep		
2"	EA.	0.267
3"	"	0.320
4"	"	0.320
Straight T		
4" x 2"	EA.	0.500
4" x 3"	"	0.500
4"	"	0.571
Sanitary T		
3" x 2"	EA.	0.500
3"	"	0.500

PLUMBING	UNIT	MAN/HOURS
15410.08 EXTRA HEAVY SOIL PIPE		
4" x 2"	EA.	0.571
4" x 3"	"	0.571
4"	"	0.615
4" x 2"	"	0.571
4" x 3"	"	0.571
4"	"	0.615
Combination Y and 1/8 bend, 4"	"	0.615
Double wye, 4"	"	0.615
Tapped sanitary T, 4" x 2"	"	0.667
P trap		
2"	EA.	0.320
4"	"	0.533
Dandy		
4", with 3" brass plug	EA.	0.444
6", 4" brass plug	"	0.500
15410.09 SERVICE WEIGHT PIPE		
Service weight pipe, single hub		
2" x 5'	EA.	0.160
3" x 5'	"	0.170
4" x 5'	"	0.178
5" x 5'	"	0.190
6" x 5'	"	0.200
8" x 5'	"	0.211
10" x 5'	"	0.222
12" x 5'	"	0.250
Double hub		
2" x 5'	EA.	0.200
3" x 5'	"	0.216
4" x 5'	"	0.229
5" x 5'	"	0.250
6" x 5'	"	0.267
10" x 5'	"	0.286
12' x 5'	"	0.308
Single hub		
2" x 10'	EA.	0.200
3" x 10'	"	0.216
4" x 10'	"	0.229
5" x 10'	"	0.250
6" x 10'	"	0.267
8" x 10'	"	0.286
10" x 10'	"	0.308
12" x 10'	"	0.333
Shorty		
2" x 42"	EA.	0.160
3" x 42"	"	0.170
4" x 42"	"	0.178
5" x 42"	"	0.190

PLUMBING	UNIT	MAN/ HOURS
15410.09 SERVICE WEIGHT PIPE		
6" x 42"	EA.	0.200
8" x 42"	"	0.211
10" x 42"	"	0.222
Soil plug		
2"	EA.	0.267
3"	"	0.286
4"	"	0.308
5"	"	0.320
6"	"	0.333
1/5 bend		
3"	EA.	0.320
4"	"	0.364
1/6 bend		
2"	EA.	0.267
3"	"	0.320
4"	"	0.364
1/8 bend		
2"	EA.	0.267
3"	"	0.320
4"	"	0.364
5"	"	0.381
6"	"	0.400
8"	"	0.500
10"	"	0.571
12"	"	0.667
1/16 bend		
2"	EA.	0.267
3"	"	0.320
4"	"	0.364
5"	"	0.381
6"	"	0.400
1/4 bend		
2"	EA.	0.267
3"	"	0.320
4"	"	0.364
5"	"	0.381
6"	"	0.400
8"	"	0.500
10"	"	0.571
12"	"	0.667
1/4 bend, long		
2" x 12"	EA.	0.286
4" x 18"	"	0.400
4" x 12"	"	0.400
Sweep		
2"	EA.	0.267
3"	"	0.320
4"	"	0.364
5"	"	0.381

PLUMBING	UNIT	MAN/ HOURS
15410.09 SERVICE WEIGHT PIPE		
6"	EA.	0.400
8"	"	0.500
Reducing long sweep		
3" x 2"	EA.	0.320
4" x 3"	"	0.400
Straight T		
2"	EA.	0.500
3"	"	0.533
4" x 2"	"	0.571
4" x 3"	"	0.615
4"	"	0.667
Sanitary T		
2"	EA.	0.500
3" x 2"	"	0.533
3"	"	0.571
4" x 2"	"	0.615
4" x 3"	"	0.667
4"	"	0.667
5"	"	0.727
6"	"	0.727
Wye		
2"	EA.	0.400
3" x 2"	"	0.444
3"	"	0.444
4" x 2"	"	0.471
4" x 3"	"	0.471
4"	"	0.471
5" x 2"	"	0.500
5" x 3"	"	0.500
5" x 4"	"	0.500
5"	"	0.500
6" x 2"	"	0.500
6" x 3"	"	0.500
6" x 4"	"	0.533
6" x 5"	"	0.533
6"	"	0.571
8" x 4"	"	0.571
8" x 6"	"	0.615
8"	"	0.667
10" x 4"	"	0.667
10" x 6"	"	0.727
10" x 8"	"	0.800
Service wye		
10"	EA.	1.000
12" x 4"	"	1.000
12" x 6"	"	1.000
12" x 8"	"	1.143
12"	"	1.333
Combination wye and 1/8 bend		

PLUMBING	UNIT	MAN/ HOURS
15410.09 SERVICE WEIGHT PIPE		
2"	EA.	0.533
3" x 2"	"	0.571
3"	"	0.571
4" x 2"	"	0.615
4" x 3"	"	0.615
4"	"	0.667
5" x 4"	"	0.667
6" x 4"	"	0.800
8"	"	1.000
Straight cross, 4"	"	0.667
Sanitary cross		
2"	EA.	0.571
3"	"	0.615
3" x 2"	"	0.615
4"	"	0.667
4" x 3"	"	0.667
4" x 2"	"	0.727
Double wye		
2"	EA.	0.571
3" x 2"	"	0.615
3"	"	0.615
4" x 2"	"	0.667
4" x 3"	"	0.667
4"	"	0.667
5"	"	0.800
6" x 4"	"	1.000
Combinatioin double wye and 1/8 bend		
2"	EA.	0.571
3" x 2"	"	0.615
3"	"	0.615
4" x 3"	"	0.667
4"	"	0.667
Tapped sanitary T		
2" x 1-1/2"	EA.	0.571
2" x 2"	"	0.571
3" x 1-1/2"	"	0.615
3" x 2"	"	0.615
4" x 1-1/2"	"	0.667
4" x 2"	"	0.667
Tapped straight T		
3" x 1-1/2"	EA.	0.615
3" x 2"	"	0.615
4" x 2"	"	0.667
4" x 1-1/2"	"	0.667
Tapped Y		
4" x 2"	EA.	0.667
Tapped sanitary cross		
2" x 1-1/2"	EA.	0.571
2" x 2"	"	0.615

PLUMBING	UNIT	MAN/ HOURS
15410.09 SERVICE WEIGHT PIPE		
3" x 1-1/2"	EA.	0.615
3" x 2"	"	0.615
3" x 3"	"	0.615
4" x 1-1/2"	"	0.667
4" x 2"	"	0.667
Cleanout, dandy, with brass plug		
2", 1-1/2" plug	EA.	0.571
3", 2" plug	"	0.615
4", 3" plug	"	0.667
5", 4" plug	"	0.727
6", 4" plug	"	0.800
8", 6" plug	"	1.000
Reducer, 3" x 2"	"	0.571
4" x		
2"	EA.	0.615
3"	"	0.615
5" x		
2"	EA.	0.667
3"	"	0.667
4"	"	0.667
6" x		
2"	EA.	0.800
3"	"	0.800
4"	"	0.800
5"	"	0.800
8" x		
4"	EA.	1.000
6"	"	1.000
10" x		
4"	EA.	1.143
6"	"	1.143
8"	"	1.143
12" x		
4"	EA.	1.333
6"	"	1.333
8"	"	1.333
10"	"	1.333
P trap		
2"	EA.	0.571
3"	"	0.615
4"	"	0.667
5"	"	0.800
6"	"	1.000
15410.10 COPPER PIPE		
Type "K" copper		
1/2"	L.F.	0.025
3/4"	"	0.027

PLUMBING	UNIT	MAN/ HOURS
15410.10 COPPER PIPE		
1"	L.F.	0.029
1-1/4"	"	0.031
1-1/2"	"	0.033
2"	"	0.036
2-1/2"	"	0.040
3"	"	0.042
4"	"	0.044
DWV, copper		
1-1/4"	L.F.	0.033
1-1/2"	"	0.036
2"	"	0.040
3"	"	0.044
4"	"	0.050
6"	"	0.057
Refrigeration tubing, copper, sealed		
1/8"	L.F.	0.032
3/16"	"	0.033
1/4"	"	0.035
5/16"	"	0.036
3/8"	"	0.038
1/2"	"	0.040
7/8"	"	0.046
1-1/8"	"	0.053
1-3/8"	"	0.062
Type "L" copper		
1/4"	L.F.	0.024
3/8"	"	0.024
1/2"	"	0.025
3/4"	"	0.027
1"	"	0.029
1-1/4"	"	0.031
1-1/2"	"	0.033
2"	"	0.036
2-1/2"	"	0.040
3"	"	0.042
3-1/2"	"	0.043
4"	"	0.044
Type "M" copper		
1/2"	L.F.	0.025
3/4"	"	0.027
1"	"	0.029
1-1/4"	"	0.031
2"	"	0.036
2-1/2"	"	0.040
3"	"	0.042
4"	"	0.044

PLUMBING	UNIT	MAN/ HOURS
15410.11 COPPER FITTINGS		
Coupling, with stop		
1/4"	EA.	0.267
3/8"	"	0.320
1/2"	"	0.348
5/8"	"	0.400
3/4"	"	0.444
1"	"	0.471
3"	"	0.800
4"	"	1.000
Reducing coupling		
1/4" x 1/8"	EA.	0.320
3/8" x 1/4"	"	0.348
1/2" x		
3/8"	EA.	0.400
1/4"	"	0.400
1/8"	"	0.400
3/4" x		
3/8"	EA.	0.444
1/2"	"	0.444
1" x		
3/8"	EA.	0.500
1" x 1/2"	"	0.500
1" x 3/4"	"	0.500
1-1/4" x		
1/2"	EA.	0.533
3/4"	"	0.533
1"	"	0.533
1-1/2" x		
1/2"	EA.	0.571
3/4"	"	0.571
1"	"	0.571
1-1/4"	"	0.571
2" x		
1/2"	EA.	0.667
3/4"	"	0.667
1"	"	0.667
1-1/4"	"	0.667
1-1/2"	"	0.667
2-1/2" x		
1"	EA.	0.800
1-1/4"	"	0.800
1-1/2"	"	0.800
2"	"	0.800
3" x		
1-1/2"	EA.	1.000
2"	"	1.000
2-1/2"	"	1.000
4" x		
2"	EA.	1.143

PLUMBING	UNIT	MAN/ HOURS
15410.11 COPPER FITTINGS		
2-1/2"	EA.	1.143
3"	"	1.143
Slip coupling		
1/8"	EA.	0.267
1/2"	"	0.320
3/4"	"	0.400
1"	"	0.444
1-1/4"	"	0.500
1-1/2"	"	0.533
2"	"	0.667
2-1/2"	"	0.667
3"	"	0.800
4"	"	1.000
Coupling with drain		
1/2"	EA.	0.400
3/4"	"	0.444
1"	"	0.500
Reducer		
3/8" x 1/4"	EA.	0.320
1/2" x 3/8"	"	0.320
3/4" x		
1/4"	EA.	0.364
3/8"	"	0.364
1/2"	"	0.364
1" x		
1/2"	EA.	0.400
3/4"	"	0.400
1-1/4" x		
1/2"	EA.	0.444
3/4"	"	0.444
1"	"	0.444
1-1/2" x		
1/2"	EA.	0.500
3/4"	"	0.500
1"	"	0.500
1-1/4"	"	0.500
2" x		
1/2"	EA.	0.571
3/4"	"	0.571
1"	"	0.571
1-1/4"	"	0.571
1-1/2"	"	0.571
2-1/2" x		
1"	EA.	0.667
1-1/4"	"	0.667
1-1/2"	"	0.667
2"	"	0.667
3" x		
1-1/4"	EA.	0.800

PLUMBING	UNIT	MAN/ HOURS
15410.11 COPPER FITTINGS		
1-1/2"	EA.	0.800
2"	"	0.800
2-1/2"	"	0.800
4" x		
2"	EA.	1.000
3"	"	1.000
Female adapters		
1/4"	EA.	0.320
3/8"	"	0.364
1/2"	"	0.400
3/4"	"	0.444
1"	"	0.444
1-1/4"	"	0.500
1-1/2"	"	0.500
2"	"	0.533
2-1/2"	"	0.571
3"	"	0.667
4"	"	0.800
Increasing female adapters		
1/8" x		
3/8"	EA.	0.320
1/2"	"	0.320
1/4" x 1/2"	"	0.348
3/8" x 1/2"	"	0.364
1/2" X		
3/4"	EA.	0.400
1"	"	0.400
3/4" X		
1"	EA.	0.444
1-1/4"	"	0.444
1" x		
1-1/4"	EA.	0.444
1-1/2"	"	0.444
1-1/4" x		
1-1/2"	EA.	0.500
2"	"	0.500
1-1/2" x 2"	"	0.533
Reducing female adapters		
3/8" x 1/4"	EA.	0.364
1/2" x		
1/4"	EA.	0.400
3/8"	"	0.400
3/4" x 1/2"	"	0.444
1" x		
1/2"	EA.	0.444
3/4"	"	0.444
1-1/4" x		
1/2"	EA.	0.500
3/4"	"	0.500

PLUMBING	UNIT	MAN/ HOURS
15410.11 COPPER FITTINGS		
1"	EA.	0.500
1-1/2" x		
1"	EA.	0.533
1-1/4"	"	0.533
2" x		
1"	EA.	0.571
1-1/4"	"	0.571
1-1/2"	"	0.571
Female fitting adapters		
1/2"	EA.	0.400
3/4"	"	0.400
3/4" x 1/2"	"	0.421
1"	"	0.444
1-1/4"	"	0.471
1-1/2"	"	0.500
2"	"	0.533
Male adapters		
1/4"	EA.	0.364
3/8"	"	0.364
3"	"	0.667
4"	"	0.800
Increasing male adapters		
3/8" x 1/2"	EA.	0.364
1/2" x		
3/4"	EA.	0.400
1"	"	0.400
3/4" x		
1"	EA.	0.421
1-1/4"	"	0.421
1" x 1-1/4"	"	0.444
1-1/2" x		
3/4"	EA.	0.471
1"	"	0.471
1-1/4"	"	0.471
2" x		
1"	EA.	0.500
1-1/4"	"	0.500
1-1/2"	"	0.500
2" x 2-1/2"	"	0.533
Reducing male adapters		
1/2" x		
1/4"	EA.	0.400
3/8"	"	0.400
3/4" x 1/2"	"	0.421
1" x		
1/2"	EA.	0.444
3/4"	"	0.444
1-1/4" x		
3/4"	EA.	0.471

PLUMBING	UNIT	MAN/ HOURS
15410.11 COPPER FITTINGS		
1"	EA.	0.471
1-1/2" x		
3/4"	EA.	0.500
1"	"	0.500
1-1/4'	"	0.500
2" x		
3/4"	EA.	0.533
1"	"	0.533
1-1/4"	"	0.533
1-1/2"	"	0.533
2-1/2" x		
2"	EA.	0.615
Fitting x male adapters		
1/2"	EA.	0.400
3/4"	"	0.421
1"	"	0.444
1-1/4"	"	0.471
1-1/2"	"	0.500
2"	"	0.533
90 ells		
1/8"	EA.	0.320
1/4"	"	0.320
3/8"	"	0.364
1/2"	"	0.400
3/4"	"	0.421
1"	"	0.444
1-1/4"	"	0.471
1-1/2"	"	0.500
2"	"	0.533
2-1/2"	"	0.615
3"	"	0.667
4"	"	0.800
Reducing 90 ell		
3/8" x 1/4"	EA.	0.364
1/2" x		
1/4"	EA.	0.400
3/8"	"	0.400
3/4" x 1/2"	"	0.421
1" x		
1/2"	EA.	0.444
3/4"	"	0.444
1-1/4" x		
1/2"	EA.	0.471
3/4"	"	0.471
1"	"	0.471
1-1/2" x		
1/2"	EA.	0.500
3/4"	"	0.500
1"	"	0.500

PLUMBING	UNIT	MAN/ HOURS
15410.11 COPPER FITTINGS		
x 1-1/4"	EA.	0.500
2" x		
3/4"	EA.	0.533
x 1"	"	0.533
x 1-1/4"	"	0.533
x 1-1/2"	"	0.533
2-1/2" x		
1-1/2"	EA.	0.615
x 2"	"	0.615
3" x		
2"	EA.	0.667
x 2-1/2"	"	0.667
Street ells, copper		
1/4"	EA.	0.320
3/8"	"	0.364
1/2"	"	0.400
3/4"	"	0.421
1"	"	0.444
1-1/4"	"	0.471
1-1/2"	"	0.500
2"	"	0.533
2-1/2"	"	0.615
3"	"	0.667
4"	"	0.800
Female, 90 ell		
1/2"	EA.	0.400
3/4"	"	0.421
1"	"	0.444
1-1/4"	"	0.471
1-1/2"	"	0.500
2"	"	0.533
Female increasing, 90 ell		
3/8" x 1/2"	EA.	0.364
1/2" x		
3/4"	EA.	0.400
1"	"	0.400
3/4" x 1"	"	0.421
1" x 1-1/4"	"	0.444
1-1/4" x 1-1/2"	"	0.471
Female reducing, 90 ell		
1/2" x 3/8"	EA.	0.400
3/4" x 1/2"	"	0.421
1" x		
1/2"	EA.	0.444
x 3/4"	"	0.444
1-1/4" x		
3/4"	EA.	0.471
x 1"	"	0.471
x 1/2"	"	0.471
1-1/2" x		
1"	EA.	0.500
x 1-1/4"	"	0.500
2" x 1-1/2"	"	0.533
Male, 90 ell		
1/4"	EA.	0.320
3/8"	"	0.364
1/2"	"	0.400
3/4"	"	0.421
1"	"	0.444
1-1/4"	"	0.471
1-1/2"	"	0.500
2"	"	0.533
Male, increasing 90 ell		
1/2" x		
3/4"	EA.	0.400
1"	"	0.400
3/4" x 1"	"	0.421
1" x 1-1/4"	"	0.444
1-1/4" x 1-1/2"	"	0.471
Male, reducing 90 ell		
1/2" x 3/8"	EA.	0.400
3/4" x 1/2"	"	0.421
1" x		
1/2"	EA.	0.444
3/4"	"	0.444
1-1/4" x 1"	"	0.471
Drop ear ells		
1/2"	EA.	0.400
Female drop ear ells		
1/2"	EA.	0.400
1/2" x 3/8"	"	0.400
3/4"	"	0.421
Female flanged sink ell		
1/2"	EA.	0.400
45 ells		
1/4"	EA.	0.320
3/8"	"	0.364
3"	"	0.667
4"	"	0.800
45 street ell		
1/4"	EA.	0.320
3/8"	"	0.364
1/2"	"	0.400
3/4"	"	0.421
1"	"	0.444
1-1/2"	"	0.500
2"	"	0.533
2-1/2"	"	0.615

15 MECHANICAL

PLUMBING	UNIT	MAN/ HOURS
15410.11 COPPER FITTINGS		
3"	EA.	0.667
4"	"	0.800
Tee		
1/8"	EA.	0.320
1/4"	"	0.320
3/8"	"	0.364
3"	"	0.667
4"	"	0.800
Caps		
1/4"	EA.	0.320
3/8"	"	0.364
3"	"	0.667
4"	"	0.800
Test caps		
1/2"	EA.	0.400
3/4"	"	0.421
1"	"	0.444
1-1/4"	"	0.471
1-1/2"	"	0.500
2"	"	0.533
3"	"	0.667
Flush bushing		
1/4" x 1/8"	EA.	0.320
1/2" x		
1/4"	EA.	0.400
3/8"	"	0.400
3/4" x		
3/8"	EA.	0.421
1/2"	"	0.421
1" x		
1/2"	EA.	0.444
3/4"	"	0.444
1-1/4" x		
1/2"	EA.	0.471
3/4"	"	0.471
x 1"	"	0.471
1-1/2" x		
1/2"	EA.	0.500
1-1/4"	"	0.500
2" x		
1"	EA.	0.533
1-1/4"	"	0.533
1-1/2"	"	0.533
Female flush bushing		
1/2" x		
1/2" x 1/8"	EA.	0.400
1/4"	"	0.400
Union		
1/4"	EA.	0.320

PLUMBING	UNIT	MAN/ HOURS
15410.11 COPPER FITTINGS		
3/8"	EA.	0.364
3"	"	0.667
Female		
1/2"	EA.	0.400
3/4"	"	0.421
Male		
1/2"	EA.	0.400
3/4"	"	0.421
1"	"	0.444
45 degree wye		
1/2"	EA.	0.400
3/4"	"	0.421
1"	"	0.444
1" x 3/4" x 3/4"	"	0.444
1-1/4"	"	0.471
1-1/4" x 1" x 1"	"	0.471
Twin ells		
1" x 3/4" x 3/4"	EA.	0.444
1" x 1" x 1"	"	0.444
1-1/4" x 1" x 1"	"	0.471
Companion flanges, 125#		
2" x 6"	EA.	0.533
2-1/2"	"	0.615
3" x 7-1/2"	"	0.667
4" x 9"	"	0.800
90 union ells, male		
1/2"	EA.	0.400
3/4"	"	0.421
1"	"	0.444
DWV fittings, coupling with stop		
1-1/4"	EA.	0.471
1-1/2"	"	0.500
1-1/2" x 1-1/4"	"	0.500
2"	"	0.533
2" x 1-1/4"	"	0.533
2" x 1-1/2"	"	0.533
3"	"	0.667
3" x 1-1/2"	"	0.667
3" x 2"	"	0.667
4"	"	0.800
Slip coupling		
1-1/2"	EA.	0.500
2"	"	0.533
3"	"	0.667
90 ells		
1-1/2"	EA.	0.500
1-1/2" x 1-1/4"	"	0.500
2"	"	0.533
2" x 1-1/2"	"	0.533

15 MECHANICAL

PLUMBING	UNIT	MAN/HOURS
15410.11 COPPER FITTINGS		
3"	EA.	0.667
4"	"	0.800
Street, 90 elbows		
1-1/2"	EA.	0.500
2"	"	0.533
3"	"	0.667
4"	"	0.800
Female, 90 elbows		
1-1/2"	EA.	0.500
2"	"	0.533
Male, 90 elbows		
1-1/2"	EA.	0.500
2"	"	0.533
90 with side inlet		
3" x 3" x 1"	EA.	0.667
3" x 3" x 1-1/2"	"	0.667
3" x 3" x 2"	"	0.667
45 ells		
1-1/4"	EA.	0.471
1-1/2"	"	0.500
2"	"	0.533
3"	"	0.667
4"	"	0.800
Street, 45 ell		
1-1/2"	EA.	0.500
2"	"	0.533
3"	"	0.667
60 ell		
1-1/2"	EA.	0.500
2"	"	0.533
3"	"	0.667
22-1/2 ell		
1-1/2"	EA.	0.500
2"	"	0.533
3"	"	0.667
11-1/4 ell		
1-1/2"	EA.	0.500
2"	"	0.533
3"	"	0.667
Wye		
1-1/4"	EA.	0.471
1-1/2"	"	0.500
2"	"	0.533
2" x 1-1/2" x 1-1/2"	"	0.533
2" x 1-1/2" x 2"	"	0.533
2" x 1-1/2" x 2"	"	0.533
3"	"	0.667
3" x 3" x 1-1/2"	"	0.667
3" x 3" x 2"	"	0.667

PLUMBING	UNIT	MAN/HOURS
15410.11 COPPER FITTINGS		
4"	EA.	0.800
4" x 4" x 2"	"	0.800
4" x 4" x 3"	"	0.800
Sanitary tee		
1-1/4"	EA.	0.471
1-1/2"	"	0.500
2"	"	0.533
2" x 1-1/2" x 1-1/2"	"	0.533
2" x 1-1/2" x 2"	"	0.533
2" x 2" x 1-1/2"	"	0.533
3"	"	0.667
3" x 3" x 1-1/2"	"	0.667
3" x 3" x 2"	"	0.667
4"	"	0.800
4" x 4" x 3"	"	0.800
Female sanitary tee		
1-1/2"	EA.	0.500
Long turn tee		
1-1/2"	EA.	0.500
2"	"	0.533
3" x 1-1/2"	"	0.667
Double wye		
1-1/2"	EA.	0.500
2"	"	0.533
2" x 2" x 1-1/2" x 1-1/2"	"	0.533
3"	"	0.667
3" x 3" x 1-1/2" x 1-1/2"	"	0.667
3" x 3" x 2" x 2"	"	0.667
4" x 4" x 1-1/2" x 1-1/2"	"	0.800
Double sanitary tee		
1-1/2"	EA.	0.500
2"	"	0.533
2" x 2" x 1-1/2"	"	0.533
3"	"	0.667
3" x 3" x 1-1/2" x 1-1/2"	"	0.667
3" x 3" x 2" x 2"	"	0.667
4" x 4" x 1-1/2" x 1-1/2"	"	0.800
Long		
2" x 1-1/2"	EA.	0.533
Twin elbow		
1-1/2"	EA.	0.500
2"	"	0.533
2" x 1-1/2" x 1-1/2"	"	0.533
Spigot adapter, manoff		
1-1/2" x 2"	EA.	0.500
1-1/2" x 3"	"	0.500
2"	"	0.533
2" x 3"	"	0.533
2" x 4"	"	0.533

15 MECHANICAL

PLUMBING	UNIT	MAN/ HOURS
15410.11 COPPER FITTINGS		
3"	EA.	0.667
3" x 4"	"	0.667
4"	"	0.800
No-hub adapters		
1-1/2" x 2"	EA.	0.500
2"	"	0.533
2" x 3"	"	0.533
3"	"	0.667
3" x 4"	"	0.667
4"	"	0.800
Fitting reducers		
1-1/2" x 1-1/4"	EA.	0.500
2" x 1-1/2"	"	0.533
3" x 1-1/2"	"	0.667
3" x 2"	"	0.667
Slip joint (Desanco)		
1-1/4"	EA.	0.471
1-1/2"	"	0.500
1-1/2" x 1-1/4"	"	0.500
Street x slip joint (Desanco)		
1-1/2"	EA.	0.500
1-1/2" x 1-1/4"	"	0.500
Flush bushing		
1-1/2" x 1-1/4"	EA.	0.500
2" x 1-1/2"	"	0.533
3" x 1-1/2"	"	0.667
3" x 2"	"	0.667
Male hex trap bushing		
1-1/4" x 1-1/2"	EA.	0.471
1-1/2"	"	0.500
1-1/2" x 2"	"	0.500
2"	"	0.533
Round trap bushing		
1-1/2"	EA.	0.500
2"	"	0.533
Female adapter		
1-1/4"	EA.	0.471
1-1/2"	"	0.500
1-1/2" x 2"	"	0.500
2"	"	0.533
2" x 1-1/2"	"	0.533
3"	"	0.667
Fitting x female adapter		
1-1/2"	EA.	0.500
2"	"	0.533
Male adapters		
1-1/4"	EA.	0.471
1-1/4" x 1-1/2"	"	0.471
1-1/2"	"	0.500

PLUMBING	UNIT	MAN/ HOURS
15410.11 COPPER FITTINGS		
1-1/2" x 2"	EA.	0.500
2"	"	0.533
2" x 1-1/2"	"	0.533
3"	"	0.667
Male x slip joint adapters		
1-1/2" x 1-1/4"	EA.	0.500
Dandy cleanout		
1-1/2"	EA.	0.500
2"	"	0.533
3"	"	0.667
End cleanout, flush pattern		
1-1/2" x 1"	EA.	0.500
2" x 1-1/2"	"	0.533
3" x 2-1/2"	"	0.667
Copper caps		
1-1/2"	EA.	0.500
2"	"	0.533
Closet flanges		
3"	EA.	0.667
4"	"	0.800
Drum traps, with cleanout		
1-1/2" x 3" x 6"	EA.	0.500
P-trap, swivel, with cleanout		
1-1/2"	EA.	0.500
P-trap, solder union		
1-1/2"	EA.	0.500
2"	"	0.533
With cleanout		
1-1/2"	EA.	0.500
2"	"	0.533
2" x 1-1/2"	"	0.533
Swivel joint, with cleanout		
1-1/2" x 1-1/4"	EA.	0.500
1-1/2"	"	0.500
2" x 1-1/2"	"	0.533
Estabrook TY, with inlets		
3", with 1-1/2" inlet	EA.	0.667
Fine thread adapters		
1/2"	EA.	0.400
1/2" x 1/2" IPS	"	0.400
1/2" x 3/4" IPS	"	0.400
1/2" x male	"	0.400
1/2" x female	"	0.400
15410.14 BRASS I.P.S. FITTINGS		
Fittings, iron pipe size, 45 deg ell		
1/8"	EA.	0.320
1/4"	"	0.320

15 MECHANICAL

PLUMBING

15410.14 BRASS I.P.S. FITTINGS

	UNIT	MAN/ HOURS
3/8"	EA.	0.364
1/2"	"	0.400
3/4"	"	0.421
1"	"	0.444
1-1/4"	"	0.471
1-1/2"	"	0.500
2"	"	0.533
90 deg ell		
1/8"	EA.	0.320
1/4"	"	0.320
3/8"	"	0.364
1/2"	"	0.400
3/4"	"	0.421
1"	"	0.444
1-1/4"	"	0.471
1-1/2"	"	0.500
2"	"	0.533
2-1/2"	"	0.615
90 deg ell, reducing		
1/4" x 1/8"	EA.	0.320
3/8" x 1/8"	"	0.364
3/8" x 1/4"	"	0.364
1/2" x 1/4"	"	0.400
1/2" x 3/8"	"	0.400
3/4" x 1/2"	"	0.421
1" x 3/8"	"	0.444
1" x 1/2"	"	0.444
1" x 3/4"	"	0.444
1-1/4" x 3/4"	"	0.471
1-1/4" x 1"	"	0.471
1-1/2" x 1/2"	"	0.500
1-1/2" x 3/4"	"	0.500
1-1/2" x 1"	"	0.500
2" x 1"	"	0.533
2" x 1-1/4"	"	0.533
2" x 1-1/2"	"	0.533
Street ell, 45 deg		
1/2"	EA.	0.400
3/4"	"	0.421
2"	"	0.533
90 deg		
1/8"	EA.	0.320
1/4"	"	0.320
3/8"	"	0.364
1/2"	"	0.400
3/4"	"	0.421
1"	"	0.444
1-1/4"	"	0.471
1-1/2"	"	0.500

PLUMBING

15410.14 BRASS I.P.S. FITTINGS

	UNIT	MAN/ HOURS
2"	EA.	0.533
Tee, 1/8"	"	0.320
1/4"	"	0.320
3/8"	"	0.364
1/2"	"	0.400
3/4"	"	0.421
1"	"	0.444
1-1/4"	"	0.471
1-1/2"	"	0.500
2"	"	0.533
2-1/2"	"	0.615
Tee, reducing, 3/8" x		
1/4"	EA.	0.364
1/2"	"	0.364
1/2" x		
1/4"	EA.	0.400
3/8"	"	0.400
3/4"	"	0.400
3/4" x		
1/4"	EA.	0.421
1/2"	"	0.421
1"	"	0.421
1" x		
1/2"	EA.	0.444
3/4"	"	0.444
1-1/4" x		
1/2"	EA.	0.471
1"	"	0.471
1-1/2" x		
1/2"	EA.	0.500
3/4"	"	0.500
1"	"	0.500
1-1/2"	"	0.500
2" x		
1/2"	EA.	0.533
3/4"	"	0.533
1"	"	0.533
1-1/4"	"	0.533
1-1/2"	"	0.533
2-1/2" x 2"	"	0.615
Tee, reducing		
1/2" x 3/8" x 1/2"	EA.	0.400
3/4" x 1/2" x 1/2"	"	0.421
3/4" x 1/2" x 3/4"	"	0.421
1" x 1/2" x 1/2"	"	0.444
1" x 1/2" x 3/4"	"	0.444
1" x 3/4" x 1/2"	"	0.444
1" x 3/4" x 3/4"	"	0.444
1-1/4" x 1/2" x 1-1/4"	"	0.471

PLUMBING

15410.14 BRASS I.P.S. FITTINGS

PLUMBING	UNIT	MAN/ HOURS
1-1/4" x 1" x 1"	EA	0.471
1-1/4" x 1-1/4" x 3/4"	"	0.471
1-1/2" x 1-1/4" x 1-1/4"	"	0.500
Union		
1/8"	EA	0.320
1/4"	"	0.320
3/8"	"	0.364
1/2"	"	0.400
3/4"	"	0.421
1"	"	0.444
1-1/4"	"	0.471
1-1/2"	"	0.500
2"	"	0.533
Brass face bushing		
3/8" x 1/4"	EA	0.364
1/2" x 3/8"	"	0.400
3/4" x 1/2"	"	0.421
1" x 3/4"	"	0.444
1-1/4" x 1"	"	0.471
Hex bushing, 1/4" x 1/8"	"	0.320
1/2" x		
1/4"	EA	0.400
3/8"	"	0.400
5/8" x		
1/8"	EA	0.400
1/4"	"	0.400
3/4" x		
1/8"	EA	0.421
1/4"	"	0.421
3/8"	"	0.421
1/2"	"	0.421
1" x		
1/4"	EA	0.444
3/8"	"	0.444
1/2"	"	0.444
3/4"	"	0.444
1-1/4" x		
3/8"	EA	0.471
1/2"	"	0.471
3/4"	"	0.471
1"	"	0.471
1-1/2" x		
1/4"	EA	0.500
1/2"	"	0.500
3/4"	"	0.500
1"	"	0.500
1-1/4"	"	0.500
2" x		
1/2"	EA	0.533

PLUMBING

15410.14 BRASS I.P.S. FITTINGS

PLUMBING	UNIT	MAN/ HOURS
3/4"	EA	0.533
1"	"	0.533
1-1/4"	"	0.533
1-1/2"	"	0.533
2-1/2" x		
1"	EA	0.615
2"	"	0.615
3" x		
1-1/2"	EA	0.667
2"	"	0.667
2-1/2"	"	0.667
4" x		
2"	EA	0.800
3"	"	0.800
Caps		
1/8"	EA	0.320
1/4"	"	0.320
3/8"	"	0.364
1/2"	"	0.400
3/4"	"	0.421
1"	"	0.444
1-1/4"	"	0.471
1-1/2"	"	0.500
2"	"	0.533
Couplings		
1/8"	EA	0.320
1/4"	"	0.320
3/8"	"	0.364
1/2"	"	0.400
3/4"	"	0.421
1"	"	0.444
1-1/4"	"	0.471
1-1/2"	"	0.500
2"	"	0.533
2-1/2"	"	0.615
Couplings, reducing, 1/4" x 1/8"	"	0.320
3/8" x		
1/8"	EA	0.364
1/4"	"	0.364
1/2" x		
1/8"	EA	0.400
1/4"	"	0.400
3/8"	"	0.400
3/4" x		
1/4"	EA	0.421
3/8"	"	0.421
1/2"	"	0.421
1/2"	"	0.421
3/4"	"	0.421

15 MECHANICAL

PLUMBING	UNIT	MAN/ HOURS
15410.14 BRASS I.P.S. FITTINGS		
1-1/4" x		
1/2"	EA.	0.471
3/4"	"	0.471
1"	"	0.471
1-1/2" x		
3/4"	EA.	0.500
1"	"	0.500
1-1/4"	"	0.500
2" x		
3/4"	EA.	0.533
1"	"	0.533
1-1/4"	"	0.533
1-1/2"	"	0.533
2-1/2" x 1-1/2"	"	0.615
Square head plug, solid		
1/8"	EA.	0.320
1/4"	"	0.320
3/8"	"	0.364
1/2"	"	0.400
3/4"	"	0.421
Cored		
1/2"	EA.	0.400
3/4"	"	0.421
1"	"	0.444
1-1/4"	"	0.471
1-1/2"	"	0.500
2"	"	0.533
3"	"	0.667
4"	"	0.800
Countersunk		
1/2"	EA.	0.400
3/4"	"	0.421
1-1/2"	"	0.500
2"	"	0.533
Locknut		
3/4"	EA.	0.421
1"	"	0.444
1-1/4"	"	0.471
2"	"	0.533
Close standard red nipple, 1/8"	"	0.320
1/8" x		
1-1/2"	EA.	0.320
2"	"	0.320
2-1/2"	"	0.320
3"	"	0.320
3-1/2"	"	0.320
4"	"	0.320
4-1/2"	"	0.320
5"	"	0.320

PLUMBING	UNIT	MAN/ HOURS
15410.14 BRASS I.P.S. FITTINGS		
5-1/2"	EA.	0.320
6"	"	0.320
1/4" x close	"	0.320
1/4" x		
1-1/2"	EA.	0.320
2"	"	0.320
2-1/2"	"	0.320
3"	"	0.320
3-1/2"	"	0.320
4"	"	0.320
4-1/2"	"	0.320
5"	"	0.320
5-1/2"	"	0.320
6"	"	0.320
3/8" x close	"	0.364
3/8" x		
1-1/2"	EA.	0.364
2"	"	0.364
2-1/2"	"	0.364
3"	"	0.364
3-1/2"	"	0.364
4"	"	0.364
4-1/2"	"	0.364
5"	"	0.364
5-1/2"	"	0.364
6"	"	0.364
1/2" x close	"	0.400
1/2" x		
1-1/2"	EA.	0.400
2"	"	0.400
2-1/2"	"	0.400
3"	"	0.400
3-1/2"	"	0.400
4"	"	0.400
4-1/2"	"	0.400
5"	"	0.400
5-1/2"	"	0.400
6"	"	0.400
7-1/2"	"	0.400
8"	"	0.400
3/4" x close	"	0.421
3/4" x		
1-1/2"	EA.	0.421
2"	"	0.421
2-1/2"	"	0.421
3"	"	0.421
3-1/2"	"	0.421
4"	"	0.421
4-1/2"	"	0.421

PLUMBING

15410.14 BRASS I.P.S. FITTINGS

Description	UNIT	MAN/HOURS
5"	EA.	0.421
5-1/2"	"	0.421
6"	"	0.421
1" x close	"	0.444
1" x		
2"	EA.	0.444
2-1/2"	"	0.444
3"	"	0.444
3-1/2"	"	0.444
4"	"	0.444
4-1/2"	"	0.444
5"	"	0.444
5-1/2"	"	0.444
6"	"	0.444
1-1/4" x close	"	0.471
1-1/4" x		
2"	EA.	0.471
2-1/2"	"	0.471
3"	"	0.471
3-1/2"	"	0.471
4"	"	0.471
4-1/2"	"	0.471
5"	"	0.471
5-1/2"	"	0.471
6"	"	0.471
1-1/2" x close	"	0.500
1-1/2" x		
2"	EA.	0.500
2-1/2"	"	0.500
3"	"	0.500
3-1/2"	"	0.500
4-1/2"	"	0.500
5"	"	0.500
5-1/2"	"	0.500
6"	"	0.500
2" x close	"	0.533
2"		
2-1/2"	EA.	0.533
3"	"	0.533
3-1/2"	"	0.533
4"	"	0.533
4-1/2"	"	0.533
5"	"	0.533
5-1/2"	"	0.533
6"	"	0.533
2-1/2" x close	"	0.615
2-1/2" x		
3"	EA.	0.615
3-1/2"	"	0.615

PLUMBING

15410.14 BRASS I.P.S. FITTINGS

Description	UNIT	MAN/HOURS
4"	EA.	0.615
2-1/2" x 5"	"	0.615
3"		
Close	EA.	0.667

15410.15 BRASS FITTINGS

Description	UNIT	MAN/HOURS
Compression fittings, union		
3/8"	EA.	0.133
1/2"	"	0.133
5/8"	"	0.133
Union elbow		
3/8"	EA.	0.133
1/2"	"	0.133
5/8"	"	0.133
Union tee		
3/8"	EA.	0.133
1/2"	"	0.133
5/8"	"	0.133
Male connector		
3/8"	EA.	0.133
1/2"	"	0.133
5/8"	"	0.133
Female connector		
3/8"	EA.	0.133
1/2"	"	0.133
5/8"	"	0.133
Brass flare fittings, union		
3/8"	EA.	0.129
1/2"	"	0.129
5/8"	"	0.129
90 deg elbow union		
3/8"	EA.	0.129
1/2"	"	0.129
5/8"	"	0.129
Three way tee		
3/8"	EA.	0.216
1/2"	"	0.216
5/8"	"	0.216
Cross		
3/8"	EA.	0.286
1/2"	"	0.286
5/8"	"	0.286
Male connector, half union		
3/8"	EA.	0.129
1/2"	"	0.129
5/8"	"	0.129
Female connector, half union		
3/8"	EA.	0.129
1/2"	"	0.129

PLUMBING	UNIT	MAN/ HOURS
15410.15 BRASS FITTINGS		
5/8"	EA.	0.129
Long forged nut		
3/8"	EA.	0.129
1/2"	"	0.129
5/8"	"	0.129
Short forged nut		
3/8"	EA.	0.129
1/2"	"	0.129
5/8"	"	0.129
Compression elbow		
1/4"	EA.	0.129
5/16"	"	0.129
3/4"	"	0.129
Nut		
1/8"	EA.	
1/4"	"	
5/16"	"	
3/8"	"	
1/2"	"	
5/8"	"	
3/4"	"	
7/8"	"	
Sleeve		
1/8"	EA.	0.160
1/4"	"	0.160
5/16"	"	0.160
3/8"	"	0.160
1/2"	"	0.160
5/8"	"	0.160
3/4"	"	0.160
7/8"	"	0.160
Tee		
1/4"	EA.	0.229
5/16"	"	0.229
Male tee		
5/16" x 1/8"	EA.	0.229
Female union		
1/8" x 1/8"	EA.	0.200
1/4" x 3/8"	"	0.200
3/8" x 1/4"	"	0.200
3/8" x 1/2"	"	0.200
5/8" x 1/2"	"	0.229
Male union, 1/4"		
1/4" x 1/4"	EA.	0.200
3/8"	"	0.200
1/2"	"	0.200
5/16" x		
1/8"	EA.	0.200
1/4"	"	0.200

PLUMBING	UNIT	MAN/ HOURS
15410.15 BRASS FITTINGS		
3/8"	EA.	0.200
3/8" x		
1/8"	EA.	0.200
1/4"	"	0.200
1/2"	"	0.200
3/4"	"	0.200
1/2" x		
1/4"	EA.	0.229
3/8"	"	0.229
5/8" x		
3/8"	EA.	0.229
1/2"	"	0.229
3/4"	"	0.229
1/2"	"	0.229
7/8" x		
1/2"	EA.	0.229
3/4"	"	0.229
Female elbow, 1/4" x 1/4"	"	0.229
5/16" x		
1/8"	EA.	0.229
1/4"	"	0.229
3/8" x		
3/8"	EA.	0.229
1/2"	"	0.229
Male elbow, 1/8" x 1/8"	"	0.229
3/16" x 1/4"	"	0.229
1/4" x		
1/8"	EA.	0.229
1/4"	"	0.229
3/8"	"	0.229
5/16" x		
1/8"	EA.	0.229
1/4"	"	0.229
3/8"	"	0.229
3/8" x		
1/8"	EA.	0.229
1/4"	"	0.229
3/8"	"	0.229
1/2"	"	0.229
1/2" x		
1/4"	EA.	0.267
3/8"	"	0.267
1/2"	"	0.267
5/8" x		
3/8"	EA.	0.267
1/2"	"	0.267
3/4"	"	0.267
3/4" x		
1/2"	EA.	0.267

15 MECHANICAL

PLUMBING	UNIT	MAN/HOURS
15410.15 BRASS FITTINGS		
3/4"	EA.	0.267
7/8" x		
3/4"	EA.	0.267
Union		
1/8"	EA.	0.229
3/16"	"	0.229
1/4"	"	0.229
5/16"	"	0.229
3/8"	"	0.229
3/4"	"	0.229
7/8"	"	0.229
Reducing union		
3/8" x 1/4"	EA.	0.267
5/8" x		
3/8"	EA.	0.267
1/2"	"	0.267
15410.17 CHROME PLATED FITTINGS		
Fittings		
90 ell		
1/2"	EA.	0.200
3/4"	"	0.200
45 ell		
3/8"	EA.	0.200
1/2"	"	0.200
Tee		
3/8"	EA.	0.267
1/2"	"	0.267
Coupling		
3/8"	EA.	0.200
1/2"	"	0.200
Union		
3/8"	EA.	0.200
1/2"	"	0.200
Tee		
1/2" x 3/8" x 3/8"	EA.	0.267
1/2" x 3/8" x 1/2"	"	0.267
15410.18 GLASS PIPE		
Glass pipe		
1-1/2" dia.	L.F.	0.160
2" dia.	"	0.178
3" dia.	"	0.200
4" dia.	"	0.229
6" dia.	"	0.267

PLUMBING	UNIT	MAN/HOURS
15410.30 PVC/CPVC PIPE		
PVC schedule 40		
1/2" pipe	L.F.	0.033
3/4" pipe	"	0.036
1" pipe	"	0.040
1-1/4" pipe	"	0.044
1-1/2" pipe	"	0.050
2" pipe	"	0.057
2-1/2" pipe	"	0.067
3" pipe	"	0.080
4" pipe	"	0.100
6" pipe	"	0.200
8" pipe	"	0.267
Fittings, 1/2"		
90 deg ell	EA.	0.100
45 deg ell	"	0.100
Tee	"	0.114
Reducing insert	"	0.133
Threaded	"	0.100
Male adapter	"	0.133
Female adapter	"	0.100
Union	"	0.160
Cap	"	0.133
Flange	"	0.160
3/4"		
90 deg elbow	EA.	0.133
45 deg elbow	"	0.133
Tee	"	0.160
Reducing insert	"	0.114
Threaded	"	0.133
1"		
90 deg elbow	EA.	0.160
45 deg elbow	"	0.160
Tee	"	0.178
Reducing insert	"	0.160
Threaded	"	0.178
Male adapter	"	0.200
Female adapter	"	0.200
Union	"	0.267
Cap	"	0.160
Flange	"	0.267
1-1/4"		
90 deg elbow	EA.	0.229
45 deg elbow	"	0.229
Tee	"	0.267
Reducing insert	"	0.267
Threaded	"	0.267
Male adapter	"	0.267
Female adapter	"	0.267
Union	"	0.320

PLUMBING	UNIT	MAN/HOURS
15410.30 PVC/CPVC PIPE		
Cap	EA.	0.267
Flange	"	0.320
1-1/2"		
90 deg elbow	EA.	0.229
45 deg elbow	"	0.229
Tee	"	0.267
Reducing insert	"	0.267
Threaded	"	0.267
Male adapter	"	0.267
Female adapter	"	0.267
Union	"	0.400
Cap	"	0.267
Flange	"	0.400
2"		
90 deg elbow	EA.	0.267
45 deg elbow	"	0.267
Tee	"	0.320
Reducing insert	"	0.320
Threaded	"	0.320
Male adapter	"	0.320
Female adapter	"	0.320
Union	"	0.500
Cap	"	0.320
Flange	"	0.500
2-1/2"		
90 deg elbow	EA.	0.500
45 deg elbow	"	0.500
Tee	"	0.533
Reducing insert	"	0.533
Threaded	"	0.533
Male adapter	"	0.533
Female adapter	"	0.533
Union	"	0.667
Cap	"	0.500
Flange	"	0.667
3"		
90 deg elbow	EA.	0.667
45 deg elbow	"	0.667
Tee	"	0.727
Reducing insert	"	0.667
Threaded	"	0.667
Male adapter	"	0.667
Female adapter	"	0.667
Union	"	0.800
Cap	"	0.667
Flange	"	0.800
4"		
90 deg elbow	EA.	0.800
45 deg elbow	"	0.800

PLUMBING	UNIT	MAN/HOURS
15410.30 PVC/CPVC PIPE		
Tee	EA.	0.889
Reducing insert	"	0.800
Threaded	"	0.800
Male adapter	"	0.800
Female adapter	"	0.800
Union	"	1.000
Cap	"	0.800
Flange	"	1.000
PVC schedule 80 pipe		
1-1/2" pipe	L.F.	0.050
2" pipe	"	0.057
3" pipe	"	0.080
4" pipe	"	0.100
Fittings, 1-1/2"		
90 deg elbow	EA.	0.267
45 deg elbow	"	0.267
Tee	"	0.400
Reducing insert	"	0.267
Threaded	"	0.267
Male adapter	"	0.267
Female adapter	"	0.267
Union	"	0.400
Cap	"	0.267
Flange	"	0.400
2"		
90 deg elbow	EA.	0.320
45 deg elbow	"	0.320
Tee	"	0.500
Reducing insert	"	0.320
Threaded	"	0.320
Male adapter	"	0.320
Female adapter	"	0.320
2-1/2"		
90 deg elbow	EA.	0.500
45 deg elbow	"	0.500
Tee	"	0.667
Reducing insert	"	0.500
Threaded	"	0.500
Male adapter	"	0.500
Female adapter	"	0.500
Union	"	0.667
Cap	"	0.500
Flange	"	0.667
3"		
90 deg elbow	EA.	0.667
45 deg elbow	"	0.667
Tee	"	0.800
Reducing insert	"	0.667
Threaded	"	0.667

15 MECHANICAL

PLUMBING	UNIT	MAN/HOURS
15410.30 PVC/CPVC PIPE		
Male adapter	EA.	0.667
Female adapter	"	0.667
Union	"	0.800
Cap	"	0.667
Flange	"	0.800
4"		
90 deg elbow	EA.	0.800
45 deg elbow	"	0.800
Tee	"	1.000
Reducing insert	"	0.800
Threaded	"	0.800
Male adapter	"	0.800
Union	"	1.000
Cap	"	0.800
Flange	"	1.000
CPVC schedule 40		
1/2" pipe	L.F.	0.033
3/4" pipe	"	0.036
1" pipe	"	0.040
1-1/4" pipe	"	0.044
1-1/2" pipe	"	0.050
2" pipe	"	0.057
Fittings, CPVC, schedule 80		
1/2", 90 deg ell	EA.	0.080
Tee	"	0.133
3/4", 90 deg ell	"	0.080
Tee	"	0.133
1", 90 deg ell	"	0.089
Tee	"	0.145
1-1/4", 90 deg ell	"	0.089
Tee	"	0.145
1-1/2", 90 deg ell	"	0.160
Tee	"	0.200
2", 90 deg ell	"	0.160
Tee	"	0.200
Polypropylene, acid resistant, DWV pipe		
Schedule 40		
1-1/2" pipe	L.F.	0.057
2" pipe	"	0.067
3" pipe	"	0.080
4" pipe	"	0.100
6" pipe	"	0.200
Fittings		
1-1/2"		
1/4 bend	EA.	0.200
1/8 bend	"	0.200
Sanitary tee	"	0.400
Cleanout with plug	"	0.400
Wye	"	0.400

PLUMBING	UNIT	MAN/HOURS
15410.30 PVC/CPVC PIPE		
Combination wye & 1/8 bend	EA.	0.400
P-trap	"	0.500
Hub adapter	"	0.200
Coupling	"	0.200
2"		
1/4 bend	EA.	0.229
1/8 bend	"	0.229
Sanitary tee	"	0.471
Cleanout with plug	"	0.471
Wye	"	0.471
Combination wye & 1/8 bend	"	0.471
P-trap	"	0.667
Hub adapter	"	0.267
Mechanical joint adapter	"	0.267
Coupling	"	0.267
3"		
1/4 bend	EA.	0.267
1/8 bend	"	0.267
Sanitary tee	"	0.533
Cleanout with plug	"	0.533
Wye	"	0.533
Combination wye & 1/8 bend	"	0.533
P-trap	"	0.800
Hub adapter	"	0.267
Mechanical joint adapter	"	0.267
Coupling	"	0.267
4"		
1/4 bend	EA.	0.400
1/8 bend	"	0.400
Sanitary tee	"	0.800
Cleanout with plug	"	0.800
Wye	"	0.800
Combination wye & 1/8 bend	"	0.800
P-trap	"	1.333
Hub adapter	"	0.400
Mechanical joint adapter	"	0.400
Coupling	"	0.400
6"		
1/4 bend	EA.	0.667
1/8 bend	"	0.667
Sanitary tee	"	1.333
Cleanout with plug	"	1.333
Wye	"	1.333
Hub adapter	"	0.667
Coupling	"	0.667
Polyethylene pipe and fittings		
SDR-21		
3" pipe	L.F.	0.100
4" pipe	"	0.133

PLUMBING

15410.30 PVC/CPVC PIPE

PLUMBING	UNIT	MAN/HOURS
6" pipe	L.F.	0.200
8" pipe	"	0.229
10" pipe	"	0.267
12" pipe	"	0.320
14" pipe	"	0.400
16" pipe	"	0.500
18" pipe	"	0.615
20" pipe	"	0.800
22" pipe	"	0.889
24" pipe	"	1.000
Fittings, 3"		
90 deg elbow	EA.	0.400
45 deg elbow	"	0.400
Tee	"	0.667
45 deg wye	"	0.667
Reducer	"	0.500
Flange assembly	"	0.400
4"		
90 deg elbow	EA.	0.500
45 deg elbow	"	0.500
Tee	"	0.800
45 deg wye	"	0.800
Reducer	"	0.667
Flange assembly	"	0.500
8"		
90 deg elbow	EA.	1.000
45 deg elbow	"	1.000
Tee	"	1.600
45 deg wye	"	1.600
Reducer	"	1.333
Flange assembly	"	1.000
10"		
90 deg elbow	EA.	1.333
45 deg elbow	"	1.333
Tee	"	2.000
45 deg wye	"	2.000
Reducer	"	1.600
Flange assembly	"	1.333
12"		
90 deg elbow	EA.	1.600
45 deg elbow	"	1.600
Tee	"	2.667
45 deg wye	"	2.667
Reducer	"	2.000
Flange assembly	"	1.600
14"		
90 deg elbow	EA.	2.000
45 deg elbow	"	2.000
Tee	"	3.200

PLUMBING

15410.30 PVC/CPVC PIPE

PLUMBING	UNIT	MAN/HOURS
45 deg wye	EA.	3.200
Reducer	"	2.667
Flange assembly	"	2.000
16"		
90 deg elbow	EA.	2.000
45 deg elbow	"	2.000
Tee	"	3.200
45 deg wye	"	3.200
Reducer	"	2.667
Flange assembly	"	2.000
18"		
90 deg elbow	EA.	2.667
45 deg elbow	"	2.667
Tee	"	4.000
45 deg wye	"	4.000
Reducer	"	2.667
Flange assembly	"	2.667
20"		
90 deg elbow	EA.	2.667
45 deg elbow	"	2.667

15410.33 ABS DWV PIPE

PLUMBING	UNIT	MAN/HOURS
Schedule 40 ABS		
1-1/2" pipe	L.F.	0.040
2" pipe	"	0.044
3" pipe	"	0.057
4" pipe	"	0.080
6" pipe	"	0.100
Fittings		
1/8 bend		
1-1/2"	EA.	0.160
2"	"	0.200
3"	"	0.267
4"	"	0.320
6"	"	0.400
Tee, sanitary		
1-1/2"	EA.	0.267
2"	"	0.320
3"	"	0.400
4"	"	0.500
6"	"	0.667
Tee, sanitary reducing		
2 x 1-1/2 x 1-1/2	EA.	0.320
2 x 1-1/2 x 2	"	0.333
2 x 2 x 1-1/2	"	0.364
3 x 3 x 1-1/2	"	0.400
3 x 3 x 2	"	0.444
4 x 4 x 1-1/2	"	0.500
4 x 4 x 2	"	0.571

PLUMBING	UNIT	MAN/ HOURS
15410.33 ABS DWV PIPE		
4 x 4 x 3	EA.	0.615
6 x 6 x 4	"	0.667
Wye		
1-1/2"	EA.	0.229
2"	"	0.320
3"	"	0.400
4"	"	0.500
6"	"	0.667
Reducer		
2 x 1-1/2	EA.	0.200
3 x 1-1/2	"	0.267
3 x 2	"	0.267
4 x 2	"	0.320
4 x 3	"	0.320
6 x 4	"	0.400
P-trap		
1-1/2"	EA.	0.267
2"	"	0.296
3"	"	0.348
4"	"	0.400
6"	"	0.500
Double sanitary, tee		
1-1/2"	EA.	0.320
2"	"	0.400
3"	"	0.500
4"	"	0.667
Long sweep, 1/4 bend		
1-1/2"	EA.	0.160
2"	"	0.200
3"	"	0.267
4"	"	0.400
Wye, standard		
1-1/2"	EA.	0.267
2"	"	0.320
3"	"	0.400
4"	"	0.500
Wye, reducing		
2 x 1-1/2 x 1-1/2	EA.	0.267
2 x 2 x 1-1/2	"	0.320
4 x 4 x 2	"	0.500
4 x 4 x 3	"	0.533
Double wye		
1-1/2"	EA.	0.320
2"	"	0.400
3"	"	0.500
4"	"	0.667
2 x 2 x 1-1/2 x 1-1/2	"	0.400
3 x 3 x 2 x 2	"	0.500
4 x 4 x 3 x 3	"	0.667

PLUMBING	UNIT	MAN/ HOURS
15410.33 ABS DWV PIPE		
Combination wye and 1/8 bend		
1-1/2"	EA.	0.267
2"	"	0.320
3"	"	0.400
4"	"	0.500
2 x 2 x 1-1/2	"	0.320
3 x 3 x 1-1/2	"	0.400
3 x 3 x 2	"	0.400
4 x 4 x 2	"	0.500
4 x 4 x 3	"	0.500
15410.35 PLASTIC PIPE		
Fiberglass reinforced pipe		
2" pipe	L.F.	0.062
3" pipe	"	0.067
4" pipe	"	0.073
6" pipe	"	0.080
8" pipe	"	0.133
10" pipe	"	0.160
12" pipe	"	0.200
Fittings		
90 deg elbow, flanged		
2"	EA.	0.800
3"	"	0.889
4"	"	1.000
6"	"	1.333
8"	"	1.600
10"	"	2.000
12"	"	2.667
45 deg elbow, flanged		
2"	EA.	0.667
3"	"	0.800
4"	"	1.000
6"	"	1.333
8"	"	1.600
10"	"	2.000
12"	"	2.667
Tee, flanged		
2"	EA.	1.000
3"	"	1.143
4"	"	1.333
6"	"	1.600
8"	"	2.000
10"	"	2.667
12"	"	4.000
Wye, flanged		
2"	EA.	1.000
3"	"	1.143
4"	"	1.333

PLUMBING	UNIT	MAN/ HOURS
15410.35 PLASTIC PIPE		
6"	EA.	1.600
8"	"	2.000
10"	"	2.667
12"	"	4.000
Concentric reducer, flanged		
2"	EA.	0.667
4"	"	0.800
6"	"	1.143
8"	"	1.600
10"	"	2.000
12"	"	2.667
Adapter, bell x male or female		
2"	EA.	0.667
3"	"	0.727
4"	"	0.800
6"	"	1.143
8"	"	1.600
10"	"	2.000
12"	"	2.667
Nipples		
2" x 6"	EA.	0.080
2" x 12"	"	0.100
3" x 8"	"	0.123
3" x 12"	"	0.133
4" x 8"	"	0.133
4" x 12"	"	0.160
6" x 12"	"	0.200
8" x 18"	"	0.200
8" x 24"	"	0.229
10" x 18"	"	0.267
10" x 24"	"	0.320
12" x 18"	"	0.364
12" x 24"	"	0.400
Sleeve coupling		
2"	EA.	0.667
3"	"	0.800
4"	"	1.143
6"	"	1.600
8"	"	2.000
10"	"	2.667
Flanges		
2"	EA.	0.667
3"	"	0.800
4"	"	1.143
6"	"	1.600
8"	"	2.000
10"	"	2.667
12"	"	2.667

PLUMBING	UNIT	MAN/ HOURS
15410.70 STAINLESS STEEL PIPE		
Stainless steel, schedule 40, threaded		
1/2" pipe	L.F.	0.114
3/4" pipe	"	0.118
1" pipe	"	0.123
1-1/2" pipe	"	0.133
2" pipe	"	0.145
2-1/2" pipe	"	0.160
3" pipe	"	0.178
4" pipe	"	0.200
Fittings, 1/2"		
90 deg ell	EA.	1.000
45 deg ell	"	1.000
Tee	"	1.333
Cap	"	0.500
Reducer	"	0.667
Union	"	1.000
Flange	"	1.000
3/4"		
90 deg ell	EA.	1.000
45 deg ell	"	1.000
Tee	"	1.333
Cap	"	0.500
Reducer	"	0.667
Union	"	1.000
Flange	"	1.000
1"		
90 deg ell	EA.	1.000
45 deg ell	"	1.000
Tee	"	1.333
Cap	"	0.500
Reducer	"	1.000
Union	"	1.000
Flange	"	1.000
1-1/4"		
90 deg ell	EA.	1.000
45 deg ell	"	1.000
Tee	"	1.333
Cap	"	0.500
Union	"	1.000
Flange	"	1.000
1-1/2"		
90 deg ell	EA.	1.333
45 deg ell	"	1.333
Tee	"	1.600
Cap	"	0.667
Reducer	"	0.800
Union	"	1.143
Flange	"	1.143
2"		

PLUMBING	UNIT	MAN/ HOURS
15410.70 STAINLESS STEEL PIPE		
90 deg ell	EA.	1.600
45 deg ell	"	1.600
Tee	"	2.667
Cap	"	0.800
Reducer	"	1.000
Union	"	1.600
Flange	"	1.600
Type 304, sch 10 pipe		
1" pipe	L.F.	0.100
1-1/4" pipe	"	0.133
1-1/2" pipe	"	0.145
2" pipe	"	0.178
2-1/2" pipe	"	0.200
3" pipe	"	0.229
4" pipe	"	0.267
6" pipe	"	0.320
Fittings, 1"		
90 deg elbow	EA.	1.600
45 deg elbow	"	1.600
Tee	"	2.667
Reducer	"	1.600
Flange	"	1.000
Cap	"	1.000
1-1/4"		
90 deg elbow	EA.	1.600
45 deg elbow	"	1.600
Tee	"	2.667
Reducer	"	1.600
Flange	"	1.000
Cap	"	1.000
1-1/2"		
90 deg elbow	EA.	1.333
45 deg elbow	"	1.333
Tee	"	2.667
Reducer	"	2.000
Flange	"	1.333
Cap	"	1.333
2"		
90 deg elbow	EA.	1.600
45 deg elbow	"	1.600
Tee	"	4.000
Reducer	"	2.667
Flange	"	2.000
Cap	"	2.000
2-1/2"		
90 deg elbow	EA.	1.600
45 deg elbow	"	1.600
Tee	"	4.000
Reducer	"	2.667

PLUMBING	UNIT	MAN/ HOURS
15410.70 STAINLESS STEEL PIPE		
Flange	EA.	2.000
Cap	"	2.000
3"		
90 deg elbow	EA.	2.000
45 deg elbow	"	2.000
Tee	"	5.333
Reducer	"	4.000
Flange	"	2.667
Cap	"	2.667
4"		
90 deg elbow	EA.	2.000
45 deg elbow	"	2.000
Reducer	"	4.000
Flange	"	2.667
Cap	"	2.667
6"		
90 deg elbow	EA.	2.667
45 deg elbow	"	2.667
Tee	"	11.429
Reducer	"	4.000
Flange	"	4.000
Cap	"	4.000
Type 304 tubing		
.035 wall		
1/4"	L.F.	0.044
3/8"	"	0.050
1/2"	"	0.057
5/8"	"	0.067
3/4"	"	0.080
7/8"	"	0.089
1"	"	0.100
.049 wall		
1/4"	L.F.	0.047
3/8"	"	0.053
1/2"	"	0.062
5/8"	"	0.073
3/4"	"	0.089
7/8"	"	0.100
1"	"	0.114
.065 wall		
1/4"	L.F.	0.053
3/8"	"	0.067
1/2"	"	0.073
5/8"	"	0.089
3/4"	"	0.114
7/8"	"	0.133
1"	"	0.160
Type 316 tubing		
.035 wall		

PLUMBING

15410.70 STAINLESS STEEL PIPE

	UNIT	MAN/ HOURS
1/4"	L.F.	0.044
3/8"	"	0.050
1/2"	"	0.057
5/8"	"	0.067
3/4"	"	0.080
7/8"	"	0.089
1"	"	0.100
.049 wall		
1/4"	L.F.	0.053
3/8"	"	0.067
1/2"	"	0.073
5/8"	"	0.089
3/4"	"	0.114
7/8"	"	0.133
1"	"	0.160
.065 wall		
1/4"	L.F.	0.053
3/8"	"	0.067
1/2"	"	0.073
5/8"	"	0.089
3/4"	"	0.114
7/8"	"	0.133
1"	"	0.160
Fittings, 1/4"		
90 deg elbow	EA.	0.160
Union tee	"	0.267
Union	"	0.267
Male connector	"	0.200
3/8"		
90 deg elbow	EA.	0.200
Union tee	"	0.308
Union	"	0.308
Male connector	"	0.200
1/2"		
90 deg elbow	EA.	0.211
Union tee	"	0.333
Union	"	0.333
Male connector	"	0.200
5/8"		
90 deg elbow	EA.	0.267
Union tee	"	0.400
Union	"	0.400
Male connector	"	0.267
3/4"		
90 deg elbow	EA.	0.267
Union tee	"	0.400
Union	"	0.400
Male connector	"	0.267
7/8"		

PLUMBING

15410.70 STAINLESS STEEL PIPE

	UNIT	MAN/ HOURS
90 deg elbow	EA.	0.286
Union tee	"	0.444
Union	"	0.444
Male connector	"	0.286
1"		
90 deg elbow	EA.	0.364
Union tee	"	0.500
Union	"	0.500
Male conector	"	0.400
Type 316 valves		
Gate valves		
1/4"	EA.	0.267
3/8"	"	0.320
1/2"	"	0.348
3/4"	"	0.400
1"	"	0.533
Globe valves		
1/4"	EA.	0.267
3/8"	"	0.320
1/2"	"	0.348
3/4"	"	0.400
1"	"	0.533
Check valves		
1/4"	EA.	0.267
3/8"	"	0.320
1/2"	"	0.348
3/4"	"	0.400
1"	"	0.533

15410.80 STEEL PIPE

	UNIT	MAN/ HOURS
Black steel, extra heavy pipe, threaded		
1/2" pipe	L.F.	0.032
3/4" pipe	"	0.032
1" pipe	"	0.040
1-1/2" pipe	"	0.044
2-1/2" pipe	"	0.100
3" pipe	"	0.133
4" pipe	"	0.160
5" pipe	"	0.200
6" pipe	"	0.200
8" pipe	"	0.267
10" pipe	"	0.320
12" pipe	"	0.400
Fittings, malleable iron, threaded, 1/2" pipe		
90 deg ell	EA.	0.267
45 deg ell	"	0.267
Tee	"	0.400
Reducing tee	"	0.400
Cap	"	0.160

15 MECHANICAL

PLUMBING	UNIT	MAN/ HOURS
15410.80 STEEL PIPE		
Coupling	EA.	0.320
Union	"	0.267
Nipple, 4" long	"	0.267
3/4" pipe		
90 deg ell	EA.	0.267
45 deg ell	"	0.400
Tee	"	0.400
Reducing tee	"	0.267
Cap	"	0.160
Coupling	"	0.267
Union	"	0.267
Nipple, 4" long	"	0.267
1" pipe		
90 deg ell	EA.	0.320
45 deg ell	"	0.320
Tee	"	0.444
Reducing tee	"	0.444
Cap	"	0.160
Coupling	"	0.320
Union	"	0.320
Nipple, 4" long	"	0.320
1-1/2" pipe		
90 deg ell	EA.	0.400
45 deg ell	"	0.400
Tee	"	0.571
Reducing tee	"	0.571
Cap	"	0.200
Coupling	"	0.400
Union	"	0.400
Nipple, 4" long	"	0.400
2-1/2" pipe		
90 deg ell	EA.	1.000
45 deg ell	"	1.000
Tee	"	1.333
Reducing tee	"	1.333
Cap	"	0.500
Coupling	"	1.333
Union	"	1.333
Nipple, 4" long	"	1.333
3" pipe		
90 deg ell	EA.	1.333
45 deg ell	"	1.333
Tee	"	2.000
Reducing tee	"	2.000
Cap	"	0.667
Coupling	"	1.333
Union	"	1.333
Nipple, 4" long	"	1.333
4" pipe		

PLUMBING	UNIT	MAN/ HOURS
15410.80 STEEL PIPE		
90 deg ell	EA.	1.600
45 deg ell	"	1.600
Tee	"	2.667
Reducing tee	"	2.667
Cap	"	2.667
Coupling	"	0.800
Union	"	2.667
Nipple, 4" long	"	2.667
6" pipe		
90 deg ell	EA.	1.600
45 deg ell	"	1.600
Tee	"	2.667
Reducing tee	"	2.667
Cap	"	0.800
8" pipe		
90 deg ell	EA.	3.200
45 deg ell	"	3.200
Tee	"	5.000
Reducing tee	"	4.211
Cap	"	1.600
10" pipe		
90 deg ell	EA.	4.000
45 deg ell	"	6.667
Tee	"	5.000
Reducing tee	"	2.000
Cap	"	2.000
12" pipe		
90 deg ell	EA.	5.000
45 deg ell	"	5.000
Tee	"	6.667
Reducing tee	"	6.667
Cap	"	2.667
Butt welded, 1/2" pipe		
90 deg ell	EA.	0.267
45 deg ell	"	0.267
Tee	"	0.400
3/4" pipe		
90 deg ell	EA.	0.267
45 deg. ell	"	0.267
Tee	"	0.400
1" pipe		
90 deg ell	EA.	0.320
45 deg ell	"	0.320
Tee	"	0.444
1-1/2" pipe		
90 deg ell	EA.	0.400
45 deg. ell	"	0.400
Tee	"	0.571
Reducing tee	"	0.571

PLUMBING	UNIT	MAN/ HOURS
15410.80 STEEL PIPE		
Cap	EA.	0.320
2-1/2" pipe		
90 deg. ell	EA.	0.800
45 deg. ell	"	0.800
Tee	"	1.143
Reducing tee	"	1.143
Cap	"	0.400
3" pipe		
90 deg ell	EA.	1.000
45 deg. ell	"	1.000
Tee	"	1.333
Reducing tee	"	1.333
Cap	"	0.667
4" pipe		
90 deg ell	EA.	1.333
45 deg. ell	"	1.333
Tee	"	2.000
Reducing tee	"	2.000
Cap	"	0.667
6" pipe		
90 deg. ell	EA.	1.600
45 deg. ell	"	1.600
Tee	"	2.667
Reducing tee	"	2.667
Cap	"	0.800
8" pipe		
90 deg. ell	EA.	2.667
45 deg. ell	"	2.667
Tee	"	4.000
Reducing tee	"	4.000
Cap	"	1.600
10" pipe		
90 deg ell	EA.	2.667
45 deg. ell	"	2.667
Tee	"	4.000
Reducing tee	"	4.000
Cap	"	2.000
12" pipe		
90 deg. ell	EA.	3.200
45 deg. ell	"	3.200
Tee	"	5.714
Reducing tee	"	5.714
Cap	"	2.000
90 deg. ell	"	0.267
45 deg. ell	"	0.267
Tee	"	0.400
Reducing tee	"	0.400
3/4" pipe		
90 deg. ell	EA.	0.267

PLUMBING	UNIT	MAN/ HOURS
15410.80 STEEL PIPE		
45 deg. ell	EA.	0.267
Tee	"	0.400
Reducing tee	"	0.400
1" pipe		
90 deg. ell	EA.	0.320
45 deg. ell	"	0.320
Tee	"	0.444
Reducing tee	"	0.444
1-1/2" pipe		
90 deg. ell	EA.	0.400
45 deg. ell	"	0.400
Tee	"	0.571
Reducing tee	"	0.571
2-1/2" pipe		
90 deg. ell	EA.	0.800
45 deg. ell	"	0.800
Tee	"	1.143
Reducing tee	"	1.143
3" pipe		
90 deg. ell	EA.	1.000
45 deg. ell	"	1.000
Tee	"	1.600
Reducing tee	"	1.600
4" pipe		
90 deg. ell	EA.	1.333
45 deg. ell	"	1.333
Tee	"	2.000
Reducing tee	"	2.000
6" pipe		
90 deg. ell	EA.	1.333
45 deg. ell	"	1.333
Tee	"	2.000
Reducing tee	"	2.000
8" pipe		
90 deg. ell	EA.	2.667
45 deg. ell	"	2.667
Tee	"	4.000
Reducing tee	"	4.000
15410.82 GALVANIZED STEEL PIPE		
Galvanized pipe		
1/2" pipe	L.F.	0.080
3/4" pipe	"	0.100
1" pipe	"	0.114
1-1/4" pipe	"	0.133
1-1/2" pipe	"	0.160
2" pipe	"	0.200
2-1/2" pipe	"	0.267
3" pipe	"	0.286

PLUMBING	UNIT	MAN/ HOURS
15410.82 GALVANIZED STEEL PIPE		
4" pipe	L.F.	0.333
6" pipe	"	0.667
90 degree ell, 150 lb malleable iron, galvanized		
1/2"	EA.	0.160
3/4"	"	0.200
1"	"	0.211
1-1/4"	"	0.235
1-1/2"	"	0.267
2"	"	0.320
2-1/2"	"	0.500
3"	"	0.615
4"	"	0.667
5"	"	0.800
6"	"	0.800
45 degree ell, 150 lb m.i., galv.		
1/2"	EA.	0.160
3/4"	"	0.200
1"	"	0.211
1-1/4"	"	0.235
1-1/2"	"	0.267
2"	"	0.320
2-1/2"	"	0.500
3"	"	0.615
4"	"	0.800
5"	"	0.800
6"	"	1.000
Tees, straight, 150 lb m.i., galv.		
1/2"	EA.	0.200
3/4"	"	0.229
1"	"	0.267
1-1/4"	"	0.320
1-1/2"	"	0.400
2"	"	0.500
2-1/2"	"	0.667
3"	"	0.800
4"	"	1.000
5"	"	1.143
6"	"	1.333
Tees, reducing, out, 150 lb m.i., galv.		
1/2"	EA.	0.200
3/4"	"	0.229
1"	"	0.267
1-1/4"	"	0.320
1-1/2"	"	0.400
2"	"	0.500
2-1/2"	"	0.667
3"	"	0.800
4"	"	1.000
5"	"	1.143

PLUMBING	UNIT	MAN/ HOURS
15410.82 GALVANIZED STEEL PIPE		
6"	EA.	1.333
Couplings, straight, 150 lb m.i., galv.		
1/2"	EA.	0.160
3/4"	"	0.178
1"	"	0.200
1-1/4"	"	0.229
1-1/2"	"	0.267
2"	"	0.320
2-1/2"	"	0.500
3"	"	0.667
4"	"	0.727
5"	"	0.800
6"	"	0.800
Couplings, reducing, 150 lb m.i., galv		
1/2"	EA.	0.160
3/4"	"	0.178
1"	"	0.200
1-1/4"	"	0.229
1-1/2"	"	0.267
2"	"	0.320
2-1/2"	"	0.500
3"	"	0.667
4"	"	0.727
5"	"	0.800
6"	"	0.800
Caps, 150 lb m.i., galv.		
1/2"	EA.	0.080
3/4"	"	0.084
1"	"	0.089
1-1/4"	"	0.094
1-1/2"	"	0.100
2"	"	0.114
2-1/2"	"	0.145
3"	"	0.200
4"	"	0.250
5"	"	0.308
6"	"	0.400
Unions, 150 lb m.i., galv.		
1/2"	EA.	0.200
3/4"	"	0.229
1"	"	0.267
1-1/4"	"	0.320
1-1/2"	"	0.400
2"	"	0.444
2-1/2"	"	0.533
3"	"	0.667
Nipples, galvanized steel, 4" long		
1/2"	EA.	0.100
3/4"	"	0.107

PLUMBING	UNIT	MAN/ HOURS
15410.82 GALVANIZED STEEL PIPE		
1"	EA.	0.114
1-1/4"	"	0.123
1-1/2"	"	0.133
2"	"	0.145
2-1/2"	"	0.160
3"	"	0.200
4"	"	0.267
90 degree reducing ell, 150 lb m.i., galv.		
3/4" x 1/2"	EA.	0.160
1" x 3/4"	"	0.178
1-1/4" x 1"	"	0.200
1-1/4" x 3/4"	"	0.229
1-1/4" x 1/2"	"	0.267
1-1/2" x 1-1/4"	"	0.267
1-1/2" x 1"	"	0.267
1-1/2" x 3/4"	"	0.267
2" x 1-1/2"	"	0.348
2" x 1-1/4"	"	0.348
2" x 1"	"	0.348
2" x 3/4"	"	0.348
2-1/2" x 2"	"	0.500
2-1/2" x 1-1/2"	"	0.500
3" x 2-1/2"	"	0.667
3" x 2"	"	0.667
4" x 3"	"	0.727
Square head plug (C.I.)		
1/2"	EA.	0.089
3/4"	"	0.100
1"	"	0.107
1-1/4"	"	0.114
1-1/2"	"	0.123
2"	"	0.133
2-1/2"	"	0.178
3"	"	0.200
4"	"	0.267
5"	"	0.320
6"	"	0.400
Screwed flanges, galv.		
1"	EA.	0.400
1-1/4"	"	0.444
1-1/2"	"	0.500
2"	"	0.500
2-1/2"	"	0.533
3"	"	0.727
4"	"	1.000
5"	"	1.333
6"	"	1.333

PLUMBING	UNIT	MAN/ HOURS
15430.23 CLEANOUTS		
Cleanout, wall		
2"	EA.	0.533
3"	"	0.533
4"	"	0.667
6"	"	0.800
8"	"	1.000
Floor		
2"	EA.	0.667
3"	"	0.667
4"	"	0.800
6"	"	1.000
8"	"	1.143
15430.24 GREASE TRAPS		
Grease traps, cast iron, 3" pipe		
35 gpm, 70 lb capacity	EA.	8.000
50 gpm, 100 lb capacity	"	10.000
15430.25 HOSE BIBBS		
Hose bibb		
1/2"	EA.	0.267
3/4"	"	0.267
15430.60 VALVES		
Gate valve, 125 lb, bronze, soldered		
1/2"	EA.	0.200
3/4"	"	0.200
1"	"	0.267
1-1/2"	"	0.320
2"	"	0.400
2-1/2"	"	0.500
Threaded		
1/4", 125 lb	EA.	0.320
1/2"		
125 lb	EA.	0.320
150 lb	"	0.320
300 lb	"	0.320
3/4"		
125 lb	EA.	0.320
150 lb	"	0.320
300 lb	"	0.320
1"		
125 lb	EA.	0.320
150 lb	"	0.320
300 lb	"	0.400
1-1/2"		
125 lb	EA.	0.400

PLUMBING	UNIT	MAN/ HOURS
15430.60 VALVES		
150 lb	EA.	0.400
300 lb	"	0.444
2"		
125 lb	EA.	0.571
150 lb	"	0.571
300 lb	"	0.667
Cast iron, flanged		
2", 150 lb	EA.	0.667
2-1/2"		
125 lb	EA.	0.667
150 lb	"	0.667
250 lb	"	0.667
3"		
125 lb	EA.	0.800
150 lb	"	0.800
250 lb	"	0.800
4"		
125 lb	EA.	1.143
150 lb	"	1.143
250 lb	"	1.143
6"		
125 lb	EA.	1.600
250 lb	"	1.600
8"		
125 lb	EA.	2.000
250 lb	"	2.000
OS&Y, flanged		
2"		
125 lb	EA.	0.667
250 lb	"	0.667
2-1/2"		
125 lb	EA.	0.667
250 lb	"	0.800
3"		
125 lb	EA.	0.800
250 lb	"	0.800
4"		
125 lb	EA.	1.333
250 lb	"	1.333
6"		
125 lb	EA.	1.600
250 lb	"	1.600
Check valve, bronze, soldered, 125 lb		
1/2"	EA.	0.200
3/4"	"	0.200
1"	"	0.267
1-1/4"	"	0.320
1-1/2"	"	0.320
2"	"	0.400

PLUMBING	UNIT	MAN/ HOURS
15430.60 VALVES		
Threaded		
1/2"		
125 lb	EA.	0.267
150 lb	"	0.267
200 lb	"	0.267
3/4"		
125 lb	EA.	0.320
150 lb	"	0.320
200 lb	"	0.320
1"		
125 lb	EA.	0.400
150 lb	"	0.400
200 lb	"	0.400
Flow check valve, cast iron, threaded		
1"	EA.	0.320
1-1/4"	"	0.400
1-1/2"		
125 lb	EA.	0.400
150 lb	"	0.400
200 lb	"	0.444
2"		
125 lb	EA.	0.444
150 lb	"	0.444
200 lb	"	0.500
2-1/2"		
125 lb	EA.	0.667
250 lb	"	0.800
3"		
125 lb	EA.	0.800
250 lb	"	1.000
4"		
125 lb	EA.	1.143
250 lb	"	1.333
6"		
125 lb	EA.	1.600
250 lb	"	1.600
Vertical check valve, bronze, 125 lb, threaded		
1/2"	EA.	0.320
3/4"	"	0.364
1"	"	0.400
1-1/4"	"	0.444
1-1/2"	"	0.500
2"	"	0.571
Cast iron, flanged		
2-1/2"	EA.	0.800
3"	"	1.000
4"	"	1.333
6	"	1.600
8"	"	2.000

PLUMBING	UNIT	MAN/ HOURS
15430.60 VALVES		
10"	EA.	2.667
12"	"	3.200
Globe valve, bronze, soldered, 125 lb		
1/2"	EA.	0.229
3/4"	"	0.250
1"	"	0.267
1-1/4"	"	0.286
1-1/2"	"	0.333
2"	"	0.400
Threaded		
1/2"		
125 lb	EA.	0.267
150 lb	"	0.267
300 lb	"	0.267
3/4"		
125 lb	EA.	0.320
150 lb	"	0.320
300 lb	"	0.320
1"		
125 lb	EA.	0.400
150 lb	"	0.400
300 lb	"	0.400
1-1/4"		
125 lb	EA.	0.400
150 lb	"	0.400
300 lb	"	0.400
1-1/2"		
125 lb	EA.	0.444
150 lb	"	0.444
300 lb	"	0.444
2"		
125 lb	EA.	0.533
150 lb	"	0.533
300 lb	"	0.533
Cast iron flanged		
2-1/2"		
125 lb	EA.	0.800
250 lb	"	0.800
3"		
125 lb	EA.	1.000
250 lb	"	1.000
4"		
125 lb	EA.	1.333
250 lb	"	1.333
6"		
125 lb	EA.	1.600
250 lb	"	1.600
8"		
125 lb	EA.	2.000

PLUMBING	UNIT	MAN/ HOURS
15430.60 VALVES		
250 lb	EA.	2.000
Butterfly valve, cast iron, wafer type		
2"		
150 lb	EA.	0.571
200 lb	"	0.667
2-1/2"		
150 lb	EA.	0.667
200 lb	"	0.727
3"		
150 lb	EA.	0.800
200 lb	"	0.889
4"		
150 lb	EA.	1.143
200 lb	"	1.333
6"		
150 lb	EA.	1.600
200 lb	"	1.600
8"		
150 lb	EA.	1.778
200 lb	"	2.000
10"		
150 lb	EA.	2.000
200 lb	"	2.667
Ball valve, bronze, 250 lb, threaded		
1/2"	EA.	0.320
3/4"	"	0.320
1"	"	0.400
1-1/4"	"	0.444
1-1/2"	"	0.500
2"	"	0.571
Angle valve, bronze, 150 lb, threaded		
1/2"	EA.	0.286
3/4"	"	0.320
1"	"	0.320
1-1/4"	"	0.400
1-1/2"	"	0.444
Balancing valve, with meter connections, circuit setter		
1/2"	EA.	0.320
3/4"	"	0.364
1"	"	0.400
1-1/4"	"	0.444
1-1/2"	"	0.533
2"	"	0.667
2-1/2"	"	0.800
3"	"	1.000
4"	"	1.333
Balancing valve, straight type		
1/2"	EA.	0.320
3/4"	"	0.320

15 MECHANICAL

PLUMBING

15430.60 VALVES

PLUMBING	UNIT	MAN/ HOURS
Angle type		
1/2"	EA.	0.320
3/4"	"	0.320
Square head cock, 125 lb, bronze body		
1/2"	EA.	0.267
3/4"	"	0.320
1"	"	0.364
1-1/4"	"	0.400
Radiator temp control valve, with control and sensor		
1/2" valve	EA.	0.500
1" valve	"	0.500
Pressure relief valve, 1/2", bronze		
Low pressure	EA.	0.320
High pressure	"	0.320
Press and temperature relief valve		
Bronze, 3/4"	EA.	0.320
Cast iron, 3/4"		
High pressure	EA.	0.320
Temperature relief	"	0.320
Pressure & temp relief valve	"	0.320
Pressure reducing valve, bronze, threaded, 250 lb		
1/2"	EA.	0.500
3/4"	"	0.500
1"	"	0.500
1-1/4"	"	0.571
1-1/2"	"	0.667
Pressure regulating valve, bronze, class 300		
1"	EA.	0.500
1-1/2"	"	0.615
2"	"	0.800
3"	"	1.143
4"	"	1.600
5"	"	2.000
6"	"	2.667
Solar water temperature regulating valve		
3/4"	EA.	0.667
1"	"	0.800
1-1/4"	"	0.889
1-1/2"	"	1.000
2"	"	1.143
2-1/2"	"	2.000
Tempering valve, threaded		
3/4"	EA.	0.267
1"	"	0.320
1-1/4"	"	0.400
1-1/2"	"	0.400
2"	"	0.500
2-1/2"	"	0.667
3"	"	0.800

PLUMBING

15430.60 VALVES

PLUMBING	UNIT	MAN/ HOURS
4"	EA.	1.143
Thermostatic mixing valve, threaded		
1/2"	EA.	0.286
3/4"	"	0.320
1"	"	0.348
1-1/2"	"	0.400
2"	"	0.500
Sweat connection		
1/2"	EA.	0.286
3/4"	"	0.320
Mixing valve, sweat connection		
1/2"	EA.	0.286
3/4"	"	0.320
Liquid level gauge, aluminum body		
3/4"	EA.	0.320
125 psi, pvc body		
3/4"	EA.	0.320
150 psi, crs body		
3/4"	EA.	0.320
1"	"	0.320
175 psi, bronze body, 1/2"	"	0.286

15430.65 VACUUM BREAKERS

PLUMBING	UNIT	MAN/ HOURS
Vacuum breaker, atmospheric, threaded connection		
3/4"	EA.	0.320
1"	"	0.320
Anti-siphon, brass		
3/4"	EA.	0.320
1"	"	0.320
1-1/4"	"	0.400
1-1/2"	"	0.444
2"	"	0.500
Air eliminators, purger, cast iron, threaded		
1"	EA.	0.320
1-1/4"	"	0.400
1-1/2"	"	0.400
2"	"	0.500
2-1/2"	"	1.000
3"	"	1.333
Airtrol fitting, 3/4"	"	0.320
Air eliminator, air vents, 1/4"	"	0.320
Air vent for hot water	"	0.286

15430.68 STRAINERS

PLUMBING	UNIT	MAN/ HOURS
Strainer, Y pattern, 125 psi, cast iron body, threaded		
3/4"	EA.	0.286
1"	"	0.320

15 MECHANICAL

PLUMBING	UNIT	MAN/ HOURS
15430.68 STRAINERS		
1-1/4"	EA.	0.400
1-1/2"	"	0.400
2"	"	0.500
250 psi, brass body, threaded		
3/4"	EA.	0.320
1"	"	0.320
1-1/4"	"	0.400
1-1/2"	"	0.400
2"	"	0.500
Cast iron body, threaded		
3/4"	EA.	0.320
1"	"	0.320
1-1/4"	"	0.400
1-1/2"	"	0.400
2"	"	0.500
15430.70 DRAINS, ROOF & FLOOR		
Floor drain, cast iron, with cast iron top		
2"	EA.	0.667
3"	"	0.667
4"	"	0.667
6"	"	0.800
Roof drain, cast iron		
2"	EA.	0.667
3"	"	0.667
4"	"	0.667
5"	"	0.800
6"	"	0.800
15430.80 TRAPS		
Bucket trap, threaded		
3/4"	EA.	0.500
1"	"	0.533
1-1/4"	"	0.615
1-1/2"	"	0.727
Inverted bucket steam trap, threaded		
3/4"	EA.	0.500
1"	"	0.500
1-1/4"	"	0.444
1-1/2"	"	0.667
With stainless interior		
1/2"	EA.	0.500
3/4"	"	0.500
1"	"	0.500
1-1/4"	"	0.571
Brass interior		
3/4"	EA.	0.500

PLUMBING	UNIT	MAN/ HOURS
15430.80 TRAPS		
1"	EA.	0.533
1-1/4"	"	0.571
Cast steel body, threaded, high temperature		
3/4"	EA.	0.500
1"	"	0.571
1-1/4	"	0.615
1-1/2"	"	0.667
2"	"	0.800
Float trap, 15 psi		
3/4"	EA.	0.500
1"	"	0.533
1-1/4"	"	0.571
1-1/2"	"	0.667
2"	"	0.800
30 psi		
3/4"	EA.	0.500
1"	"	0.533
1-1/4"	"	0.667
1-1/2"	"	0.800
75 psi		
3/4"	EA.	0.500
1"	"	0.533
1-1/4"	"	0.571
1-1/2"	"	0.667
125 psi		
3/4"	EA.	0.500
1"	"	0.533
1-1/4	"	0.571
1-1/2	"	0.667
Float and thermostatic trap, 15 psi		
3/4"	EA.	0.500
1"	"	0.533
1-1/4"	"	0.571
1-1/2"	"	0.667
2"	"	0.800
30 psi		
3/4"	EA.	0.500
1"	"	0.533
1-1/4"	"	0.571
1-1/2"	"	0.667
75 psi		
3/4"	EA.	0.500
1"	"	0.533
1-1/4"	"	0.571
1-1/2"	"	0.667
Steam trap, cast iron body, threaded, 125 psi		
3/4"	EA.	0.500
1"	"	0.533
1-1/4"	"	0.571

15 MECHANICAL

PLUMBING

15430.80 TRAPS

	UNIT	MAN/HOURS
1-1/2"	EA.	0.667
Thermostatic trap, low pressure, angle type, 25 psi		
1/2"	EA.	0.500
3/4"	"	0.500
1"	"	0.533
50 psi		
1/2"	EA.	0.500
3/4"	"	0.500
1"	"	0.533
Cast iron body, threaded, 125 psi		
3/4"	EA.	0.500
1"	"	0.571
1-1/4"	"	0.615
1-1/2"	"	0.667
Low pressure, 25 psi, swivel type, 1/2"	"	0.500
Straightway type, 3/4"	"	0.500
Vertical type, 1/2"	"	0.500
Medium pressure, 50 psi, angle type, 1/2"	"	0.500
High pressure, 125 psi, angle type		
1/2"	EA.	0.500
3/4"	"	0.500
1"	"	0.571
Straightway type		
1/2"	EA.	0.500
Thermo disc trap		
3/4"	EA.	0.500
1"	"	0.533
Drip pan ell, cast iron		
2-1/2"	EA.	1.000
3"	"	1.143
4"	"	1.333
5"	"	1.600
6"	"	2.000
8"	"	2.667
Steel		
2-1/2"	EA.	1.000
3"	"	1.143
4"	"	1.333
5"	"	1.600
6"	"	2.000
8"	"	2.667

PLUMBING FIXTURES

15440.10 BATHS

	UNIT	MAN/HOURS
Bath tub, 5' long		
Minimum	EA.	2.667
Average	"	4.000
Maximum	"	8.000
6' long		
Minimum	EA.	2.667
Average	"	4.000
Maximum	"	8.000
Square tub, whirlpool, 4'x4'		
Minimum	EA.	4.000
Average	"	8.000
Maximum	"	10.000
5'x5'		
Minimum	EA.	4.000
Average	"	8.000
Maximum	"	10.000
6'x6'		
Minimum	EA.	4.000
Average	"	8.000
Maximum	"	10.000
For trim and rough-in		
Minimum	EA.	2.667
Average	"	4.000
Maximum	"	8.000

15440.12 DISPOSALS & ACCESSORIES

	UNIT	MAN/HOURS
Continuous feed		
Minimum	EA.	1.600
Average	"	2.000
Maximum	"	2.667
Batch feed, 1/2 hp		
Minimum	EA.	1.600
Average	"	2.000
Maximum	"	2.667
Hot water dispenser		
Minimum	EA.	1.600
Average	"	2.000
Maximum	"	2.667
Epoxy finish faucet	"	1.600
Lock stop assembly	"	1.000
Mounting gasket	"	0.667
Tailpipe gasket	"	0.667
Stopper assembly	"	0.800
Switch assembly, on/off	"	1.333
Tailpipe gasket washer	"	0.400
Stop gasket	"	0.444
Tailpipe flange	"	0.400
Tailpipe	"	0.500

15 MECHANICAL

PLUMBING FIXTURES	UNIT	MAN/ HOURS
15440.15 FAUCETS		
Kitchen		
Minimum	EA.	1.333
Average	"	1.600
Maximum	"	2.000
Bath		
Minimum	EA.	1.333
Average	"	1.600
Maximum	"	2.000
Lavatory, domestic		
Minimum	EA.	1.333
Average	"	1.600
Maximum	"	2.000
Hospital, patient rooms		
Minimum	EA.	2.000
Average	"	2.667
Maximum	"	4.000
Operating room		
Minimum	EA.	2.000
Average	"	2.667
Maximum	"	4.000
Washroom		
Minimum	EA.	1.333
Average	"	1.600
Maximum	"	2.000
Handicapped		
Minimum	EA.	1.600
Average	"	2.000
Maximum	"	2.667
Shower		
Minimum	EA.	1.333
Average	"	1.600
Maximum	"	2.000
For trim and rough-in		
Minimum	EA.	1.600
Average	"	2.000
Maximum	"	4.000
15440.18 HYDRANTS		
Wall hydrant		
8" thick	EA.	1.333
12" thick	"	1.600
18" thick	"	1.778
24" thick	"	2.000
Ground hydrant		
2' deep	EA.	1.000
4' deep	"	1.143
6' deep	"	1.333
8' deep	"	2.000

PLUMBING FIXTURES	UNIT	MAN/ HOURS
15440.20 LAVATORIES		
Lavatory, counter top, porcelain enamel on cast iron		
Minimum	EA.	1.600
Average	"	2.000
Maximum	"	2.667
Wall hung, china		
Minimum	EA.	1.600
Average	"	2.000
Maximum	"	2.667
Handicapped		
Minimum	EA.	2.000
Average	"	2.667
Maximum	"	4.000
For trim and rough-in		
Minimum	EA.	2.000
Average	"	2.667
Maximum	"	4.000
15440.30 SHOWERS		
Shower, fiberglass, 36"x34"x84"		
Minimum	EA.	5.714
Average	"	8.000
Maximum	"	8.000
Steel, 1 piece, 36"x36"		
Minimum	EA.	5.714
Average	"	8.000
Maximum	"	8.000
Receptor, molded stone, 36"x36"		
Minimum	EA.	2.667
Average	"	4.000
Maximum	"	6.667
For trim and rough-in		
Minimum	EA.	3.636
Average	"	4.444
Maximum	"	8.000
15440.40 SINKS		
Service sink, 24"x29"		
Minimum	EA.	2.000
Average	"	2.667
Maximum	"	4.000
Kitchen sink, single, stainless steel, single bowl		
Minimum	EA.	1.600
Average	"	2.000
Maximum	"	2.667
Double bowl		
Minimum	EA.	2.000
Average	"	2.667
Maximum	"	4.000
Porcelain enamel, cast iron, single bowl		

15 MECHANICAL

PLUMBING FIXTURES	UNIT	MAN/ HOURS
15440.40 SINKS		
Minimum	EA.	1.600
Average	"	2.000
Maximum	"	2.667
Double bowl		
Minimum	EA.	2.000
Average	"	2.667
Maximum	"	4.000
Mop sink, 24"x36"x10"		
Minimum	EA.	1.600
Average	"	2.000
Maximum	"	2.667
Washing machine box		
Minimum	EA.	2.000
Average	"	2.667
Maximum	"	4.000
For trim and rough-in		
Minimum	EA.	2.667
Average	"	4.000
Maximum	"	5.333
15440.50 URINALS		
Urinal, flush valve, floor mounted		
Minimum	EA.	2.000
Average	"	2.667
Maximum	"	4.000
Wall mounted		
Minimum	EA.	2.000
Average	"	2.667
Maximum	"	4.000
For trim and rough-in		
Minimum	EA.	2.000
Average	"	4.000
Maximum	"	5.333
15440.60 WATER CLOSETS		
Water closet flush tank, floor mounted		
Minimum	EA.	2.000
Average	"	2.667
Maximum	"	4.000
Handicapped		
Minimum	EA.	2.667
Average	"	4.000
Maximum	"	8.000
Bowl, with flush valve, floor mounted		
Minimum	EA.	2.000
Average	"	2.667
Maximum	"	4.000
Wall mounted		

PLUMBING FIXTURES	UNIT	MAN/ HOURS
15440.60 WATER CLOSETS		
Minimum	EA.	2.000
Average	"	2.667
Maximum	"	4.000
For trim and rough-in		
Minimum	EA.	2.000
Average	"	2.667
Maximum	"	4.000
15440.70 WATER HEATERS		
Water heater, electric		
6 gal	EA.	1.333
10 gal	"	1.333
15 gal	"	1.333
20 gal	"	1.600
30 gal	"	1.600
40 gal	"	1.600
52 gal	"	2.000
66 gal	"	2.000
80 gal	"	2.000
100 gal	"	2.667
120 gal	"	2.667
Oil fired		
20 gal	EA.	4.000
50 gal	"	5.714
15440.90 MISCELLANEOUS FIXTURES		
Electric water cooler		
Floor mounted	EA.	2.667
Wall mounted	"	2.667
Wash fountain		
Wall mounted	EA.	4.000
Circular, floor supported	"	8.000
Deluge shower and eye wash	"	4.000
15440.95 FIXTURE CARRIERS		
Water fountain, wall carrier		
Minimum	EA.	0.800
Average	"	1.000
Maximum	"	1.333
Lavatory, wall carrier		
Minimum	EA.	0.800
Average	"	1.000
Maximum	"	1.333
Sink, industrial, wall carrier		
Minimum	EA.	0.800
Average	"	1.000

15 MECHANICAL

PLUMBING FIXTURES	UNIT	MAN/ HOURS
15440.95 FIXTURE CARRIERS		
Maximum	EA.	1.333
Toilets, water closets, wall carrier		
Minimum	EA.	0.800
Average	"	1.000
Maximum	"	1.333
Floor support		
Minimum	EA.	0.667
Average	"	0.800
Maximum	"	1.000
Urinals, wall carrier		
Minimum	EA.	0.800
Average	"	1.000
Maximum	"	1.333
Floor support		
Minimum	EA.	0.667
Average	"	0.800
Maximum	"	1.000
15450.30 PUMPS		
In-line pump, bronze, centrifugal		
5 gpm, 20' head	EA.	0.500
20 gpm, 40' head	"	0.500
50 gpm		
50' head	EA.	1.000
100' head	"	1.000
70 gpm, 100' head	"	1.333
100 gpm, 80' head	"	1.333
250 gpm, 150' head	"	2.000
Cast iron, centrifugal		
50 gpm, 200' head	EA.	1.000
100 gpm		
100' head	EA.	1.333
200' head	"	1.333
200 gpm		
100' head	EA.	2.000
200' head	"	2.000
Centrifugal, close coupled, c.i., single stage		
50 gpm, 100' head	EA.	1.000
100 gpm, 100' head	"	1.333
Base mounted		
50 gpm, 100' head	EA.	1.000
100 gpm, 50' head	"	1.333
200 gpm, 100' head	"	2.000
300 gpm, 175' head	"	2.000
Suction diffuser, flanged, strainer		
3" inlet, 2-1/2" outlet	EA.	1.000
3" outlet	"	1.000
4" inlet		

PLUMBING FIXTURES	UNIT	MAN/ HOURS
15450.30 PUMPS		
3" outlet	EA.	1.333
4" outlet	"	1.333
6" inlet		
4" outlet	EA.	1.600
5" outlet	"	1.600
6" Outlet	"	1.600
8" inlet		
6" outlet	EA.	2.000
8" outlet	"	2.000
10" inlet		
8" outlet	EA.	2.667
Vertical turbine		
Single stage, C.I., 3550 rpm, 200 gpm, 50'head	EA.	2.667
Multi stage, 3550 rpm		
50 gpm, 100' head	EA.	2.000
100 gpm		
100' head	EA.	2.000
200 gpm		
50' head	EA.	2.667
100' head	"	2.667
Bronze		
Single stage, 3550 rpm, 100 gpm, 50' head	EA.	2.000
Multi stage, 3550 rpm, 50 gpm, 100' head	"	2.000
100 gpm		
100' head	EA.	2.000
200 gpm		
50' head	EA.	2.667
100' head	"	2.667
Sump pump, bronze, 1750 rpm, 25 gpm		
20' head	EA.	10.000
150' head	"	13.333
50 gpm		
100' head	EA.	10.000
100 gpm		
50' head	EA.	10.000
Condensate pump, simplex		
1000 sf EDR, 2 gpm	EA.	6.667
2000 sf EDR, 3 gpm	"	6.667
4000 sf EDR, 6 gpm	"	7.273
6000 sf EDR, 9 gpm	"	7.273
Duplex, bronze		
8000 sf EDR, 12 gpm	EA.	7.273
10,000 sf EDR, 15 gpm	"	10.000
15,000 sf EDR, 23 gpm	"	11.429
20,000 sf EDR, 30 gpm	"	16.000
25,000 sf EDR, 38 gpm	"	16.000
30,000 sf EDR, 45 gpm	"	17.778
40,000 sf EDR, 60 gpm	"	10.000
50,000 sf EDR, 75 gpm	"	11.429

15 MECHANICAL

PLUMBING FIXTURES

PLUMBING FIXTURES	UNIT	MAN/ HOURS
15450.30 PUMPS		
75,000 sf EDR, 112 gpm	EA.	13.333
100,000 sf EDR, 150 gpm	"	20.000
15450.40 STORAGE TANKS		
Hot water storage tank, cement lined		
10 gallon	EA.	2.667
70 gallon	"	4.000
200 gallon	"	5.714
900 gallon	"	10.000
1100 gallon	"	10.000
2000 gallon	"	10.000
15480.10 SPECIAL SYSTEMS		
Air compressor, air cooled, two stage		
5.0 cfm, 175 psi	EA.	16.000
10 cfm, 175 psi	"	17.778
20 cfm, 175 psi	"	19.048
50 cfm, 125 psi	"	21.053
80 cfm, 125 psi	"	22.857
Single stage, 125 psi		
1.0 cfm	EA.	11.429
1.5 cfm	"	11.429
2.0 cfm	"	11.429
Automotive compressor, hose reel, air and water, 50' hose	"	6.667
Lube equipment, 3 reel, with pumps	"	32.000
Tire changer		
Truck	EA.	11.429
Passenger car	"	6.154
Air hose reel, includes, 50' hose	"	6.154
Hose reel, 5 reel, motor oil, gear oil, lube, air & water	"	32.000
Water hose reel, 50' hose	"	6.154
Pump, air operated, for motor or gear oil, fits 55 gal drum	"	0.800
For chassis lube	"	0.800
Fuel dispensing pump, lighted dial, one product		
One hose	EA.	6.667
Two hose	"	6.667
Two products, two hose	"	6.667

HEATING & VENTILATING

HEATING & VENTILATING	UNIT	MAN/ HOURS
15555.10 BOILERS		
Cast iron, gas fired, hot water		
115 mbh	EA.	20.000
175 mbh	"	21.818
235 mbh	"	24.000
940 mbh	"	48.000
1600 mbh	"	60.000
3000 mbh	"	80.000
6000 mbh	"	120
Steam		
115 mbh	EA.	20.000
175 mbh	"	21.818
235 mbh	"	24.000
940 mbh	"	48.000
1600 mbh	"	60.000
3000 mbh	"	80.000
6000 mbh	"	120
Electric, hot water		
115 mbh	EA.	12.000
175 mbh	"	12.000
235 mbh	"	12.000
940 mbh	"	24.000
1600 mbh	"	48.000
3000 mbh	"	60.000
6000 mbh	"	80.000
Steam		
115 mbh	EA.	12.000
175 mbh	"	12.000
235 mbh	"	12.000
940 mbh	"	24.000
1600 mbh	"	48.000
3000 mbh	"	60.000
6000 mbh	"	80.000
Oil fired, hot water		
115 mbh	EA.	16.000
175 mbh	"	18.462
235 mbh	"	21.818
940 mbh	"	40.000
1600 mbh	"	48.000
3000 mbh	"	60.000
6000 mbh	"	120
Steam		
115 mbh	EA.	16.000
175 mbh	"	18.462
235 mbh	"	21.818
940 mbh	"	40.000
1600 mbh	"	48.000
3000 mbh	"	60.000
6000 mbh	"	120

HEATING & VENTILATING

15610.10 FURNACES	UNIT	MAN/HOURS
Electric, hot air		
40 mbh	EA.	4.000
60 mbh	"	4.211
80 mbh	"	4.444
100 mbh	"	4.706
125 mbh	"	4.848
160 mbh	"	5.000
200 mbh	"	5.161
400 mbh	"	5.333
Gas fired hot air		
40 mbh	EA.	4.000
60 mbh	"	4.211
80 mbh	"	4.444
100 mbh	"	4.706
125 mbh	"	4.848
160 mbh	"	5.000
200 mbh	"	5.161
400 mbh	"	5.333
Oil fired hot air		
40 mbh	EA.	4.000
60 mbh	"	4.211
80 mbh	"	4.444
100 mbh	"	4.706
125 mbh	"	4.848
160 mbh	"	5.000
200 mbh	"	5.161
400 mbh	"	5.333

REFRIGERATION

15670.10 CONDENSING UNITS	UNIT	MAN/HOURS
Air cooled condenser, single circuit		
3 ton	EA.	1.333
5 ton	"	1.333
7.5 ton	"	3.810
20 ton	"	4.000
25 ton	"	4.000
30 ton	"	4.000
40 ton	"	5.714
50 ton	"	5.714
60 ton	EA.	5.000
With low ambient dampers		
3 ton	EA.	2.000
5 ton	"	2.000
7.5 ton	"	4.000
20 ton	"	5.333
25 ton	"	5.333
30 ton	"	5.333
40 ton	"	6.667
50 ton	"	7.273
60 ton	"	7.273
Dual circuit		
10 ton	EA.	4.000
15 ton	"	5.714
20 ton	"	5.714
25 ton	"	5.714
30 ton	"	5.714
40 ton	"	6.667
50 ton	"	6.667
60 ton	"	6.667
80 ton	"	8.889
100 ton	"	8.889
120 ton	"	8.889
With low ambient dampers		
15 ton	EA.	5.714
20 ton	"	5.714
25 ton	"	5.714
30 ton	"	5.714
40 ton	"	6.667
50 ton	"	6.667
60 ton	"	6.667
80 ton	"	8.889
100 ton	"	8.889
120 ton	"	8.889

15680.10 CHILLERS	UNIT	MAN/HOURS
Chiller, reciprocal		
Air cooled, remote condenser, starter		
20 ton	EA.	8.000
25 ton	"	8.000
30 ton	"	8.000
40 ton	"	12.000
50 ton	"	13.333
60 ton	"	14.118
80 ton	"	21.818
100 ton	"	24.000
120 ton	"	26.667
150 ton	"	30.000
180 ton	"	34.286

REFRIGERATION

15680.10 CHILLERS

	UNIT	MAN/HOURS
200 ton	EA.	40.000
Water cooled, with starter		
20 ton	EA.	8.000
25 ton	"	8.000
30 ton	"	12.000
40 ton	"	12.000
50 ton	"	13.333
60 ton	"	14.118
80 ton	"	21.818
100 ton	"	24.000
120 ton	"	26.667
150 ton	"	30.000
180 ton	"	34.286
200 ton	"	40.000
Packaged, air cooled, with starter		
20 ton	EA.	6.000
25 ton	"	6.000
30 ton	"	6.000
40 ton	"	6.000
50 ton	"	8.000
60 ton	"	8.000
80 ton	"	12.000
100 ton	"	12.000
120 ton	"	12.000
Heat recovery, air cooled, with starter		
40 ton	EA.	12.000
50 ton	"	12.000
60 ton	"	16.000
75 ton	"	24.000
100 ton	"	24.000
Water cooled, with starter		
40 ton	EA.	12.000
50 ton	"	12.000
60 ton	"	16.000
75 ton	"	24.000
100 ton	"	26.667
Centrifugal, single bundle condenser, with starter		
80 ton	EA.	34.286
130 ton	"	40.000
160 ton	"	43.636
180 ton	"	48.000
230 ton	"	53.333
280 ton	"	60.000
360 ton	"	60.000
460 ton	"	80.000
560 ton	"	85.714
670 ton	"	96.000

REFRIGERATION

15710.10 COOLING TOWERS

	UNIT	MAN/HOURS
Cooling tower, propeller type		
100 ton	EA.	8.000
200 ton	"	12.000
300 ton	"	20.000
400 ton	"	24.000
600 ton	"	34.286
800 ton	"	48.000
1000 ton	"	60.000
Centrifugal		
100 ton	EA.	8.000
200 ton	"	12.000
300 ton	"	20.000
400 ton	"	24.000
600 ton	"	34.286
800 ton	"	48.000
1000 ton	"	60.000

HEAT TRANSFER

15780.10 COMPUTER ROOM A/C

	UNIT	MAN/HOURS
Air cooled, alarm, high efficiency filter, elec. heat		
3 ton	EA.	6.154
5 ton	"	6.667
7.5 ton	"	8.000
10 ton	"	10.000
15 ton	"	11.429
Steam heat		
3 ton	EA.	6.154
5 ton	"	6.667
7.5 ton	"	8.000
10 ton	"	10.000
15 ton	"	11.429
Hot water heat		
3 ton	EA.	6.154
5 ton	"	6.667
7.5 ton	"	8.000
10 ton	"	10.000
15 ton	"	11.429
Air cooled condenser, low ambient damper		
3 ton	EA.	1.600
5 ton	"	2.000

HEAT TRANSFER	UNIT	MAN/ HOURS
15780.10 COMPUTER ROOM A/C		
7.5 ton	EA.	4.000
10 ton	"	5.714
15 ton	"	4.706
Water cooled, high efficiency filter, alarm, elec. heat		
3 ton	EA.	5.714
5 ton	"	6.667
7.5 ton	"	10.000
10 ton	"	11.429
15 ton	"	13.333
Steam heat		
3 ton	EA.	5.714
5 ton	"	6.667
7.5 ton	"	10.000
10 ton	"	11.429
15 ton	"	13.333
Hot water heat		
3 ton	EA.	5.714
5 ton	"	6.667
7.5 ton	"	10.000
10 ton	"	11.429
15 ton	"	13.333
Chilled water, alarm, high eff. filter, elec. heat		
7.5 ton	EA.	7.273
10 ton	"	8.889
15 ton	"	10.000
Steam heat		
7.5 ton	EA.	7.273
10 ton	"	8.889
15 ton	"	10.000
Hot water heat		
7.5 ton	EA.	7.273
10 ton	"	8.889
15 ton	"	10.000
15780.20 ROOFTOP UNITS		
Packaged, single zone rooftop unit, with roof curb		
2 ton	EA.	8.000
3 ton	"	8.000
4 ton	"	10.000
5 ton	"	13.333
7.5 ton	"	16.000
15830.10 RADIATION UNITS		
Baseboard radiation unit		
1.7 mbh/lf	L.F.	0.320
2.1 mbh/lf	"	0.400
Enclosure only		
Two tier	L.F.	0.133

HEAT TRANSFER	UNIT	MAN/ HOURS
15830.10 RADIATION UNITS		
Three tier	L.F.	0.133
Copper element only, 3/4" dia.		
Two tier	L.F.	0.200
Three tier	"	0.267
Fin-tube, 16 ga, sloping cover, 1-1/4" steel		
One tier	L.F.	0.267
Two tier	"	0.320
2" steel		
Two tier	L.F.	0.320
Three tier	"	0.400
1-1/4" copper		
Two tier	L.F.	0.267
18 ga flat cover, 1-1/4" steel		
One tier	L.F.	0.267
Two tier	"	0.320
Three tier	"	0.400
2" steel		
One tier	L.F.	0.267
Two tier	"	0.320
Three tier	"	0.400
1-1/4" copper		
One tier	L.F.	0.267
Two tier	"	0.320
Three tier	"	0.400
15830.20 FAN COIL UNITS		
Fan coil unit, 2 pipe, complete		
200 cfm ceiling hung	EA.	2.667
Floor mounted	"	2.000
300 cfm, ceiling hung	"	3.200
Floor mounted	"	2.667
400 cfm, ceiling hung	"	3.810
Floor mounted	"	2.667
500 cfm, ceiling hung	"	4.000
Floor mounted	"	3.077
600 cfm, ceiling hung	"	4.420
Floor mounted	"	3.636
800 cfm, ceiling hung	"	5.000
Floor mounted	"	3.810
1000 cfm, ceiling hung	"	5.714
Floor mounted	"	4.211
1200 cfm ceiling hung	"	6.667
Floor mounted	"	5.000
15830.70 UNIT HEATERS		
Steam unit heater, horizontal		
12,500 btuh, 200 cfm	EA.	1.333
17,000 btuh, 300 cfm	"	1.333

15 MECHANICAL

HEAT TRANSFER

15830.70 UNIT HEATERS

	UNIT	MAN/ HOURS
40,000 btuh, 500 cfm	EA.	1.333
60,000 btuh, 700 cfm	"	1.333
70,000 btuh, 1000 cfm	"	2.000
Vertical		
12,500 btuh, 200 cfm	EA.	1.333
17,000 btuh, 300 cfm	"	1.333
40,000 btuh, 500 cfm	"	1.333
60,000 btuh, 700 cfm	"	1.333
70,000 btuh, 1000 cfm	"	1.333
Gas unit heater, horizontal		
27,400 btuh	EA.	3.200
38,000 btuh	"	3.200
56,000 btuh	"	3.200
82,200 btuh	"	3.200
103,900 btuh	"	5.000
125,700 btuh	"	5.000
133,200 btuh	"	5.000
149,000 btuh	"	5.000
172,000 btuh	"	5.000
190,000 btuh	"	5.000
225,000 btuh	"	5.000
Hot water unit heater, horizontal		
12,500 btuh, 200 cfm	EA.	1.333
17,000 btuh, 300 cfm	"	1.333
25,000 btuh, 500 cfm	"	1.333
30,000 btuh, 700 cfm	"	1.333
50,000 btuh, 1000 cfm	"	2.000
60,000 btuh, 1300 cfm	"	2.000
Vertical		
12,500 btuh, 200 cfm	EA.	1.333
17,000 btuh, 300 cfm	"	1.333
25,000 btuh, 500 cfm	"	1.333
30,000 btuh, 700 cfm	"	1.333
50,000 btuh, 1000 cfm	"	1.333
60,000 btuh, 1300 cfm	"	1.333
Cabinet unit heaters, ceiling, exposed, hot water		
200 cfm	EA.	2.667
300 cfm	"	3.200
400 cfm	"	3.810
600 cfm	"	4.211
800 cfm	"	5.000
1000 cfm	"	5.714
1200 cfm	"	6.667
2000 cfm	"	8.889

AIR HANDLING

15855.10 AIR HANDLING UNITS

	UNIT	MAN/ HOURS
Air handling unit, medium pressure, single zone		
1500 cfm	EA.	5.000
3000 cfm	"	8.889
4000 cfm	"	10.000
5000 cfm	"	10.667
6000 cfm	"	11.429
7000 cfm	"	12.308
8500 cfm	"	13.333
10,500 cfm	"	16.000
12,500 cfm	"	17.778
15,500 cfm	"	22.857
17,500 cfm	"	26.667
20,500 cfm	"	32.000
25,000 cfm	"	40.000
31,500 cfm	"	53.333
Rooftop air handling units		
4950 cfm	EA.	8.889
7370 cfm	"	11.429
9790 cfm	"	13.333
14,300 cfm	"	11.429
21,725 cfm	"	11.429
33,000 cfm	"	13.333

15870.20 EXHAUST FANS

	UNIT	MAN/ HOURS
Belt drive roof exhaust fans		
640 cfm, 2618 fpm	EA.	1.000
940 cfm, 2604 fpm	"	1.000
1050 cfm, 3325 fpm	"	1.000
1170 cfm, 2373 fpm	"	1.000
2440 cfm, 4501 fpm	"	1.000
2760 cfm, 4950 fpm	"	1.000
3890 cfm, 6769 fpm	"	1.000
2380 cfm, 3382 fpm	"	1.000
2880 cfm, 3859 fpm	"	1.000
3200 cfm, 4173 fpm	"	1.333
3660 cfm, 3437 fpm	"	1.333
4070 cfm, 3694 fpm	"	1.333
5030 cfm, 3251 fpm	"	1.333
5830 cfm, 6932 fpm	"	1.600
6380 cfm, 3817 fpm	"	1.600
8460 cfm, 6721 fpm	"	1.600
10,970 cfm, 5906 fpm	"	2.000
12,470 cfm, 6620 fpm	"	2.667
7000 cfm, 3449 fpm	"	2.000
13,000 cfm, 5456 fpm	"	2.000
11,250 cfm, 4854 fpm	"	2.000
18,490 cfm, 7405 fpm	"	3.636
11,300 cfm, 3232 fpm	"	3.478
18,330 cfm, 4488 fpm	"	3.478

15 MECHANICAL

AIR HANDLING	UNIT	MAN/ HOURS
15870.20 EXHAUST FANS		
21,720 cfm, 5131 fpm	EA.	3.478
31,110 cfm, 6965 fpm	"	4.000
Direct drive fans		
60 to 390 cfm	EA.	1.000
145 to 590 cfm	"	1.000
295 to 860 cfm	"	1.000
235 to 1300 cfm	"	1.000
415 to 1630 cfm	"	1.000
590 to 2045 cfm	"	1.000
805 cfm, 3235 fpm	"	1.000
1455 cfm, 4360 fpm	"	1.000
1385 cfm, 3655 fpm	"	1.000
2260 cfm, 4930 fpm	"	1.000
1720 cfm, 3870 fpm	"	1.000
2700 cfm, 5220 fpm	"	1.000
Terminal blenders and cooling		
400 cfm	EA.	1.600
800 cfm	"	1.600
1200 cfm	"	2.000
2000 cfm	"	2.000

AIR DISTRIBUTION	UNIT	MAN/ HOURS
15890.10 METAL DUCTWORK		
Rectangular duct		
Galvanized steel		
Minimum	Lb.	0.073
Average	"	0.089
Maximum	"	0.133
Aluminum		
Minimum	Lb.	0.160
Average	"	0.200
Maximum	"	0.267
Fittings		
Minimum	EA.	0.267
Average	"	0.400
Maximum	"	0.800

AIR DISTRIBUTION	UNIT	MAN/ HOURS
15890.30 FLEXIBLE DUCTWORK		
Flexible duct, 1.25" fiberglass		
5" dia.	L.F.	0.040
6" dia.	"	0.044
7" dia.	"	0.047
8" dia.	"	0.050
10" dia.	"	0.057
12" dia.	"	0.062
14" dia.	"	0.067
16" dia.	"	0.073
Flexible duct connector, 3" wide fabric	"	0.133
15895.10 ROOF CURBS		
8" high, insulated, with liner and raised can		
15" x 15"	EA.	0.400
17" x 17"	"	0.400
19" x 19"	"	0.400
21" x 21"	"	0.400
25" x 25"	"	0.500
28" x 28"	"	0.533
32" x 32"	"	0.571
36" x 36"	"	0.571
40" x 40"	"	0.571
44" x 44"	"	0.615
48" x 48"	"	0.615
52" x 52"	"	0.667
56" x 56"	"	0.667
60" x 60"	"	0.800
64" x 64"	"	0.800
68" x 68"	"	0.889
72" x 72"	"	1.000
15910.10 DAMPERS		
Horizontal parallel aluminum backdraft damper		
12" x 12"	EA.	0.200
16" x 16"	"	0.229
20" x 20"	"	0.286
24" x 24"	"	0.400
28" x 28"	"	0.444
32" x 32"	"	0.500
36" x 36"	"	0.571
40" x 40"	"	0.667
44" x 44"	"	0.727
48" x 48"	"	0.800
"Up", parallel dampers		
12" x 12"	EA.	0.200
16" x 16"	"	0.229
20" x 20"	"	0.286
24" x 24"	"	0.400
28" x 28"	"	0.444

15 MECHANICAL

AIR DISTRIBUTION	UNIT	MAN/ HOURS
15910.10 DAMPERS		
32" x 32"	EA.	0.500
36" x 36"	"	0.571
40" x 40"	"	0.667
44" x 44"	"	0.727
48" x 48"	"	0.800
"Down", parallel dampers		
12" x 12"	EA.	0.200
16" x 16"	"	0.229
20" x 20"	"	0.286
24" x 24"	"	0.400
28" x 28"	"	0.444
32" x 32"	"	0.500
36" x 36"	"	0.571
40" x 40"	"	0.667
44" x 44"	"	0.727
48" x 48"	"	0.800
Fire damper, 1.5 hr rating		
12" x 12"	EA.	0.400
16" x 16"	"	0.400
20" x 20"	"	0.400
24" x 24"	"	0.400
28" x 28"	"	0.571
32" x 32"	"	0.667
36" x 36"	"	0.800
40" x 40"	"	0.889
44" x 44"	"	1.000
48" x 48"	"	1.143
15940.10 DIFFUSERS		
Ceiling diffusers, round, baked enamel finish		
6" dia.	EA.	0.267
8" dia.	"	0.333
10" dia.	"	0.333
12" dia.	"	0.333
14" dia.	"	0.364
16" dia.	"	0.364
18" dia.	"	0.400
20" dia.	"	0.400
Rectangular		
6x6"	EA.	0.267
9x9"	"	0.400
12x12"	"	0.400
15x15"	"	0.400
18x18"	"	0.400
21x21"	"	0.500
24x24"	"	0.500
Lay in, flush mounted, perforated face, with grid		
6x6/24x24	EA.	0.320

AIR DISTRIBUTION	UNIT	MAN/ HOURS
15940.10 DIFFUSERS		
8x8/24x24	EA.	0.320
9x9/24x24	"	0.320
10x10/24x24	"	0.320
12x12/24x24	"	0.320
15x15/24x24	"	0.320
18x6/24x24	"	0.320
18x18/24x24	"	0.320
Two-way slot diffuser with balancing damper, 4'	"	0.800
15940.20 RELIEF VENTILATORS		
Intake ventilator, aluminum, with screen, no curbs		
12" x 12"	EA.	0.667
16" x 16"	"	0.800
20" x 20"	"	0.800
30" x 30"	"	1.143
36" x 36"	"	1.333
42" x 42"	"	1.333
48" x 48"	"	1.600
15940.40 REGISTERS AND GRILLES		
Lay in flush mounted, perforated face, return		
6x6/24x24	EA.	0.320
8x8/24x24	"	0.320
9x9/24x24	"	0.320
10x10/24x24	"	0.320
12x12/24x24	"	0.320
Rectangular, ceiling return, single deflection		
10x10	EA.	0.400
12x12	"	0.400
14x14	"	0.400
16x8	"	0.400
16x16	"	0.400
18x8	"	0.400
20x20	"	0.400
24x12	"	0.400
24x18	"	0.400
36x24	"	0.444
36x30	"	0.444
Wall, return air register		
12x12	EA.	0.200
16x16	"	0.200
18x18	"	0.200
20x20	"	0.200
24x24	"	0.200
Ceiling, return air grille		
6x6	EA.	0.267
8x8	"	0.320
10x10	"	0.320
Ceiling, exhaust grille, aluminum egg crate		

AIR DISTRIBUTION	UNIT	MAN/ HOURS
15940.40 REGISTERS AND GRILLES		
6x6	EA.	0.267
8x8	"	0.320
10x10	"	0.320
12x12	"	0.400
14x14	"	0.400
16x16	"	0.400
18x18	"	0.400
15940.80 PENTHOUSE LOUVERS		
Penthouse louvers		
12" high, extruded aluminum, 4" louver		
6' perimeter	EA.	2.000
8' perimeter	"	2.000
10' perimeter	"	2.000
12' perimeter	"	2.000
14' perimeter	"	2.667
16' perimeter	"	3.200
18' perimeter	"	4.444
20' perimeter	"	5.333
16" high x 4' perimeter	"	2.000
6' perimeter	"	2.000
8' perimeter	"	2.000
10' perimeter	"	2.000
12' perimeter	"	2.000
14' perimeter	"	2.667
16' perimeter	"	3.200
18' perimeter	"	4.444
20' perimeter	"	5.333
22' perimeter	"	6.667
24' perimeter	"	8.889
20" high x 4' perimeter	"	2.000
6' perimeter	"	2.000
8' perimeter	"	2.000
10' perimeter	"	2.000
12' perimeter	"	2.000
14' perimeter	"	2.667
16' perimeter	"	3.200
18' perimeter	"	4.444
20' perimeter	"	5.333
22' perimeter	"	6.667
24' perimeter	"	8.889
24" high x 4' perimeter	"	2.000
6' perimeter	"	2.000
8' perimeter	"	2.000
10' perimeter	"	2.000
12' perimeter	"	2.000
16' perimeter	"	3.200
18' perimeter	"	4.444

AIR DISTRIBUTION	UNIT	MAN/ HOURS
15940.80 PENTHOUSE LOUVERS		
20' perimeter	EA.	5.333
22' perimeter	"	6.667
24' perimeter	"	8.889

CONTROLS	UNIT	MAN/ HOURS
15950.10 HVAC CONTROLS		
Pressure gauge, direct reading gage cock and syphon	EA.	0.500
Control valve, 1", modulating		
2-way	EA.	0.667
3-way	"	1.000
Self contained control valve with sensing element, 3/4"	"	0.500
Control dampers, round		
6" dia.	EA.	0.320
8" dia	"	0.320
10" dia	"	0.320
12" dia	"	0.320
12" dia	"	0.400
18" dia	"	0.400
20" dia	"	0.400
Rectangular, parallel blade standard leakage		
12" x 12"	EA.	0.400
16" x 16"	"	0.400
20" x 20"	"	0.400
28" x 28"	"	0.500
32" x 32"	"	0.500
36" x 36"	"	0.667
40" x 40"	"	0.800
44" x 44"	"	1.000
48" x 48"	"	1.143
48" x 52"	"	1.333
48" x 56"	"	1.333
48" x 60"	"	1.333
48" x 64"	"	1.333
48" x 68"	"	1.333
48" x 72"	"	1.333
Low leakage		
12" x 12"	EA.	0.400
16" x 16"	"	0.400
20" x 20"	"	0.400

CONTROLS	UNIT	MAN/ HOURS
15950.10 HVAC CONTROLS		
24" x 24"	EA.	0.400
28" x 28"	"	0.500
32" x 32"	"	0.571
36" x 36"	"	0.667
40" x 40"	"	0.800
44" x 44"	"	1.000
48" x 48"	"	1.143
48" x 56"	"	1.333
48" x 60"	"	1.333
48" x 64"	"	1.333
48" x 68"	"	1.333
48" x 72"	"	1.333
Rectangular, opposed horizontal blade		
12" x 12"	EA.	0.400
16" x 16"	"	0.400
20" x 20"	"	0.400
24" x 24"	"	0.400
28" x 28"	"	0.500
32" x 32"	"	0.533
36" x 36"	"	0.667
40" x 40"	"	0.800
44" x 44"	"	1.000
48" x 48"	"	1.143
48" x 52"	"	1.143
48" x 56"	"	1.333
48" x 60"	"	1.333
48" x 64"	"	1.333
48" x 68"	"	1.333
48" x 72"	"	1.333

16 ELECTRICAL

BASIC MATERIALS	UNIT	MAN/ HOURS
16050.30 BUS DUCT		
Bus duct, 100a, plug-in		
10', 600v	EA.	2.759
With ground	"	4.211
10', 277/480v	"	2.759
With ground	"	4.211
Cable tap box	"	2.500
End closure	"	0.400
Edgewise hanger	"	0.727
Flatwise hanger	"	0.727
Outside elbow	"	0.800
Inside elbow	"	0.800
Outside tee	"	1.100
Inside tee	"	1.100
Outlet cover	"	0.400
Wall flange	"	0.400
Circuit breakers, with enclosure		
1 pole		
15a-60a	EA.	1.000
70a-100a	"	1.250
2 pole		
15a-60a	EA.	1.100
70a-100a	"	1.301
3 pole		
15a-60a	EA.	1.159
70a-100a	"	1.509
Bus duct, copper feeder duct, 277/480v, 4 wire		
800a	L.F.	0.400
1000a	"	0.500
1200a	"	0.533
1350a	"	0.615
1600a	"	0.727
2000a	"	0.800
2500a	"	0.851
3000a	"	0.952
Weatherproof		
800a	L.F.	0.444
1000a	"	0.533
1350a	"	0.667
1600a	"	0.727
2000a	"	0.833
2500a	"	0.899
3000a	"	0.976
4000a	"	1.509
5000a	"	1.818
Plug-in feeder duct, 277/480v, 4 wire		
400a	L.F.	0.400
600a	"	0.444
800a	"	0.500
1000a	"	0.500

BASIC MATERIALS	UNIT	MAN/ HOURS
16050.30 BUS DUCT		
1200a	L.F.	0.533
1350a	"	0.615
1600a	"	0.727
2000a	"	0.800
2500a	"	0.851
3000a	"	0.952
Copper flanged ends, 277/480v, 4 wire		
225a	EA.	2.500
400a	"	2.759
600a	"	2.963
800a	"	3.077
1000a	"	3.200
1200a	"	3.200
1350a	"	3.333
1600a	"	3.478
2000a	"	3.478
2500a	"	3.636
3000a	"	3.810
4000a	"	4.444
5000a	"	4.706
Bus duct, copper elbows, 277/480v-4w		
225a-1000a	EA.	2.105
1200a-3000a	"	2.500
4000a-5000a	"	2.963
Tees, 277/480v-4w		
225a-1000a	EA.	2.222
1200a-3000a	"	2.581
4000a-5000a	"	2.963
Crosses, 277/480v-4w		
225a-1000a	EA.	2.222
1200a-3000a	"	2.581
4000a-5000a	"	2.963
Copper end closures, 277/480v-4w		
225a-1000a	EA.	0.899
1200a-3000a	"	1.194
4000a-5000a	"	1.667
Tap boxes, 277/480v-4w		
225a	EA.	3.478
400a	"	4.444
600a	"	7.273
800a	"	8.000
1000a	"	10.000
1200a	"	11.004
1350a	"	13.008
1600a	"	14.011
2000a	"	16.985
2500a	"	22.989
3000a	"	27.972
4000a	"	37.915

16 ELECTRICAL

BASIC MATERIALS	UNIT	MAN/ HOURS
16050.30 BUS DUCT		
5000a	EA.	44.944
Circuit breaker, adapter cubicle		
225a	EA.	1.509
200a	"	1.600
600a	"	1.702
800a	"	1.818
1000a	"	1.905
1200a	"	2.000
1600a	"	2.105
2000a	"	2.222
Transformer taps, 1 phase 277/480v		
600a	EA.	7.273
800a	"	8.000
1000a	"	10.000
1200a	"	11.004
1350a	"	13.008
1600a	"	14.011
2000a	"	16.985
2500a	"	22.989
3000a	"	27.972
4000a	"	37.915
5000a	"	45.977
3 phase, 227/480v, 3 wire		
600a	EA.	7.273
800a	"	8.000
1000a	"	10.000
1200a	"	11.004
1350a	"	13.008
1600a	"	14.011
2000a	"	16.985
2500a	"	22.989
3000a	"	27.972
4000a	"	37.915
5000a	"	45.977
3 phase, 4 wire, 277/480v		
600a	EA.	7.273
800a	"	8.000
1000a	"	10.000
1200a	"	11.004
1350a	"	13.008
1600a	"	14.011
2000a	"	16.985
2500a	"	22.989
3000a	"	27.972
4000a	"	
5000a	"	45.977
Transformer connection, 4 wire, 277/480v		
600a	EA.	2.759
800a	"	2.857

BASIC MATERIALS	UNIT	MAN/ HOURS
16050.30 BUS DUCT		
1000a	EA.	2.963
1200a	"	3.077
1350a	"	3.200
1600a	"	3.333
2000a	"	3.478
2500a	"	3.636
3000a	"	3.810
4000a	"	4.444
5000a	"	4.706
Unfused reducers, 4 wire, 277/480v		
400a	EA.	2.500
600a	"	3.810
800a	"	4.706
1000a	"	5.000
1200a	"	5.333
1350a	"	5.714
1600a	"	6.154
2000a	"	6.154
2500a	"	6.667
3000a	"	7.273
4000a	"	8.753
5000a	"	10.796
400a	"	2.759
600a	"	4.000
800a	"	5.000
1000a	"	5.333
1200a	"	5.714
1350a	"	6.154
1600a	"	6.667
2000a	"	7.273
2500a	"	7.273
3000a	"	8.247
4000a	"	9.744
5000a	"	11.994
Circuit breaker reducers, 4 wire, 277/480v		
400a	EA.	2.222
600a	"	3.478
800a	"	4.211
1000a	"	4.706
1350a	"	5.333
1600a	"	5.714
2000a	"	6.154
2500a	"	6.667
3000a	"	5.714
4000a	"	7.273
5000a	"	10.000
Expansion fittings, 4 wire, 277/480v		
225a	EA.	2.500
400a	"	3.810

16 ELECTRICAL

BASIC MATERIALS	UNIT	MAN/ HOURS
16050.30 BUS DUCT		
600a	EA.	4.706
800a	"	5.000
1000a	"	5.333
1200a	"	5.714
1350a	"	6.154
1600a	"	6.154
2000a	"	6.667
2500a	"	7.273
3000a	"	8.753
4000a	"	10.796
5000a	"	11.994
Wall flanges		
225a-2500a	EA.	4.000
3000a-5000a	"	6.154
Weather seals	"	1.000
Roof flanges	"	4.000
Fire barriers	"	1.509
Spring hangers	"	1.739
Sway brace collars	"	1.250
Hook sticks		
8'	EA.	
14'	"	
Fusible switches, 240v, 3 phase		
30a	EA.	1.000
60a	"	1.250
100a	"	1.509
200a	"	2.105
400a	"	4.000
600a	"	6.154
208v, 4 wire		
30a	EA.	1.194
60a	"	1.356
100a	"	1.818
200a	"	2.759
400a	"	5.000
600a	"	8.000
600v		
30a	EA.	1.000
60a	"	1.250
100a	"	1.509
200a	"	2.105
400a	"	4.000
600a	"	6.154
800a	"	6.667
1000a	"	8.000
1200a	"	11.004
1600a	"	11.994
480v, 4 wire		
30a	EA.	1.194

BASIC MATERIALS	UNIT	MAN/ HOURS
16050.30 BUS DUCT		
60a	EA.	1.356
100a	"	1.818
200a	"	2.759
400a	"	5.000
600a	"	8.000
800a	"	8.247
1000a	"	11.004
1200a	"	11.994
1600a	"	14.011
Fusible combination starters, 600v, 3 phase		
Size 0	EA.	1.290
Size 1	"	1.600
Size 2	"	1.818
Size 3	"	2.581
Circuit breaker combination starters, 600v, 3 phase		
Size 0	EA.	1.290
Size 1	"	1.600
Size 2	"	1.818
Size 3	"	2.581
Fusible combination contactors, 600v, 3 phase		
30a	EA.	1.290
60a	"	1.600
100a	"	1.818
200a	"	2.581
Circuit breaker, combination contactors, 600v, 3 phase		
30a	EA.	1.290
60a	"	1.600
100a	"	1.818
200a	"	2.581
Fusible contactor electrically held, 480v, 4 wire		
30a	EA.	1.290
60a	"	1.600
100a	"	1.818
200a	"	2.759
Mechanically held		
30a	EA.	1.290
60a	"	1.600
100a	"	1.860
200a	"	2.759
Circuit breakers, 240v, 3 phase		
15a-60a	EA.	1.159
70a-100a	"	1.600
600v, 3 phase		
15a-60a	EA.	1.159
125a-225a	"	2.286
250a-400a	"	4.211
500a-600a	"	5.333
700a-800a	"	8.000
900a-1000a	"	10.000

16 ELECTRICAL

BASIC MATERIALS

16050.30 BUS DUCT

	UNIT	MAN/HOURS
1200a-1600a	EA.	11.004
120/208v, 4 wire		
15a-60a	EA.	1.290
70a-100a	"	1.860
277/480v, 4 wire		
15a-60a	EA.	1.290
70a-100a	"	1.905
125a-225a	"	2.759
250a-400a	"	5.000
500a-600a	"	8.000
700a-800a	"	8.000
900a-1000a	"	11.004
1200a-1600a	"	14.011
600v, 3 phase, 65,000 aic.		
60a	EA.	1.159
70a-100a	"	1.600
125a-225a	"	2.286
250a-400a	"	4.211
500a-600a	"	6.154
700a-800a	"	8.000
900a-1000a	"	10.000
277/480v, 4 wire, 65,000 aic.		
15a-60a	EA.	1.290
70a-100a	"	1.905
125a-225a	"	2.759
250a-400a	"	5.000
500a-600a	"	8.000
700a-800a	"	8.000
900a-1000a	"	10.349
600v, 3 phase, current limiting		
15a-60a	EA.	1.159
70a-100a	"	1.600
125a-225a	"	2.286
250a-400a	"	4.211
500a-600a	"	6.154
700a-800a	"	8.000
900a-1000a	"	10.000
277/480v, 4 wire, current limiting		
15a-60a	EA.	1.290
70a-100a	"	40.000
125a-225a	"	2.759
250a-400a	"	5.000
500a-600a	"	8.000
700a-800a	"	8.000
900a-1000a	"	11.004
Capacitors, 3 phase, 240v		
5 kvar	EA.	5.000
7.5 kvar	"	6.154
10 kvar	"	8.097

BASIC MATERIALS

16050.30 BUS DUCT

	UNIT	MAN/HOURS
15 kvar	EA.	9.195
480v		
2.5 kvar	EA.	2.759
5 kvar	"	4.706
7.5 kvar	"	5.714
10 kvar	"	8.097
15 kvar	"	8.999
20 kvar	"	11.494
25 kvar	"	13.008
30 kvar	"	14.210
Transformers, 3 phase, 480v		
1.0 kva	EA.	1.739
1.5 kva	"	2.000
2 kva	"	2.500
3 kva	"	2.759
5 kva	"	4.000
7.5 kva	"	5.000
10 kva	"	5.333

16110.12 CABLE TRAY

	UNIT	MAN/HOURS
Cable tray, 6"	L.F.	0.059
Ventilated cover	"	0.030
Solid cover	"	0.030
Flat 90	EA.	0.500
Outside 90	"	0.500
Inside 90	"	0.500
Flat 45	"	0.500
Outside 45	"	0.500
Inside 45	"	0.500
Adjustable elbow	"	0.500
Support riser	"	0.500
Adjustable riser	"	0.500
Tee	"	1.739
Cross	"	1.818
Blind end	"	0.290
Expansion joint	"	0.500
Box connector	"	2.500
Standard dropout	"	0.500
2"	"	0.615
3"	"	0.800
4"	"	1.000
Cable tray, 9"	L.F.	0.070
Ventilated cover	"	0.040
Solid cover	"	0.040
Flat 90	EA.	0.533
Outside 90	"	0.533
Inside 90	"	0.533
Flat 45	"	0.533
Outside 45	"	0.533

BASIC MATERIALS	UNIT	MAN/ HOURS
16110.12 CABLE TRAY		
Inside 45	EA.	0.533
Adjustable elbow	"	0.533
Support riser	"	0.533
Adjustable riser	"	0.533
Tee	"	1.818
Cross	"	1.818
Blind end	"	0.320
Expansion joint	"	0.533
Box connector	"	2.581
Standard dropout	"	0.533
2"	"	0.667
3"	"	0.667
4"	"	1.096
Cable tray, 12"	L.F.	0.080
Ventilated cover	"	0.050
Solid cover	"	0.050
Flat 90	EA.	0.533
Outside 90	"	0.533
Inside 90	"	0.533
Flat 45	"	0.533
Outside 45	"	0.533
Inside 45	"	0.533
Adjustable elbow	"	0.533
Support riser	"	0.533
Adjustable riser	"	0.533
Tee	"	2.000
Cross	"	2.000
Blind end	"	0.348
Expansion joint	"	0.615
Box connector	"	2.759
Standard dropout	"	0.615
2"	"	0.727
3"	"	0.899
4"	"	1.159
Cable tray, 18"	L.F.	0.100
Ventilated cover	"	0.059
Solid cover	"	0.059
Flat 90	EA.	0.727
Outside 90	"	0.727
Inside 90	"	0.727
Flat 45	"	0.727
Outside 45	"	0.727
Inside 45	"	0.727
Adjustable elbow	"	0.727
Support riser	"	0.727
Adjustable riser	"	0.727
Tee	"	2.105
Cross	"	2.105
Blind end	"	0.400

BASIC MATERIALS	UNIT	MAN/ HOURS
16110.12 CABLE TRAY		
Expansion joint	EA.	0.727
Box connector	"	2.963
Standard dropout	"	0.667
2"	"	0.727
3"	"	0.952
4"	"	1.194
Cable tray, 24"	L.F.	0.123
Ventilated cover	"	0.070
Solid cover	"	0.070
Flat 90	EA.	0.727
Outside 90	"	0.727
Inside 90	"	0.727
Flat 45	"	0.727
Outside 45	"	0.727
Inside 45	"	0.727
Adjustable elbow	"	0.727
Support riser	"	0.727
Adjustable riser	"	0.727
Tee	"	2.222
Cross	"	2.222
Blind end	"	0.444
Expansion joint	"	0.727
Box connector	"	3.478
Standard dropout	"	0.667
2"	"	0.800
3"	"	0.976
4"	"	1.250
Cable tray, 36"	L.F.	0.145
Ventilated cover	"	0.080
Solid cover	"	0.080
Flat 90	EA.	0.851
Outside 90	"	0.851
Inside 90	"	0.851
Flat 45	"	0.851
Outside 45	"	0.851
Inside 45	"	0.851
Adjustable elbow	"	0.851
Support riser	"	0.851
Adjustable riser	"	0.851
Tee	"	2.286
Cross	"	2.963
Blind end	"	0.471
Expansion joint	"	0.727
Box connector	"	3.810
Standard dropout	"	0.727
2"	"	0.800
3"	"	1.000
4"	"	1.290
Reducers		

BASIC MATERIALS	UNIT	MAN/ HOURS
16110.12 CABLE TRAY		
9" - 6"	EA.	0.500
12" - 9"	"	0.500
18" - 12"	"	0.615
24" - 18"	"	0.727
36" - 18"	"	0.800
36" - 24"	"	0.899
Conduit dropouts		
3/4"	EA.	0.348
1"	"	0.348
1-1/4"	"	0.400
1-1/2"	"	0.500
2"	"	0.533
2-1/2"	"	0.727
3"	"	0.800
Wall brackets		
6"	EA.	0.145
9"	"	0.145
12"	"	0.200
18"	"	0.250
24"	"	0.296
36"	"	0.400
16110.15 FIBERGLASS CABLE TRAY		
Fiberglass cable tray, 6"	L.F.	0.040
Tray cover	"	0.030
Horizontal		
90	EA.	0.276
45	"	0.276
30	"	0.276
Inside		
90	EA.	0.276
45	"	0.276
30	"	0.276
Horizontal tee	"	0.727
Horizontal cross	"	1.096
Splice plate	"	0.145
Floor flange	"	0.348
Panel flange	"	1.250
End plate	"	0.145
Nylon rivet	"	0.050
Barrier strip	"	0.050
Hold down clamp	"	0.050
Drop out	"	0.533
Cover stand off	"	0.050
Wall bracket	"	0.200
Sealer	"	
Outside		
90	EA.	0.276

BASIC MATERIALS	UNIT	MAN/ HOURS
16110.15 FIBERGLASS CABLE TRAY		
45	EA.	0.276
30	"	0.276
Fiberglass cable tray, 9"	L.F.	0.050
Tray cover	"	0.040
Horizontal		
90	EA.	0.276
45	"	0.276
30	"	0.276
Inside		
90	EA.	0.276
45	"	0.276
30	"	0.276
Horizontal tee	"	0.727
Horizontal cross	"	1.096
Splice plate	"	0.145
Floor flange	"	0.348
Panel flange	"	1.250
End plate	"	0.145
Nylon rivet	"	0.050
Barrier strap	"	0.050
Hold down clamp	"	0.050
Drop out	"	0.533
Cover stand off	"	0.050
Wall bracket	"	0.200
Sealer	"	
Outside		
90	EA.	0.276
45	"	0.276
30	"	0.276
Fiberglass cable tray, 12"	L.F.	0.059
Tray cover	"	0.050
Horizontal		
90	EA.	0.348
45	"	0.348
30	"	0.348
Inside		
90	EA.	0.348
45	"	0.348
Inside 30	"	0.348
Horizontal tee	"	0.952
Horizontal cross	"	1.290
Splice plate	"	0.160
Floor flange	"	0.364
Panel flange	"	1.600
End plate	"	0.160
Nylon rivet	"	0.050
Barrier strip	"	0.050
Hold down clamp	"	0.050
Drop out	"	0.533

BASIC MATERIALS	UNIT	MAN/ HOURS
16110.15 FIBERGLASS CABLE TRAY		
Cover stand off	EA.	0.050
Wall bracket	"	0.200
Sealer	"	
Outside		
90	EA.	0.348
45	"	0.348
30	"	0.348
Fiberglass cable tray, 18"	L.F.	0.070
Tray cover	"	0.059
Horizontal		
90	EA.	0.400
45	"	0.400
30	"	0.400
Inside		
90	EA.	0.400
45	"	0.400
30	"	0.400
Horizontal tee	"	1.538
Horizontal cross	"	1.356
Splice plate	"	0.160
Floor flange	"	0.381
Panel flange	"	1.702
End plate	"	0.170
Nylon rivet	"	0.050
Barrier strip	"	0.050
Hold down clamp	"	0.050
Drop out	"	0.533
Cover stand off	"	0.050
Wall bracket	"	0.200
Sealer	"	
Outside		
90	EA.	0.400
45	"	0.400
30	"	0.400
Fiberglass cable tray, 24"	L.F.	0.080
Tray cover	"	0.070
Horizontal		
90	EA.	0.500
45	"	0.500
30	"	0.500
Inside		
90	EA.	0.500
45	"	0.500
30	"	0.500
Horizontal tee	"	1.000
Horizontal cross	"	1.509
Splice plate	"	0.160
Floor flange	"	0.400
Panel flange	"	1.818

BASIC MATERIALS	UNIT	MAN/ HOURS
16110.15 FIBERGLASS CABLE TRAY		
End plate	EA.	0.170
Nylon Rivet	"	0.050
Barrier strip	"	0.050
Hold down clamp	"	0.050
Drop out	"	0.533
Cover stand off	"	0.050
Wall bracket	"	0.200
Sealer	"	
Outside		
90	EA.	0.500
45	"	0.500
30	"	0.500
Fiberglass cable tray, 30"	L.F.	0.089
Tray cover	"	0.080
Horizontal		
90	EA.	0.615
45	"	0.615
30	"	0.615
Inside		
90	EA.	0.615
45	"	0.615
30	"	0.615
Horizontal tee	"	1.096
Horizontal cross	"	1.600
Splice plate	"	0.160
Floor flange	"	0.400
Panel flange	"	1.905
End plate	"	0.200
Nylon rivet	"	0.050
Barrier strip	"	0.050
Hold down clamp	"	0.050
Dropout	"	0.615
Cover stand off	"	0.050
Wall bracket	"	0.200
Sealer	"	
Outside		
90	EA.	0.615
45	"	0.615
30	"	0.615
Fibeglass cable tray, 36"	L.F.	0.100
Tray cover	"	0.089
Horizontal		
90	EA.	0.667
45	"	0.667
30	"	0.667
Inside		
90	EA.	0.667
45	"	0.667
30	"	0.667

BASIC MATERIALS	UNIT	MAN/ HOURS
16110.15 FIBERGLASS CABLE TRAY		
Horizontal tee	EA.	1.194
horizontal cross	"	1.702
Splice plate	"	0.160
Floor flange	"	0.444
Panel flange	"	2.105
End plate	"	0.200
Nylon rivet	"	0.050
Barrier strip	"	0.050
Hold down clamp	"	0.050
Drop out	"	0.667
5Cover stand off	"	0.050
Wall bracket	"	0.200
Sealer	"	
Outside		
90	EA.	0.667
45	"	0.667
30	"	0.667
Reducers		
12" - 6"	EA.	0.296
12" - 9"	"	2.963
18" - 6"	"	0.348
18" - 9"	"	0.348
18" - 12"	"	0.400
24" - 6"	"	0.400
24" - 9"	"	0.400
24" - 12"	"	0.400
24" - 18"	"	0.444
30" - 9"	"	0.400
30" - 12"	"	0.400
30" - 18"	"	0.444
30" - 24"	"	0.500
36" - 9"	"	0.500
36" - 12"	"	0.500
36" - 18"	"	0.533
36" - 24"	"	0.615
36" - 30"	"	0.727
16110.20 CONDUIT SPECIALTIES		
Rod beam clamp, 1/2"	EA.	0.050
Hanger rod		
3/8"	L.F.	0.040
1/2"	"	0.050
All thread rod		
1/4"	L.F.	0.030
3/8"	"	0.040
1/2"	"	0.050
5/8"	"	0.080
Hanger channel, 1-1/2"		
No holes	EA.	0.030

BASIC MATERIALS	UNIT	MAN/ HOURS
16110.20 CONDUIT SPECIALTIES		
Holes	EA.	0.030
Channel strap		
1/2"	EA.	0.050
3/4"	"	0.050
1"	"	0.050
1-1/4"	"	0.080
1-1/2"	"	0.080
2"	"	0.080
2-1/2"	"	0.123
3"	"	0.123
3-1/2"	"	0.123
4"	"	0.145
5"	"	0.145
6"	"	0.145
Conduit penetrations, roof and wall, 8" thick		
1/2"	EA.	0.615
3/4"	"	0.615
1"	"	0.800
1-1/4"	"	0.800
1-1/2"	"	0.800
2"	"	1.600
2-1/2"	"	1.600
3"	"	1.600
3-1/2"	"	2.000
4"	"	2.000
Plastic duct bank conduit spacer, 3" separation		
2"	EA.	0.050
3"	"	0.050
4"	"	0.050
5"	"	0.050
6"	"	0.050
Intermediate, 3" separation		
2"	EA.	0.050
3"	"	0.050
4"	"	0.050
5"	"	0.050
6"	"	0.050
Base with 1-1/2" separation		
2"	EA.	0.050
3"	"	0.160
4"	"	0.160
5"	"	0.160
6"	"	0.160
Intermediate, 1-1/2" separation		
2"	EA.	0.160
3"	"	0.160
3-1/2"	"	0.160
4"	"	0.160
5"	"	0.160

BASIC MATERIALS	UNIT	MAN/ HOURS
16110.20 CONDUIT SPECIALTIES		
6"	EA.	0.160
OD beam clamp, 1/4"	"	0.200
Threaded rod couplings		
1/4"	EA.	0.050
3/8"	"	0.050
1/2"	"	0.050
5/8"	"	0.050
3/4"	"	0.050
Hex nuts		
1/4"	EA.	0.050
3/8"	"	0.050
1/2"	"	0.050
5/8"	"	0.050
3/4"	"	0.050
Square nuts		
1/4"	EA.	0.050
3/8"	"	0.050
3/8"	"	0.050
5/8"	"	0.050
3/4"	"	0.050
Flat washers		
1/4"	EA.	
3/8"	"	
1/2"	"	
5/8"	"	
3/4"	"	
Lockwashers		
1/4"	EA.	
3/8"	"	
1/2"	"	
5/8"	"	
3/4"	"	
Channel closure strip	L.F.	0.133
Channel end cap	EA.	0.133
Li-channel trapeze hangers		
12" long	EA.	0.145
18" long	"	0.145
24" long	"	0.145
30" long	"	0.250
36" long	"	0.250
42" long	"	0.296
Channel spring nuts		
1/4"	EA.	0.059
3/8"	"	0.080
1/2"	"	0.100
Fireproofing, for conduit penetrations		
1/2"	EA.	0.500
3/4"	"	0.500
1"	"	0.500

BASIC MATERIALS	UNIT	MAN/ HOURS
16110.20 CONDUIT SPECIALTIES		
1-1/4"	EA.	0.727
1-1/2"	"	0.727
2"	"	0.727
2-1/2"	"	0.899
3"	"	0.899
3-1/2"	"	1.250
4"	"	1.509
16110.21 ALUMINUM CONDUIT		
Aluminum conduit		
1/2"	L.F.	0.030
3/4"	"	0.040
1"	"	0.050
1-1/4"	"	0.059
1-1/2"	"	0.080
2"	"	0.089
2-1/2"	"	0.100
3"	"	0.107
3-1/2"	"	0.123
4"	"	0.145
5"	"	0.182
6"	"	0.200
90 deg. elbow		
1/2"	EA.	0.190
3/4"	"	0.250
1"	"	0.308
1-1/4"	"	0.381
1-1/2"	"	0.400
2"	"	0.444
2-1/2"	"	0.571
3"	"	0.667
3-1/2"	"	0.800
4"	"	0.889
5"	"	1.143
6"	"	2.222
Coupling		
1/2"	EA.	0.050
3/4"	"	0.059
1"	"	0.080
1-1/4"	"	0.089
1-1/2"	"	0.100
2"	"	0.107
2-1/2"	"	0.123
3"	"	0.123
3-1/2"	"	0.145
4"	"	0.160
5"	"	0.160
6"	"	0.190

BASIC MATERIALS	UNIT	MAN/ HOURS
16110.22 EMT CONDUIT		
EMT conduit		
1/2"	L.F.	0.030
3/4"	"	0.040
1"	"	0.050
1-1/4"	"	0.059
1-1/2"	"	0.080
2"	"	0.089
2-1/2"	"	0.100
3"	"	0.123
3-1/2"	"	0.145
4"	"	0.182
90 deg. elbow		
1/2"	EA.	0.089
3/4"	"	0.100
1"	"	0.107
1-1/4"	"	0.123
1-1/2"	"	0.145
2"	"	0.190
2-1/2"	"	0.211
3"	"	0.242
3-1/2"	"	0.258
4"	"	0.286
Connector, steel compression		
1/2"	EA.	0.089
3/4"	"	0.089
1"	"	0.089
1-1/4"	"	0.107
1-1/2"	"	0.145
2"	"	0.190
2-1/2"	"	0.250
3"	"	0.286
3-1/2"	"	0.308
4"	"	0.333
Coupling, steel, compression		
1/2"	EA.	0.059
3/4"	"	0.059
1"	"	0.059
1-1/4"	"	0.089
1-1/2"	"	0.107
2"	"	0.145
2-1/2"	"	0.222
3"	"	0.250
3-1/2"	"	0.286
4"	"	0.308
1 hole strap, steel		
1/2"	EA.	0.040
3/4"	"	0.040
1"	"	0.040
1-1/4"	"	0.050

BASIC MATERIALS	UNIT	MAN/ HOURS
16110.22 EMT CONDUIT		
1-1/2"	EA.	0.050
2"	"	0.050
2-1/2"	"	0.059
3"	"	0.059
3-1/2"	"	0.059
4"	"	0.059
Connector, steel set screw		
1/2"	EA.	114
3/4"	"	0.070
1"	"	0.070
1-1/4"	"	0.107
1-1/2"	"	0.145
2"	"	0.182
2-1/2"	"	0.242
3"	"	0.267
3-1/2"	"	0.296
4"	"	0.348
Insulated throat		
1/2"	EA.	0.070
3/4"	"	0.070
1"	"	0.070
1-1/4"	"	0.107
1-1/2"	"	0.145
2"	"	0.182
2-1/2"	"	0.242
3"	"	0.267
3-1/2"	"	0.296
4"	"	0.348
Connector, die cast set screw		
1/2"	EA.	0.059
3/4"	"	0.059
1"	"	0.059
1-1/4"	"	0.089
1-1/2"	"	0.107
2"	"	0.145
2-1/2"	"	0.200
3"	"	0.222
3-1/2"	"	0.250
4"	"	0.286
Insulated throat		
1/2"	EA.	0.059
3/4"	"	0.059
1"	"	
1-1/4"	"	0.089
1-1/2"	"	0.107
2"	"	0.145
2-1/2"	"	0.200
3"	"	0.222
3-1/2"	"	0.250

BASIC MATERIALS	UNIT	MAN/ HOURS
16110.22 EMT CONDUIT		
4"	EA.	0.286
Coupling, steel set screw		
1/2"	EA.	0.040
3/4"	"	0.040
1"	"	0.040
1-1/4"	"	0.050
1-1/2"	"	0.080
2"	"	0.107
2-1/2"	"	0.160
3"	"	0.190
3-1/2"	"	0.222
4"	"	0.250
Diecast set screw		
1/2"	EA.	0.040
3/4"	"	0.040
1"	"	0.040
1-1/4"	"	0.050
1-1/2"	"	0.080
2"	"	0.107
2-1/2"	"	0.160
3"	"	0.186
3-1/2"	"	0.222
4"	"	0.250
1 hole malleable straps		
1/2"	EA.	0.040
3/4"	"	0.040
1"	"	0.040
1-1/4"	"	0.050
1-1/2"	"	0.050
2"	"	0.050
2-1/2"	"	0.059
3"	"	0.059
3-1/2"	"	0.059
4"	"	0.059
EMT to rigid compression coupling		
1/2"	EA.	0.100
3/4"	"	0.100
1"	"	0.150
Set screw couplings		
1/2"	EA.	0.100
3/4"	"	0.100
1"	"	0.145
Set screw offset connectors		
1/2"	EA.	0.100
3/4"	"	0.100
1"	"	0.145
Compression offset connectors		
1/2"	EA.	0.100
3/4"	"	0.100

BASIC MATERIALS	UNIT	MAN/ HOURS
16110.22 EMT CONDUIT		
1"	EA.	0.145
Type "LB" set screw condulets		
1/2"	EA.	0.229
3/4"	"	0.381
1"	"	0.381
1-1/4"	"	0.444
1-1/2"	"	0.533
2"	"	0.615
2-1/2"	"	0.727
3"	"	1.000
3-1/2"	"	1.333
4"	"	1.600
Type "T" set screw condulets		
1/2"	EA.	0.296
3/4"	"	0.444
1"	"	0.444
1-1/4"	"	0.533
1-1/2"	"	0.615
2"	"	0.667
Type "C" set screw condulets		
1/2"	EA.	0.250
3/4"	"	0.296
1"	"	0.381
1-1/4"	"	0.444
1-1/2"	"	0.533
2"	"	0.381
Type "LL" setscrew condulets		
1/2"	EA.	0.250
3/4"	"	0.296
1"	"	0.381
1-1/4"	"	0.444
1-1/2"	"	0.533
2"	"	0.615
Type "LR" set screw condulets		
1/2"	EA.	0.250
3/4"	"	0.296
1"	"	0.381
1-1/4"	"	0.444
1-1/2"	"	0.533
2"	"	0.615
Type "LB" compression condulets		
1/2"	EA.	0.296
3/4"	"	0.500
1"	"	0.500
Type "T" compression condulets		
1/2"	EA.	0.400
3/4"	"	0.444
1"	"	0.615
Condulet covers		

BASIC MATERIALS	UNIT	MAN/ HOURS
16110.22 EMT CONDUIT		
1/2"	EA.	0.123
3/4"	"	0.123
1"	"	0.123
1-1/4"	"	0.123
1-1/2"	"	0.145
2"	"	0.145
2-1/2"	"	0.145
3"	"	0.182
3-1/2"	"	0.182
4"	"	0.182
Clamp type entrance caps		
1/2"	EA.	0.250
3/4"	"	0.296
1"	"	0.023
1-1/4"	"	0.533
1-1/2"	"	0.615
2"	"	0.899
2-1/2"	"	1.000
3"	"	1.509
3-1/2"	"	1.739
4"	"	2.222
Slip fitter type entrance caps		
1/2"	EA.	0.250
3/4"	"	0.296
1"	"	0.400
1-1/4"	"	0.533
1-1/2"	"	0.615
2"	"	0.899
2-1/2"	"	1.000
3"	"	1.509
3-1/2"	"	1.739
4"	"	2.222
16110.23 FLEXIBLE CONDUIT		
Flexible conduit, steel		
3/8"	L.F.	0.030
1/2	"	0.030
3/4"	"	0.040
1"	"	0.040
1-1/4"	"	0.050
1-1/2"	"	0.059
2"	"	0.080
2-1/2"	"	0.089
3"	"	0.107
Flexible conduit, liquid tight		
3/8"	L.F.	0.030
1/2"	"	0.030
3/4"	"	0.040
1"	"	0.040

BASIC MATERIALS	UNIT	MAN/ HOURS
16110.23 FLEXIBLE CONDUIT		
1-1/4"	L.F.	0.050
1-1/2"	"	0.059
2"	EA.	0.080
2-1/2"	"	0.089
3"	"	0.107
4"	"	0.145
Connector, straight		
3/8"	EA.	0.080
1/2"	"	0.080
3/4"	"	0.089
1"	"	0.100
1-1/4"	"	0.107
1-1/2"	"	0.123
2"	"	0.145
2-1/2"	"	0.182
3"	"	0.190
Straight insulated throat connectors		
3/8"	EA.	0.123
1/2"	"	0.123
3/4"	"	0.145
1"	"	0.145
1-1/4"	"	0.182
1-1/2"	"	0.211
2"	"	0.229
2-1/2"	"	0.267
3"	"	0.333
4"	"	0.421
90 deg connectors		
3/8"	EA.	0.148
1/2"	"	0.148
3/4"	"	0.170
1"	"	0.182
1-1/4"	"	0.229
1-1/2"	"	0.250
2"	"	0.267
2-1/2"	"	0.333
3"	"	0.381
4"	"	0.444
90 degree insulated throat connectors		
3/8"	EA.	0.145
1/2"	"	0.145
3/4"	"	0.170
1"	"	0.178
1-1/4"	"	0.229
1-1/2"	"	0.250
2"	"	0.267
2-1/2"	"	0.333
3"	"	0.381
4"	"	0.444

BASIC MATERIALS	UNIT	MAN/ HOURS
16110.23 FLEXIBLE CONDUIT		
Flexible aluminum conduit		
3/8"	L.F.	0.030
1/2"	"	0.030
3/4"	"	0.040
1"	"	0.040
1-1/4"	"	0.050
1-1/2"	"	0.059
2"	"	0.080
2-1/2"	"	0.089
3"	"	0.107
3-1/2"	"	0.123
4"	"	0.145
Connector, straight		
3/8"	EA.	0.100
1/2"	"	0.100
3/4"	"	0.107
1"	"	0.123
1-1/4"	"	0.145
1-1/2"	"	0.182
2"	"	0.190
2-1/2"	"	0.222
3"	"	0.276
4"	"	0.348
Straight insulated throat connectors		
3/8"	EA.	0.089
1/2"	"	0.080
3/4"	"	0.089
1"	"	0.100
1-1/4"	"	0.107
1-1/2"	"	0.123
2"	"	0.145
2-1/2"	"	0.182
3"	"	0.190
3-1/2"	"	0.222
4"	"	0.276
90 deg connectors		
3/8"	EA.	0.145
1/2"	"	0.145
3/4"	"	0.145
1"	"	0.170
1-1/4"	"	0.182
1-1/2"	"	0.200
2"	"	0.211
2-1/2"	"	0.229
3"	"	0.267
90 deg insulated throat connectors		
3/8"	EA.	0.145
1/2"	"	0.145
3/4"	"	0.145

BASIC MATERIALS	UNIT	MAN/ HOURS
16110.23 FLEXIBLE CONDUIT		
1"	EA.	0.170
1-1/4"	"	0.182
1-1/2"	"	0.200
2"	"	0.211
2-1/2"	"	0.229
3"	"	0.267
3-1/2"	"	0.333
4"	"	0.421
3/4"	"	0.050
1"	"	0.050
1-1/4"	"	0.059
1-1/2"	"	0.070
16110.24 GALVANIZED CONDUIT		
Galvanized rigid steel conduit		
1/2"	L.F.	0.040
3/4"	"	0.050
1"	"	0.059
1-1/4"	"	0.080
1-1/2"	"	0.089
2"	"	0.100
2-1/2"	"	0.145
3"	"	0.182
3-1/2"	"	0.190
4"	"	0.211
5"	"	0.286
6"	"	0.381
90 degree ell		
1/2"	EA.	0.250
3/4"	"	0.308
1"	"	0.381
1-1/4"	"	0.444
1-1/2"	"	0.500
2"	"	0.533
2-1/2"	"	0.667
3"	"	0.889
3-1/2"	"	1.000
4"	"	1.333
5"	"	2.222
6"	"	3.333
Couplings, with set screws		
1/2"	EA.	0.050
3/4"	"	0.059
1"	"	0.080
1-1/4"	"	0.100
1-1/2"	"	0.123
2"	"	0.145
2-1/2"	"	0.190
3"	"	0.250

BASIC MATERIALS	UNIT	MAN/HOURS
16110.24 GALVANIZED CONDUIT		
3-1/2"	EA.	0.286
4"	"	0.308
5"	"	0.444
6"	"	0.500
Split couplings		
1/2"	EA.	0.190
3/4"	"	0.250
1"	"	0.276
1-1/4"	"	0.308
1-1/2"	"	0.381
2"	"	0.571
2-1/2"	"	0.571
3"	"	0.727
3-1/2"	"	1.000
4"	"	1.333
5"	"	1.633
6"	"	2.051
Erickson couplings		
1/2"	EA.	0.444
3/4"	"	0.500
1"	"	0.615
1-1/4"	"	0.889
1-1/2"	"	1.000
2"	"	1.333
2-1/2"	"	1.860
3"	"	2.105
3-1/2"	"	2.500
4"	"	2.667
5"	"	2.963
6"	"	3.200
Seal fittings		
1/2"	EA.	0.667
3/4"	"	0.800
1"	"	1.000
1-1/4"	"	1.143
1-1/2"	"	1.333
2"	"	1.600
2-1/2"	"	1.905
3"	"	2.105
3-1/2"	"	2.500
4"	"	2.963
5"	"	4.444
6"	"	5.000
Entrance fitting, (weather head), threaded		
1/2"	EA.	0.444
3/4"	"	0.500
1"	"	0.571
1-1/4"	"	0.727
1-1/2"	"	0.800

BASIC MATERIALS	UNIT	MAN/HOURS
16110.24 GALVANIZED CONDUIT		
2"	EA.	0.889
2-1/2"	"	1.000
3"	"	1.333
3-1/2"	"	1.739
4"	"	2.500
5"	"	3.478
6"	"	4.444
Locknuts		
1/2"	EA.	0.050
3/4"	"	0.050
1"	"	0.050
1-1/4"	"	0.050
1-1/2"	"	0.059
2"	"	0.059
2-1/2"	"	0.080
3"	"	0.080
3-1/2"	"	0.080
4"	"	0.089
5"	"	0.089
6"	"	0.089
Plastic conduit bushings		
1/2"	EA.	0.123
3/4"	"	0.145
1"	"	0.190
1-1/4"	"	0.222
1-1/2"	"	0.250
2"	"	0.308
2-1/2"	"	0.500
3"	"	0.667
3-1/2"	"	0.800
4"	"	0.889
5"	"	1.143
6"	"	1.600
Conduit bushings, steel		
1/2"	EA.	0.123
3/4"	"	0.145
1"	"	0.190
1-1/4"	"	0.222
1-1/2"	"	0.250
2"	"	0.308
2-1/2"	"	0.500
3"	"	0.667
3-1/2"	"	0.800
4"	"	0.889
5"	"	1.143
6"	"	1.600
Pipe cap		
1/2"	EA.	0.050
3/4"	"	0.050

BASIC MATERIALS	UNIT	MAN/HOURS
16110.24 GALVANIZED CONDUIT		
1"	EA.	0.050
1-1/4"	"	0.080
1-1/2"	"	0.080
2"	"	0.080
2-1/2"	"	0.089
3"	"	0.089
3-1/2"	"	0.089
4"	"	0.107
5"	"	0.145
6"	"	0.200
GRS elbows, 36" radius		
2"	EA.	0.667
2-1/2"	"	0.808
3"	"	1.053
3-1/2"	"	1.250
4"	"	1.509
5"	"	2.500
6"	"	3.810
42" radius		
2"	EA.	0.808
2-1/2"	"	1.000
3"	"	1.250
3-1/2"	"	1.509
4"	"	1.739
5"	"	2.857
6"	"	4.000
48" radius		
2"	EA.	0.930
2-1/2"	"	1.127
3"	"	1.429
3-1/2"	"	1.739
4"	"	2.162
5"	"	3.077
6"	"	4.444
Threaded couplings		
1/2"	EA.	0.050
3/4"	"	0.059
1"	"	0.080
1-1/4"	"	0.089
1-1/2"	"	0.100
2"	"	0.107
2-1/2"	"	0.123
3"	"	0.145
3-1/2"	"	0.145
4"	"	0.160
5"	"	0.182
6"	"	0.190
Threadless couplings		
1/2"	EA.	0.100

BASIC MATERIALS	UNIT	MAN/HOURS
16110.24 GALVANIZED CONDUIT		
3/4"	EA.	0.123
1"	"	0.145
1-1/4"	"	0.190
1-1/2"	"	0.250
2"	"	0.308
2-1/2"	"	0.500
3"	"	0.615
3-1/2"	"	0.808
4"	"	1.000
5"	"	1.250
6"	"	5.333
Threadless connectors		
1/2"	EA.	1.000
3/4"	"	0.123
1"	"	0.145
1-1/4"	"	0.190
1-1/2"	"	0.250
2"	"	0.308
2-1/2"	"	0.500
3"	"	0.615
3-1/2"	"	0.808
4"	"	1.000
5"	"	1.250
6"	"	1.509
Setscrew connectors		
1/2"	EA.	0.080
3/4"	"	0.089
1"	"	0.100
1-1/4"	"	0.123
1-1/2"	"	0.145
2"	"	0.190
2-1/2"	"	0.250
3"	"	0.308
3-1/2"	"	0.381
4"	"	0.500
5"	"	0.615
6"	"	0.808
Clamp type entrance caps		
1/2"	EA.	0.308
3/4"	"	0.381
1"	"	0.444
1-1/4"	"	0.011
1-1/2"	"	0.615
2"	"	0.727
3-1/2"	"	0.941
3"	"	1.127
3-1/2"	"	1.379
4"	"	2.424
"LB" condulets		

BASIC MATERIALS	UNIT	MAN/ HOURS
16110.24 GALVANIZED CONDUIT		
1/2"	EA.	0.308
3/4"	"	0.381
1"	"	0.444
1-1/4"	"	0.500
1-1/2"	"	0.615
2"	"	0.727
2-1/2"	"	1.000
3"	"	1.379
3-1/2"	"	1.739
4"	"	2.105
"T" condulets		
1/2"	EA.	0.381
3/4"	"	0.444
1"	"	0.500
1-1/4"	"	0.571
1-1/2"	"	0.615
2"	"	0.727
2-1/2"	"	1.127
3"	"	1.509
3-1/2"	"	1.860
4"	"	2.222
"X" condulets		
1/2"	EA.	0.444
3/4"	"	0.500
1"	"	0.571
1-1/4"	"	0.615
1-1/2"	"	0.667
2"	"	0.879
Blank steel condulet covers		
1/2"	EA.	0.100
3/4"	"	0.100
1"	"	0.100
1-1/4"	"	0.123
1-1/2"	"	0.123
2"	"	0.123
2-1/2"	"	0.145
3"	"	0.145
3-1/2"	"	0.145
4"	"	0.200
Solid condulet gaskets		
1/2"	EA.	0.050
3/4"	"	0.050
1"	"	0.050
1-1/4"	"	0.080
1-1/2"	"	0.080
2"	"	0.080
2-1/2"	"	0.100
3"	"	0.100
3-1/2"	"	0.100

BASIC MATERIALS	UNIT	MAN/ HOURS
16110.24 GALVANIZED CONDUIT		
4"	EA.	0.145
One-hole malleable straps		
1/2"	EA.	0.040
3/4"	"	0.040
1"	"	0.040
1-1/4"	"	0.050
1-1/2"	"	0.050
2"	"	0.050
2-1/2"	"	0.059
3"	"	0.059
3-1/2"	"	0.059
4"	"	0.080
5"	"	0.080
6"	"	0.080
One-hole steel straps		
1/2"	EA.	0.040
3/4"	"	0.040
1"	"	0.040
1-1/4"	"	0.050
1-1/2"	"	0.050
2"	"	0.050
2-1/2"	"	0.059
3"	"	0.059
3-1/2"	"	0.059
4"	"	0.080
Bushed chase nipples		
1/2"	EA.	0.059
3/4"	"	0.070
1"	"	0.089
1-1/4"	"	0.010
1-1/2"	"	0.123
2"	"	0.145
2-1/2"	"	0.145
3"	"	0.182
3-1/2"	"	0.250
4"	"	0.296
Offset nipples		
1/2"	EA.	0.059
3/4"	"	0.070
1"	"	0.089
1-1/4"	"	0.107
1-1/2"	"	0.123
2"	"	0.145
3"	"	0.182
Short elbows		
1/2"	EA.	0.145
3/4"	"	0.200
1"	"	0.250

16 ELECTRICAL

BASIC MATERIALS	UNIT	MAN/ HOURS
16110.24 GALVANIZED CONDUIT		
1-1/4"	EA.	0.296
1-1/2"	"	0.348
2"	"	0.400
Pulling elbows, female to female		
1/2"	EA.	0.250
3/4"	"	0.296
1"	"	0.400
1-1/4"	"	0.533
1-1/2"	"	0.727
2"	"	0.851
Grounding locknuts		
1/2"	EA.	0.080
3/4"	"	0.080
1"	"	0.080
1-1/4"	"	0.089
1-1/2"	"	0.089
2"	"	0.089
2-1/2"	"	0.100
3"	"	0.100
3-1/2"	"	0.100
4"	"	0.145
Insulated grounding metal bushings		
1/2"	EA.	0.190
3/4"	"	0.222
1"	"	0.250
1-1/4"	"	0.308
1-1/2"	"	0.381
2"	"	0.444
2-1/2"	"	0.667
3"	"	0.808
3-1/2"	"	0.941
4"	"	1.053
5"	"	1.569
6"	"	1.739
Nipples		
1/2" x		
4"	EA.	0.145
6"	"	0.145
8"	"	0.145
10"	"	0.145
12"	"	0.145
3/4" x		
4"	EA.	0.145
6"	"	0.145
8"	"	0.145
10"	"	0.145
12"	"	0.145
1" x		
4"	EA.	0.145

BASIC MATERIALS	UNIT	MAN/ HOURS
16110.24 GALVANIZED CONDUIT		
6"	EA.	0.145
8"	"	0.145
10"	"	0.145
12"	"	0.145
1-1/4" x		
4"	EA.	0.250
6"	"	0.250
8"	"	0.250
10"	"	0.250
12"	"	0.250
1-1/2" x		
4"	EA.	0.250
6"	"	0.250
8"	"	0.250
10"	"	0.250
12"	"	0.250
2" x		
4"	EA.	0.250
6"	"	0.250
8"	"	0.250
10"	"	0.250
12"	"	0.250
2-1/2" x		
6"	EA.	0.300
8"	"	0.300
10"	"	0.300
12"	"	0.300
3" x		
6"	EA.	0.300
8"	"	0.300
10"	"	0.300
12"	"	0.300
3-1/2" x		
6"	EA.	0.300
8"	"	0.300
10"	"	0.300
12"	"	0.300
4" x		
8"	EA.	0.400
10"	"	0.400
10"	"	0.400
12"	"	0.400
5" x		
8"	EA.	0.400
12"	"	0.400
6" x		
8"	EA.	0.400
10"	"	0.400
12"	"	0.400

BASIC MATERIALS	UNIT	MAN/ HOURS
16110.25 PLASTIC CONDUIT		
PVC conduit, schedule 40		
1/2"	L.F.	0.030
3/4"	"	0.030
1"	"	0.040
1-1/4"	"	0.040
1-1/2"	"	0.050
2"	"	0.050
2-1/2"	"	0.059
3"	"	0.059
3-1/2"	"	0.080
4"	"	0.080
5"	"	0.089
6"	"	0.100
Couplings		
1/2"	EA.	0.050
3/4"	"	0.050
1"	"	0.050
1-1/4"	"	0.059
1-1/2"	"	0.059
2"	"	0.059
2-1/2"	"	0.059
3"	"	0.080
3-1/2"	"	0.080
4"	"	0.100
5"	"	0.100
6"	"	0.100
90 degree elbows		
1/2"	EA.	0.100
3/4"	"	0.123
1"	"	0.123
1-1/4"	"	0.145
1-1/2"	"	0.190
2"	"	0.222
2-1/2"	"	0.250
3"	"	0.308
3-1/2"	"	0.381
4"	"	0.500
5"	"	0.615
6"	"	0.727
Terminal adapters		
1/2"	EA.	0.100
3/4"	"	0.100
1"	"	0.100
1-1/4"	"	0.160
1-1/2"	"	0.160
2"	"	0.160
2-1/2"	"	0.222
3"	"	0.222
3-1/2"	"	0.222

BASIC MATERIALS	UNIT	MAN/ HOURS
16110.25 PLASTIC CONDUIT		
4"	EA.	0.381
5"	"	0.381
6"	"	0.381
End bells		
1"	EA.	1.000
1-1/4"	"	0.160
1-1/2"	"	0.160
2"	"	0.160
2-1/2"	"	0.222
3"	"	0.222
3-1/2"	"	0.222
4"	"	0.381
5"	"	0.381
6"	"	0.381
Lb conduit body		
1/2"	EA.	0.190
3/4"	"	0.190
1	"	0.190
1-1/4"	"	0.308
1-1/2"	"	0.308
2"	"	0.308
2-1/2"	"	0.444
3"	"	0.533
3-1/2"	"	0.615
4"	"	0.727
Direct burial, conduit		
2"	L.F.	0.050
3"	"	0.059
4"	"	0.080
5"	"	0.089
6"	"	0.100
Encased burial conduit		
2"	L.F.	0.050
3"	"	0.059
4"	"	0.080
5"	"	0.089
6"	"	0.100
"EB" and "DB" duct, 90 degree elbows		
1-1/2"	EA.	0.145
2"	"	0.229
3"	"	0.381
3-1/2"	"	0.444
4"	"	0.533
5"	"	0.667
6"	"	0.899
45 degree elbows		
1-1/2"	EA.	0.229
2"	"	0.229
3"	"	0.381

16 ELECTRICAL

BASIC MATERIALS	UNIT	MAN/ HOURS
16110.25 PLASTIC CONDUIT		
3-1/2"	EA.	0.444
4"	"	0.533
5"	"	0.667
6"	"	0.899
Couplings		
1-1/2"	EA.	0.059
2"	"	0.059
3"	"	0.080
3-1/2"	"	0.080
4"	"	0.100
5"	"	0.100
6"	"	0.160
Bell ends		
1-1/2"	EA.	0.160
2"	"	0.160
3"	"	0.222
3-1/2"	"	0.222
4"	"	0.381
5"	"	0.381
6"	"	0.381
Female adapters, 1-1/2"	"	0.200
5 degree couplings		
1-1/2"	EA.	0.070
2"	"	0.070
3"	"	0.100
4"	"	0.145
5"	"	0.145
6"	"	0.145
45 degree elbows		
1/2"	EA.	0.123
3/4"	"	0.145
1"	"	0.145
1-1/4"	"	0.182
1-1/2"	"	0.229
2"	"	0.267
2-1/2"	"	0.296
3"	"	0.381
3-1/2"	"	0.444
4"	"	0.615
5"	"	0.727
6"	"	0.899
Female adapters		
1/2"	EA.	0.123
3/4"	"	0.123
1"	"	0.123
1-1/4"	"	0.200
1-1/2"	"	0.200
2"	"	0.200
2-1/2"	"	0.267

BASIC MATERIALS	UNIT	MAN/ HOURS
16110.25 PLASTIC CONDUIT		
3"	EA.	0.267
3-1/2"	"	0.267
4"	"	0.444
5"	"	0.444
6"	"	0.444
Expansion couplings		
1/2"	EA.	0.123
3/4"	"	0.123
1"	"	0.145
1-1/4"	"	0.200
1-1/2"	"	0.200
2"	"	0.200
2-1/2"	"	0.296
3"	"	0.296
3-1/2"	"	0.296
4"	"	0.444
5"	"	0.444
6"	"	0.444
Plugs		
2"	EA.	0.200
3"	"	0.296
3-1/2"	"	0.296
4"	"	0.444
5"	"	0.444
6"	"	0.500
PVC cement		
1 pint	EA.	
1 quart	"	
1 gallon	"	
Type "T" condulets		
1/2"	EA.	0.296
3/4"	"	0.296
1"	"	0.296
1-1/4"	"	0.500
1-1/2"	"	0.500
2"	"	0.500
EB & DB female adapters		
2"	EA.	0.250
3"	"	0.381
3-1/2"	"	0.615
4"	"	0.727
5"	"	1.000
6"	"	0.016
16110.27 PLASTIC COATED CONDUIT		
Rigid steel conduit, plastic coated		
1/2"	L.F.	0.050
3/4"	"	0.059

BASIC MATERIALS	UNIT	MAN/ HOURS
16110.27 PLASTIC COATED CONDUIT		
1"	L.F.	0.080
1-1/4"	"	0.100
1-1/2"	"	0.123
2"	"	0.145
2-1/2"	"	0.190
3"	"	0.222
3-1/2"	"	0.250
4"	"	0.308
5"	"	0.381
90 degree elbows		
1/2"	EA.	0.308
3/4"	"	0.381
1"	"	0.444
1-1/4"	"	0.500
1-1/2"	"	0.615
2"	"	0.800
2-1/2"	"	1.143
3"	"	1.333
3-1/2"	"	1.633
4"	"	2.000
5"	"	2.500
Couplings		
1/2"	EA.	0.059
3/4"	"	0.080
1"	"	0.089
1-1/4"	"	0.107
1-1/2"	"	0.123
2"	"	0.145
2-1/2"	"	0.182
3"	"	0.190
3-1/2"	"	0.200
4"	"	0.222
5"	"	0.250
1 hole conduit straps		
3/4"	EA.	0.050
1"	"	0.050
1-1/4"	"	0.059
1-1/2"	"	0.059
2"	"	0.059
3"	"	0.080
3-1/2"	"	0.080
4"	"	0.100
"L.B." condulets with covers		
1/2"	EA.	0.500
3/4"	"	0.500
1"	"	0.615
1-1/4"	"	0.727
1-1/2"	"	0.879
2"	"	1.000

BASIC MATERIALS	UNIT	MAN/ HOURS
16110.27 PLASTIC COATED CONDUIT		
2-1/2"	EA.	1.379
3"	"	1.739
3-1/2"	"	2.162
4"	"	2.500
"T" condulets with covers		
1/2"	EA.	0.571
3/4"	"	0.615
1"	"	0.667
1-1/4"	"	0.808
1-1/2"	"	0.941
2"	"	1.053
2-1/2"	"	1.509
3-1/2"	"	2.222
4"	"	2.667
5"	"	1.905
16110.28 STEEL CONDUIT		
Intermediate metal conduit (IMC)		
1/2"	L.F.	0.030
3/4"	"	0.040
1"	"	0.050
1-1/4"	"	0.059
1-1/2"	"	0.080
2"	"	0.089
2-1/2"	"	0.119
3"	"	0.145
3-1/2"	"	0.182
4"	"	0.190
90 degree ell		
1/2"	EA.	0.250
3/4"	"	0.308
1"	"	0.381
1-1/4"	"	0.444
1-1/2"	"	0.500
2"	"	0.571
2-1/2"	"	0.667
3"	"	0.889
3-1/2"	"	1.143
4"	"	1.333
Couplings		
1/2"	EA.	0.050
3/4"	"	0.059
1"	"	0.080
1-1/4"	"	0.089
1-1/2"	"	0.100
2"	"	0.107
2-1/2"	"	0.123
3"	"	0.145

16 ELECTRICAL

BASIC MATERIALS	UNIT	MAN/ HOURS
16110.28 STEEL CONDUIT		
3-1/2"	EA.	0.145
4"	"	0.160
16110.32 FLEXIBLE WIRING SYSTEMS		
Single circuit cables		
5'	EA.	0.059
10'	"	0.100
15'	"	0.145
20'	"	0.200
25'	"	0.267
30'	"	0.296
40'	"	0.400
Two circuit cables		
5'	EA.	0.059
10'	"	0.100
15'	"	0.145
20'	"	0.200
25'	"	0.267
30'	"	0.296
40'	"	0.400
Two wire switch and receptacle cables		
5'	EA.	0.059
10'	"	0.100
15'	"	0.145
20'	"	0.200
25'	"	0.267
30'	"	0.296
40'	"	0.348
Three wire switch		
5'	EA.	0.059
10'	"	0.100
15'	"	0.145
20'	"	0.200
25'	"	0.267
30'	"	0.296
40'	"	0.348
Distribution boxes		
2 circuit	EA.	0.533
3 circuit	"	0.667
4 circuit	"	0.800
6 circuit	"	1.096
12 circuit	"	2.222
18 circuit	"	2.963
Tap boxes		
1 single pole switch	EA.	0.400
2 single pole switches	"	0.533
1 3 way switch	"	0.444
1 4 way switch	"	0.533
1 receptacle	"	0.400

BASIC MATERIALS	UNIT	MAN/ HOURS
16110.32 FLEXIBLE WIRING SYSTEMS		
2 receptacles	EA.	0.533
4 receptacles	"	0.800
1 clock	"	0.400
2 clocks	"	0.533
4 clocks	"	0.800
Dust cap	"	0.100
Cable coupler	"	0.200
Reversing connector	"	0.296
16110.35 WIREMOLD		
Wiremold raceway with fittings, surface mounted		
#200	L.F.	0.030
#500	"	0.030
#700	"	0.040
#800	"	0.040
Fittings, #200, 90 degree flat elbow	EA.	0.050
Internal elbow	"	0.050
Extension adapter	"	0.059
#200, #500, #700		
Single pole switch and box	EA.	0.400
Duplex receptacle with box	"	0.308
#500, #700		
90 deg. flat elbow	EA.	0.080
Internal elbow	"	0.080
Junction box	"	0.133
Fixture box	"	0.133
Shallow switch and receptacle box	"	0.080
#800		
90 deg. flat elbow	EA.	0.080
Internal elbow	"	0.080
Junction box	"	0.123
#1500 series		
Raceway	L.F.	0.050
Wire clip	EA.	0.040
2 hole clip	"	0.059
Flat elbow	"	0.200
Internal elbow	"	0.200
External elbow	"	0.200
Adapter fitting	"	0.050
Duplex receptacle box	"	0.145
#2000 series		
Raceway	L.F.	0.050
Wire clip	EA.	0.040
Coupling	"	0.050
Supporting clip	"	0.040
Entrance end	"	0.145
Flat elbow	"	0.145
Tee	"	0.200
Single pole switch	"	0.200

BASIC MATERIALS	UNIT	MAN/ HOURS
16110.35 WIREMOLD		
#2100 series		
Raceway	L.F.	0.050
Cover	"	0.010
Wireclip	EA.	0.040
Coupling	"	0.050
Entrance end	"	0.200
Flat elbow	"	0.200
Tee	"	0.200
Device box	"	0.145
#2220 series		
Raceway	L.F.	0.100
Cover	"	0.020
Wire cup	EA.	0.040
Coupling	"	0.050
Entrance end	"	0.200
Flat elbow	"	0.200
Wall box connection	"	0.200
External elbow	"	0.145
#2600 series		
Raceway	L.F.	0.100
Wireclip	EA.	0.040
Bushing	"	0.040
Flat elbow	"	0.200
Junction box	"	0.200
#3000 series		
Raceway	L.F.	0.100
Wire clip	EA.	0.040
Coupling	"	0.040
Supporting clip	"	0.050
Device bracket	"	0.059
Entrance end	"	0.200
Flat elbow	"	0.250
Tee	"	0.348
Device box	"	0.200
Circuit breaker housing	"	0.250
Panel connection	"	0.500
Conduit connector	"	0.250
#4000 series		
Raceway	L.F.	0.100
Cover	"	0.050
Divider	"	0.050
Wire clip	EA.	0.050
Coupling	"	0.059
Device bracket	"	0.059
Entrance end	"	0.308
Flat elbow	"	0.250
Tee	"	0.308
Internal elbow	"	0.308
External elbow	"	0.308

BASIC MATERIALS	UNIT	MAN/ HOURS
16110.35 WIREMOLD		
Panel connector	EA.	0.727
#6000 series		
Raceway	L.F.	0.148
Cover	"	0.050
Wire cup	EA.	0.050
Coupling	"	0.100
Device bracket	"	0.100
Entrance end	"	0.444
Flat elbow	"	0.400
Tee	"	0.500
Internal elbow	"	0.400
External elbow	"	0.444
Panel connector	"	0.899
Telepower poles		
# 21TP2	EA.	0.727
# 21TP4	"	1.000
# 30TP4	"	1.000
Fittings		
Telephone	EA.	0.200
T-bar clamp	"	0.200
C-hanger clamp	"	0.200
Tap-off fitting	"	0.200
Take-off connector	"	0.727
16110.40 UNDERFLOOR DUCT		
Underfloor blank duct, insert duct		
7/8"	L.F.	0.050
1-3/8"	"	0.050
1-7/8"	"	0.050
Box opening plugs	EA.	0.145
Duct end plugs	"	0.145
Sleeve couplings	"	0.348
Expansion couplings	"	0.348
Vertical elbow	"	0.348
Offset elbow	"	0.348
Horizontal elbow	"	0.348
Adjustable elbow	"	0.348
Cabinet connector	"	1.194
Y-take off	"	0.615
Underfloor duct leveling legs	"	0.145
Conduit adapters		
1/2"	EA.	0.250
3/4"	"	0.250
1"	"	0.250
1-1/4"	"	0.296
2"	"	0.348
Reducer bushings		
1-1/4" x 3/4"	EA.	0.200

BASIC MATERIALS	UNIT	MAN/ HOURS
16110.40 UNDERFLOOR DUCT		
1-1/4" x 1"	EA.	0.200
2" x 1-1/2"	"	0.250
Support couplers		
1 standard	EA.	0.250
2 standard	"	0.296
3 standard	"	0.348
Supports		
1 duct	EA.	0.145
2 duct	"	0.170
3 duct	"	0.190
4 duct	"	0.250
5 duct	"	0.296
Single level junction box		
1 standard	EA.	0.800
2 standard	"	1.509
3 standard	"	2.963
4 standard	"	4.000
Two level junction boxes		
1 standard	EA.	1.000
2 standard	"	2.000
Sealing compound	"	
Insert adapters	"	0.145
Ellipsoids	"	0.348
Insert closing cap	"	0.145
Marker Screws	"	0.145
Access boxes	"	1.000
Closing caps	"	0.145
Afterset markers	"	0.145
Cell markers	"	0.145
Tie down straps	"	0.145
Plastic grommets	"	0.400
Metal grommets	"	0.400
Receptacle		
Duplex, 20a	EA.	0.500
Single		
30a	EA.	0.500
50a	"	0.615
Double duplex	"	0.615
Single, 20a	"	0.444
Double single	"	0.500
Twist lock	"	0.444
1 conduit opening	"	0.444
2 conduit openings	"	0.500
1 bushed opening	"	0.444
2 bushed openings	"	0.500
Amphenol connector		
1"	EA.	0.500
2"	"	0.533
5"	"	0.615

BASIC MATERIALS	UNIT	MAN/ HOURS
16110.40 UNDERFLOOR DUCT		
Standpipes		
Aluminum	EA.	0.250
Brass	"	0.250
Abandonment plates		
Aluminum	EA.	0.250
Brass	"	0.250
Split bell caps		
Aluminum	EA.	0.250
Brass	"	0.250
Flush floor receptacles		
Aluminum	EA.	0.727
Brass	"	0.727
Flush floor telephone		
Aluminum	EA.	0.500
Brass	"	0.500
Super underfloor duct blank duct		
1/2"	L.F.	0.059
7/8"	"	0.059
1-3/8"	"	0.059
1-7/8"	"	0.059
Box opening plugs	EA.	0.145
End plugs	"	0.145
Conduit adapters	"	0.727
Sleeve coupling	"	0.444
Expansion coupling	"	0.444
Reducing coupling	"	0.444
Vertical elbow	"	0.444
Offset elbow	"	0.444
Horizontal elbow	"	0.444
Adjustable elbow	"	0.444
Cabinet connector	"	1.509
Super underfloor duct Y-take off	"	0.727
Leveling legs	"	0.145
Support couplers		
1 super	EA.	0.250
2 super	"	0.296
1 super, 1 standard	"	0.296
2 super, 2 standard	"	0.348
Single level junction boxes		
1 super	EA.	1.509
2 super	"	2.000
4 super	"	3.478
Double level junction boxes		
1 super	EA.	1.194
2 super	"	1.509

BASIC MATERIALS

16110.50 WALL DUCT

	UNIT	MAN/HOURS
Lay-in wall duct, 10"	L.F.	0.059
Horizontal elbow	EA.	0.727
Edgewise elbow	"	0.727
Tee	"	0.899
Cross	"	1.096
Cabinet connector	"	1.600
Reverse elbow	"	0.533
Sweep elbow	"	0.727
Partition	"	0.059
Straight tunnel	"	0.276
Elbow tunnel	"	0.348
Tee kit	"	0.400
Ceiling dropout	"	1.000
Coupling device	"	0.145
End cap	"	0.276
Lay-in wall duct, 18"	L.F.	0.080
Horizontal elbow	EA.	0.800
Edgeware elbow	"	1.000
Tee	"	1.096
Cross	"	1.250
Reverse elbow	"	0.727
Sweep elbow	"	0.800
Partition	"	0.100
Straight tunnel	"	0.400
Elbow tunnel	"	0.500
Tee kit	"	0.533
Ceiling dropout	"	1.096
Coupling device	"	0.200
Reducer coupling	"	0.400
Cabinet connector	"	2.000
End cap	"	0.400

16110.60 TRENCH DUCT

	UNIT	MAN/HOURS
Trench duct, with cover		
9"	L.F.	0.170
12"	"	0.200
18"	"	0.267
24"	"	0.348
30"	"	0.400
36"	"	0.571
Tees		
9"	EA.	1.739
12"	"	2.000
18"	"	2.222
24"	"	2.500
30"	"	2.963
36"	"	3.376
Vertical elbows		

BASIC MATERIALS

16110.60 TRENCH DUCT

	UNIT	MAN/HOURS
9"	EA.	0.800
12"	"	1.096
18"	"	1.356
24"	"	1.667
30"	"	2.000
36"	"	2.500
Cabinet connectors		
9"	EA.	2.000
12"	"	2.105
18"	"	2.424
24"	"	2.500
30"	"	2.759
36"	"	2.963
End closers		
9"	EA.	0.615
12"	"	0.667
18"	"	0.800
24"	"	1.096
30"	"	1.290
36"	"	1.455
Horizontal elbows		
9"	EA.	1.509
12"	"	1.739
18"	"	2.105
24"	"	2.500
30"	"	2.857
36"	"	3.200
Crosses		
9"	EA.	2.000
12"	"	2.222
18"	"	2.500
24"	"	2.759
30"	"	3.200
36"	"	3.478

16110.80 WIREWAYS

	UNIT	MAN/HOURS
Wireway, hinge cover type		
2-1/2" x 2-1/2"		
1' section	EA.	0.154
2'	"	0.190
3'	"	0.250
5'	"	0.381
10'	"	0.667
4" x 4"		
1'	EA.	0.250
2'	"	0.250
3'	"	0.308
4'	"	0.308
10'	"	0.800

16 ELECTRICAL

BASIC MATERIALS	UNIT	MAN/ HOURS
16110.80 WIREWAYS		
6" x 6"		
1'	EA.	0.381
2'	"	0.381
3'	"	0.444
4'	"	0.444
5'	"	0.571
10'	"	0.889
8" x 8"		
1'	EA.	0.444
2'	"	0.444
3'	"	0.500
4'	"	0.500
5'	"	0.615
12" x 12"		
1'	EA.	0.615
2'	"	0.615
3'	"	0.727
4'	"	0.727
5'	"	0.889
Fittings		
2-1/2" x 2-1/2"		
Drop hanger	EA.	0.123
Bracket hanger	"	0.123
Panel adapter	"	0.500
End plate	"	0.123
U-connector	"	0.123
Tee	"	0.200
Cross	"	0.250
90 degree elbow	"	0.200
Sweep elbow	"	0.200
45 degree elbow	"	0.200
Lay-in adapter	"	0.145
4" x 4"		
Drop hanger	EA.	0.145
Bracket hanger	"	0.145
Panel adapter	"	0.615
End plate	"	0.145
U-connector	"	0.145
Tee	"	0.250
Cross	"	0.348
90 degree elbow	"	0.250
Sweep elbow	"	0.250
45 degree elbow	"	0.250
Lay-in adapter	"	0.200
6" x 6"		
Drop hanger	EA.	0.145
Bracket hanger	"	0.145
Reducing bushing	"	0.145
Panel adapter	"	0.727

BASIC MATERIALS	UNIT	MAN/ HOURS
16110.80 WIREWAYS		
End plate	EA.	0.145
U-connector	"	0.145
Tee	"	0.250
Cross	"	0.348
90 degree elbow	"	0.250
Sweep elbow	"	0.250
45 degree elbow	"	0.250
Lay-in adapter	"	0.250
8" x 8"		
Drop hanger	EA.	0.145
Bracket hanger	"	0.145
Reducing bushing	"	0.145
Panel adapter	"	0.899
End plate	"	0.145
U-connector	"	0.145
Tee	"	0.400
Cross	"	0.444
90 degree elbow	"	0.400
Sweep elbow	"	0.400
45 degree elbow	"	0.400
Lay-in adapter	"	0.250
10" x 10"		
Drop hanger	EA.	0.145
Bracket hanger	"	0.145
Reducing bushing	"	0.145
Panel adapter	"	1.000
End plate	"	0.145
U-connector	"	0.145
Tee	"	0.444
Cross	"	0.500
90 degree elbow	"	0.444
Sweep elbow	"	0.444
45 degree elbow	"	0.444
Lay-in adapter	"	0.296
12" x 12"		
Drop hanger	EA.	0.145
Bracket hanger	"	0.145
Reducing bushing	"	0.145
Panel adapter	"	1.250
End plate	"	0.145
U-connector	"	0.615
Tee	"	5.333
Cross	"	0.615
90 degree elbow	"	0.615
Sweep elbow	"	0.615
45 degree elbow	"	0.615
Lay-in adapter	"	0.533
Raintight wireway, 4" x 4"		
1' section	EA.	0.400

BASIC MATERIALS	UNIT	MAN/ HOURS
16110.80 WIREWAYS		
5'	EA.	0.400
10'	"	1.000
Fittings		
90 degree elbow	EA.	0.400
Tee	"	0.444
Cross	"	0.500
Panel adapter	"	0.727
End plate	"	0.145
Gusset bracket	"	0.145
6" x 6"		
1' section	EA.	0.500
5'	"	0.727
10'	"	1.509
Fittings		
90 degree elbow	EA.	0.500
Tee	"	0.500
Cross	"	0.727
Panel adapter	"	1.000
End plate	"	0.145
Gusset bracket	"	0.145
16120.41 ALUMINUM CONDUCTORS		
Type XHHW, stranded aluminum, 600v		
#8	L.F.	0.005
#6	"	0.006
#4	"	0.008
#2	"	0.009
1/0	"	0.011
2/0	"	0.012
3/0	"	0.014
4/0	"	0.015
300 MCM	"	0.020
350 MCM	"	0.023
400 MCM	"	0.028
500 MCM	"	0.033
600 MCM	"	0.040
700 MCM	"	0.047
750 MCM	"	0.052
THW, stranded		
#8	L.F.	0.005
#6	"	0.006
#4	"	0.008
#3	"	0.009
#1	"	0.010
1/0	"	0.011
2/0	"	0.012
3/0	"	0.012
4/0	"	0.015
250 MCM	"	0.018

BASIC MATERIALS	UNIT	MAN/ HOURS
16120.41 ALUMINUM CONDUCTORS		
300 MCM	L.F.	0.020
350 MCM	"	0.023
400 MCM	"	0.028
500 MCM	"	0.033
600 MCM	"	0.040
700 MCM	"	0.047
750 MCM	"	0.052
XLP, stranded		
#6	L.F.	0.005
#4	"	0.008
#2	"	0.009
#1	"	0.010
1/0	"	0.011
2/0	"	0.012
3/0	"	0.014
4/0	"	0.015
250 MCM	"	0.016
300 MCM	"	0.020
350 MCM	"	0.023
400 MCM	"	0.028
500 MCM	"	0.033
600 MCM	"	0.040
700 MCM	"	0.047
750 MCM	"	0.052
1000 MCM	"	0.057
Bare stranded aluminum wire		
#4	L.F.	0.008
#2	"	0.009
1/0	"	0.011
2/0	"	0.012
3/0	"	0.014
4/0	"	0.015
Triplex XLP cable		
#4	L.F.	0.015
#2	"	0.020
1/0	"	0.030
4/0	"	0.048
Aluminum quadruplex XLP cable		
#4	L.F.	0.018
#2	"	0.023
1/0	"	0.076
2/0	"	0.042
4/0	"	0.064
Triplexed URD-XLP cable		
#6	L.F.	0.011
#4	"	0.014
#2	"	0.018
1/0	"	0.028
2/0	"	0.033

16 ELECTRICAL

BASIC MATERIALS

16120.41 ALUMINUM CONDUCTORS

	UNIT	MAN/HOURS
3/0	L.F.	0.040
4/0	"	0.047
250 MCM	"	0.055
350 MCM	"	0.057
Type S.E.U. cable		
#8/3	L.F.	0.025
#6/3	"	0.028
#4/3	"	0.035
#2/3	"	0.038
#1/3	"	0.040
1/0-3	"	0.042
2/0-3	"	0.044
3/0-3	"	0.052
4/0-3	"	0.057
Type S.E.R. cable with ground		
#8/3	L.F.	0.028
#6/3	"	0.035
#4/3	"	0.038
#2/3	"	0.040
#1/3	"	0.044
1/0-3	"	0.050
2/0-3	"	0.055
3/0-3	"	0.059
4/0-3	"	0.067
#6/4	"	0.038
#4/4	"	0.044
#2/4	"	0.044
#1/4	"	0.050
1/0-4	"	0.052
2/0-4	"	0.057
3/0-4	"	0.064
4/0-4	"	0.076

16120.43 COPPER CONDUCTORS

	UNIT	MAN/HOURS
Copper conductors, type THW, solid		
#14	L.F.	0.004
#12	"	0.005
#10	"	0.006
Stranded		
#14	L.F.	0.004
#12	"	0.005
#10	"	0.006
#8	"	0.008
#6	"	0.009
#4	"	0.010
#3	"	0.010
#2	"	0.012
#1	"	0.014
1/0	"	0.016

BASIC MATERIALS

16120.43 COPPER CONDUCTORS

	UNIT	MAN/HOURS
2/0	L.F.	0.020
3/0	"	0.025
4/0	"	0.028
250 MCM	"	0.030
300 MCM	"	0.033
350 MCM	"	0.040
400 MCM	"	0.044
500 MCM	"	0.052
600 MCM	"	0.059
750 MCM	"	0.067
1000 MCM	"	0.076
THHN-THWN, solid		
#14	L.F.	0.004
#12	"	0.005
#10	"	0.006
Stranded		
#14	L.F.	0.004
#12	"	0.005
#10	"	0.006
#8	"	0.008
#6	"	0.009
#4	"	0.010
#2	"	0.012
#1	"	0.014
1/0	"	0.016
2/0	"	0.020
3/0	"	0.025
4/0	"	0.028
250 MCM	"	0.030
350 MCM	"	0.040
XHHW		
#14	L.F.	0.004
#10	"	0.006
#8	"	0.008
#6	"	0.009
#4	"	0.009
#2	"	0.011
#1	"	0.014
1/0	"	0.016
2/0	"	0.019
3/0	"	0.025
XLP, 600v		
#12	L.F.	0.005
#10	"	0.006
#8	"	0.008
#6	"	0.009
#4	"	0.010
#3	"	0.011
#2	"	0.012

BASIC MATERIALS	UNIT	MAN/ HOURS
16120.43 COPPER CONDUCTORS		
#1	L.F.	0.014
1/0	"	0.016
2/0	"	0.020
3/0	"	0.026
4/0	"	0.028
250 MCM	"	0.030
300 MCM	"	0.033
350 MCM	"	0.039
400 MCM	"	0.044
500 MCM	"	0.052
600 MCM	"	0.059
750 MCM	"	0.067
1000 MCM	"	0.076
Bare solid wire		
#14	L.F.	0.004
#12	"	0.005
#10	"	0.006
#8	"	0.008
#6	"	0.009
#4	"	0.010
#2	"	0.012
Bare stranded wire		
#8	L.F.	0.008
#6	"	0.010
#4	"	0.010
#2	"	0.011
#1	"	0.014
1/0	"	0.018
2/0	"	0.020
3/0	"	0.025
4/0	"	0.028
250 MCM	"	0.030
300 MCM	"	0.033
350 MCM	"	0.040
400 MCM	"	0.044
500 MCM	"	0.052
Type "BX" solid armored cable		
#14/2	L.F.	0.025
#14/3	"	0.028
#14/4	"	0.031
#12/2	"	0.028
#12/3	"	0.031
#12/4	"	0.035
#10/2	"	0.031
#10/3	"	0.035
#10/4	"	0.040
#8/2	"	0.035
#8/3	"	0.040
Steel type, metal clad cable, solid, with ground		

BASIC MATERIALS	UNIT	MAN/ HOURS
16120.43 COPPER CONDUCTORS		
#14/2	L.F.	0.018
#14/3	"	0.020
#14/4	"	0.023
#12/2	"	0.020
#12/3	"	0.025
#12/4	"	0.030
#10/2	"	0.023
#10/3	"	0.028
#10/4	"	0.033
Metal clad cable, stranded, with ground		
#8/2	L.F.	0.028
#8/3	"	0.035
#8/4	"	0.042
#6/2	"	0.030
#6/3	"	0.038
#6/4	"	0.044
#4/2	"	0.040
#4/3	"	0.044
#4/4	"	0.055
#3/3	"	0.050
#3/4	"	0.059
#2/3	"	0.057
#2/4	"	0.067
#1/3	"	0.076
#4	"	0.035
16120.45 FLAT CONDUCTOR CABLE		
Flat conductor cable, with shield, 3 conductor		
#12 awg	L.F.	0.059
#10 awg	"	0.059
4 conductor		
#12 awg	L.F.	0.080
#10 awg	"	0.080
Transition boxes		
#12 awg	L.F.	0.089
#10 awg	"	0.089
Flat conductor cable communication, with shield		
10 conductor	L.F.	0.059
16 conductor	"	0.070
24 conductor	"	0.100
Power and communication heads, duplex receptacle	EA.	0.800
Double duplex receptacle	"	0.952
Telephone	"	0.800
Receptacle and telephone	"	0.952
Blank cover	"	0.145
Transition boxes		
Surface	EA.	0.727
Flush	"	1.000

BASIC MATERIALS

16120.45 FLAT CONDUCTOR CABLE

BASIC MATERIALS	UNIT	MAN/ HOURS
Flat conductor cable fittings		
End caps	EA.	0.145
Insulators	"	0.296
Splice connectors	"	0.444
Tap connectors	"	0.444
Cable connectors	"	0.444
Terminal blocks	"	0.615
Tape	"	

16120.47 SHEATHED CABLE

BASIC MATERIALS	UNIT	MAN/ HOURS
Non-metallic sheathed cable		
Type NM cable with ground		
#14/2	L.F.	0.015
#12/2	"	0.016
#10/2	"	0.018
#8/2	"	0.020
#6/2	"	0.025
#14/3	"	0.026
#12/3	"	0.027
#10/3	"	0.027
#8/3	"	0.028
#6/3	"	0.028
#4/3	"	0.032
#2/3	"	0.035
Type U.F. cable with ground		
#14/2	L.F.	0.016
#12/2	"	0.019
#10/2	"	0.020
#8/2	"	0.023
#6/2	"	0.027
#14/3	"	0.020
#12/3	"	0.022
#10/3	"	0.025
#8/3	"	0.028
#6/3	"	0.032
Type S.F.U. cable, 3 conductor		
#8	L.F.	0.028
#6	"	0.031
#3	"	0.040
#2	"	0.044
#1	"	0.050
#1/0	"	0.055
#2/0	"	0.064
#3/0	"	0.070
#4/0	"	0.076
Type SER cable, 4 conductor		
#6	L.F.	0.036
#4	"	0.039
#3	"	0.044

BASIC MATERIALS

16120.47 SHEATHED CABLE

BASIC MATERIALS	UNIT	MAN/ HOURS
#2	L.F.	0.048
#1	"	0.055
#1/0	"	0.064
#2/0	"	0.067
#3/0	"	0.076
#4/0	"	0.084
Flexible cord, type STO cord		
#18/2	L.F.	0.004
#18/3	"	0.005
#18/4	"	0.006
#16/2	"	0.004
#16/3	"	0.004
#16/4	"	0.005
#14/2	"	0.005
#14/3	"	0.006
#14/4	"	0.007
#12/2	"	0.006
#12/3	"	0.007
#12/4	"	0.008
#10/2	"	0.007
#10/3	"	0.008
#10/4	"	0.009
#8/2	"	0.008
#8/3	"	0.009
#8/4	"	0.010
Flexible cord, copper, type STO cord		

16130.10 FLOOR BOXES

BASIC MATERIALS	UNIT	MAN/ HOURS
Adjustable floor boxes, steel	EA.	0.533
Cast bronze round	"	0.727
1 gang	"	0.800
2 gang	"	0.952
3 gang	"	1.000
Aluminum round	"	0.727
1 gang	"	0.800
2 gang	"	0.952
3 gang	"	1.000
Steel plate single recept	"	0.145
Duplex receptacle	"	0.182
Twist lock receptacle	"	0.182
Plug, 3/4"	"	0.145
1" plug	"	0.145
Carpet flange	"	0.145
Adjustable bronze plates for round cast boxes		
1/2" plug	EA.	0.145
3/4" plug	"	0.145
1" plug	"	0.145
1-1/4" plug	"	0.182
2" plug	"	0.200

BASIC MATERIALS	UNIT	MAN/ HOURS
16130.10 FLOOR BOXES		
Combination plug	EA.	0.200
Duplex receptacle plug	"	0.200
Adjustable aluminum plates for round cast boxes		
1/2" plug	EA.	0.145
3/4" plug	"	0.145
1" plug	"	0.145
1-1/4" plug	"	0.182
2" plug	"	0.200
Combination plug	"	0.200
Duplex receptacle plug	"	0.200
Adjustable bronze plates for gang type boxes		
1/2" plug	EA.	0.145
3/4" plug	"	0.145
1" plug	"	0.145
1-1/4" plug	"	0.182
2" plug	"	0.200
Carpet plate		
1 gang	EA.	0.145
2 gang	"	0.145
3 gang	"	0.200
Adjustable aluminum plates for gang type boxes		
1/2" plug	EA.	0.145
3/4" plug	"	0.145
1" plug	"	0.145
1-1/4" plug	"	0.182
2" plug	"	0.200
Duplex recept	"	0.200
Carpet plate		
1 gang	EA.	0.145
2 gang	"	0.145
3 gang	"	0.200
4 gang carpet plate	"	0.571
Telephone	"	0.500
Floor box nozzles, horizontal		
Duplex recept	EA.	0.533
Single recept	"	0.533
Double duplex recept	"	0.727
Vertical with duplex recept	"	0.615
Double duplex recept	"	0.727
Floor box bell nozzles split bell	"	0.250
One piece bell	"	0.250
Floor box standpipe		
1/2" x 3"	EA.	0.145
1/2" x 1"	"	0.145
Poke thru floor outlets		
2" floor	EA.	1.000
3" floor	"	1.194
4" floor	"	1.290
7" floor	"	1.509

BASIC MATERIALS	UNIT	MAN/ HOURS
16130.10 FLOOR BOXES		
9" floor	EA.	1.600
11" floor	"	1.818
13" floor	"	2.000
16130.40 BOXES		
Round cast box, type SEH		
1/2"	EA.	0.348
3/4"	"	0.421
SEHC		
1/2"	EA.	0.348
3/4"	"	0.421
SEHL		
1/2"	EA.	0.348
3/4"	"	0.444
SEHT		
1/2"	EA.	0.421
3/4"	"	0.500
SEHX		
1/2"	EA.	0.500
3/4"	"	0.615
Blank cover	"	0.145
1/2", hub cover	"	0.145
Cover with gasket	"	0.178
Rectangle, type FS boxes		
1/2"	EA.	0.348
3/4"	"	0.400
1"	"	0.500
FSA		
1/2"	EA.	0.348
3/4"	"	0.400
FSC		
1/2"	EA.	0.348
3/4"	"	0.421
1"	"	0.500
FSL		
1/2"	EA.	0.348
3/4"	"	0.400
FSR		
1/2"	EA.	0.348
3/4"	"	0.400
FSS		
1/2"	EA.	0.348
3/4"	"	0.400
FSLA		
1/2"	EA.	0.348
3/4"	"	0.400
FSCA		
1/2"	EA.	0.348

16 ELECTRICAL

BASIC MATERIALS	UNIT	MAN/ HOURS
16130.40 BOXES		
3/4"	EA.	0.400
FSCC		
1/2"	EA.	0.400
3/4"	"	0.500
FSCT		
1/2"	EA.	0.400
3/4"	"	0.500
1"	"	0.571
FST		
1/2"	EA.	0.500
3/4"	"	0.571
FSX		
1/2"	EA.	0.615
3/4"	"	0.727
FSCD boxes		
1/2"	EA.	0.615
3/4"	"	0.727
Rectangle, type FS, 2 gang boxes		
1/2"	EA.	0.348
3/4"	"	0.400
1"	"	0.500
FSC, 2 gang boxes		
1/2"	EA.	0.348
3/4"	"	0.400
1"	"	0.500
FSS, 2 gang boxes		
3/4"	EA.	0.400
FS, tandem boxes		
1/2"	EA.	0.400
3/4"	"	0.444
FSC, tandem boxes		
1/2"	EA.	0.400
3/4"	"	0.444
FS, three gang boxes		
3/4"	EA.	0.444
1"	"	0.500
FSS, three gang boxes, 3/4"	"	0.500
Weatherproof cast aluminum boxes, 1 gang, 3 outlets		
1/2"	EA.	0.400
3/4"	"	0.500
2 gang, 3 outlets		
1/2"	EA.	0.500
3/4"	"	0.533
1 gang, 4 outlets		
1/2"	EA.	0.615
3/4"	"	0.727
2 gang, 4 outlets		
1/2"	EA.	0.615
3/4"	"	0.727

BASIC MATERIALS	UNIT	MAN/ HOURS
16130.40 BOXES		
1 gang, 5 outlets		
1/2"	EA.	0.727
3/4"	"	0.800
2 gang, 5 outlets		
1/2"	EA.	0.727
3/4"	"	0.800
2 gang, 6 outlets		
1/2"	EA.	0.851
3/4"	"	0.899
2 gang, 7 outlets		
1/2"	EA.	1.000
3/4"	"	1.096
Weatherproof and type FS box covers, blank, 1 gang	"	0.145
Tumbler switch, 1 gang	"	0.145
1 gang, single recept	"	0.145
Duplex recept	"	0.145
Despard	"	0.145
Red pilot light	"	0.145
SW and		
Single recept	EA.	0.200
Duplex recept	"	0.200
2 gang		
Blank	EA.	0.182
Tumbler switch	"	0.182
Single recept	"	0.182
Duplex recept	"	0.182
3 gang		
Blank	EA.	0.200
Tumbler switch	"	0.200
4 gang		
Tumbler switch	EA.	0.250
Explosion proof boxes type E		
1/2"	EA.	0.348
3/4"	"	0.400
1"	"	0.500
1-1/4"	"	0.571
1-1/2"	"	0.615
Type L.B.		
1/2"	EA.	0.400
3/4"	"	0.500
1"	"	0.571
1-1/4"	"	0.667
1-1/2"	"	0.727
2"	"	0.800
Type C		
1/2"	EA.	0.400
3/4"	"	0.500
1"	"	0.571
1-1/4"	"	0.667

BASIC MATERIALS	UNIT	MAN/ HOURS
16130.40 BOXES		
1-1/2"	EA.	0.727
2"	"	0.800
Type CA		
1/2"	EA.	0.571
3/4"	"	0.727
Type L		
1/2"	EA.	0.400
3/4"	"	0.500
1"	"	0.571
1-1/4"	"	0.667
1-1/2"	"	0.727
2"	"	0.800
Type N		
1/2"	EA.	0.400
3/4"	"	0.500
1"	"	0.615
1-1/4"	"	0.667
Type T		
1/2"	EA.	0.533
3/4"	"	0.727
1"	"	0.851
1-1/4"	"	1.000
1-1/2"	"	1.159
2"	"	1.290
Type TA		
1/2"	EA.	0.727
3/4"	"	0.800
Type X		
1/2"	EA.	0.727
3/4"	"	0.851
1"	"	1.000
1-1/4"	"	1.159
1-1/2"	"	1.290
2"	"	1.455
With union hubs		
1/2"	EA.	0.727
3/4"	"	0.800
Box covers		
Surface	EA.	0.200
Sealing	"	0.200
Dome	"	0.200
1/2" nipple	"	0.200
3/4" nipple	"	0.200
16130.45 EXPLOSION PROOF FITTINGS		
Flexible couplings with female unions		
1/2" x 18"	EA.	0.200
3/4" x 18"	"	0.276

BASIC MATERIALS	UNIT	MAN/ HOURS
16130.45 EXPLOSION PROOF FITTINGS		
1" x 18"	EA.	0.348
1-1/4" x 18"	"	0.421
1-1/2" x 18"	"	0.500
2" x 18"	"	0.571
1/2" x 24"	"	0.250
3/4" x 24"	"	0.296
1" x 24"	"	0.400
1-1/4" x 24"	"	0.444
1-1/2" x 24"	"	0.571
2" x 24"	"	0.615
Female seal-offs		
1/2"	EA.	0.571
3/4"	"	0.667
1"	"	0.727
1-1/4"	"	0.851
1-1/2"	"	1.000
2"	"	1.159
2-1/2"	"	1.739
3"	"	2.162
4"	"	2.667
Conduit plugs		
1/2"	EA.	0.145
3/4"	"	0.145
1"	"	0.145
1-1/4"	"	0.250
1-1/2"	"	0.250
2"	"	0.296
2-1/2"	"	0.296
3"	"	0.348
4"	"	0.348
Sealing cement		
1 pound	EA.	
5 pound	"	
Fibre		
1 ounce	EA.	
8 ounce	"	
Male unions		
1/2"	EA.	0.200
3/4"	"	0.242
1"	"	0.276
1-1/4"	"	0.296
1-1/2"	"	0.348
2"	"	0.421
2-1/2"	"	0.500
3"	"	0.727
4"	"	0.899
Female unions		
1/2"	EA.	0.200
3/4"	"	0.242

16 ELECTRICAL

BASIC MATERIALS	UNIT	MAN/ HOURS
16130.45 EXPLOSION PROOF FITTINGS		
1"	EA.	0.276
1-1/4"	"	0.296
1-1/2"	"	0.348
2"	"	0.421
2-1/2"	"	0.500
3"	"	0.727
4"	"	0.899
Male elbows		
1/2"	EA.	0.250
3/4"	"	0.296
1"	"	0.348
1-1/4"	"	0.444
Female elbows		
1/2"	EA.	0.250
3/4"	"	0.296
1"	"	0.348
1-1/4"	"	0.444
Pulling elbows		
1/2"	EA.	0.348
3/4"	"	0.444
1"	"	0.500
1-1/4"	"	0.615
1-1/2"	"	0.727
2"	"	1.905
2-1/2"	"	2.500
3"	"	2.963
3-1/2"	"	3.478
4"	"	4.211
Male expansion couplings		
1/2"	EA.	0.250
3/4"	"	0.296
1"	"	0.444
Female expansion couplings		
1/2"	EA.	0.250
3/4"	"	0.296
1"	"	0.444
16130.60 PULL AND JUNCTION BOXES		
4"		
Octagon box	EA.	0.114
Box extension	"	0.059
Plaster ring	"	0.059
Cover blank	"	0.059
Square box	"	0.114
Box extension	"	0.059
Plaster ring	"	0.059
Cover blank	"	0.059
4-11/16"		
Square box	EA.	0.114

BASIC MATERIALS	UNIT	MAN/ HOURS
16130.60 PULL AND JUNCTION BOXES		
Box extension	EA.	0.059
Plaster ring	"	0.059
Cover blank	"	0.059
Switch and device boxes		
2 gang	EA.	0.114
3 gang	"	0.114
4 gang	"	0.160
Device covers		
2 gang	EA.	0.059
3 gang	"	0.059
4 gang	"	0.059
Handy box	"	0.114
Extension	"	0.059
Switch cover	"	0.059
Switch box with knockout	"	0.145
Weatherproof cover, spring type	"	0.080
Cover plate, dryer receptacle 1 gang plastic	"	0.100
For 4" receptacle, 2 gang	"	0.100
Duplex receptacle cover plate, plastic	"	0.059
4", vertical bracket box, 1-1/2" with		
RMX clamps	EA.	0.145
BX clamps	"	0.145
4", octagon device cover		
1 switch	EA.	0.059
1 duplex recept	"	0.059
4", octagon swivel hanger box, 1/2" hub	"	0.059
3/4" hub	"	0.059
4" octagon adjustable bar hangers		
18-1/2"	EA.	0.050
26-1/2"	"	0.050
With clip		
18-1/2"	EA.	0.050
26-1/2"	"	0.050
4", square face bracket boxes, 1-1/2"		
RMX	EA.	0.145
BX	"	0.145
4" square to round plaster rings	"	0.059
2 gang device plaster rings	"	0.059
Surface covers		
1 gang switch	EA.	0.059
2 gang switch	"	0.059
1 single recept	"	0.059
1 20a twist lock recept	"	0.059
1 30a twist lock recept	"	0.059
1 duplex recept	"	0.059
2 duplex recept	"	0.059
Switch and duplex recept	"	0.059
4-11/16" square to round plaster rings	"	0.059
2 gang device plaster rings	"	0.059

16 ELECTRICAL

BASIC MATERIALS

16130.60 PULL AND JUNCTION BOXES

BASIC MATERIALS	UNIT	MAN/ HOURS
Surface covers		
1 gang switch	EA.	0.059
2 gang switch	"	0.059
1 single recept	"	0.059
1 20a twist lock recept	"	0.059
1 30a twist lock recept	"	0.059
1 duplex recept	"	0.059
2 duplex recept	"	0.059
Switch and duplex recept	"	0.059
4" plastic round boxes, ground straps		
box only	EA.	0.145
Box w/clamps	"	0.200
Box w/16" bar	"	0.229
Box w/24" bar	"	0.250
4" plastic round box covers		
blank cover	EA.	0.059
Plaster ring	"	0.059
4" plastic square boxes		
box only	EA.	0.145
Box w/clamps	"	0.200
Box w/hanger	"	0.250
Box w/nails and clamp	"	0.250
4" plastic square box covers		
Blank cover	EA.	0.059
1 gang ring	"	0.059
2 gang ring	"	0.059
Round ring	"	0.059

16130.65 PULL BOXES AND CABINETS

BASIC MATERIALS	UNIT	MAN/ HOURS
Galvanized pull boxes, screw cover		
4x4x4	EA.	0.190
4x6x4	"	0.190
6x6x4	"	0.190
6x8x4	"	0.190
8x8x4	"	0.250
8x10x4	"	0.242
8x12x4	"	0.250
Screw cover		
10x10x4	EA.	0.308
12x12x6	"	0.444
12x15x6	"	0.444
12x18x6	"	0.500
15x18x6	"	0.571
18x24x6	"	0.615
18x30x6	"	0.727
24x36x6	"	0.727
Cast iron junction box, unflanged		
6x6x4		

BASIC MATERIALS

16130.65 PULL BOXES AND CABINETS

BASIC MATERIALS	UNIT	MAN/ HOURS
3/4" tap	EA.	0.500
1" tap	"	0.500
Two 1/2" taps	"	0.500
3/4" taps	"	0.500
6" adapter plate	"	0.348
6" exterior collar	"	0.348
Screw cover cabinet		
12x12x4	EA.	0.615
12x16x4	"	0.615
12x16x6	"	0.615
12x18x4	"	0.667
12x18x6	"	0.667
18x18x4	"	1.000
18x18x6	"	1.000
18x24x6	"	1.143
24x24x6	"	1.333
24x36x6	"	1.667
36x48x6	"	2.500
NEMA 3R, rain tight screw cover enclosures		
6x6x4	EA.	0.211
8x6x4	"	0.296
8x8x4	"	0.296
10x8x4	"	0.400
10x10x4	"	0.400
12x8x4	"	0.444
12x12x4	"	0.444
15x12x4	"	0.533
8x8x6	"	0.400
10x8x6	"	0.444
10x10x6	"	0.444
12x8x6	"	0.533
12x10x6	"	0.548
12x12x6	"	0.548
18x12x6	"	0.702

16130.80 RECEPTACLES

BASIC MATERIALS	UNIT	MAN/ HOURS
Contractor grade duplex receptacles, 15 a 120v		
Duplex	EA.	0.200
125 volt, 20a, duplex, grounding type, standard grade	"	0.200
Ground fault interrupter type	"	0.296
250 volt, 20a, 2 pole, single receptacle, ground type	"	0.200
120/208v, 4 pole, single receptacle, twist lock		
20a	EA.	0.348
50a	"	0.348
125/250v, 3 pole, flush receptacle		
30a	EA.	0.296
50a	"	0.296
60a	"	0.348
277 v, 20a, 2 pole, grounding type, twist lock	"	0.200

16 ELECTRICAL

BASIC MATERIALS

BASIC MATERIALS	UNIT	MAN/HOURS
16130.80 RECEPTACLES		
Dryer receptacle, 250v, 30a/50a, 3 wire	EA.	0.296
Clock receptacle, 2 pole, grounding type	"	0.200
125v, 20a single recept. grounding type		
Standard grade	EA.	0.200
Specification	"	0.200
Hospital	"	0.200
Isolated ground orange	"	0.250
Duplex		
Specification grade	EA.	0.200
Hospital	"	0.200
Isolated ground orange	"	0.250
250v, 20a, duplex, 2 pole, grounding, spec. grade	"	0.200
Combination recepts, 20a, 125v and 250v, duplex	"	0.200
GFI hospital grade recepts, 20a, 125v, duplex	"	0.296
125/250v, 3 pole, 3 wire surface recepts		
30a	EA.	0.296
50a	"	0.296
60a	"	0.348
Cord set, 3 wire, 6' cord		
30a	EA.	0.296
50a	"	0.296
125/250v, 3 pole, 3 wire cap		
30a	EA.	0.400
50a	"	0.400
60a	"	0.444
16198.10 ELECTRIC MANHOLES		
Precast, handhole, 4' deep		
2'x2'	EA.	3.478
3'x3'	"	5.556
4'x4'	"	10.256
Power manhole, complete, precast, 8' deep		
4'x4'	EA.	14.035
6'x6'	"	20.000
8'x8'	"	21.053
6' deep, 9' x 12'	"	25.000
Cast in place, power manhole, 8' deep		
4'x4'	EA.	14.035
6'x6'	"	20.000
8'x8'	"	21.053
16199.10 UTILITY POLES & FITTINGS		
Wood pole, creosoted		
25'	EA.	2.353
30'	"	2.963
35'	"	3.478
40'	"	3.791

BASIC MATERIALS	UNIT	MAN/HOURS
16199.10 UTILITY POLES & FITTINGS		
45'	EA.	6.957
50'	"	7.207
55'	"	7.547
Treated, wood preservative, 6"x6"		
8'	EA.	0.500
10'	"	0.800
12'	"	0.889
14'	"	1.333
16'	"	1.600
18'	"	2.000
20'	"	2.000
Aluminum, brushed, no base		
8'	EA.	2.000
10'	"	2.667
15'	"	2.759
20'	"	3.200
25'	"	3.810
30'	"	4.396
35'	"	5.000
40'	"	6.250
Steel, no base		
10'	EA.	2.500
15'	"	2.963
20'	"	3.810
25'	"	4.520
30'	"	5.096
35'	"	6.250
Concrete, no base		
13'	EA.	5.517
16'	"	7.273
18'	"	8.791
25'	"	10.000
30'	"	12.121
35'	"	14.035
40'	"	16.000
45'	"	17.021
50'	"	18.182
55'	"	19.048
60'	"	20.000
Pole line hardware		
Wood crossarm		
4'	EA.	1.333
8'	"	1.667
10'	"	2.051
Angle steel brace		
1 piece	EA.	0.250
2 piece	"	0.348
Eye nut, 5/8"	"	0.050
Bolt (14-16"), 5/8"	"	0.200

16 ELECTRICAL

BASIC MATERIALS

16199.10 UTILITY POLES & FITTINGS

	UNIT	MAN/HOURS
Transformer, ground connection	EA.	0.250
Stirrup	"	0.308
Secondary lead support	"	0.400
Spool insulator	"	0.200
Guy grip, preformed		
7/16"	EA.	0.145
1/2"	"	0.145
Hook	"	0.250
Strain insulator	"	0.364
Wire		
5/16"	L.F.	0.005
7/16"	"	0.006
1/2"	"	0.008
Soft drawn ground, copper, #8	"	0.008
Ground clamp	EA.	0.308
Perforated strapping for conduit, 1-1/2"	L.F.	0.145
Hot line clamp	EA.	0.800
Lightning arrester		
3kv	EA.	1.000
10kv	"	1.600
30kv	"	2.000
36kv	"	2.500
Fittings		
Plastic molding	L.F.	0.145
Molding staples	EA.	0.050
Ground wires staples	"	0.030
Copper butt plate	"	0.296
Anchor bond clamp	"	0.145
Guy wire		
1/4"	L.F.	0.030
3/8"	"	0.050
Guy grip		
1/4"	EA.	0.050
3/8"	"	0.050

POWER GENERATION

16210.10 GENERATORS

	UNIT	MAN/HOURS
Diesel generator, with auto transfer switch		
30kw	EA.	30.769
50kw	"	30.769
75kw	EA.	42.105
100kw	"	47.059
125kw	"	50.000
150kw	"	57.143
175kw	"	66.667
200kw	"	80.000
250kw	"	88.889
300kw	"	100
350kw	"	114
400kw	"	133
450kw	"	145
500kw	"	160
600kw	"	200
750kw	"	200

16230.10 CAPACITORS

	UNIT	MAN/HOURS
Three phase capacitors		
240v		
1.5 kvar	EA.	2.500
2.5 kvar	"	3.200
3.0 kvar	"	4.000
4 kvar	"	5.000
5 kvar	"	5.333
6 kvar	"	5.714
7.5 kvar	"	6.154
10 kvar	"	8.000
15 kvar	"	9.501
20 kvar	"	11.994
25 kvar	"	13.008
40 kvar	"	18.018
50 kvar	"	20.997
60 kvar	"	21.505
75 kvar	"	25.000
100 kvar	"	29.963
480v		
1.5 kvar	EA.	2.500
2.5 kvar	"	3.200
3 kvar	"	4.000
4 kvar	"	5.000
5 kvar	"	5.333
6 kvar	"	5.714
7.5 kvar	"	6.154
10 kvar	"	8.000
12.5 kvar	"	9.501
15 kvar	"	11.994
18 kvar	"	12.500
20 kvar	"	13.008
22.5 kvar	"	13.491
25 kvar	"	14.842

16 ELECTRICAL

POWER GENERATION	UNIT	MAN/ HOURS
16230.10 CAPACITORS		
30 kvar	EA.	14.842
35 kvar	"	16.000
40 kvar	"	18.018
45 kvar	"	20.000
50 kvar	"	20.997
60 kvar	"	21.978
70 kvar	"	24.024
75 kvar	"	25.000
80 kvar	"	27.027
90 kvar	"	28.986
100 kvar	"	29.963
125 kvar	"	33.058
150 kvar	"	37.037
16320.10 TRANSFORMERS		
Floor mounted, single phase, int. dry, 480v-120/240v		
3 kva	EA.	1.818
5 kva	"	3.077
7.5 kva	"	3.478
10 kva	"	3.810
15 kva	"	4.301
25 kva	"	7.547
37.5 kva	"	9.412
50 kva	"	10.256
75 kva	"	10.667
100 kva	"	11.594
Three phase, 480v-120/280v		
15 kva	EA.	6.015
30 kva	"	9.412
45 kva	"	10.811
75 kva	"	10.959
112.5 kva	"	12.698
150 kva	"	13.559
225 kva	"	15.385
Single phase, dry type, 2400v		
167 kva	EA.	22.472
250 kva	"	29.963
333 kva	"	22.472
5000v		
167 kva	EA.	22.472
250 kva	"	29.963
333 kva	"	37.559
8660v		
167 kva	EA.	27.491
250 kva	"	34.934
333 kva	"	67.797
1500v		
167 kva	EA.	27.491

POWER GENERATION	UNIT	MAN/ HOURS
16320.10 TRANSFORMERS		
250 kva	EA.	34.934
333 kva	"	42.553
Three phase, dry type transformer, 2400v		
225 kva	EA.	25.000
300 kva	"	27.491
500 kva	"	42.553
750 kva	"	52.632
5000v		
225.0 kva	EA.	25.000
300 kva	"	27.491
500 kva	"	42.553
750 kva	"	52.632
8660v		
225.0 kva	EA.	29.963
300 kva	"	32.520
500 kva	"	47.619
750 kva	"	57.554
1500v		
225 kva	EA.	29.963
300 kva	"	32.520
500 kva	"	47.619
750 kva	"	57.554
Buck boost transformers		
.25 kva	EA.	1.000
.50 kva	"	1.250
.75 kva	"	1.509
1.00 kva	"	1.739
1.50 kva	"	2.000
2.00 kva	"	2.500
3.00 kva	"	2.963
16350.10 CIRCUIT BREAKERS		
Molded case, 240v, 15-60a, bolt-on		
1 pole	EA.	0.250
2 pole	"	0.348
70-100a, 2 pole	"	0.533
15-60a, 3 pole	"	0.400
70-100a, 3 pole	"	0.615
480v, 2 pole		
15-60a	EA.	0.296
70-100a	"	0.400
3 pole		
15-60a	EA.	0.400
70-100a	"	0.444
70-225a	"	0.615
Draw out air circuit breakers		
600a	EA.	16.000
800a	"	18.182
1600a	"	24.242

POWER GENERATION	UNIT	MAN/ HOURS
16350.10 CIRCUIT BREAKERS		
2000a	EA.	27.586
3000a	"	32.000
4000a	"	38.095
Load center circuit breakers, 240v		
1 pole, 10-60a	EA.	0.250
2 pole		
10-60a	EA.	0.400
70-100a	"	0.667
110-150a	"	0.727
3 pole		
10-60a	EA.	0.500
70-100a	"	0.727
Load center, G.F.I. breakers, 240v		
1 pole, 15-30a	EA.	0.296
2 pole, 15-30a	"	0.400
Key operated breakers, 240v, 1 pole, 10-30a	"	0.296
Tandem breakers, 240v		
1 pole, 15-30a	EA.	0.400
2 pole, 15-30a	"	0.533
Bolt-on, G.F.I. breakers, 240v, 1 pole, 15-30a	"	0.348
Enclosed breaker, 120v, 1 pole, 15-50a, NEMA 1	"	0.800
240v, 2 pole		
15-60a, NEMA 1	EA.	1.250
70-100a, NEMA 1	"	1.739
3 pole		
15-60a, NEMA 1	EA.	1.509
70-100a, NEMA 1	"	2.222
Enclosed circuit breakers		
120v, 1 pole, NEMA 3R, 15-50a	EA.	0.899
240v, 2 pole, NEMA 3R		
15-60a	EA.	1.250
70-100a	"	1.739
3 pole, NEMA 3R		
15-60a	EA.	1.509
70-100a	"	2.222
480v, NEMA 1		
1 pole, 15-50a	EA.	0.800
2 pole, 15-60a	"	1.250
70-100a	"	1.509
3 pole, NEMA 1		
15-60a	EA.	1.509
70-100a	"	2.222
480v, 1 pole, 15-50a, NEMA 3R	"	1.000
2 pole		
2 pole, 15-60a, NEMA 3R	EA.	1.250
70-100a, NEMA 3R	"	1.739
3 pole		
15-60a, NEMA 3R	EA.	1.509
70-100a, NEMA 3R	"	2.222

POWER GENERATION	UNIT	MAN/ HOURS
16350.10 CIRCUIT BREAKERS		
70-100a, NEMA 3R	EA.	1.250
70-100a, NEMA 1	"	1.739
3 pole		
15-60a, NEMA 1	EA.	1.739
70-100a, NEMA 1	"	2.222
Enclosed breakers, 600v, 2 phase, NEMA 3R		
15-60a	EA.	1.250
70-100a	"	1.739
3 phase, NEMA 3R		
15-60a	EA.	1.509
70-100a	"	2.222
600v, 3 phase, NEMA 1		
125a	EA.	2.222
150a	"	2.963
175a	"	2.963
200a	"	2.963
225a	"	2.963
250a	"	6.154
300a	"	6.154
350a	"	6.154
400a	"	6.154
500a	"	9.744
600a	"	9.744
700a	"	10.753
800a	"	10.753
900a	"	15.009
100a	"	15.009
1200a	"	18.519
1400a	"	18.519
1600a	"	24.024
1800a	"	29.963
2000a	"	29.963
600v, 3 phase, NEMA 3R		
125-225a	EA.	2.222
250-400a	"	5.714
500-600a	"	9.744
700-800a	"	11.004
900-1000a	"	15.009
1000-1200a	"	19.002
1400-1600a	"	24.024
1800-2000a	"	29.963
16360.10 SAFETY SWITCHES		
Fused, 3 phase, 30 amp, 600v, heavy duty		
NEMA 1	EA.	1.143
NEMA 3r	"	1.143
NEMA 4	"	1.600
NEMA 12	"	1.739

16 ELECTRICAL

POWER GENERATION

16360.10 SAFETY SWITCHES

	UNIT	MAN/ HOURS
60a		
NEMA 1	EA.	1.143
NEMA 3r	"	1.143
NEMA 4	"	1.600
NEMA 12	"	1.739
100a		
NEMA 1	EA.	1.739
NEMA 3r	"	1.739
NEMA 4	"	2.000
NEMA 12	"	2.500
200a		
NEMA 1	EA.	2.500
NEMA 3r	"	2.500
NEMA 4	"	2.759
NEMA 12	"	3.478
400a		
NEMA 1	EA.	5.517
NEMA 3r	"	5.517
NEMA 4	"	5.755
NEMA 12	"	7.018
600a		
NEMA 1	EA.	8.000
NEMA 3r	"	8.000
NEMA 4	"	8.989
NEMA 12	"	12.308
Non-fused, 240-600v, heavy duty, 3 phase, 30 amp		
NEMA 1	EA.	1.143
NEMA 3r	"	1.143
NEMA 4	"	1.739
NEMA 12	"	1.739
60a		
NEMA1	EA.	1.143
NEMA 3r	"	1.143
NEMA 4	"	1.739
NEMA 12	"	1.739
100a		
NEMA 1	EA.	1.739
NEMA 3r	"	1.739
NEMA 4	"	2.500
NEMA 12	"	2.500
200a, NEMA 1	"	2.500
600a, NEMA 12	"	12.308
Bolt-on hubs		
3/4" - 1-1/2"	EA.	0.250
2"	"	0.296
2-1/2"	"	0.296
3"	"	0.348
3-1/2"	"	0.400
4"	"	0.400

POWER GENERATION

16360.10 SAFETY SWITCHES

	UNIT	MAN/ HOURS
Watertight hubs		
1/2"	EA.	0.250
3/4"	"	0.296
1"	"	0.400
1-1/4"	"	0.444
1-1/2"	"	0.471
2"	"	0.500
2-1/2"	"	0.533
3"	"	0.615
3-1/2"	"	0.800
4"	"	0.851
Non-fused, 600v, 3 pole, NEMA 7		
600a	EA.	2.222
100a	"	3.200
225a	"	4.000
NEMA 9		
60a	EA.	2.500
100a	"	3.333
225a	"	4.211
Fusible bolted pressure switches, 600v/3 pole, NEMA 1		
800a	EA.	16.000
1200a	"	21.978
1600a	"	25.000
2000a	"	29.963
2500a	"	34.934
3000a	"	44.944
4000a	"	51.948
Non-fusible		
800a	EA.	14.493
1200a	"	20.000
1600a	"	22.989
2000a	"	27.972
2500a	"	34.934
3000a	"	44.944
4000a	"	51.948
Fusible load interrupter switches, 4.16 kv, NEMA 1		
200a	EA.	29.963
600a	"	70.175
Fusible load interrupter switch, 13.8 kv		
NEMA 1, 600a	EA.	100
NEMA 3R, 600a	"	100
4.16 kv, NEMA 3R		
200a	EA.	29.963
600a	"	70.175
Non-fused load interrupter switch, 4.16 kv, NEMA 1		
200a	EA.	29.963
600a	"	70.175
13.8 kv, NEMA 1, 600a	"	100
4.16 kv, NEMA 3R		

16 ELECTRICAL

POWER GENERATION	UNIT	MAN/ HOURS
16360.10 SAFETY SWITCHES		
200a	EA.	29.963
600a	"	70.175
13.8 kv, NEMA 3R, 600a	"	100
Interrupter switch accessories, strip heater	"	
Cable lugs	"	
Key interlock	"	
Auxiliary switch	"	
5 kv lightning arrester		
5 kva	EA.	
15 kv	"	
16365.10 FUSES		
Fuse, one-time, 250v		
30a	EA.	0.050
60a	"	0.050
100a	"	0.050
200a	"	0.050
400a	"	0.050
600a	"	0.050
600v		
30a	EA.	0.050
60a	"	0.050
100a	"	0.050
200a	"	0.050
400a	"	0.050
Fusetron, 600v		
200a	EA.	0.050
400a	"	0.050
Fuse, amp-trap, K1, 250v		
30a	EA.	0.050
60a	"	0.050
100a	"	0.050
200a	"	0.050
400a	"	0.050
600a	"	0.050
600v		
30a	EA.	0.050
60a	"	0.050
100a	"	0.050
200a	"	0.050
400a	"	0.050
K5, 250v		
30a	EA.	0.050
60a	"	0.050
100a	"	0.050
200a	"	0.050
400a	"	0.050
600a	"	0.050
600v		

POWER GENERATION	UNIT	MAN/ HOURS
16365.10 FUSES		
30a	EA.	0.050
60a	"	0.050
100a	"	0.050
200a	"	0.050
400a	"	0.050
600a	"	0.050
J, 600v		
30a	EA.	0.050
60a	"	0.050
100a	"	0.050
200a	"	0.050
400a	"	0.050
L, 600v		
1200a	EA.	0.400
1600a	"	0.400
2000a	"	0.400
2500a	"	0.400
3000a	"	0.400
4000a	"	0.400
5000a	"	0.400
Fuse cl-ay 250v		
600a	EA.	0.296
1200a	"	0.296
1600a	"	0.296
2000a	"	0.296
600v		
1200a	EA.	0.296
1600a	"	0.296
2000a	"	0.296
Reducers, 600v		
60a-30a	EA.	0.145
100a-30a	"	0.145
100a-60a	"	0.145
200a-60a	"	0.250
200a-100a	"	0.250
400a-100a	"	0.348
400a-200a	"	0.348
600a-100a	"	40.000
600a-200a	"	40.000
600a-400a	"	40.000
16395.10 GROUNDING		
Ground rods, copper clad, 1/2" x		
6'	EA.	0.667
8'	"	0.727
10'	"	1.000
5/8" x		
5'	EA.	0.615
6'	"	0.727

16 ELECTRICAL

POWER GENERATION

16395.10 GROUNDING	UNIT	MAN/HOURS
8'	EA.	1.000
10'	"	1.250
3/4" x		
8'	EA.	0.727
10'	"	0.800
Ground rod clamp		
5/8"	EA.	0.123
3/4"	"	0.123
Coupling, on threaded rods, 3/4"	"	0.050
Ground receptacles	"	0.250
Bus bar, copper, 2" x 1/4"	L.F.	0.145
Copper braid, 1" x 1/8", for door ground	EA.	0.100
Brazed connection for		
#6 wire	EA.	0.500
#2 wire	"	0.800
#2/0 wire	"	1.000
#4/0 wire	"	1.143
Ground rod couplings		
1/2"	EA.	0.100
5/8"	"	0.100
Ground rod, driving stud		
1/2"	EA.	0.100
5/8"	"	0.100
3/4"	"	0.100
Ground rod clamps, #8-2 to		
1" pipe	EA.	0.200
2" pipe	"	0.250
3" pipe	"	0.296
5" pipe	"	0.348
6" piep	"	0.444
#4-4/0 to		
1" pipe	EA.	0.200
2" pipe	"	0.250
3" pipe	"	0.296
3" pipe	"	0.348
8 pipe	"	0.444
8 pipe	"	0.667
10 pipe	"	0.952
12 pipe	"	1.290

SERVICE AND DISTRIBUTION

16425.10 SWITCHBOARDS	UNIT	MAN/HOURS
Switchboard, 90" high, no main disconnect, 208/120v		
400a	EA.	7.921
600a	"	8.000
1000a	"	8.000
1200a	"	10.000
1600a	"	11.940
2000a	"	14.035
2500a	"	16.000
240/480v		
600a	EA.	8.163
800a	"	8.163
1600a	"	11.940
2000a	"	14.035
2500a	"	16.000
3000a	"	27.586
4000a	"	29.630
Main breaker sections, 600v		
1200a, GFI	EA.	16.985
1600a, GFI	"	19.512
2000a, GFI	"	20.000
2500a, GFI	"	25.000
3000a, GFI	"	29.963
4000a, GFI	"	34.934
Switchboard meter sections, 600v		
400a	EA.	8.000
600a	"	10.000
800a	"	11.004
1000a	"	13.491
2000a	"	16.000
2500a	"	20.000
3000a	"	25.000
4000a	"	29.963
Insulated case, draw out compartment, 208/120v		
800a	EA.	2.500
1600a	"	2.963
2000a	"	3.478
2500a	"	3.478
3000a	"	4.000
4000a	"	4.790
Accessories for power trip breakers		
Shunt trip	EA.	0.500
Key interlock	"	2.222
Lifting and transport truck	"	4.494
Lifting device	"	1.333
Bus duct connection, 3 phase, 4 wire		
225a	EA.	3.333
400a	"	2.963
600a	"	3.333
800a	"	4.000

16 ELECTRICAL

SERVICE AND DISTRIBUTION

16425.10 SWITCHBOARDS	UNIT	MAN/HOURS
2500a	EA.	6.015
3000a	"	7.477
4000a	"	8.791
Provision for mounting current transformers		
800a & below primary	EA.	2.963
1000 to 1500a primary	"	2.963
2000 to 6000a primary	"	2.963
Provision for mounting potential transformers		
2000a max	EA.	3.810
Switchboard instruments		
Voltmeter	EA.	1.000
Ammeter, incoming line	"	1.000
Wattmeter	"	1.000
Varmeter	"	1.000
Power factor meter	"	1.000
Frequency meter	"	1.000
Recording voltmeter	"	2.000
Wattmeter	"	2.000
Power factor meter	"	2.000
Frequency meter	"	2.000
Instrument phase select switch	"	0.500
Enclosure, 90" high, 3 phase, 4 wire		
1000a	EA.	6.838
1200a	"	7.018
1600a	"	8.602
2000a	"	13.333
5500a	"	15.686
3000a	"	18.182
4000a	"	23.529
Circuit breakers, 600v, 100a, frame		
15-30a, 1 pole	EA.	0.296
15-60a, 2 pole	"	0.348
70-100a, 2 pole	"	0.400
15-60a, 3 pole	"	0.444
70-100a, 3 pole	"	0.500
Bolt on breakers, 600v, 225a frame, 110-225a		
2 pole	EA.	0.615
3 pole	"	1.096
400a frame, 250-400a, 2 pole	"	1.250
800a frame		
450-600a, 2 pole	EA.	1.905
700-800a, 2 pole	"	2.500
450-600a, 3 pole	"	4.211
700-800a, 3 pole	"	4.444
Bolt on branch breakers, 600v		
1000-2000a, 2 pole	EA.	5.333
2500a, 2 pole	"	10.753
1000-2000a, 3 pole	"	8.000
2500a, 3 pole	"	11.004

SERVICE AND DISTRIBUTION

16425.10 SWITCHBOARDS	UNIT	MAN/HOURS
3000a, 3 pole	EA.	20.000
Metal clad substation switch board, selector switch		
600a, 5kv	EA.	42.105
600a, 15kv	"	47.904
Fused switch, 600a		
5kv	EA.	34.934
15kv	"	34.934
1200a		
5kv	EA.	40.000
15kv	"	40.000
Oil cutout switch		
5 kv	EA.	15.009
15 kv	"	18.018
Liquid air terminal section	"	8.000
Dry air terminal section	"	8.502
Auxillary compartment	"	29.963

16430.20 METERING	UNIT	MAN/HOURS
Outdoor wp meter sockets, 1 gang, 240v, 1 phase		1.509
Includes sealing ring 100a	EA.	1.509
150a	"	1.778
200a	"	2.000
Die cast hubs, 1-1/4"	"	0.320
1-1/2"	"	0.320
2"	"	0.320
Indoor meter center, main switch single phase, 240v		
400a	EA.	8.000
600a	"	11.004
800a	"	11.696
Main breaker		
400a	EA.	8.000
600a	"	11.004
800a	"	11.696
1000a	"	16.000
Terminal box	"	16.495
1600a	"	18.018
Terminal box		
800a	EA.	10.000
1600a	"	18.018
Main switch, three phase, 208v		
400a	EA.	8.502
600a	"	11.994
800a	"	13.491
Main breaker		
400a	EA.	8.502
600a	"	11.994
800a	"	13.491
1000a	"	16.985
1200a	"	18.018

16 ELECTRICAL

SERVICE AND DISTRIBUTION

16430.20 METERING

SERVICE AND DISTRIBUTION	UNIT	MAN/HOURS
1600a	EA.	20.997
Terminal box		
800a	EA.	13.008
1600a	"	20.997
Indoor meter center		
2 meters	EA.	5.000
3 meters	"	6.154
4 meters	"	7.273
5 meters	"	8.000
6 meters	"	8.999
Plug on breakers, single phase, 208v		
60a	EA.	0.250
70a	"	0.250
80a	"	0.250
90a	"	0.250
100a	"	0.348
Indoor meter center, single phase, 125a breakers		
3 meters	EA.	6.154
4 meters	"	7.273
5 meters	"	8.000
6 meters	"	8.502
7 meters	"	10.000
8 meters	"	11.004
10 meters	"	11.994
150a breakers		
3 meters	EA.	6.154
4 meters	"	7.273
6 meters	"	8.000
7 meters	"	10.000
8 meters	"	11.004
200a breakers		
3 meters	EA.	6.154
4 meters	"	7.273
6 meters	"	8.000
7 meters	"	10.000
8 meters	"	11.004
Indoor meter center, three phase, 125a breakers		
3 meters	EA.	6.154
4 meters	"	7.273
5 meters	"	8.000
6 meters	"	8.999
7 meters	"	10.000
8 meters	"	11.004
10 meters	"	11.994
150a breakers		
3 meters	EA.	6.154
4 meters	"	7.273
6 meters	"	8.502
7 meters	"	11.004

SERVICE AND DISTRIBUTION

16430.20 METERING

SERVICE AND DISTRIBUTION	UNIT	MAN/HOURS
8 meters	EA.	11.994
200a breakers		
3 meters	EA.	6.667
4 meters	"	7.273
6 meters	"	8.999
7 meters	"	11.004
8 meters	"	11.994
NEMA 3R, meter center, main switch, 1 phase, 240v		
400a	EA.	8.000
600a	"	10.000
800a	"	11.004
Main breaker		
400a	EA.	8.000
600a	"	10.000
800a	"	7.273
1000a	"	15.009
1200a	"	16.000
Terminal box		
225a	EA.	7.273
800a	"	11.494
1600a	"	18.018
NEMA 3R, three phase, 280v		
400a	EA.	8.502
600a	"	11.994
800a	"	13.008
Main breaker		
400a	EA.	8.502
600a	"	11.994
800a	"	13.008
1000a	"	16.985
1200a	"	18.018
Terminal box		
225a	EA.	8.000
800a	"	13.008
1600a	"	20.997
NEMA 3R meter center, single phase, 208v, 100a		
2 meters	EA.	5.000
3 meters	"	6.154
4 meters	"	7.273
5 meters	"	8.000
6 meters	"	8.999
4 meters	"	7.273
6 meters	"	8.000
7 meters	"	10.000
8 meters	"	11.004
125a, 3 meters	"	6.154
4 meters	"	7.273
6 meters	"	8.239
7 meters	"	10.000

SERVICE AND DISTRIBUTION

16430.20 METERING

SERVICE AND DISTRIBUTION	UNIT	MAN/ HOURS
8 meters	EA.	11.004
150a, 3 meters	"	6.154
4 meters	"	7.273
6 meters	"	8.502
7 meters	"	10.000
8 meters	"	11.004
NEMA 3R center, 3 phase, 208v, 125a breakers		
3 meters	EA.	6.154
4 meters	"	7.273
6 meters	"	8.502
7 meters	"	10.000
8 meters	"	11.004
150a		
3 meters	EA.	6.667
4 meters	"	7.273
6 meters	"	8.999
7 meters	"	10.499
8 meters	"	11.494
200a		
3 meters	EA.	6.667
4 meters	"	7.273
6 meters	"	8.999
7 meters	"	11.004
8 meters	"	11.494
NEMA 3R, center plug-on breakers, 208v, 1 phase		
60a	EA.	2.500
70a	"	0.250
90a	"	0.250
100a	"	0.348
125a	"	0.400

16460.10 TRANSFORMERS

SERVICE AND DISTRIBUTION	UNIT	MAN/ HOURS
Pad mounted, single phase, dry type, 480v-120/240v		
15 kva	EA.	8.000
25 kva	"	8.989
37.5 kva	"	10.000
50 kva	"	10.959
3 phase		
225 kva	EA.	25.000
300 kva	"	30.769
500 kva	"	38.095
750 kva	"	47.059
1000 kva	"	50.000
1500 kva	"	57.143
Substation transformers, outdoor, 5 kv - 208v		
112.5 kva	EA.	21.978
150 kva	"	24.024
225 kva	"	27.972
300 kva	"	29.963

SERVICE AND DISTRIBUTION

16460.10 TRANSFORMERS

SERVICE AND DISTRIBUTION	UNIT	MAN/ HOURS
500 kva	EA.	44.944
750 kva	"	55.172
1000 kva	"	65.041
15 kv, 208v		
112 kva	EA.	27.972
150 kva	"	29.963
225 kva	"	34.934
300 kva	"	40.000
500 kva	"	50.000
750 kva	"	60.150
1000 kva	"	70.175
5kv, 480v		
112kva	EA.	21.978
150 kva	"	24.024
225 kva	"	27.972
300 kva	"	29.963
500 kva	"	44.944
750 kva	"	55.172
1000 kva	"	65.041
1500 kva	"	74.766
2000 kva	"	89.888
2500 kva	"	110
15 kv, 480v		
112.5 kva	EA.	27.972
150 kva	"	29.963
225 kva	"	34.934
300 kva	"	40.000
500 kva	"	50.000
750 kva	"	60.150
1000 kva	"	70.175
1500 kva	"	80.000
2000 kva	"	89.888
2500 kva	"	119
Pad mounted 3 phase, 15 kv outdoor		
50 kva	EA.	10.256
75 kva	"	11.765
112 kva	"	12.903
150 kva	"	14.545
225 kva	"	15.385
300 kva	"	17.021
500 kva	"	27.586
750 kva	"	36.364
1000 kva	"	44.444
1500 kva	"	53.333
Dry type, for power gear, 5 kv indoor		
75 kva	EA.	16.000
112.5 kva	"	18.605
150 kva	"	21.053
225 kva	"	23.529

SERVICE AND DISTRIBUTION

16460.10 TRANSFORMERS

	UNIT	MAN/HOURS
300 kva	EA.	25.000
500 kva	"	27.586
750 kva	"	36.364

16470.10 PANELBOARDS

	UNIT	MAN/HOURS
Indoor load center, 1 phase 240v main lug only		
30a - 2 spaces	EA.	2.000
100a - 8 spaces	"	2.424
150a - 16 spaces	"	2.963
200a - 24 spaces	"	3.478
200a - 42 spaces	"	4.000
Main circuit breaker		
100a - 8 spaces	EA.	2.424
100a - 16 spaces	"	2.759
150a - 16 spaces	"	2.963
150a - 24 spaces	"	3.200
200a - 24 spaces	"	3.478
200a - 42 spaces	"	3.636
3 phase, 480/277v, main lugs only, 120a, 30 circuits	"	3.478
277/480v, 4 wire, flush surface		
225a, 30 circuits	EA.	4.000
400a, 30 circuits	"	5.000
600a, 42 circuits	"	6.015
208/120v, main circuit breaker, 3 phase, 4 wire		
100a		
12 circuits	EA.	5.096
20 circuits	"	6.299
30 circuits	"	7.018
225a		
30 circuits	EA.	7.767
42 circuits	"	9.524
400a		
30 circuits	EA.	14.815
42 circuits	"	16.000
600a, 42 circuits	"	18.182
120/208v, flush, 3 ph., 4 wire, main only		
100a		
12 circuits	EA.	5.096
20 circuits	"	6.299
30 circuits	"	7.018
225a		
30 circuits	EA.	7.767
42 circuits	"	9.524
400a		
30 circuits	EA.	14.815
42 circuits	"	16.000
600a, 42 circuits	"	18.182
Panelboard accessories		

SERVICE AND DISTRIBUTION

16470.10 PANELBOARDS

	UNIT	MAN/HOURS
Grounding bus	EA.	0.348
Handle lock device	"	0.145
Factory assembled panel		
1 pole space	EA.	0.348
2 pole space	"	0.145
3 pole space	"	0.133
Panelboards 1 phase, 240/120v main circuit breaker		
Single phase, 3 wire, 120/240v flush		
100a, 20 circuits	EA.	3.478
225a, 30 circuits	"	4.000
240/120v, main lugs only	"	
100a		
8 circuits	EA.	2.963
12 circuits	"	2.963
20 circuits	"	2.963
225a		
24 circuits	EA.	3.478
30 circuits	"	3.810
42 circuits	"	3.810
Distribution panelboards, 3 ph, main breaker		
225a	EA.	16.000
400a	"	18.018
600a	"	21.978
800a	"	24.024
1000a	"	27.972
1200a	"	29.963
Single phase		
225a	EA.	14.011
400a	"	16.000
600a	"	20.000
800a	"	24.024
1000a	"	27.972
1200a	"	29.963
Fusible distribution panelboards, 3 phase, 600v		
100a	EA.	14.011
200a	"	16.000
400a	"	20.000
600a	"	24.024
800a	"	27.972
Single phase		
100a	EA.	11.994
200a	"	14.011
400a	"	18.018
600a	"	21.978
800a	"	25.974
Hospital panels, operating room		
3kv - 208v	EA.	6.154
3kv - 277v	"	6.154
5kv - 208v	"	6.154

16 ELECTRICAL

SERVICE AND DISTRIBUTION	UNIT	MAN/ HOURS
16470.10 PANELBOARDS		
5kv - 277v	EA.	6.154
Coronary care		
3kv - 208v	EA.	7.273
3kv - 277v	"	7.273
5kv - 208v	"	7.273
5kv - 277v	"	7.273
Intensive care		
3kv - 208v	EA.	8.000
3kv - 277v	"	8.000
5kv - 208v	"	8.000
5kv - 277v	"	8.000
15kv - 208v	"	11.994
15kv - 277v	"	11.994
25kv - 208v	"	16.000
25kv - 277v	"	16.000
Explosion proof, 240v, m.l.b. 20a, single phase		
6 breakers	EA.	11.004
8 breakers	"	11.747
10 breakers	"	12.500
12 breakers	"	13.245
14 breakers	"	14.011
16 breakers	"	14.011
18 breakers	"	15.504
20 breakers	"	16.260
22 breakers	"	16.985
24 breakers	"	17.738
16480.10 MOTOR CONTROLS		
Motor generator set, 3 phase, 480/277v, w/controls		
10kw	EA.	27.586
15kw	"	30.769
20kw	"	32.000
25kw	"	34.783
30kw	"	36.364
40kw	"	38.095
50kw	"	40.000
60kw	"	44.444
75kw	"	50.000
100kw	"	61.538
125kw	"	66.667
150kw	"	66.667
200kw	"	72.727
250kw	"	72.727
300kw	"	80.000
2 pole, 230 volt starter, w/NEMA-1		
1 hp, 9 amp, size 00	EA.	1.000
2 hp, 18amp, size 0	"	1.000
3 hp, 27amp, size 1	"	1.000
5 hp, 45amp, size 1p	"	1.000

SERVICE AND DISTRIBUTION	UNIT	MAN/ HOURS
16480.10 MOTOR CONTROLS		
7-1/2 hp, 45a, size 2	EA.	1.000
15 hp, 90a, size 3	"	1.000
2 pole, w/NEMA-4 enclosure		
2 hp, 18a, size 1	EA.	1.600
5 hp, 45amp, size 1p	"	1.600
7-1/2 hp, 45a, size 2	"	1.600
3 pole, 2 hp, 9a, 200-575v starter		
W/NEMA-1, size 00	EA.	1.333
W/NEMA-4 enclosure, size 00	"	1.739
5hp, 18a		
W/NEMA-1 enclosure, size 0	EA.	1.333
W/NEMA-4 enclosure, size 0	"	1.739
7.5-10hp, 27a		
7.5-10hp 27a, w/NEMA-1 enclosure, size 1	EA.	1.333
W/NEMA-4 enclosure size 1	"	1.739
10-25hp, 45a		
W/NEMA-1 enclosure, size 2	EA.	1.333
W/NEMA-4 enclosure, size 2	"	1.739
25-50hp, 90a		
W/NEMA-1 enclosure, size 3	EA.	1.739
W/NEMA-4 enclosure, size 3	"	2.500
40-100hp, 135a		
W/NEMA-1 enclosure, size 4	EA.	2.500
W/NEMA-4 enclosure, size 4	"	3.478
75-200hp, 270a		
W/NEMA-1 enclosure, size 5	EA.	5.517
W/NEMA-4 enclosure, size 5	"	7.018
Magnetic starter accessories		
On-off-auto selector switch kit	EA.	0.320
With pilot light	"	0.348
Control center main lug only, 208v, 3 phase		
600a	EA.	11.994
1200a	"	16.000
Main circuit breakers, 208v, 3 phase		
400a	EA.	10.000
600a	"	14.011
800a	"	16.000
1000a	"	18.018
1200a	"	20.000
Non-reversing starters		
Size 1	EA.	0.727
Size 2	"	1.250
Size 3	"	1.509
Size 4	"	1.739
Reversing starters		
Size 1	EA.	0.727
Size 2	"	1.096
Fusible switch, non-revolving starters		
Size 1	EA.	0.727

SERVICE AND DISTRIBUTION

16480.10 MOTOR CONTROLS

SERVICE AND DISTRIBUTION	UNIT	MAN/ HOURS
Size 2	EA.	1.250
Size 3	"	1.509
Size 4	"	1.739
Reversing starters		
Size 1	EA.	0.727
Size 2	"	1.096
Two speed, non-reversing starter		
Size 1	EA.	0.727
Size 2	"	1.096
Magnetic starter, 600v, 2 pole, NEMA 3R		
Size 0, 2 hp	EA.	1.000
Size 1, 5hp	"	1.096
NEMA 3R		
Size 2, 7.5 hp	EA.	1.143
Size 3, 15 hp	"	1.194
NEMA 7		
Size 0, 2 hp	EA.	1.739
Size 1, 5 hp	"	2.000
Size 2, 7.5 hp	"	2.222
NEMA 12		
Size 0, 2 hp	EA.	1.509
Size 1, 5 hp	"	1.739
Size 2, 7.5 hp	"	2.000
Size 3, 15 hp	"	2.222
3 pole, NEMA 1		
Size 00	EA.	
Size 0	"	
Size 1	"	
Size 2	"	
Size 3	"	
Size 4	"	
Size 5	"	
Size 6	"	10.000
Size 7	"	11.994
Size 8	"	16.000
NEMA 4		
Size 00	EA.	
Size 0	"	
Size 1	"	
Size 2	"	
Size 3	"	
Size 4	"	
Size 5	"	
Size 6	"	14.011
Size 7	"	16.000
Size 8	"	20.000
NEMA 3R		
Size 0	EA.	1.250
Size 1	"	1.356

SERVICE AND DISTRIBUTION

16480.10 MOTOR CONTROLS

SERVICE AND DISTRIBUTION	UNIT	MAN/ HOURS
Size 2	EA.	1.818
Size 3	"	1.905
Size 4	"	2.759
NEMA 7		
Size 0	EA.	2.000
Size 1	"	2.162
Size 2	"	2.222
Size 3	"	2.759
Size 4	"	4.444
Size 5	"	9.744
Size 6	"	16.000
NEMA 12		
Size 00	EA.	1.739
Size 0	"	1.860
Size 1	"	1.905
Size 2	"	2.000
Size 3	"	2.500
Size 4	"	3.478
Size 5	"	7.273
Size 6	"	12.500
Size 7	"	14.011
Size 8	"	20.000
Reversing magnetic starters, 600v, 3 pole, NEMA 1		
Size 00	EA.	1.250
Size 0	"	1.290
Size 1	"	1.356
Size 2	"	1.509
Size 3	"	1.739
Size 4	"	2.000
Size 5	"	5.333
Size 6	"	9.501
Size 7	"	11.004
Size 8	"	18.018
NEMA 4		
Size 0	EA.	1.739
Size 4	"	1.818
Size 2	"	1.905
Size 3	"	2.000
Size 4	"	2.500
Size 5	"	7.273
Size 6	"	12.500
Size 7	"	15.009
Size 8	"	20.000
NEMA 7		
Size 0	EA.	2.000
Size 1	"	2.222
Size 2	"	2.500
Size 3	"	2.963
NEMA 12		

SERVICE AND DISTRIBUTION	UNIT	MAN/ HOURS
16480.10 MOTOR CONTROLS		
Size 0	EA.	1.739
Size 1	"	1.905
Size 2	"	2.000
Size 3	"	2.222
Size 4	"	2.500
Size 5	"	7.273
Size 6	"	14.011
Size 7	"	16.000
Size 8	"	20.000
Electrically held lighting contactors, NEMA 1, 20a		
2 pole	EA.	1.000
3 pole	"	1.250
4 pole	"	1.509
6 pole	"	2.000
8 pole	"	2.500
10 pole	"	2.963
12 pole	"	3.478
30a		
2 pole	EA.	1.000
3 pole	"	1.250
4 pole	"	1.509
5 pole	"	1.739
60a		
2 pole	EA.	1.000
3 pole	"	1.250
4 pole	"	1.509
5 pole	"	1.739
100a		
2 pole	EA.	1.250
3 pole	"	1.739
4 pole	"	2.222
5 pole	"	2.759
200a		
2 pole	EA.	2.759
3 pole	"	2.500
4 pole	"	3.200
300a		
2 pole	EA.	4.211
3 pole	"	5.333
400a		
2 pole	EA.	4.211
3 pole	"	5.333
600a		
2 pole	EA.	6.667
3 pole	"	9.249
800a		
2 pole	EA.	8.000
3 pole	"	11.004
Mechanically held lighting contactors, NEMA 1, 20a		

SERVICE AND DISTRIBUTION	UNIT	MAN/ HOURS
16480.10 MOTOR CONTROLS		
2 pole	EA.	1.000
3 pole	"	1.250
4 pole	"	1.509
6 pole	"	2.000
8 pole	"	2.500
10 pole	"	2.963
30a		
2 pole	EA.	1.000
3 pole	"	1.250
4 pole	"	1.509
5 pole	"	1.739
60a		
2 pole	EA.	1.000
3 pole	"	1.250
4 pole	"	1.509
5 pole	"	1.739
100a		
2 pole	EA.	1.250
3 pole	"	1.739
4 pole	"	2.000
5 pole	"	2.500
200a		
2 pole	EA.	1.739
3 pole	"	2.500
4 pole	"	3.200
300a		
2 pole	EA.	4.211
3 pole	"	5.333
400a		
2 pole	EA.	4.211
3 pole	"	5.333
600a		
2 pole	EA.	6.667
3 pole	"	8.889
800a		
2 pole	EA.	8.000
3 pole	"	11.429
AC relays, control type open, 15a, 600v		
2 pole	EA.	1.000
3 pole	"	1.250
4 pole	"	1.509
6 pole	"	2.000
8 pole	"	2.500
10 pole	"	2.963
12 pole	"	3.478

16 ELECTRICAL

SERVICE AND DISTRIBUTION	UNIT	MAN/HOURS
16490.10 SWITCHES		
Oil switches, medium voltage, bus components		
Switches, 277/120v, toggle device only	EA.	1.600
With oil 35kv, g&w gram 44, 4 way switch	"	8.000
Weatherproof enclosure		
3 way switch	EA.	10.000
4 way switch	"	10.959
Fused interrupter load, 35kv		
20A		
1 pole	EA.	16.000
2 pole	"	17.021
3 way	"	17.021
4 way	"	18.182
30a, 1 pole	"	16.000
3 way	"	17.021
4 way	"	18.182
Weatherproof switch, including box & cover, 20a		
1 pole	EA.	16.000
2 pole	"	17.021
3 way	"	18.182
4 way	"	18.182
3 way, oil switch, 15kv enclosure	"	11.940
Pedestal for 35kv double breaker switch	"	5.000
Bus terminal connector, 2	"	2.500
2 to 3	"	2.500
Support connector, 3	"	1.600
Tee connector, 2 to 3	"	2.000
Flexible bus stud connector	"	1.739
End cap 3	"	1.333
Weldment connection, 3	"	1.000
Plate switch, 1 gang	"	0.050
Start stop stations, manual motor starters	"	0.727
Lockout switch	"	0.250
Forward-reverse switch	"	0.727
On-off switch	"	0.727
Open-close switch	"	0.727
Forward-reverse-stop switch	"	1.000
Standard 3 button switch any standard legend	"	1.000
Standard 3 button with lockout	"	1.000
Manual motor starters, tog, 115/230v		
Size 1 gp	EA.	1.000
Size 2	"	1.000
Button		
Size 0	EA.	1.000
Size 1	"	1.000
Size 2	"	1.000
3-phase		
Size 0	EA.	1.333
Size 1	"	1.333
Time & float switches	"	1.600

SERVICE AND DISTRIBUTION	UNIT	MAN/HOURS
16490.10 SWITCHES		
Astronomical time switch, 40a, 240v	EA.	1.000
Timer switch 0-5 minute, with box	"	0.500
Single pole/single throw time, 277v, NEMA-1	"	0.727
Single toggle switch, 20a, 120v, with pilot	"	0.250
3-way toggle	"	0.296
Photo electric switches		
1000 watt		
105-135v	EA.	0.727
208-277v	"	0.727
3000 watt, 105-130v		
Double throw	EA.	1.000
Single throw	"	1.000
Double pole/single throw, 210-250v	"	1.333
Dimmer switch and switch plate		
600 w	EA.	0.308
1000 w	"	0.348
Dimmer switch incandescent		
1500w	EA.	0.702
2000w	"	0.748
Fluorescent		
12 lamps	EA.	0.500
20 lamps	"	0.552
30 lamps	"	0.602
40 lamps	"	0.702
Time clocks with skip, 40a, 120v		
SPST	EA.	0.748
SPDT	"	0.748
DPST	"	0.748
DPDT	"	1.000
SPST	"	1.000
Astronomic time clocks with skip, 40a, 120v		
DPST	EA.	0.748
SPST	"	1.000
SPDT	"	0.748
Raintight time clocks, 40a, 120v		
SPDT	EA.	1.000
DPST	"	1.000
Contractor grade wall switch 15a, 120v		
Single pole	EA.	0.160
Three way	"	0.200
Four way	"	0.267
Specification grade toggle switches, 20a, 120-277v		
Single pole	EA.	0.200
Double pole	"	0.296
3 way	"	0.250
4 way	"	0.296
30a, 120-277v		
Single pole	EA.	0.200
Double pole	"	0.296

16 ELECTRICAL

SERVICE AND DISTRIBUTION

16490.10 SWITCHES

	UNIT	MAN/HOURS
3 way	EA.	0.250
Specification grade key switches, 20a, 120-277v		
Single pole	EA.	0.200
Double pole	"	0.296
3 way	"	0.250
4 way	"	0.296
Red pilot light handle switches, 20a, 120-277v		
Single pole	EA.	
Double pole	"	0.296
3 way	"	0.250
30a, 120-277v		
Single pole	EA.	0.200
Double pole	"	0.296
3 way	"	0.250
Momentary contact switches, 20a		
SPDT, ivory	EA.	0.250
SPDT, locking	"	0.296
Maintained contact switches		
SPDT ivory	EA.	0.250
DPDT ivory	"	0.250
SPDT locking	"	0.296
DPDT locking	"	0.348
Mercury switch, 3 way	"	0.250
Door switches, open on or off	"	0.500
Combination switch and pilot light, single pole	"	0.296
3 way	"	0.348
Combination switch and receptacle, single pole	"	0.296
3 way	"	0.296
Combination two switches, single pole/single pole	"	0.250
3 way	"	0.400
Switch plates, plastic ivory		
1 gang	EA.	0.080
2 gang	"	0.100
3 gang	"	0.119
4 gang	"	0.145
5 gang	"	0.160
6 gang	"	0.182
Stainless steel		
1 gang	EA.	0.080
2 gang	"	0.100
3 gang	"	0.123
4 gang	"	0.145
5 gang	"	0.160
6 gang	"	0.145
Brass		
1 gang	EA.	0.080
2 gang	"	0.100
3 gang	"	0.123
4 gang	"	0.145

SERVICE AND DISTRIBUTION

16490.10 SWITCHES

	UNIT	MAN/HOURS
5 gang	EA.	0.145
6 gang	"	0.182

16490.20 TRANSFER SWITCHES

	UNIT	MAN/HOURS
Automatic transfer switch 600v, 3 pole		
30a	EA.	3.478
60a	"	3.478
100a	"	4.762
150a	"	6.015
225a	"	8.000
260a	"	8.000
400a	"	10.000
600a	"	15.094
800a	"	18.182
1000a	"	21.053
1200a	"	22.857
1600a	"	25.000
2000a	"	29.630
2600a	"	42.105
Automatic transfer switches, 600v, 3 phase, 3000a	"	50.000

16490.80 SAFETY SWITCHES

	UNIT	MAN/HOURS
Safety switch, 600v, 3 pole, heavy duty, NEMA-1		
30a	EA.	1.000
60a	"	1.143
100a	"	1.600
200a	"	2.500
400a	"	5.517
600a	"	8.000
800a	"	10.526
1200a	"	14.286

LIGHTING

16510.05 INTERIOR LIGHTING

	UNIT	MAN/HOURS
Recessed fluorescent fixtures, 2'x2'		
2 lamp	EA.	0.727
4 lamp	"	0.727
2 lamp w/flange	"	1.000
4 lamp w/flange	"	1.000

16 ELECTRICAL

LIGHTING	UNIT	MAN/HOURS
16510.05 INTERIOR LIGHTING		
1'x4'		
2 lamp	EA.	0.667
3 lamp	"	0.667
2 lamp w/flange	"	0.727
3 lamp w/flange	"	0.727
2'x4'		
2 lamp	EA.	0.727
3 lamp	"	0.727
4 lamp	"	0.727
2 lamp w/flange	"	1.000
3 lamp w/flange	"	1.000
4 lamp w/flange	"	1.000
4'x4'		
4 lamp	EA.	1.000
6 lamp	"	1.000
8 lamp	"	1.000
4 lamp w/flange	"	1.509
6 lamp w/flange	"	1.509
8 lamp, w/flange	"	1.509
Surface mounted incandescent fixtures		
40w	EA.	0.667
75w	"	0.667
100w	"	0.667
150w	"	0.667
Pendant		
40w	EA.	0.800
75w	"	0.800
100w	"	0.800
150w	"	0.800
Contractor grade recessed down lights		
100 watt housing only	EA.	1.000
150 watt housing only	"	1.000
100 watt trim	"	0.500
150 watt trim	"	0.500
Recessed incandescent fixtures		
40w	EA.	1.509
75w	"	1.509
100w	"	1.509
150w	"	1.509
Exit lights, 120v		
Recessed	EA.	1.250
Back mount	"	0.727
Universal mount	"	0.727
Emergency battery units, 6v-120v, 50 unit	"	1.509
With 1 head	"	1.509
With 2 heads	"	1.509
Mounting bucket	"	0.727
Light track single circuit		
2'	EA.	0.500

LIGHTING	UNIT	MAN/HOURS
16510.05 INTERIOR LIGHTING		
4'	EA.	0.500
8'	"	1.000
12'	"	1.509
Fittings and accessories		
Dead end	EA.	0.145
Starter kit	"	0.250
Conduit feed	"	0.145
Straight connector	"	0.145
Center feed	"	0.145
L-connector	"	0.145
T-connector	"	0.145
X-connector	"	0.200
Cord and plug	"	0.100
Rigid corner	"	0.145
Flex connector	"	0.145
2 way connector	"	0.200
Spacer clip	"	0.050
Grid box	"	0.145
T-bar clip	"	0.050
Utility hook	"	0.145
Fixtures, square		
R-20	EA.	0.145
R-30	"	0.145
40w flood	"	0.145
40w spot	"	0.145
100w flood	"	0.145
100w spot	"	0.145
Mini spot	"	0.145
Mini flood	"	0.145
Quartz, 500w	"	0.145
R-20 sphere	"	0.145
R-30 sphere	"	0.145
R-20 cylinder	"	0.145
R-30 cylinder	"	0.145
R-40 cylinder	"	0.145
R-30 wall wash	"	0.145
R-40 wall wash	"	0.145
Explosion proof, incan., surface mounted		
100w - 200w	EA.	1.739
300w	"	1.739
500w	"	1.739
With guard		
100w-200w	EA.	2.222
300w	"	2.222
500w	"	2.222
Reflectors for incan. light fixtures, dome	"	0.250
Angle	"	0.250
Highbay	"	0.296
Explosion proof fluor. fixtures, 800 ms.		

16 ELECTRICAL

LIGHTING	UNIT	MAN/ HOURS
16510.05 INTERIOR LIGHTING		
1 lamp	EA.	2.222
2 lamp	"	2.667
3 lamp	"	2.963
4 lamp	"	3.200
Explosion proof hp sodium fixtures		
50w-70w	EA.	2.222
100w	"	2.222
150w	"	2.500
200w	"	2.500
250w	"	2.500
310w	"	2.500
400w	"	2.500
With guard		
50w-70w	EA.	2.500
100w	"	2.500
150w	"	2.759
200w	"	2.759
250w	"	2.759
310w	"	2.759
400w	"	2.759
Explosion proof metal halide fixtures		
175w	EA.	2.500
250w	"	2.500
400w	"	2.500
With guard, 175w	"	2.759
250w	"	2.759
400w	"	2.759
Energy saving rapid start fluor. lamps		
F30cw	EA.	0.100
F40 cw	"	0.100
F40 cwx	"	0.100
F30 ww	"	0.100
F40 ww	"	0.100
F40 wwx	"	0.100
Slimline		
F48 cw	EA.	0.145
F96 cwx	"	0.145
F48 ww	"	0.145
F96 ww	"	0.145
F96 wwx	"	0.145
High output	"	0.145
F96 cwx	"	0.145
F96 cw	"	0.145
Power groove, F48 cw	"	0.145
Circle		
Fc6 cw	EA.	0.100
Fc8 cw	"	0.100
Fc12 cw	"	0.100
Fc16 cw	"	0.100

LIGHTING	UNIT	MAN/ HOURS
16510.05 INTERIOR LIGHTING		
Fc6 ww	EA.	0.100
Fc8 ww	"	0.100
Fc12 ww	"	0.100
Fc16 ww	"	0.100
Incandescent lamps		
200w	EA.	0.100
300w	"	0.100
500w	"	0.100
750w	"	0.145
1000w	"	0.200
1500w	"	0.200
Energy saving reflector floodlight lamps		
25w	EA.	0.100
30w	"	0.100
50w	"	0.100
75w	"	0.100
120w	"	0.100
150w	"	0.100
200w	"	0.100
300w	"	0.100
500w	"	0.145
750w	"	0.145
Reflector spotlight		
75w	EA.	0.100
100w	"	0.100
125w	"	0.100
150w	"	0.100
250w	"	0.100
300w	"	0.100
400w	"	0.145
500w	"	0.145
1000w	"	0.200
Medium par flood lamps		
75w	EA.	0.100
100w	"	0.100
150w	"	0.100
200w	"	0.100
300w	"	0.100
500w	"	0.145
Medium par spot lamps		
75w	EA.	0.100
120w	"	0.100
150w	"	0.100
Tubular quartz lamps		
100w	EA.	0.145
150w	"	0.145
200w	"	0.145
400w	"	0.145
500w	"	0.200

LIGHTING	UNIT	MAN/ HOURS
16510.05 INTERIOR LIGHTING		
750w	EA.	0.200
1000w	"	0.250
1250w	"	0.250
1500w	"	0.250
Ballast replacements rapid start fluor		
1f-40-120v	EA.	0.727
1f-40-277v	"	0.727
1f-96-120v	"	0.727
1f-96-277v	"	0.727
2f-40-120v	"	0.727
2f-40-277v	"	0.727
2f-96-120v	"	0.727
2f-96-277v	"	0.727
Circline, 1fc6-1fc16	"	0.727
Very high output, 1500ma		
1f48-120v	EA.	0.727
1f48-277v	"	0.727
1f96-120v	"	0.727
1f96-277v	"	0.727
2f48-120v	"	0.727
2f48-277v	"	0.727
2f96-120v	"	0.727
2f96-277v	"	0.727
Mercury, multi tap		
475w	EA.	1.000
100w	"	1.000
175w	"	1.000
250w	"	1.000
400w	"	1.000
1000W	"	1.000
Metal halide, multi tap		
175w	EA.	1.000
250w	"	1.000
400w	"	1.000
1000w	"	1.000
1500w	"	1.000
High pressure sodium		
70w	EA.	1.000
100w	"	1.000
150w	"	1.000
250w	"	1.000
400w	"	1.000
1000w	"	1.000
Surface mounted fluorescent, wrap around lens		
1 lamp	EA.	0.800
2 lamps	"	0.889
4 lamps	"	1.000
Wall mounted fluorescent		
2-20w lamps	EA.	0.500

LIGHTING	UNIT	MAN/ HOURS
16510.05 INTERIOR LIGHTING		
2-30w lamps	EA.	0.500
2-40w lamps	"	0.667
Indirect, with wood shielding, 2049w lamps		
4'	EA.	1.000
8'	"	1.600
Industrial fluorescent, 2 lamp		
4'	EA.	0.727
8'	"	1.333
Strip fluorescent		
4'		
1 lamp	EA.	0.667
2 lamps	"	0.667
8'		
1 lamp	EA.	0.727
2 lamps	"	0.889
Wire guard for strip fixture, 4' long	"	0.348
Strip fluorescent, 8' long, two 4' lamps	"	1.333
With four 4' lamps	"	1.600
Wet location fluorescent, plastic housing		
4' long		
1 lamp	EA.	1.000
2 lamps	"	1.333
8' long		
2 lamps	EA.	1.600
4 lamps	"	1.739
Parabolic troffer, 2'x2'		
With 2 "U" lamps	EA.	1.000
With 3 "U" lamps	"	1.143
2'x4'		
With 2 40w lamps	EA.	1.143
With 3 40w lamps	"	1.333
With 4 40w lamps	"	1.333
1'x4'		
With 1 t12 lamp, 9 cell	EA.	0.727
With 2 t12 lamps	"	0.889
With 1 t12 lamp, 20 cell	"	0.727
With 2 t12 lamps	"	0.889
Steel sided surface fluorescent, 2'x4'		
3 lamps	EA.	1.333
4 lamps	"	1.333
Outdoor sign fluor., 1 lamp, remote ballast		
4' long	EA.	6.015
6' long	"	8.000
Recess mounted, commercial, 2'x2', 13" high		
100w	EA.	4.000
250w	"	4.494
High pressure sodium, hi-bay open		
400w	EA.	1.739
1000w	"	2.424

16 ELECTRICAL

LIGHTING	UNIT	MAN/ HOURS
16510.05 INTERIOR LIGHTING		
Enclosed		
400w	EA.	2.424
1000w	"	2.963
Metal halide hi-bay, open		
400w	EA.	1.739
1000w	"	2.424
Enclosed		
400w	EA.	2.424
1000w	"	2.963
High pressure sodium, low bay, surface mounted		
100w	EA.	1.000
150w	"	1.143
250w	"	1.333
400w	"	1.600
Metal halide, low bay, pendant mounted		
175w	EA.	1.333
250w	"	1.600
400w	"	2.222
Indirect luminare, square, metal halide, freestanding		
175w	EA.	1.000
250w	"	1.000
400w	"	1.000
High pressure sodium		
150w	EA.	1.000
250w	"	1.000
400w	"	1.000
Round, metal halide		
175w	EA.	1.000
250w	"	1.000
400w	"	1.000
High pressure sodium		
150w	EA.	1.000
250w	"	1.000
400w	"	1.000
Wall mounted, metal halide		
175w	EA.	2.500
250w	"	2.500
400w	"	3.200
High pressure sodium		
150w	EA.	2.500
250w	"	2.500
400w	"	3.200
Wall pack lithonia, high pressure sodium		
35w	EA.	0.889
55w	"	1.000
150w	"	1.600
250w	"	1.739
Low pressure sodium		
35w	EA.	1.739

LIGHTING	UNIT	MAN/ HOURS
16510.05 INTERIOR LIGHTING		
55w	EA.	2.000
Wall pack hubbell, high pressure sodium		
35w	EA.	0.889
150w	"	1.600
250w	"	1.739
Compact fluorescent		
2-7w	EA.	1.000
2-13w	"	1.333
1-18w	"	1.333
Handball & racquet ball court, 2'x2', metal halide		
250w	EA.	2.500
400w	"	2.759
High pressure sodium		
250w	EA.	2.500
400w	"	2.759
Bollard light, 42" w/found., high pressure sodium		
70w	EA.	2.581
100w	"	2.581
150w	"	2.581
Light fixture lamps		
Lamp		
20w med. bipin base, cool white, 24"	EA.	0.145
30w cool white, rapid start, 36"	"	0.145
40w cool white "U", 3"	"	0.145
40w cool white, rapid start, 48"	"	0.145
70w high pressure sodium, mogul base	"	0.200
75w slimline, 96"	"	0.200
100w		
Incandescent, 100a, inside frost	EA.	0.100
Mercury vapor, clear, mogul base	"	0.200
High pressure sodium, mogul base	"	0.200
150w		
Par 38 flood or spot, incandescent	EA.	0.100
High pressure sodium, 1/2 mogul base	"	0.200
175w		
Mercury vapor, clear, mogul base	EA.	0.200
Mercury halide, clear, mogul base	"	0.200
250w		
High pressure sodium, mogul base	EA.	0.200
Mercury vapor, clear, mogul base	"	0.200
Metal halide, clear, mogul base	"	0.200
High pressure sodium, mogul base	"	0.200
400w		
Mercury vapor, clear, mogul base	EA.	0.200
Metal halide, clear, mogul base	"	0.200
High pressure sodium, mogul base	"	0.200
1000w		
Mercury vapor, clear, mogul base	EA.	0.250
High pressure sodium, mogul base	"	0.250

16 ELECTRICAL

LIGHTING	UNIT	MAN/ HOURS
16510.30 EXTERIOR LIGHTING		
Exterior light fixtures		
Rectangle, high pressure sodium		
70w	EA.	2.500
100w	"	2.581
150w	"	2.581
250w	"	2.759
400w	"	3.478
Flood, rectangular, high pressure sodium		
70w	EA.	2.500
100w	"	2.581
150w	"	2.581
400w	"	3.478
1000w	"	4.494
Round		
400w	EA.	3.478
1000w	"	4.494
Round, metal halide		
400w	EA.	3.478
1000w	"	4.494
Light fixture arms, cobra head, 6', high press. sodium		
100w	EA.	2.000
150w	"	2.500
250w	"	2.500
400w	"	2.963
Flood, metal halide		
400w	EA.	3.478
1000w	"	4.494
1500w	"	6.015
Mercury vapor		
250w	EA.	2.759
400w	"	3.478
Incandescent		
300w	EA.	1.739
500w	"	2.000
1000w	"	3.200
16510.90 POWER LINE FILTERS		
Heavy duty power line filter, 240v		
100a	EA.	10.000
300a	"	16.000
600a	"	24.242
16600.20 CENTRAL INVERTER SYSTEMS		
Central inverter systems		
500va	EA.	2.963
1000va	"	4.000
1500va	"	5.333
2400va	"	6.667

LIGHTING	UNIT	MAN/ HOURS
16600.20 CENTRAL INVERTER SYSTEMS		
3000va	EA.	8.502
4500va	"	10.000
6000va	"	11.004
7500va	"	14.011
10,000va	"	16.000
16,600va	"	22.989
25,000va	"	34.934
16610.30 UNINTERRUPTIBLE POWER		
Uninterruptible power systems, (U.P.S.)	EA.	8.000
5 kva	"	11.004
7.5 kva	"	16.000
10 kva	"	21.978
15 kva	"	26.667
20 kva	"	24.024
25 kva	"	25.000
30 kva	"	25.974
35 kva	"	27.027
40 kva	"	27.972
45 kva	"	28.986
50 kva	"	29.963
62.5 kva	"	32.000
75 kva	"	34.934
100 kva	"	36.036
150 kva	"	50.000
200 kva	"	55.172
300 kva	"	74.766
400 kva	"	89.888
500 kva	"	110
16670.10 LIGHTNING PROTECTION		
Lightning protection		
Copper point, nickel plated, 12'		
1/2" dia.	EA.	1.000
5/8" dia.	"	1.000

COMMUNICATIONS	UNIT	MAN/ HOURS
16720.10 FIRE ALARM SYSTEMS		
Master fire alarm box, pedestal mounted	"	16.000
Master fire alarm box	"	6.015
Box light	"	0.500
Ground assembly for box	"	0.667
Bracket for pole type box	"	0.727
Pull station		
Waterproof	EA.	0.500
Manual	"	0.400
Horn, waterproof	"	1.000
Interior alarm	"	0.727
Coded transmitter, automatic	"	2.000
Control panel, 8 zone	"	8.000
Battery charger and cabinet	"	2.000
Batteries, nickel cadmium or lead calcium	"	5.000
CO2 pressure switch connection	"	0.727
Annunciator panels		
Fire detection annunciator, remote type, 8-zone	EA.	1.818
12-zone	"	2.000
16-zone	"	2.500
Fire alarm systems		
Bell	EA.	0.615
Weatherproof bell	"	0.667
Horn	"	0.727
Siren	"	2.000
Chime	"	0.615
Audio/visual	"	0.727
Strobe light	"	0.727
Smoke detector	"	0.667
Heat detection	"	0.500
Thermal detector	"	0.500
Ionization detector	"	0.533
Duct detector	"	2.759
Test switch	"	0.500
Remote indicator	"	0.571
Door holder	"	0.727
Telephone jack	"	0.296
Fireman phone	"	1.000
Speaker	"	0.800
Remote fire alarm annunciator panel		
24 zone	EA.	6.667
48 zone	"	13.008
Control panel		
12 zone	EA.	2.963
16 zone	"	4.444
24 zone	"	6.667
48 zone	"	16.000
Power supply	"	1.509
Status command	"	5.000
Printer	"	1.509

COMMUNICATIONS	UNIT	MAN/ HOURS
16720.10 FIRE ALARM SYSTEMS		
Transponder	EA.	0.899
Transformer	"	0.667
Transceiver	"	0.727
Relays	"	0.500
Flow switch	"	2.000
Tamper switch	"	2.963
End of line resistor	"	0.348
Printed ckt. card	"	0.500
Central processing unit	"	6.154
UPS backup to c.p.u.	"	8.999
Smoke detector, fixed temp. & rate of rise comb.	"	1.600
16720.50 SECURITY SYSTEMS		
Sensors		
Balanced magnetic door switch, surface mounted	EA.	0.500
With remote test	"	1.000
Flush mounted	"	1.860
Mounted bracket	"	0.348
Mounted bracket spacer	"	0.348
Photoelectric sensor, for fence		
6 beam	EA.	2.759
9 beam	"	4.255
Photoelectric sensor, 12 volt dc		
500' range	EA.	1.600
800' range	"	2.000
Capacitance wire grid kit		
Surface	EA.	1.000
Duct	"	1.600
Tube grid kit	"	0.500
Vibration sensor, 30 max per zone	"	0.500
Audio sensor, 30 max per zone	"	0.500
Inertia sensor		
Outdoor	EA.	0.727
Indoor	"	0.500
Ultrasonic transmitter, 20 max per zone		
Omni-directional	EA.	1.600
Directional	"	1.333
Transceiver		
Omni-directional	EA.	1.000
Directional	"	1.000
Passive infra-red sensor, 20 max per zone	"	1.600
Access/secure control unit, balanced magnetic switch	"	1.600
Photoelectric sensor	"	1.600
Photoelectric fence sensor	"	1.600
Capacitance sensor	"	1.739
Audio and vibration sensor	"	1.600
Inertia sensor	"	1.600
Ultrasonic sensor	"	1.739

16 ELECTRICAL

COMMUNICATIONS	UNIT	MAN/HOURS
16720.50 SECURITY SYSTEMS		
Infra-red sensor	EA.	2.000
Monitor panel, with access/secure tone, standard	"	1.739
High security	"	2.000
Emergency power indicator	"	0.500
Monitor rack with 115v power supply		
1 zone	EA.	1.000
10 zone	"	2.500
Monitor cabinet, wall mounted		
1 zone	EA.	1.000
5 zone	"	1.600
10 zone	"	1.739
20 zone	"	2.000
Floor mounted, 50 zone	"	4.000
Security system accessories		
Tamper assembly for monitor cabinet	EA.	0.444
Monitor panel blank	"	0.348
Audible alarm	"	0.500
Audible alarm control	"	0.348
Termination screw, terminal cabinet		
25 pair	EA.	1.600
50 pair	"	2.500
150 pair	"	5.000
Universal termination, cabinets & panel		
Remote test	EA.	1.739
No remote test	"	0.727
High security line supervision termination	"	1.000
Door cord for capacitance sensor, 12"	"	0.500
Insulation block kit for capacitance sensor	"	0.348
Termination block for capacitance sensor	"	0.348
Guard alert display	"	0.615
Uninterrupted power supply		
Plug-in 40kva transformer		
12 volt	EA.	0.348
18 volt	"	0.348
24 volt	"	0.348
Test relay	"	0.348
Coaxial cable, 50 ohm	L.F.	0.006
Door openers	EA.	0.500
Push buttons		
Standard	EA.	0.348
Weatherproof	"	0.444
Bells	"	0.727
Horns		
Standard	EA.	1.000
Weatherproof	"	1.250
Chimes	"	0.667
Flasher	"	0.615
Motion detectors	"	1.509
Intercom units	"	0.727

COMMUNICATIONS	UNIT	MAN/HOURS
16720.50 SECURITY SYSTEMS		
Remote annunciator	EA.	5.000
16730.20 CLOCK SYSTEMS		
Clock systems		
Single face	EA.	0.800
Double face	"	0.800
Skeleton	"	2.759
Master	"	5.000
Signal generator	"	4.000
Elapsed time indicator	"	0.800
Controller	"	0.533
Clock and speaker	"	1.096
Bell		
Standard	EA.	0.533
Weatherproof	"	0.800
Horn		
Standard	EA.	0.727
Weatherproof	"	0.952
Chime	"	0.533
Buzzer	"	0.533
Flasher	"	0.615
Control Board	"	3.478
Program unit	"	5.000
Block back box	"	0.500
Double clock back box	"	0.667
Wire guard	"	0.200
16740.10 TELEPHONE SYSTEMS		
Communication cable		
25 pair	L.F.	0.026
100 pair	"	0.029
150 pair	"	0.033
200 pair	"	0.040
300 pair	"	0.042
400 pair	"	0.044
Cable tap in manhole or junction box		
25 pair cable	EA.	3.810
50 pair cable	"	7.547
75 pair cable	"	11.268
100 pair cable	"	15.094
150 pair cable	"	22.222
200 pair cable	"	29.630
300 pair cable	"	44.444
400 pair cable	"	61.538
Cable terminations, manhole or junction box		
25 pair cable	EA.	3.756

16 ELECTRICAL

COMMUNICATIONS

COMMUNICATIONS	UNIT	MAN/ HOURS
16740.10 TELEPHONE SYSTEMS		
50 pair cable	EA.	7.477
100 pair cable	"	15.094
150 pair cable	"	22.222
200 pair cable	"	29.630
300 pair cable	"	44.444
400 pair cable	"	61.538
Telephones, standard		
1 button	EA.	2.963
2 button	"	3.478
6 button	"	5.333
12 button	"	7.619
18 button	"	8.889
Hazardous area		
Desk	EA.	7.273
Wall	"	5.000
Accessories		
Standard ground	EA.	1.600
Push button	"	1.600
Buzzer	"	1.600
Interface device	"	0.800
Long cord	"	0.800
Interior jack	"	0.400
Exterior jack	"	0.615
Hazardous area		
Selector switch	EA.	3.200
Bell	"	3.200
Horn	"	4.211
Horn relay	"	3.077
16740.30 CALL SYSTEMS		
Call systems, single bed station	EA.	0.533
Double bed station	"	0.727
Call-in cord	"	0.200
Pull cord	"	0.200
Pillow speaker	"	0.276
Dome light	"	0.533
Zone light	"	0.533
Stake station	"	0.615
Duty station	"	0.500
Utility station	"	0.615
Nurses station	"	0.533
Surgical station	"	0.727
Master station	"	2.500
Control station	"	8.000
Annunciator	"	2.000
Power supply	"	1.538
Speakers	"	0.800
Foot switch	"	0.296
Code blue systems		

COMMUNICATIONS	UNIT	MAN/ HOURS
16740.30 CALL SYSTEMS		
Bed station	EA.	0.727
Dome light	"	0.667
Zone light	"	0.727
Pull cord	"	0.250
Nurses station	"	0.533
Annunciator	"	2.000
Power supply	"	1.455
Nurse station indicator, alarm annunciators, flush		
4 circuit	EA.	4.000
6 circuit	"	8.000
12 circuit	"	12.012
Desktop		
4 circuit	EA.	3.478
6 circuit	"	3.478
12 circuit	"	5.000
16750.20 SIGNALING SYSTEMS		
Signaling systems		
4" bell	EA.	0.602
6" bell	"	0.650
10" bell	"	0.748
Buzzer		
Size 0	EA.	0.444
Size 1	"	0.444
Size 2	"	0.500
Size 3	"	0.533
Horn	"	0.615
Chime	"	0.533
Push button		
Standard	EA.	0.400
Weatherproof	"	0.500
Door opener		
Mortise	EA.	0.500
Rim	"	0.400
Transformer	"	0.444
Contractor grade doorbell chime kit		
Chime	EA.	1.000
Doorbutton	"	0.320
Transformer	"	0.500
16770.30 SOUND SYSTEMS		
Power amplifiers	EA.	3.478
Pre-amplifiers	"	2.759
Tuner	"	1.455
Horn		
Equilizer	EA.	1.600
Mixer	"	2.222
Tape recorder	"	1.860

16 ELECTRICAL

COMMUNICATIONS

16770.30 SOUND SYSTEMS

COMMUNICATIONS	UNIT	MAN/HOURS
Microphone		
Cassette Player	EA.	2.162
Record player	"	1.905
Equipment rack	"	1.290
Speaker		
Ceiling	EA.	1.356
Wall	"	4.000
Paging	"	0.800
Column	"	0.533
Single	"	0.615
Double	"	4.444
Volume control	"	0.533
Plug-in	"	0.800
Desk	"	0.400
Outlet	"	0.400
Stand	"	0.296
Console	"	8.000
Power supply	"	1.290

16780.10 ANTENNAS AND TOWERS

COMMUNICATIONS	UNIT	MAN/HOURS
Guy cable, alumaweld		
1x3, 7/32"	L.F.	0.050
1x3, 1/4"	"	0.050
1x3, 25/64"	"	0.059
1x19, 1/2"	"	0.070
1x7, 35/64"	"	0.080
1x19, 13/16"	"	0.100
Preformed alumaweld end grip		
1/4" cable	EA.	0.100
3/8" cable	"	0.100
1/2" cable	"	0.145
9/16" cable	"	0.200
5/8" cable	"	0.250
Fiberglass guy rod, white epoxy coated		
1/4" dia.	L.F.	0.145
3/8" dia	"	0.145
1/2" dia	"	0.200
5/8" dia	"	0.250
Preformed glass grip end grip, guy rod		
1/4" dia.	EA.	0.145
3/8" dia.	"	0.200
1/2" dia.	"	0.250
5/8" dia.	"	0.250
Spelter socket end grip, 1/4" dia. guy rod		
Standard strength	EA.	0.500
High performance	"	0.500
3/8" dia. guy rod		
Standard strength	EA.	0.348
High performance	"	0.500

COMMUNICATIONS

16780.10 ANTENNAS AND TOWERS

COMMUNICATIONS	UNIT	MAN/HOURS
Timber pole, Douglas Fir		
80-85 ft	EA.	19.512
90-95 ft	"	22.222
Southern yellow pine		
35-45 ft	EA.	10.959
50-55 ft	"	14.035

16780.50 TELEVISION SYSTEMS

COMMUNICATIONS	UNIT	MAN/HOURS
TV outlet, self terminating, w/cover plate	EA.	0.308
Thru splitter	"	1.600
End of line	"	1.333
In line splitter multitap		
4 way	EA.	1.818
2 way	"	1.702
Equipment cabinet	"	1.600
Antenna		
Broad band uhf	EA.	3.478
Lightning arrester	"	0.727
TV cable	L.F.	0.005
Coaxial cable rg	"	0.005
Cable drill, with replacement tip	EA.	0.500
Cable blocks for in-line taps	"	0.727
In-line taps ptu-series 36 tv system	"	1.143
Control receptacles	"	0.449
Coupler	"	2.424
Head end equipment	"	6.667
TV camera	"	1.667
TV power bracket	"	0.800
TV monitor	"	1.455
Video recorder	"	2.105
Console	"	8.502
Selector switch	"	1.379
TV controller	"	1.404

RESISTANCE HEATING

16850.10 ELECTRIC HEATING

RESISTANCE HEATING	UNIT	MAN/HOURS
Baseboard heater		
2', 375w	EA.	1.000
3', 500w	"	1.000

16 ELECTRICAL

RESISTANCE HEATING

16850.10 ELECTRIC HEATING

RESISTANCE HEATING	UNIT	MAN/ HOURS
4', 750w	EA.	1.143
5', 935w	"	1.333
6', 1125w	"	1.600
7', 1310w	"	1.818
8', 1500w	"	2.000
9', 1680w	"	2.222
10', 1875w	"	2.286
Unit heater wall mounted		
1500w	EA.	1.600
2500w	"	1.818
4000w	"	2.286
Thermostat		
Integral	EA.	0.500
Line voltage	"	0.500
Electric heater connection	"	0.250
Fittings		
Inside corner	EA.	0.400
Outside corner	"	0.400
Receptacle section	"	0.400
Blank section	"	0.400
Unit heater, wall mounted		
750w	EA.	1.600
1500w	"	1.667
2000w	"	1.739
3000w	"	2.000
Infrared heaters		
600w	EA.	1.000
2000w	"	1.194
3000w	"	2.000
4000w	"	2.500
1000w	"	1.000
2000w	"	1.000
3000w	"	1.250
4000w	"	1.290
Controller	"	0.667
Wall bracket	"	0.727
Radiant ceiling heater panels		
500w	EA.	1.000
750w	"	1.000
Unit heaters, suspended, single phase		
3.0 kw	EA.	2.759
5.0 kw	"	2.759
7.5 kw	"	3.200
10.0 kw	"	3.810
Three phase		
5 kw	EA.	2.759
7.5 kw	"	3.200
10 kw	"	3.810
15 kw	"	4.211
20 kw	EA.	5.333
25 kw	"	1.509
30 kw	"	8.000
35 kw	"	8.000
Unit heater thermostat	"	0.533
Mounting bracket	"	0.727
Relay	"	0.615
Duct heaters, three phase		
10 kw	EA.	3.810
15 kw	"	3.810
17.5 kw	"	4.000
20 kw	"	6.154

CONTROLS

16910.40 CONTROL CABLE

CONTROLS	UNIT	MAN/ HOURS
Control cable, 600v, #14 THWN, PVC jacket		
2 wire	L.F.	0.008
4 wire	"	0.010
6 wire	"	0.131
8 wire	"	0.145
10 wire	"	0.160
12 wire	"	0.182
14 wire	"	0.211
16 wire	"	0.222
18 wire	"	0.242
20 wire	"	0.250
22 wire	"	0.286
Audio cables, shielded, #24 gauge		
3 conductor	L.F.	0.004
4 conductor	"	0.006
5 conductor	"	0.007
6 conductor	"	0.009
7 conductor	"	0.011
8 conductor	"	0.012
9 conductor	"	0.014
10 conductor	"	0.015
15 conductor	"	0.018
20 conductor	"	0.023
25 conductor	"	0.027
30 conductor	"	0.030

CONTROLS	UNIT	MAN/ HOURS
16910.40 CONTROL CABLE		
40 conductor	L.F.	0.036
50 conductor	"	0.042
#22 gauge		
3 conductor	L.F.	0.004
4 conductor	"	0.006
#20 gauge		
3 conductor	L.F.	0.004
10 conductor	"	0.015
15 conductor	"	0.018
#18 gauge		
3 conductor	L.F.	0.004
4 conductor	"	0.006
Microphone cables, #24 gauge		
2 conductor	L.F.	0.004
3 conductor	"	0.005
#20 gauge		
1 conductor	L.F.	0.004
2 conductor	"	0.004
2 conductor	"	0.004
3 conductor	"	0.006
4 conductor	"	0.007
5 conductor	"	0.009
7 conductor	"	0.011
8 conductor	"	0.012
Computer cables shielded, #24 gauge		
1 pair	L.F.	0.004
2 pair	"	0.004
3 pair	"	0.006
4 pair	"	0.007
5 pair	"	0.009
6 pair	"	0.011
7 pair	"	0.012
8 pair	"	0.014
50 pair	"	0.039
Coaxial cables		
RG 6/u	L.F.	0.006
RG 6a/u	"	0.006
RG 8/u	"	0.006
RG 8a/u	"	0.006
RG 9/u	"	0.006
RG 11/u	"	0.006
RG 58/u	"	0.006
RG 59/u	"	0.006
RG 62/u	"	0.006
RG 174/u	"	0.006
RG 213/u	"	0.006
MATV and CCTV camera cables		
1 conductor	L.F.	0.004
2 conductor	"	0.005

CONTROLS	UNIT	MAN/ HOURS
16910.40 CONTROL CABLE		
4 conductor	L.F.	0.006
7 conductor	"	0.001
12 conductor	"	0.015
13 conductor	"	0.016
14 conductor	"	0.018
28 conductor	"	0.027
Fire alarm cables, #22 gauge		
6 conductor	L.F.	0.010
9 conductor	"	0.015
12 conductor	"	0.016
#18 gauge		
2 conductor	L.F.	0.005
4 conductor	"	0.007
#16 gauge		
2 conductor	L.F.	0.007
4 conductor	"	0.008
#14 gauge		
2 conductor	L.F.	0.008
#12 gauge		
2 conductor	L.F.	0.010
Plastic jacketed thermostat cable		
2 conductor	L.F.	0.004
3 conductor	"	0.005
4 conductor	"	0.006
5 conductor	"	0.008
6 conductor	"	0.009
7 conductor	"	0.012
8 conductor	"	0.013

Supporting Construction Reference Data

This section contains information, text, charts and tables on various aspects of construction. The intent is to provide the user with a better understanding of unfamiliar areas in order to be able to estimate better. This information includes actual takeoff data for some areas and also selected explanations of common construction materials, methods and common practices.

TYPICAL BUILDING COST BROKEN DOWN BY CSI FORMAT

(Commercial Construction)

Division	New Construction	Remodeling Construction
1. General Requirements	6 to 8%	Up to 30%
2. Sitework	4 to 6%	
3. Concrete	15 to 20%	
4. Masonry	8 to 12%	
5. Metals	5 to 7%	
6. Wood And Plastics	1 to 5%	
7. Thermal And Moisture Protection	4 to 6%	
8. Doors And Windows	5 to 7%	Up to 30% (Divisions 8–9)
9. Finishes	8 to 12%	
10. Specialties	6 to 10% (Divisions 10–14)	
11. Architectural Equipment		
12. Furnishings		
13. Special Construction		
14. Conveying Systems		
15. Mechanical	15 to 25%	Up to 40% (Divisions 15–16)
16. Electrical	8 to 12%	
Total Cost	100%	

CONVERSION FACTORS

Change	To	Multiply By
Atmospheres	Pounds per square inch	14.696
Atmospheres	Inches of mercury	29.92
Atmospheres	Feet of water	34
Barrels, oil	Gallons, of oil	42
Barrels, cement	Pounds of cement	376
Bags or sacks, cement	Pounds of cement	94
Btu/min.	Foot-pounds/sec.	12.96
Btu/min.	Horse-power	0.02356
Btu/min.	Kilowatts	0.01757
Btu/min.	Watts	17.57
Centimeters	Inches	0.3937
Centimeters of mercury	Atmospheres	0.01316
Centimeters of mercury	Feet of water	0.4461
Cubic inches	Cubic feet	0.00058
Cubic feet	Cubic inches	1728
cubic feet	Cubic yards	0.03703
Cubic yards	Cubic feet	27
Cubic inches	Gallons	0.00433
Cubic feet	Gallons	7.48
Feet	Inches	12
Feet	Yards	0.3333
Yards	Feet	3
Feet of water	Atmospheres	0.02950
Feet of water	Inches of mercury	0.8826
Gallons	Cubic Inches	231
Gallons	Cubic feet	0.1337
Gallons	Pounds of water	8.33
Gallons	Quarts	4
Gallons per min.	Cubic feet sec.	0.002228
Gallons per min.	Cubic feet hour	8.0208
Gallons water per min.	Tons water/24 hours	6.0086
Horse-power	Foot-lbs./sec.	550
Inches	Centimeters	2.540
Inches	Feet	0.0833
Inches	Millimeters	25.4
Inches of water	Pounds per Sq. inch	0.0361
Inches of water	Inches of mercury	0.0735
Inches of water	Ounces per square inch	0.578
Inches of water	Ounces per square foot	5.2
Inches of mercury	Inches of water	13.6
Inches of mercury	Feet of water	1.1333
Inches of mercury	Pounds per square inch	0.4914
Kilometers	Miles	0.6214
Meters	Inches	39.37
Miles	Feet	5280
Millimeters	Centimeters	0.1
Millimeters	Inches	0.03937
Ounces (fluid)	Cubic inches	1.805
Ounces	Pounds	0.0625
Pounds	Ounces	16
Pounds per square inch	Inches of water	27.72
Pounds per square inch	Feet of water	2.310
Pounds per square inch	Inches of mercury	2.04
Pounds per square inch	Atmospheres	0.0681
Quarts	Cubic Inches	67.20
Square Inches	Square feet	0.00694
Square Feet	Square inches	144
Square Feet	Square yards	0.11111
Square yards	Square feet	9
Square miles	Acres	640
Short tons	Pounds	2000
Short tons	Long tons	0.89285
Tons of water/24 hours	Gallons per minute	0.16643
Yards	Feet	3
Yards	Centimeters	91.44
Yards	Inches	36

CONVERSION CALCULATIONS

Commercial Measure

16 drams . . . . = 1 ounce
16 ounces . . . . = 1 pound
2,000 pounds . . . . = 1 ton

Long Measure

12 inches . . . . = 1 foot
3 feet . . . . = 1 yard
16½ feet . . . . = 1 rod
40 rods . . . . = 1 furlong
8 furlongs (5,280 ft) = = 1 mile
3 miles . . . . = 1 league

Square Measure

144 square inches . . . . = 1 square foot
9 square feet . . . . = 1 square yard
30¼ square yards . . . . = 1 square rod
160 square rods . . . . = 1 acre
4840 square yards . . . . = 1 acre
640 acres . . . . = 1 square mile
36 square miles . . . . = 1 township

Surveyors Measure

7.92 inches . . . . = 1 link
25 links . . . . = 1 rod
4 rods (66 ft.) . . . . = 1 chain
10 chains . . . . = 1 furlong
8 furlongs . . . . = 1 mile
1 square mile . . . . = 1 section

Cubic Measure

1728 cubic inches . . . . = 1 cubic foot
27 cubic feet . . . . = 1 cubic yard
128 cubic feet . . . . = 1 cord (wood/stone)
231 cubic inches . . . . = 1 U.S. gallon
7.48 U.S. Gallons . . . . = 1 cubic foot
2150.4 cubic inches . . . . = 1 U.S. bushel

Liquid Measure

4 fluid ounces . . . . = 1 gill
4 gills . . . . = 1 pint
2 pints . . . . = 1 quart
4 quarts . . . . = 1 gallon
9 gallons . . . . = 1 firkin
31½ gallons . . . . = 1 barrel
2 barrels . . . . = 1 hogshead

Dry Measure

2 pints . . . . = 1 quart
8 quarts . . . . = 1 peck
4 pecks . . . . = 1 bushel
2150.42 cubic inches . . . . = 1 bushel

SQUARE

$$A = a^2$$

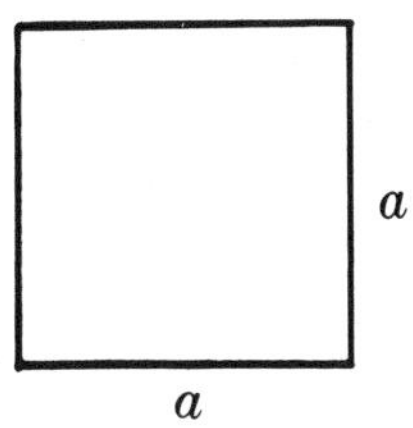

RECTANGLE

$$A = bh$$

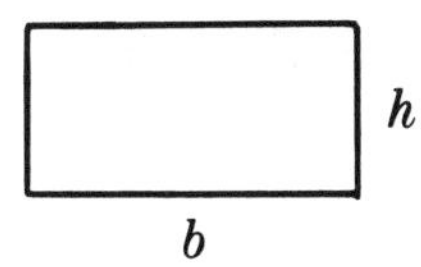

TRIANGLE

$$A = \frac{1}{2} bh$$

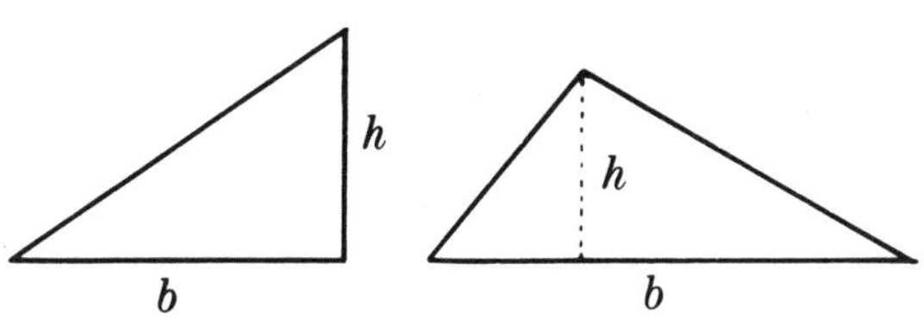

PARALLELOGRAM

$$A = bh = ab \ Sin ø$$

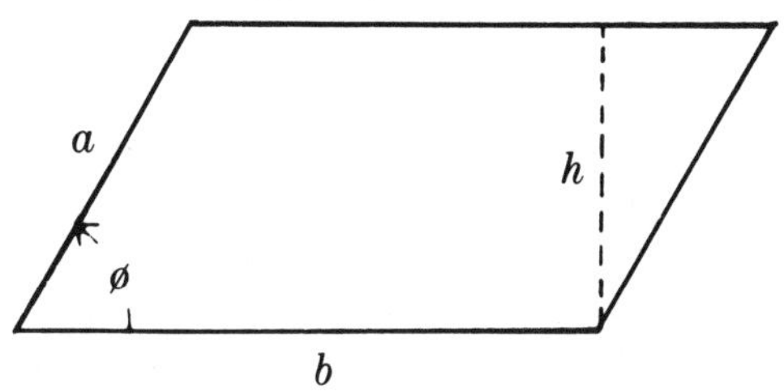

TRAPEZOID

$$A = \left(\frac{a+b}{2}\right) h$$

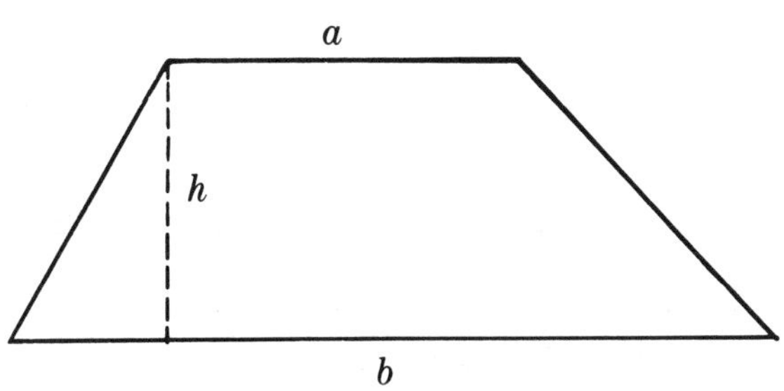

CIRCLE

$$A = \pi r^2 = \frac{\pi d^2}{4}$$

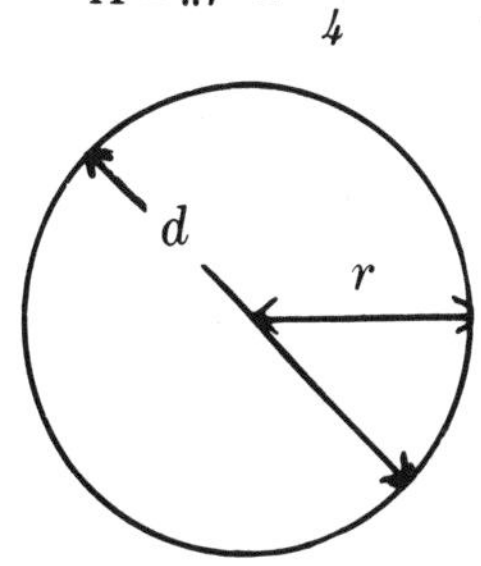

$$Circumference = C = 2\pi r = \pi d$$

ELLIPSE

$A = 0.7854\ ab$

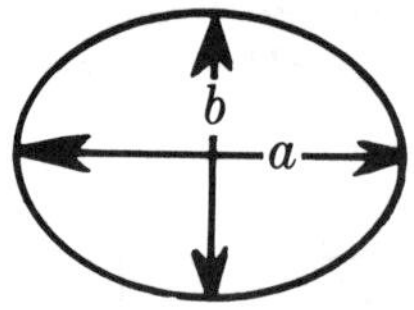

PARABOLA

$A = \frac{2}{3} bh$

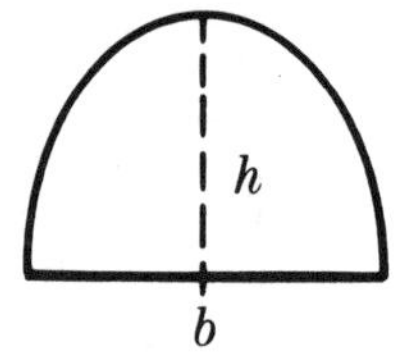

VOLUMES

CUBE

$V = a^3$

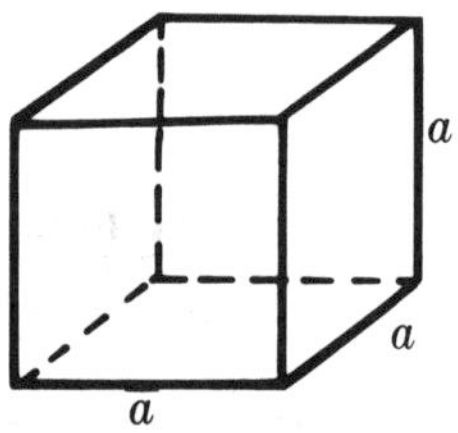

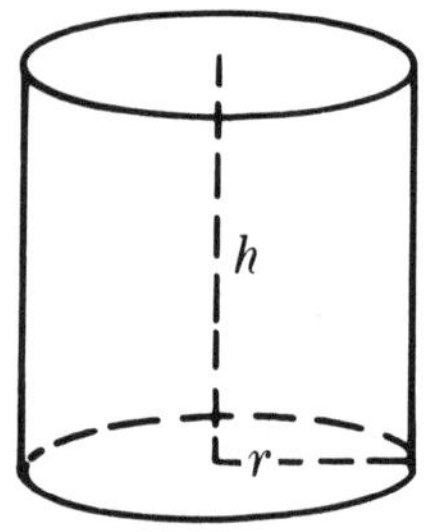

CYLINDER

$V = \pi r^3 h$

PYRAMID

$V = \frac{1}{3} (Base)\ h$

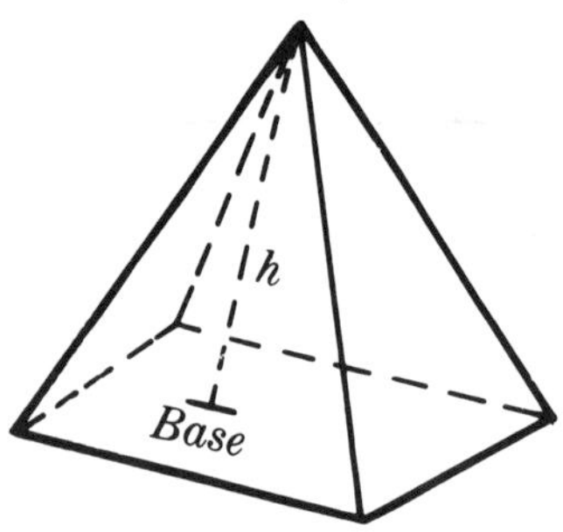

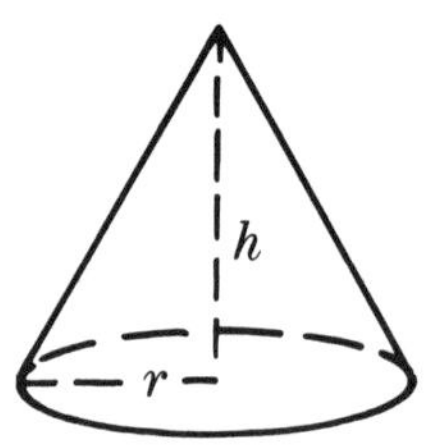

CONE

$V = \frac{1}{3} \pi r^2 h$

SPHERE

$V = \frac{4}{3} \pi r^3 = \frac{1}{6} \pi d^3$

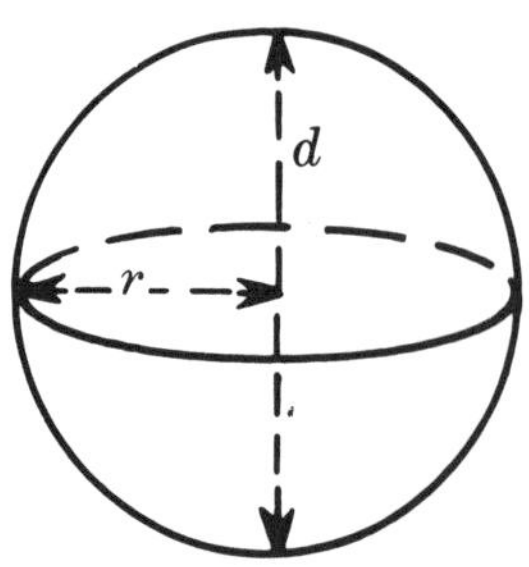

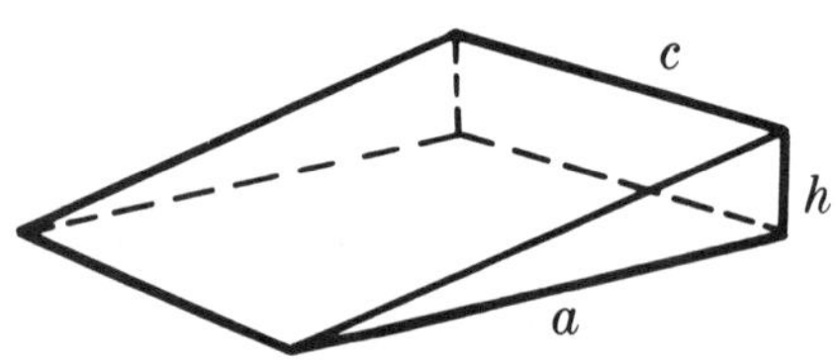

WEDGE

$V = \frac{1}{2} abc$

CONVERSION FACTORS

ENGLISH TO SI (SYSTEM INTERNATIONAL)

To Convert from	To	Multiply by
LENGTH		
Inches	Millimetres	25.4[a]
Feet	Metres	0.3048[a]
Yards	Metres	0.9144[a]
Miles (statute)	Kilometres	1.609
AREA		
Square inches	Square millimetres	645.2
Square feet	Square metres	0.0929
Square yards	Square metres	0.8361
VOLUME		
Cubic inches	Cubic millimetres	16.387
Cubic feet	Cubic metres	0.02832
Cubic yards	Cubic metres	0.7646
Gallons (U.S. liquid)[b]	Cubic metres[c]	0.003785
Gallons (Canadian liquid)[b]	Cubic metres[c]	0.004546
Ounces (U.S. liquid)[b]	Millilitres[c, d]	29.57
Quarts (U.S. liquid)[b]	Litres[c, d]	0.9464
Gallons (U.S. liquid)[b]	Litres[c]	3.785
FORCE		
Kilograms force	Newtons	9.807
Pounds force	Newtons	4.448
Pounds force	Kilograms force[d]	0.4536
Kips	Newtons	4448
Kips	Kilograms force[d]	453.6
PRESSURE, STRESS, STRENGTH (FORCE PER UNIT AREA)		
Kilograms force per sq. centimetre	Megapascals	0.09807
Pounds force per square inch (psi)	Megapascals	6895
Kips per square inch	Megapascals	6.895
Pounds force per square inch (psi)	Kilograms force per square centimetre[d]	0.07031
Pounds force per square foot	Pascals	47.88
Pounds force per square foot	Kilograms force per square metre[d]	4.882
BENDING MOMENT OR TORQUE		
Inch-pounds force	Metre-kilog. force[d]	0.01152
Inch-pounds force	Newton-metres	0.1130
Foot-pounds force	Metre-kilog. force[d]	0.1383
Foot-pounds force	Newton-metres	1.356
Metre-kilograms force	Newton-metres	9.807
MASS		
Ounce (avoirdupois)	Grams	28.35
Pounds (avoirdupois)	Kilograms	0.4536
Tons (metric)	Kilograms	1000[a]
Tons, short (2000 pounds)	Kilograms	907.2
Tons, short (2000 pounds)	Megagrams[e]	0.9072
MASS PER UNIT VOLUME		
Pounds mass per cubic foot	Kilog. per cubic metre	16.02
Pounds mass per cubic yard	Kilog. per cubic metre	0.5933
Pds. mass per gallon (U.S. liquid)[b]	Kilog. per cubic metre	119.8
Pds. mass p/gal. (Canadian liquid)[b]	Kilog. per cubic metre	99.78
TEMPERATURE		
Degrees Fahrenheit	Degrees Celsius	tK = (1F − 32)/1.8
Degrees Fahrenheit	Degrees Kelvin	tK = (1F + 459.67)/1.8
Degree Celsius	Degree Kelvin	tK = 1C + 273.15

[a] The factor given is exact.
[b] One U.S. gallon equals 0.8327 Canadian gallon.
[c] 1 litre = 1000 millilitres = 10,000 cubic centimetres = 1 cubic decimetre = 0.001 cubic metre.
[d] Metric but not SI unit.
[e] Called "tonne" in England. Called "metric ton" in other metric systems.

TRENCH BRACING

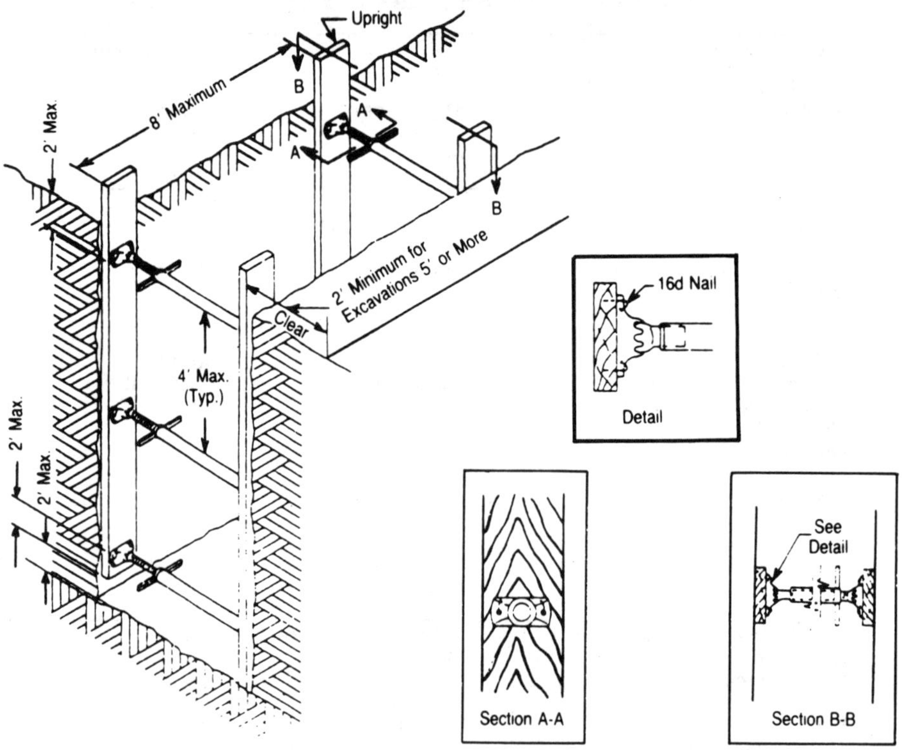

CLOSED VERTICAL SHEETING

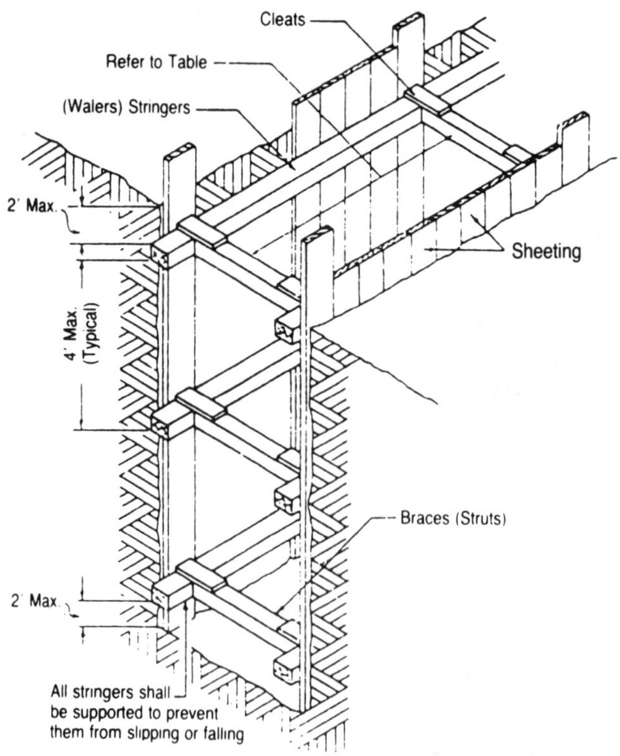

STANDARD NOMENCLATURE FOR STREET CONSTRUCTION

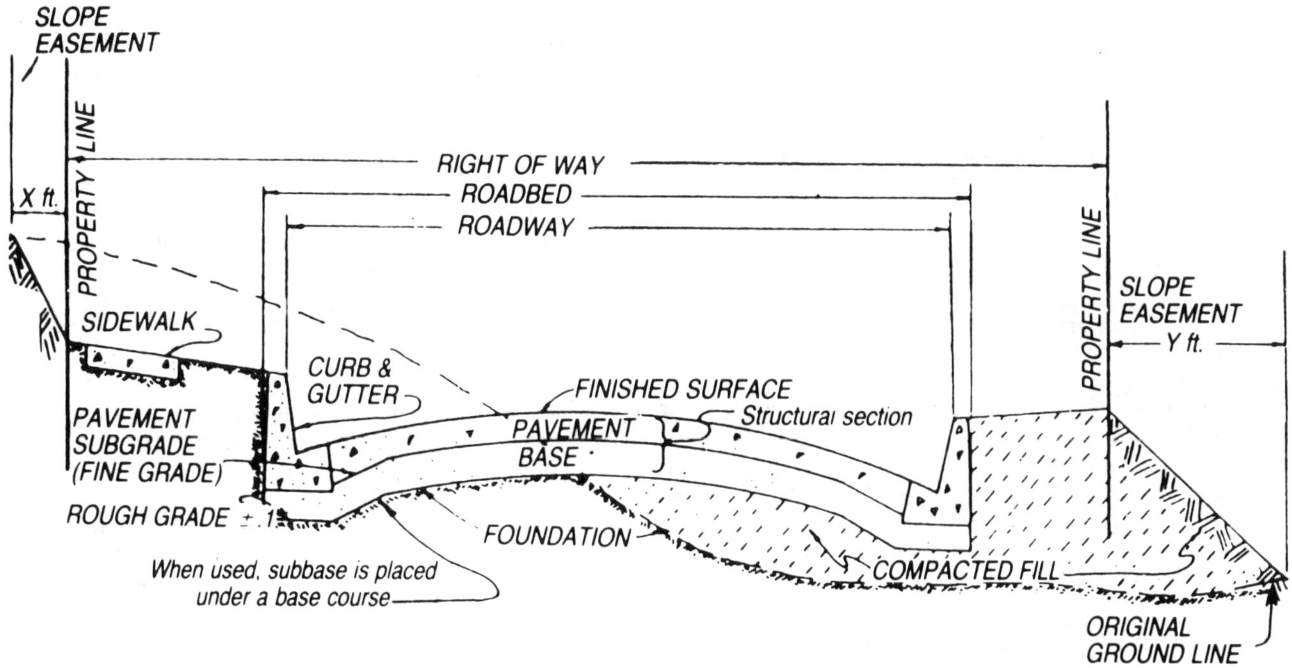

PIPE

Clay Pipe. Clay pipe is manufactured by blending various clays together, milling, mixing, extruding and firing in a kiln to obtain vitrification. The physical properties of the pipe can be changed by varying the proportions of the several clays used. The pipe is supplied in two basic styles: spigot and socket; and plain end.

Spigot and Socket Pipe has a spigot on one end and a socket on the other, and is commonly referred to as "bell and spigot" pipe. The plans generally specify the type of joint to be used from the several types of jointing methods available. This type of pipe is manufactured with matching polyurethane gaskets molded on the spigot and socket which form a tight seal when the pipe is jointed.

Plain End Pipe is without a socket on either end and is joined with special couplings. The coupling consists of a circular rubber sleeve, two stainless steel compression bands with tightening devices and a corrosion resistant shear ring. Sometimes this joint is supplied with a cardboard form, open at the top, which is filled with portland cement mortar to resist shear and prevent future corrosion of the bands.

Concrete Pipe. Unreinforced and reinforced concrete pipe is manufactured by casting in stationary or revolving metal molds. At the present time, the design practice is to specify reinforced concrete pipe for all purposes.

Unreinforced Concrete Pipe is cast in vertical steel molds, usually in pipe sizes of 21 inches or less, and is of the spigot and socket type. No steel reinforcement is used and the pipe is usually intended for use in irrigation systems and under light loading conditions.

Reinforced Concrete Pipe (RCP) is made in a number of different manufacturing processes and for a wide variety of pressure and non-pressure classes. It is available in standard sizes or it can be made to order to any diameter desired. Some of the larger diameters include diameters of 12 and 14 feet. A large variety of joint details are used with RCP. Tongue and groove joints are used for storm drain pipelines.

Reinforced concrete pipe for wastewater pipeline projects is supplied with gasketed joints and a polyvinyl chloride (PVC) plastic liner cast into the pipe.

(a) **Cast Pipe** is cast vertically in steel forms with the reinforcing cage securely held in place. The reinforcement is generally elliptical in shape to provide the maximum structural strength to resist the loads imposed on the pipe by the backfill and other stresses. Consolidation of the concrete is obtained by the use of external form vibrators.

(b) **Centrifugally Spun Pipe** is manufactured by introducing concrete into a spinning horizontal steel cylinder into which the reinforcement cage has been previously installed and which is equipped with end dams to provide the proper pipe wall thickness. The speed of rotation of the mold is increased and the centrifugal force produces a smooth, dense concrete pipe.

(c) **Pressure Pipe** may be cast or centrifugally spun pipe but it usually has a circular steel reinforcement cage (or cages) designed not only to resist the trench loading, but also the internal pressures exerted on the pipe from the fluid under pressure in the line.

Concrete Cylinder Pipe. This class of pipe is generally used for high pressure water lines and sewer force mains and is available in sizes ranging from 10 inches to 60 inches and larger in special cases.

A sheet steel cylinder is wrapped with the designed steel reinforcement and a concrete lining is centrifugally spun in the interior of the steel cylinder. An exterior coating of concrete is applied generally by the gunite process, while the cylinder is slowly rotated. These coatings vary in thickness from ½ to ¾ of an inch. The joints are commonly of the steel ring and rubber gasket type, but are generally designed for the special purpose for which the pipe line is intended.

Definition of Terms. In general, the terms used to designate types of reinforced concrete pipe refer to the process used in manufacture.

Cast RCP (Cast Reinforced Concrete Pipe). A concrete pipe having one or more cylindrical or elliptical cages of reinforcement steel embedded in it, the concrete for which is cast with the forms in a vertical position.

CSRCP (Centrifugally Spun Reinforced Concrete Pipe). A concrete pipe having one or more cylindrical or elliptical cages of reinforcement steel embedded in it, and cast in a horizontal position while the forms are spinning rapidly. This type of pipe may be designated as Spun RCP or as CCP (Centrifugal Concrete Pipe).

RCP (Reinforced Concrete Pipe). A reinforced concrete pipe manufactured by either the casting or spinning method.

Steel Reinforcement. Steel for reinforcing concrete pipe is generally furnished in large coils which will permit the use of machines to fabricate the "cages." The continuous steel rod is wound spirally at a prescribed pitch on a drum of the proper diameter. Where the rod crosses a longitudinal spacer rod, it is electrically welded to it so that the complete cage is relatively rigid.

Reinforcement cages for pipe designed for external loading are generally elliptical in shape to take full advantage of the steel in tension. Pipe to be used with relatively small external loads or pipe designed for pressure lines will have circular cages.

Reinforcement cages must be rigidly fixed in the forms so that the placement of concrete or the effects of centrifugal spinning will not result in distortion or displacement of the steel. The orientation of an elliptical cage must be marked on the forms to assure that the minor axis can be located after the concrete is placed.

CAPACITIES FOR SEPTIC TANKS SERVING AN INDIVIDUAL DWELLING

No. of bedrooms	Capacity of tank (gals.)
2 or less	750
3	900
4	1,000

CONCRETE / REINFORCING STEEL — 03210

STANDARD SIZES OF STEEL REINFORCEMENT BARS

STANDARD REINFORCEMENT BARS				
		Nominal Dimensions		
Bar Designation Number*	Nominal Weight, lb. per ft.	Diameter, in.	Cross Sectional Area, sq. in.	Perimeter, in.
3	0.376	0.375	0.11	1.178
4	0.668	0.500	0.20	1.571
5	1.043	0.625	0.31	1.963
6	1.502	0.750	0.44	2.356
7	2.044	0.875	0.60	2.749
8	2.670	1.000	0.79	3.142
9	3.400	1.128	1.00	3.544
10	4.303	1.270	1.27	3.990
11	5.313	1.410	1.56	4.430
14	7.65	1.693	2.25	5.32
18	13.60	2.257	4.00	7.09

*The bar numbers are based on the number of ⅛ inches included in the nominal diameter of the bar.

Type of Steel and ASTM Specification No.	Size Nos. Inclusive	Grade	Tensile Strength Min., psi	Yield (a) Min., psi
Billet Steel A 615	3-11	40	70,000	40,000
	3-11 14, 18	60	90,000	60,000
	11, 14, 18	75	100,000	75,000

CONCRETE / WELDED WIRE FABRIC — 03220

WELDED WIRE FABRIC – COMMON STOCK STYLES OF WELDED WIRE FABRIC

Style Designation	Steel Area sq. in. per ft.		Weight Approx. lbs. per 100 sq. ft.
	Longit.	Transv.	
Rolls			
6x6—W1.4xW1.4	.03	.03	21
6x6—W2xW2	.04	.04	29
6x6—W2.9xW2.9	.06	.06	42
6x6—W4xW4	.08	.08	58
4x4—W1.4xW1.4	.04	.04	31
4x4—W2xW2	.06	.06	43
4x4—W2.9xW2.9	.09	.09	62
4x4—W4xW4	.12	.12	86
Sheets			
6x6—W2.9xW2.9	.06	.06	42
6x6—W4xW4	.08	.08	58
6x6—W5.5xW5.5	.11	.11	80
4x4—W4xW4	.12	.12	86

METALS / WELDING | 05050

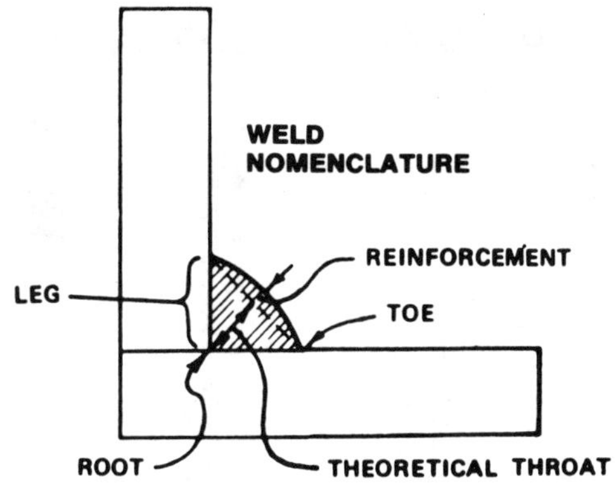

WELDED JOINTS

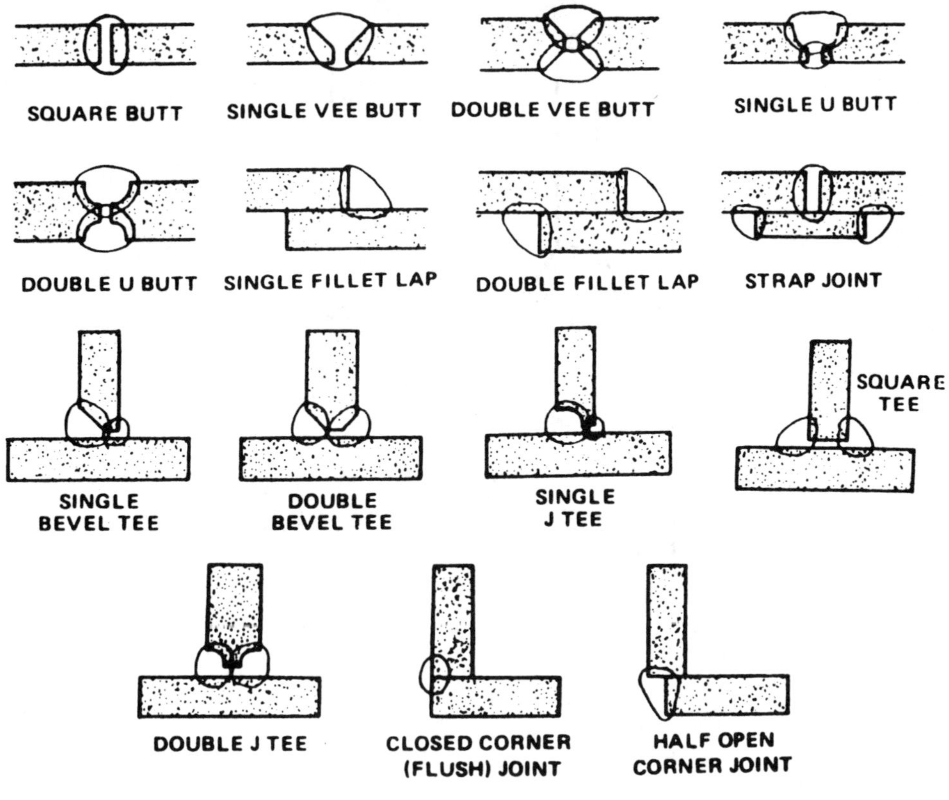

WELDING POSITIONS

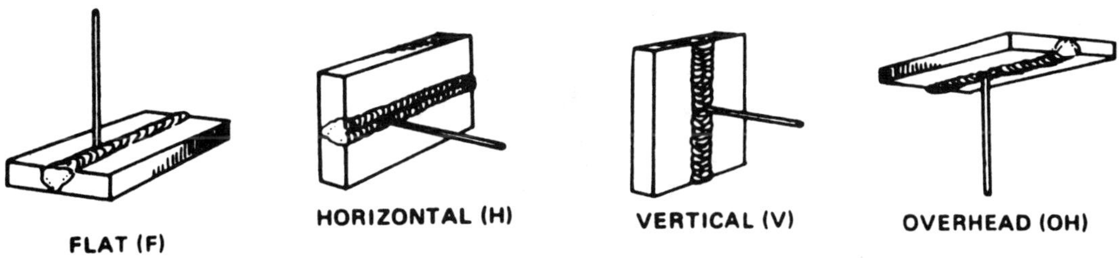

BOLTS IN COMMON USAGE

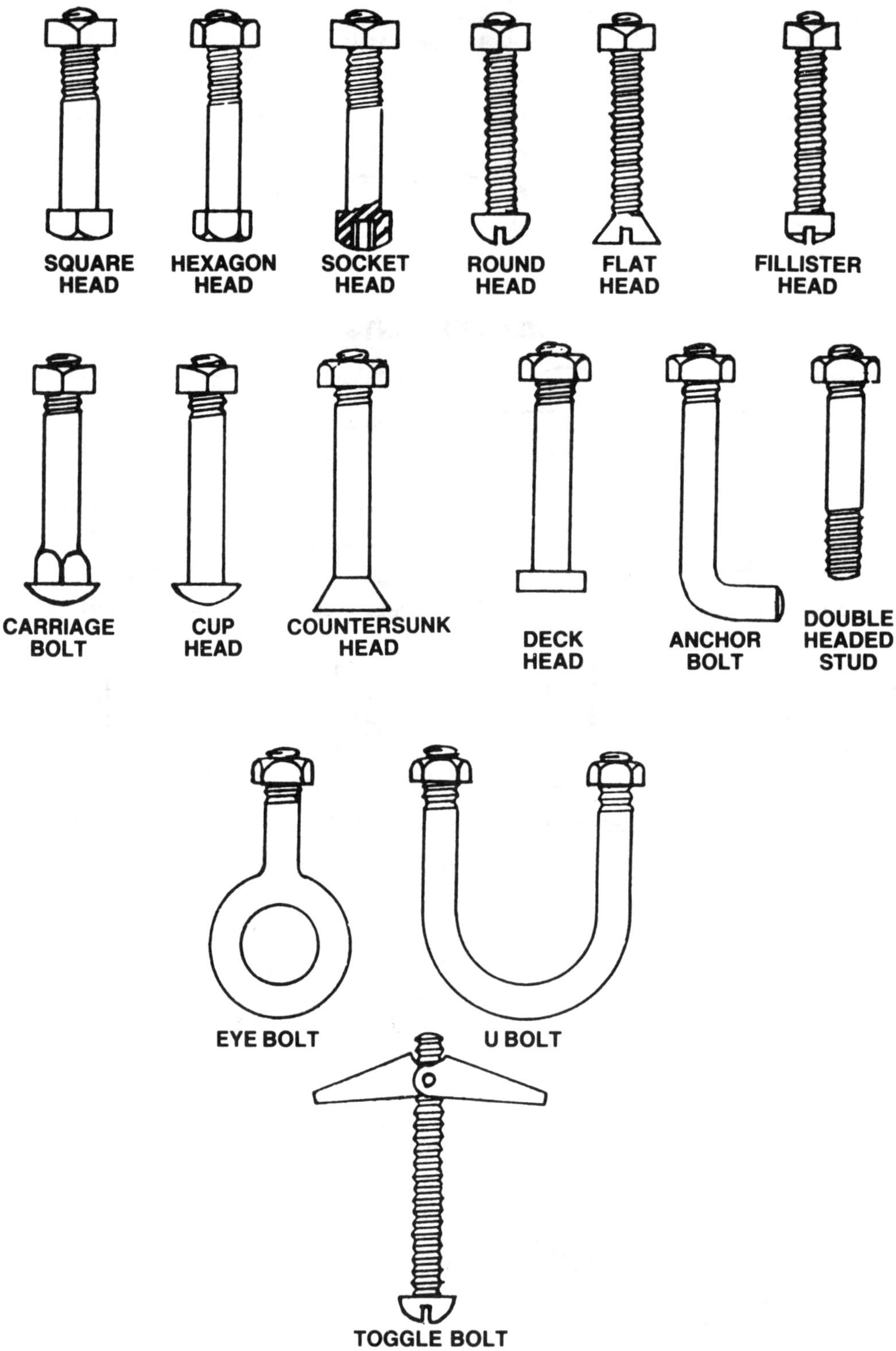

MOISTURE PROTECTION / DOWNSPOUTS 07600

DOWNSPOUT/VERTICAL LEADER CALCULATIONS

Roof Type	Slope	S.F. Roof/ Sq. In. Leader
Gravel	Less than ¼″ per foot	300
Gravel	Greater than ¼″ per foot	250
Metal or Shingle	Any	200

Alternate calculations:

$$\text{Diameter of downspout/leader} = 1.128\sqrt{\frac{\text{Area of drainage}}{\text{SF Roof/Sq. Inch}}}$$

TYPICAL MINIMUM SIZE OF VERTICAL CONDUCTORS AND LEADERS

Size of leader or conductor (Inches)	Maximum projected roof area (Square feet)
2	544
2½	987
3	1,610
4	3,460
5	6,280
6	10,200
8	22,000

TYPICAL MINIMUM SIZE OF ROOF GUTTERS

	Maximum projected roof area for gutters of various slopes			
Diameter gutter (Inches)	1/16 in. per Ft. slope (Sq. ft.)	⅛ in. per Ft. slope (Sq. ft.)	¼ in. per Ft. slope (Sq. ft.)	½ in. per Ft. slope (Sq. ft.)
3	170	240	340	480
4	360	510	720	1020
5	625	880	1250	1770
6	960	1360	1920	2770
7	1380	1950	2760	3900
8	1990	2800	3980	5600
10	3600	5100	7200	10000

PIPE WEIGHTS

CAST IRON PIPE

SERVICE WEIGHT			
Size, Inches	Weight of Pipe	Weight of Water	Total Weight-Lbs.
2″	3.8	1.45	5.3
3″	5.6	3.2	8.8
4″	7.5	5.5	13.0
5″	9.8	8.7	18.5
6″	12.4	12.5	24.9
8″	18.5	21.7	40.2

EXTRA HEAVY			
Size, Inches	Weight of Pipe	Weight of Water	Total Weight-Lbs.
2″	4.3	1.45	5.8
3″	8.3	3.2	11.5
4″	10.8	5.5	16.3
5″	13.3	8.7	22.0
6″	16.0	12.5	28.5
8″	26.5	21.7	48.2

STEEL PIPE

Pipe Size	W/40	H_2O/Lbs.	Total Lbs./L.F.
2″	3.65	1.45	5.1
2½″	5.79	2.07	7.86
3″	7.57	3.2	10.77
3½″	9.11	4.28	13.39
4″	10.8	5.51	16.31
5″	14.6	8.66	23.26
6″	18.0	12.5	30.5
8″	28.6	21.66	50.26
10″	40.5	34.15	74.65
12″			

SPRINKLER AREA CALCULATIONS

Typical maximum floor area allowed per system riser:

Light Hazard	Ordinary Hazard	Extra Hazard
52,000 S.F.	40,000-52,000 S.F.	25,000 S.F.

Typical maximum floor area coverage allowed per sprinkler head:

Light Hazard	Ordinary Hazard	Extra Hazard
130-200 S.F.	100-130 S.F.	90 S.F.

Typical maximum spacing between lines and sprinkler heads:

Light Hazard	Ordinary Hazard	Extra Hazard
12-15 feet	12-15 feet	12 feet

Note: This data is for estimating purposes only. Check all applicable codes and regulations for specific requirements.

SPRINKLER HEAD CALCULATIONS

Typical maximum quantity of sprinkler heads allowed by pipe size.

Light Hazard:

For sprinklers below ceiling:

Steel		Copper	
1 in. pipe	2 sprinklers	1 in. tube	2 sprinklers
1¼ in. pipe	3 sprinklers	1¼ in. tube	3 sprinklers
1½ in. pipe	5 sprinklers	1½ in. tube	5 sprinklers
2 in. pipe	10 sprinklers	2 in. tube	12 sprinklers
2½ in. pipe	30 sprinklers	2½ in. tube	40 sprinklers
3 in. pipe	60 sprinklers	3 in. tube	65 sprinklers
3½ in. pipe	100 sprinklers	3½ in. tube	115 sprinklers

For sprinklers above and below ceiling:

Steel		Copper	
1 in.	2 sprinklers	1 in.	2 sprinklers
1¼ in.	4 sprinklers	1¼ in.	4 sprinklers
1½ in.	7 sprinklers	1½ in.	7 sprinklers
2 in.	15 sprinklers	2 in.	18 sprinklers
2½ in.	50 sprinklers	2½ in.	65 sprinklers

Ordinary Hazard:

For sprinklers above ceiling:

Steel		Copper	
1 in. pipe	2 sprinklers	1 in. tube	2 sprinklers
1¼ in. pipe	3 sprinklers	1¼ in. tube	3 sprinklers
1½ in. pipe	5 sprinklers	1½ in. tube	5 sprinklers
2 in. pipe	10 sprinklers	2 in. tube	12 sprinklers
2½ in. pipe	20 sprinklers	2½ in. tube	25 sprinklers
3 in. pipe	40 sprinklers	3 in. tube	45 sprinklers
3½ in. pipe	65 sprinklers	3½ in. tube	75 sprinklers
4 in. pipe	100 sprinklers	4 in. tube	115 sprinklers
5 in. pipe	160 sprinklers	5 in. tube	180 sprinklers
6 in. pipe	275 sprinklers	6 in. tube	300 sprinklers

For sprinklers above and below ceiling:

Steel		Copper	
1 in.	2 sprinklers	1 in.	2 sprinklers
1¼ in.	4 sprinklers	1¼ in.	4 sprinklers
1½ in.	7 sprinklers	1½ in.	7 sprinklers
2 in.	15 sprinklers	2 in.	18 sprinklers
2½ in.	30 sprinklers	2½ in.	40 sprinklers
3 in.	60 sprinklers	3 in.	65 sprinklers

Extra Hazard:

For sprinklers below ceiling:

Steel		Copper	
1 in. pipe	1 sprinkler	1 in. tube	1 sprinkler
1¼ in. pipe	2 sprinklers	1¼ in. tube	2 sprinklers
1½ in. pipe	5 sprinklers	1½ in. tube	5 sprinklers
2 in. pipe	8 sprinklers	2 in. tube	8 sprinklers
2½ in. pipe	15 sprinklers	2½ in. tube	20 sprinklers
3 in. pipe	27 sprinklers	3 in. tube	30 sprinklers
3½ in. pipe	40 sprinklers	3½ in. tube	45 sprinklers
4 in. pipe	55 sprinklers	4 in. tube	65 sprinklers
5 in. pipe	90 sprinklers	5 in. tube	100 sprinklers
6 in. pipe	150 sprinklers	6 in. tube	170 sprinklers

Note: This data is for estimating purposes only. Check all applicable codes and regulations for specific requirements.

SPRINKLER HAZARD OCCUPANCIES

Typical Light Hazard Occupancies:

- Churches
- Clubs
- Eaves and overhangs, if combustible construction with no combustible beneath
- Educational
- Hospitals
- Institutional
- Libraries, except large stack rooms
- Museums
- Nursing or Convalescent Homes
- Office, including Data Processing
- Residential
- Restaurant seating areas
- Theaters seating areas
- Theaters and Auditoriums excluding stages and prosceniums
- Unused attics

Typical Ordinary Hazard Occupancies (Group 1):

- Automobile parking garages
- Bakeries
- Beverage manufacturing
- Canneries
- Dairy products manufacturing and processing
- Electronic plants
- Glass and glass products manufacturing
- Laundries
- Restaurant service areas

Typical Ordinary Hazard Occupancies (Group 2):

- Cereal mills
- Chemical plants — Ordinary
- Cold Storage warehouses
- Confectionery products
- Distilleries
- Leather goods mfg.
- Libraries-large stack room areas
- Mercantiles
- Machine shops
- Metal working
- Printing and publishing
- Textile mfg.
- Tobacco Products mfg.
- Wood product assembly

Typical Ordinary Hazard Occupancies (Group 3):

- Feed mills
- Paper and pulp mills
- Paper process plants
- Piers and wharves
- Repair garages
- Tire manufacturing
- Warehouses (having moderate to higher combustibility of content, such as paper, household furniture, paint, general storage, whiskey, etc.)[1]
- Wood machining

Typical Extra Hazard Occupancies (Group 1):

- Combustible Hydraulic Fluid use areas
  - Die Casting
  - Metal Extruding
- Plywood and particle board manufacting
- Printing (using inks with below 100°F [37.8°C] flash points)
- Rubber reclaiming, compounding, drying, milling, vulcanizing
- Saw Mills
- Textile picking, opening, blending, garnetting, carding, combining of cotton, synthetics, wool shoddy or burlap
- Upholstering with plastic foams

Typical Extra Hazard Occupancies (Group 2):

- Asphalt saturating
- Flammable liquids spraying
- Flow coating
- Mobile Home or Modular Building assemblies (where finished enclosure is present and has combustible interiors)
- Open Oil quenching
- Solvent cleaning
- Varnish and paint dipping

....

TYPICAL MINIMUM SIZE OF HORIZONTAL BUILDING STORM DRAINS AND BUILDING STORM SEWERS

Diameter of drain	Maximum projected area in square feet for various slopes		
	1/8 inch per feet slope	1/4 inch per feet slope	1/2 inch per feet slope
3	822	1160	1644
4	1880	2650	3760
5	3340	4720	6680
6	5350	7550	10700
8	11500	16300	23000
10	20700	29200	41400
12	33300	47000	66600
15	59500	84000	119000

MECHANICAL / PLUMBING 15400

PLUMBING BUDGETS

Budget estimates for plumbing can be determined as a percentage of total building costs depending on building type. For example:

- Apartments .... 9 to 12 percent
- Assembly .... 4 to 7 percent
- Banks .... 3 to 6 percent
- Dormitories .... 7 to 10 percent
- Factories .... 4 to 8 percent
- Hospitals .... 8 to 12 percent
- Motels .... 9 to 12 percent
- Office Buildings .... 4 to 7 percent
- Retail (small) .... 4 to 7 percent
- Retail (large) .... 3 to 6 percent
- Schools .... 3 to 6 percent
- Warehouses .... 3 to 7 percent

MECHANICAL / PLUMBING VENTS 15410

TYPICAL SIZE AND LENGTH OF PLUMBING VENTS

Diameter of soil or waste stack (in.)	Total fixture units connected to stack (dfu)	DIAMETER OF VENT PIPE										
		1¼	1	2	2½	3	4	5	6	8	10	12
1¼	2	30										
1½	8	50	150									
1½	10	30	100									
2	12	30	75	200								
2	20	26	50	150								
2½	42		30	100	300							
3	10		42	150	360	1040						
3	21		32	110	270	810						
3	53		27	94	230	680						
3	102		25	86	210	620						
4	43			35	85	250	980					
4	140			27	65	200	750					
4	320			23	55	170	640					
4	540			21	50	150	580					
5	190				28	82	320	990				
5	490				21	63	250	760				
5	940				18	53	210	670				
5	1400				16	49	190	590				
6	500					33	130	400	1000			
6	1100					26	100	310	780			
6	2000					22	84	260	660			
6	2900					20	77	240	600			
8	1800						31	95	240	940		
8	3400						24	73	190	720		
8	5600						20	62	160	610		
8	7600						18	56	140	560		

TYPICAL SIZES OF FIXTURE WATER SUPPLY PIPES

Fixture	Nominal pipe size (inches)
Bath tubs	½
Combination sink and tray	½
Drinking fountain	⅜
Dishwasher (domestic)	½
Kitchen sink, residential	½
Kitchen sink, commercial	¾
Lavatory	⅜
Laundry tray, 1, 2 or 3 compartments	½
Shower (single head)	½
Sinks (service, slop)	½
Sinks flushing rim	¾
Urinal (flash tank)	½
Urinal (direct flush valve)	¾
Water closet (tank type)	⅜
Water closet (flush valve type)	1
Hose bibs	½
Wall hydrant	½

TYPICAL VENTILATION AIR REQUIREMENTS FOR SPECIAL USES

Occupancy Classification	Required ventilation air in cfm per human occupant
Special areas	
Lockers	2* (or 30 per locker)
Wardrobes	2*
Public bathrooms	40**
Private bathrooms	25**
Swimming pools	15 (per occupant)
Exitways and corridors	1½*

*Per square foot floor area.
**Per water closet or urinal.

TYPICAL VENTILATION AIR REQUIREMENTS FOR RETAIL USES

Occupancy Classification	Required ventilation air in cfm per human occupant
Mercantile	
Sales floors and showrooms (basement & grade floors)	7
Sales and showrooms (upper floors)	7
Storage areas	5
Dressing rooms	7
Malls	7
Shipping areas	15
Elevators	7
Supermarkets	
Meat processing rooms	5
Drugs stores	
Pharmacists' work rooms	20
Specialty shops	
Pet shops	1.0*
Florists	5
Greenhouses	5

*cfm per sq. ft. floor area.

TYPICAL VENTILATION AIR REQUIREMENTS FOR RESIDENTIAL USES

Occupancy Classification	Required ventilation air in cfm per human occupant
Residential	
General living areas	5
Bedrooms	5
Kitchens	20
Basements, utility rooms	5
Mobile homes	5
Hotels, motels	
Bedrooms (single, double)	7
Living rooms (suites)	10
Corridors	5
Lobbies	7
Conference rooms (small)	20
Assembly rooms (large)	15

TYPICAL VENTILATION AIR REQUIREMENTS FOR STORAGE USES

Occupancy Classification	Required ventilation air in cfm per human occupant
Storage	
Garages, service stations, parking garages (enclosed)	1.5*
Auto repair shops	1.5**
Warehouses	
General	7

*cfm/s.f. floor area.
**Must have positive engine exhaust system.

TYPICAL VENTILATION AIR REQUIREMENTS FOR FACTORY AND INDUSTRIAL USES

Occupancy Classification	Required ventilation air in cfm per human occupant
Factory and industrial	
Metalworking & finishing	35
Automotive engine test	Require Special Exhaust Systems
Paint spray booths	
Picking, etching and plating lines	
Degreasing booths	
Sandblasting booths	
Chemicals and pharmaceuticals	
Dusty operations	30
Rooms containing potential gas emitters	20
Drying oven rooms	15
Fermentation rooms	15
Pillmaking booths	10
Packaging areas	10
Utility rooms	7
Computer rooms	7
Textiles-clothes manufacturer	15
Electronics and aerospace circuit board and soldering rooms	20
Wood products, papermaking	20
Brewing, distilling, wineries, bottling	20*
Food processing	20
Tobacco processing	20
Power plants	
Control rooms	10
Boiler rooms	35
Generator rooms	20
Sewage treatment plants	
Control rooms	10
Compressor/blower motor rooms	20
Glass and ceramic manufacturer	20
Agricultural	20

TYPICAL VENTILATION AIR REQUIREMENTS FOR BUSINESS USES

Occupancy Classification	Required ventilation air in cfm per human occupant
Business	
Banks	
(see offices)	
Vaults	5
Barber, beauty and health services	
Beauty shops (hair dressers)	25
Reducing salons	25
Sauna baths, steam rooms	5
barber shops	7
Photo studios	
Camera rooms, stages	5
Dark rooms	
Shoe repair shops	
Workrooms/trade areas	10
Offices	
General office space and showrooms	15
Conference rooms	25
Drafting/art rooms	7
Doctor's consultation rooms	10
Waiting rooms	10
Lithographing rooms	7
Diazo printing rooms	7
Computer rooms	5
Keypunch rooms	7
Communication	
TV/radio broadcasting booths, studios	30
Motion picture and TV stages	30
Pressrooms	15
Composing rooms	7
Engraving rooms	7
Telephone switchboard rooms (manual)	7
Telephone switchgear rooms (automatic)	7
Teletypewriter/facsimile rooms	5
Research institutes	
Laboratories:	
Light duty; non-chemical	15
Chemical	15
Heavy-duty	15
Radioisotope, chemical & biologically toxic	15
Machine shops	15
Dark rooms, spectroscopy rooms	10
Animal rooms	40
Veterinary hospitals	
Kennels, stalls	25
Operating rooms	25
Reception rooms	10

TYPICAL VENTILATION AIR REQUIREMENTS FOR INSTITUTIONAL USES

Occupancy Classification	Required ventilation air in cfm per human occupant
Institutional	
Prisons	
Cell blocks	7
Eating halls	15
Guard stations	7

AIR CONDITIONING

RECOMMENDED SHEET METAL GAUGES AND CONSTRUCTION FOR RECTANGULAR DUCT

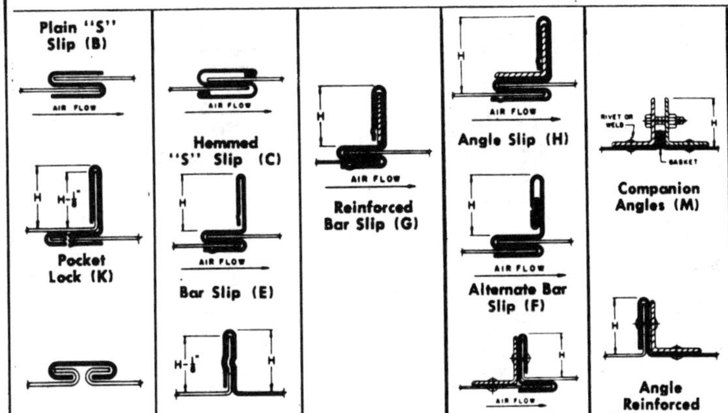

LOW PRESSURE — LOW VELOCITY = 2" W.G. MAX

Plate No.	Dimension of Longest Side of Duct	Steel Metal Gauges: Steel	Steel Metal Gauges: Aluminum	At Joints: Plain "S" Slip (B), Pocket Lock (K), Drive Slip (A)	At Joints: Hemmed "S" Slip (C), Bar Slip (E), Standing Seam (1)	At Joints: Reinforced Bar Slip (G)	At Joints: Angle Slip (H), Alternate Bar Slip (F), Angle RFD Pocket (L)	At Joints: Companion Angles (M), Angle Reinforced Standing Seam (J)	Reinforcing Between Joints
6	Thru 12"	26	24 (.020)	A-B-K	—	—	—	—	
6	13" thru 18"	24	22 (.025)	A-B-K	—	—	—	—	
7 7A	19" thru 30"	24	22 (.025)	K @ 5' cc A	C-E-@ 5' cc C-E-@ 10' cc	—	—	—	1"x1"x⅛" @ 5' cc
8	31" thru 42"	22	20 (.032)	K @ 5' cc	E-G-K @ 5' cc E-G-K @ 10' cc	—	—	—	1"x1"x⅛" @ 5' cc
9	43" thru 54"	22	20 (.032)	K @ 4' cc K @ 8' cc	E-@ 4' cc E-@ 8' cc	G -@ 4' cc G -@ 8' cc	—	—	1½"x1½"x⅛" @ 4' cc
9	55" thru 60"	20	18 (.040)	K @ 4' cc K @ 8' cc	E-@ 4' cc E-@ 8' cc	G -@ 4' cc G -@ 8' cc	—	—	1½"x1½"x⅛" @ 4' cc
10	61" thru 84"	20	18 (.040)	—	—	G -@ 4' cc G -@ 5' cc	H-@ 4' cc F-@ 4' cc L-@ 4' cc H-@ 5' cc F-@ 5' cc L-@ 5' cc	J-@ 2' cc	1½"x1½"x⅛" @ 2' cc 1½"x1½"x⅛" @ 2'-6" cc
11	85" thru 96"	18	16 (.051)	—	—	—	H-@ 4' cc L-@ 4' cc H-@ 5' cc L-@ 5' cc	M-@ 4' cc M-@ 5' cc J-@ 2' cc	1½"x1½"x3/16" @ 2' cc 1½"x1½"x3/16" @ 2'-6" cc 1½"x1½"x3/16" @ 2' cc
12	Over 96"	18	16 (.051)	—	—	—	H-@ 4' cc L-@ 4' cc H-@ 5' cc L-@ 5' cc	M-@ 4' cc M-@ 5' cc J-@ 2' cc	2"x2"x¼" @ 2' cc 2"x2"x¼" @ 2'-6" cc 2"x2"x¼" @ 2' cc

H (height dimension)—up to 42" = 1"
H (height dimension)—43" to 96" = 1½"
H (height dimension)—over 96" = 2"

AIR CONDITIONING

TYPICAL DUCT CONNECTIONS

CROSS JOINTS FOR SHEET METAL DUCTWORK

(NOT TO SCALE)

H=HEIGHT REFERRED TO IN DIMENSIONS

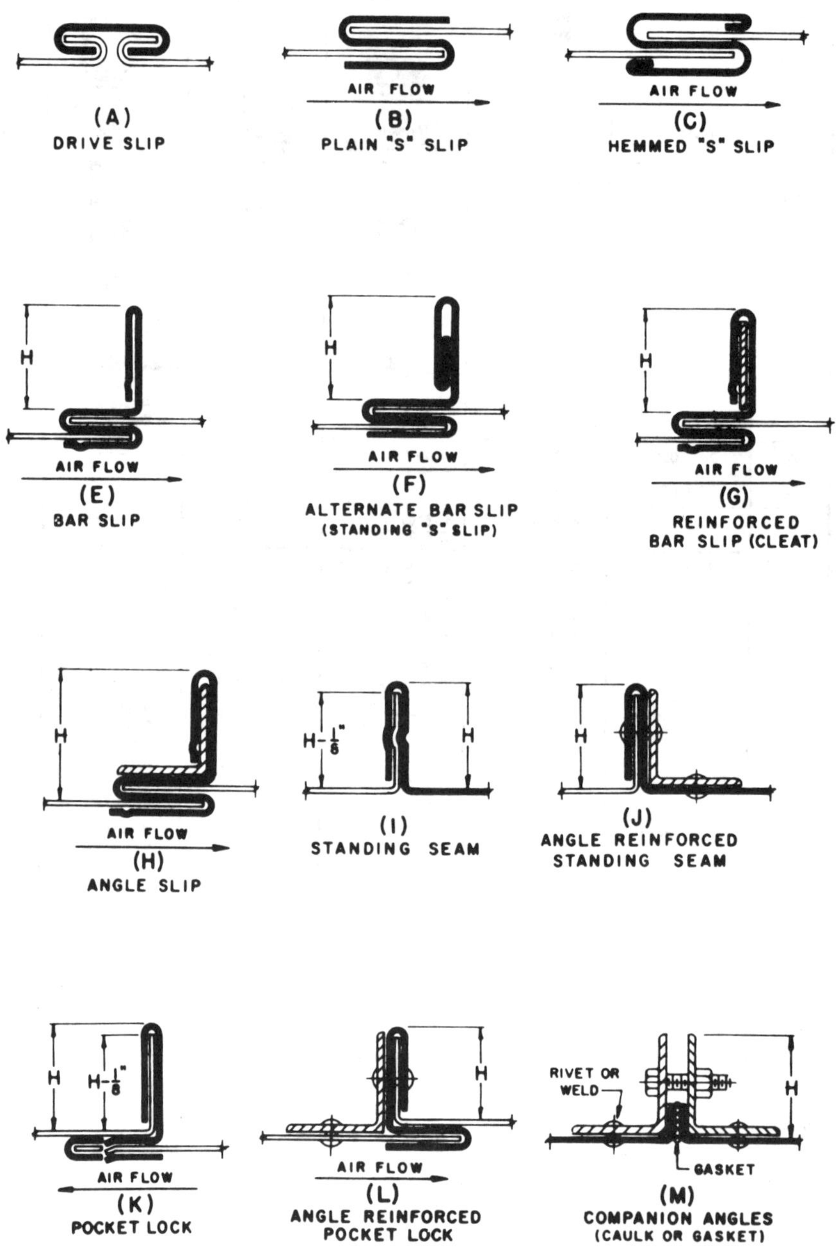

AIR CONDITIONING (Cont.)

LONGITUDINAL SEAMS FOR SHEET METAL DUCTWORK

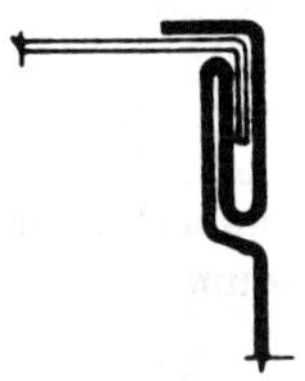

Fig. "N"
PITTSBURGH LOCK

Fig. "Z"
BUTTON PUNCH SNAP LOCK

Fig. "O"
ACME LOCK-GROOVED SEAM

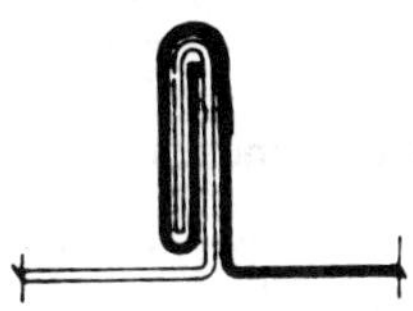

Fig. "T"
DOUBLE SEAM

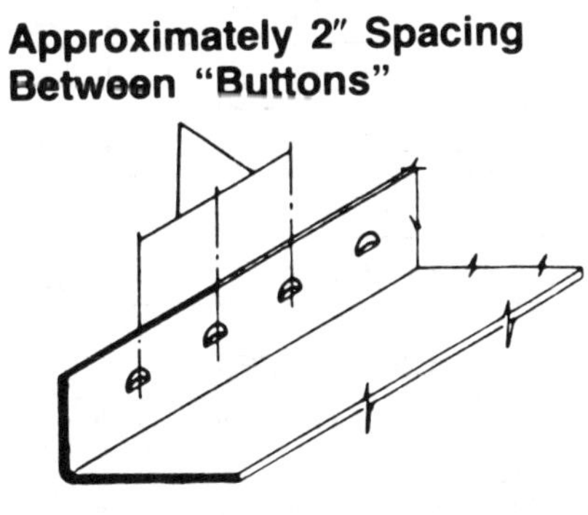

DETAIL NO. 1
MALE PIECE-SNAP LOCK

TYPICAL DUCT CONSTRUCTION SHEET METAL GAGES IN ONE- AND TWO-FAMILY DWELLINGS

Metal Gauges (duct not enclosed in partitions)		
ROUND DUCTS		
Diameter, inches	Minimum thickness galvanized sheet gage	Minimum thickness aluminum B&S gage
Less than 12	30	26
12-14	28	26
15-18	26	24
Over 18	24	22
RECTANGULAR DUCTS		
Width, inches	Minimum thickness galvanized sheet gage	Minimum thickness aluminum B&S gage
Less than 14	28	24
14-24	26	22
25-30	24	22
Over 30	22	20
Metal Gauges (duct enclosed in partitions)		
Width, inches	Minimum thickness galvanized sheet gage	Minimum thickness aluminum B&S gage
14 or less	30	26
Over 14	28	24

TYPICAL DUCT CONSTRUCTION SHEET METAL GAGES

(All uses except 1- and 2-family dwellings)

RECTANGULAR DUCTS		
Maximum side inches	Steel min. Galv. Sheet Gage	Aluminum Min. B&S Gage
Through 12	26 (0.022 in.)	24 (0.020 in.)
13 through 30	24 (0.028 in.)	22 (0.025 in.)
31 through 54	22 (0.034 in.)	20 (0.032 in.)
55 through 84	20 (0.040 in.)	18 (0.040 in.)
Over 84	18 (0.052 in.)	16 (0.051 in.)

ROUND DUCTS			
	Spiral seam duct	Longitudinal seam duct	Fittings
Diameter inches	Steel min. Galv. Sht. Gage	Steel min. Galv. Sht. Gage	Steel min. Galv. Sht. Gage
Through 12	28.(0.019 in.)	26 (0.022 in.)	26 (0.022 in.)
13 through 18	26 (0.022 in.)	24 (0.028 in.)	24 (0.028 in.)
19 through 28	24 (0.028 in.)	22 (0.034 in.)	22 (0.034 in.)
29 through 36	22 (0.034 in.)	20 (0.040 in.)	20 (0.040 in.)
37 through 52	20 (0.040 in.)	18 (0.052 in.)	18 (0.052 in.)

APPLICATIONS FOR CONDUCTORS USED FOR GENERAL WIRING

	AMBIENT TEMPERATURE								
	60°C 140°F	75°C 167°F	85°C 185°F	90°C 194°F	110°C 230°F	200°C 392°F	Dry	Dry or Wet	FEATURES
R	X						X		Code Rubber
RH		X					X		Heat Resistant
RHH				X			X		More Heat Resistant
RW	X							X	Moisture Resistant
RH-RW	X							X	Moisture and Heat Resistant
		X					X		Moisture and Heat Resistant
RHW		X						X	Moisture and Heat Resistant
RU	X						X		Latex Rubber
RUH		X					X		Heat Resistant
RUW	X							X	Moisture Resistant
T	X						X		Thermoplastic
TW	X							X	Moisture Resistant
THHN				X			X		Heat Resistant
THW		X						X	Moisture and Heat Resistant
THWN		X						X	Moisture and Heat Resistant
MI			X					X	Mineral Insulated Metal Sheathed
V			X				X		Varnished Cambric
AVA					X		X		With Asbestos
AVB				X			X		With Asbestos
AVL					X			X	With Asbestos

This table does not include special condition conductors, thickness of conductor insulation, or reference to all outer protective coverings.

GENERAL CLASSIFICATION OF INSULATIONS:

A .... Asbestos
H .... Heat Resistant
MI .... Mineral Insulation
R .... Rubber
RU .... Latex Rubber
V .... Varnished Cambric
T .... Thermoplastic
W .... (Water) Moisture Resistant

WIRE AND SHEET METAL GAGES

(In Decimals of an Inch)

Name of Gage	American Wire Gage (A.W.G.) (Corresponds to Brown & Sharpe Gage)	Birmingham Iron Wire Gage (B.W.G.)	United States Standard Gage (U.S.S.G.)	
Principal Use	Electrical Wire & Non-Ferrous Sheet Metal	Iron or Steel Wire	Ferrous Sheet Metal	
Gage No.				Gage No.
00 00000				00 00000
0 00000	.5800			0 00000
00000	.5165	.500		00000
0000	.4600	.454		0000
000	.4096	.425		000
00	.3648	.380		00
0	.3249	.340		0
1	.2893	.300		1
2	.2576	.284		2
3	.2294	.259	23.91	3
4	.2043	.238	.2242	4
5	.1819	.220	.2092	5
6	.1620	.203	.1943	6
7	.1443	.180	.1793	7
8	.1285	.165	.1644	8
9	.1144	.148	.1495	9
10	.1019	.134	.1345	10
11	.0907	.120	.1196	11
12	.0808	.109	.1046	12
13	.0720	.095	.0897	13
14	.0641	.083	.0747	14
15	.0571	.072	.0673	15
16	.0508	.065	.0598	16
17	.0453	.058	.0538	17
18	.0403	.049	.0478	18
19	.0359	.042	.0418	19
20	.0320	.035	.0359	20
21	.0285	.032	.0329	21
22	.0253	.028	.0299	22
23	.0226	.025	.0269	23
24	.0201	.022	.0239	24
25	.0179	.020	.0209	25
26	.0159	.018	.0179	26
27	.0142	.016	.0164	27
28	.0126	.014	.0149	28
29	.0113	.013	.0135	29
30	.0100	.012	.0120	30
31	.0089	.010	.0105	31
32	.0080	.009	.0097	32
33	.0071	.008	.0090	33
34	.0063	.007	.0082	34
35	.0056	.005	.0075	35
36	.0050	.004	.0067	36
37	.0045		.0064	37
38	.0040		.0060	38
39	.0035			39
40	.0031			40

Geographic Cost Modifiers

The costs as presented in this book attempt to represent national averages. Costs, however, vary among regions, states and even between adjacent localities.

In order to more closely approximate the probable costs for specific locations throughout the U.S., this table of Geographic Cost Modifiers is provided. These adjustment factors are used to modify costs obtained from this book to help account for regional variations of construction costs and to provide a more accurate estimate for specific areas. The factors are formulated by comparing costs in a specific area to the costs as presented in the Costbook pages. An example of how to use these factors is shown below. Whenever local current costs are known, whether material prices or labor rates, they should be used when more accuracy is required.

Cost obtained from Costbook pages × Location Cost Adjustment Factor = Adjusted Cost

For example, a project estimated to cost $125,000 using the Costbook pages can be adjusted to more closely approximate the cost in Los Angeles: $125,000 × 1.05 = $131,250.

Location	Factor
Alabama	0.80
Alaska	1.30
Arizona	0.89
Arkansas	0.74
California	
Los Angeles	1.05
San Francisco	1.19
Other	1.00
Colorado	0.90
Connecticut	1.01
Delaware	0.91
Florida	0.82
Georgia	
Atlanta	0.83
Other	0.75
Hawaii	1.25
Idaho	0.88
Illinois	
Chicago	1.00
Other	0.94
Indiana	0.90
Iowa	0.87
Kansas	0.81
Kentucky	0.85
Louisiana	0.82
Maine	0.85
Maryland	0.90
Massachussetts	
Boston	1.01
Other	0.96
Michigan	0.90
Minnesota	0.93
Missouri	0.86
Montana	0.87
Nebraska	0.83
Nevada	0.95
New Hampshire	0.86
New Jersey	0.97
New Mexico	0.84
New York	
New York City	1.20
Other	0.92
North Carolina	0.75
North Dakota	0.86
Ohio	0.87
Oklahoma	0.79
Oregon	0.90
Pennsylvania	
Philadelphia	1.01
Other	0.89
Rhode Island	0.97
South Carolina	0.77
South Dakota	0.76
Tennessee	0.79
Texas	0.82
Utah	0.83
Vermont	0.85
Virginia	0.81
Washington	1.01
Washington D.C.	0.95
West Virginia	0.85
Wisconsin	0.92
Wyoming	0.87

BNi Building News™

Square Foot Tables

The following Square Foot Tables list hundreds of actual projects for dozens of building types each with associated building size, total square foot building cost and percentage of project costs for total mechanical and electrical components. This data provides an overview of construction costs by building type. These costs are for actual projects. The variations within similar building types may be due, among other factors, to size, location, quality and specified components, materials and processes. Depending upon all such factors, specific building costs can vary significantly and may not necessarily fall within the range of costs as presented.

The data has been updated to reflect current construction costs and is an expanded summary of the projects and costs presented in the **Building News Square Foot 1993 Costbook.** The source of this data is **Design Cost & Data** magazine which has over 25 years of history in tracking, collecting and publishing construction cost data.

SQUARE FOOT TABLES

COMMERCIAL

AUTO DEALERSHIP

Project Size Gross S.F.	Project Cost $/S.F.	% Cost Mechanical	% Cost Electrical
7,700	73.03	16.7	9.4
16,100	40.66	10.2	15.9
20,000	42.56	12.9	23.4
26,300	43.42	12.5	22.0
43,600	34.24	19.4	13.2
53,600	58.84	12.5	11.5

BUSINESS CENTER

Project Size Gross S.F.	Project Cost $/S.F.	% Cost Mechanical	% Cost Electrical
3,900	44.33	12.0	9.2
9,900	41.72	9.1	7.6
54,400	28.32	3.4	12.2
135,000	31.18	8.2	1.5

CINEMA

Project Size Gross S.F.	Project Cost $/S.F.	% Cost Mechanical	% Cost Electrical
18,000	96.65	10.9	6.7
22,500 A	62.05	6.6	4.2

MALL/PLAZA

Project Size Gross S.F.	Project Cost $/S.F.	% Cost Mechanical	% Cost Electrical
9,700	31.29	15.0	13.3
10,500	50.77	8.0	13.5
16,300	46.29	9.4	10.6
26,900	45.72	15.0	7.0
36,000	38.24	10.0	11.0
36,300	44.19	12.4	8.2
44,720	45.93	18.3	12.4
59,100	44.06	9.8	9.5
60,000 R	45.89	10.0	9.5
64,100	46.03	22.4	18.4
66,000	53.33	13.5	11.5
67,400	38.08	21.0	15.0
73,500	55.84	14.9	6.9

For more detailed information see **Building News Square Foot 1993 Costbook.**

MALL/PLAZA (Cont.)

Project Size Gross S.F.	Project Cost $/S.F.	% Cost Mechanical	% Cost Electrical
142,000	35.21	7.1	8.0
220,000	93.65	11.0	6.4
223,700	26.97	9.0	9.3
321,200	31.92	7.3	7.4
379,900	41.61	11.2	6.2
405,100	43.35	13.9	6.0
482,000	73.32	10.5	9.4
630,000	48.79	12.2	12.4

RESTAURANT

Project Size Gross S.F.	Project Cost $/S.F.	% Cost Mechanical	% Cost Electrical
4,300 R	81.36	6.9	8.7
4,400 R	119.56	14.6	8.5
5,800	95.33	28.0	10.6
6,800 A	113.24	7.0	11.1
7,360 R	121.04	16.0	6.5
9,600	120.51	24.7	13.1
10,000 R	125.36	21.0	10.0
10,100	107.98	28.6	18.4
10,600	195.62	20.4	6.4
22,900 R	129.99	15.8	16.9

RETAIL STORE

Project Size Gross S.F.	Project Cost $/S.F.	% Cost Mechanical	% Cost Electrical
1,000	119.67	12.8	6.7
3,000 R	107.03	14.3	10.5
12,300	146.95	14.0	10.0
30,000	74.27	15.6	26.2
61,300	37.92	13.3	13.0
115,000	53.21	14.6	11.3
154,700	75.64	11.2	12.4
314,700 R	63.21	13.8	9.4

A = Addition R = Remodel

SQUARE FOOT TABLES

RESIDENTIAL

APARTMENTS

Project Size Gross S.F.	Project Cost $/S.F.	% Cost Mechanical	% Cost Electrical
3,700	44.88	14.9	4.4
13,900	60.47	7.9	7.2
19,200 R	96.81	45.3	7.5
19,700	56.85	7.4	10.4
23,700	62.16	10.6	4.3
26,500	58.78	25.2	12.8
35,100	47.67	16.4	5.6
54,000	83.67	23.3	13.1
62,700	60.74	17.0	9.0
67,300	55.32	13.8	8.4
70,600	30.76	18.1	7.8
75,300	68.79	13.2	8.1
75,600	66.89	14.5	8.9
72,200	58.36	18.4	10.9
77,600	75.89	26.9	14.3
88,100	70.32	15.3	9.3
89,500	65.21	10.7	11.1
94,100	38.87	8.5	6.8
96,000	48.44	17.0	13.3
102,000	85.17	17.8	12.5
103,200	38.45	12.1	8.9
103,600	62.30	19.1	9.0
105,200	79.70	14.9	8.9
106,200	57.47	12.8	9.3
110,900	54.36	15.6	8.4
111,800	77.33	17.3	7.4
115,900	58.60	12.5	8.6
117,200	35.50	12.9	8.0
119,000	50.16	15.6	8.4

For more detailed information see **Building News Square Foot 1993 Costbook.**

APARTMENTS (Cont.)

Project Size Gross S.F.	Project Cost $/S.F.	% Cost Mechanical	% Cost Electrical
119,400	32.55	18.6	8.7
144,300	72.69	15.0	8.6
176,300	62.23	19.1	9.0
192,300	34.86	10.1	6.0
210,900	71.90	19.1	9.5
220,200	74.29	14.8	7.5
253,900	96.60	20.6	7.7
369,500	69.84	15.8	9.0

CONDOS/TOWNHOUSES

Project Size Gross S.F.	Project Cost $/S.F.	% Cost Mechanical	% Cost Electrical
8,600	63.52	8.5	4.4
16,700	50.25	13.2	6.5
18,000	108.45	14.6	9.7
18,400	55.34	12.9	5.8
74,800	49.30	13.8	5.2
111,700	63.63	9.2	7.1
150,300	69.67	15.9	7.9
278,800	110.11	14.1	7.9
1,109,900	50.41	9.8	4.7

SINGLE-FAMILY HOMES

Project Size Gross S.F.	Project Cost $/S.F.	% Cost Mechanical	% Cost Electrical
600 R	42.47	16.3	2.0
900	57.03	33.0	3.0
2,100	66.00	8.8	4.0
2,200	141.79	23.8	3.4
2,500	69.10	6.1	7.1
2,900	66.79	7.7	2.8
3,000	46.67	9.5	6.8
3,100	104.71	15.5	4.6
3,600	65.73	8.5	3.0
3,700	91.44	9.8	3.4

A = Addition R = Remodel

SQUARE FOOT TABLES

RESIDENTIAL (Cont.)

SINGLE-FAMILY HOMES (Cont.)

Project Size Gross S.F.	Project Cost $/S.F.	% Cost Mechanical	% Cost Electrical
4,200	79.95	15.5	5.7
4,600	127.89	9.0	4.7
5,200	166.02	8.3	6.0
5,700	96.23	7.4	3.7
5,700	65.54	8.7	12.6
21,300*	46.40	26.0	4.0
22,700*	42.35	27.0	4.6
45,000*	65.57	7.8	2.5
51,458*	46.98	11.0	5.0

*Townhouses.

EDUCATIONAL

ADMINISTRATION (OFFICES)

Project Size Gross S.F.	Project Cost $/S.F.	% Cost Mechanical	% Cost Electrical
53,700	112.61	15.4	9.0

ATHLETIC FACILITY

Project Size Gross S.F.	Project Cost $/S.F.	% Cost Mechanical	% Cost Electrical
38,100	111.14	16.8	9.8
44,100	102.23	16.8	8.4
100,000	68.58	19.2	5.3
160,000	126.62	13.7	7.9
247,500	109.24	11.2	9.0
271,000	101.23	13.2	7.6
283,100	122.62	14.6	6.0

AUDITORIUM/PERFORMING ARTS

Project Size Gross S.F.	Project Cost $/S.F.	% Cost Mechanical	% Cost Electrical
9,900	175.29	17.5	29.2
17,800	164.86	11.4	16.1
29,200	161.01	16.2	10.5
62,700	119.81	13.0	12.9

For more detailed information see **Building News Square Foot 1993 Costbook.**

CLASSROOM

Project Size Gross S.F.	Project Cost $/S.F.	% Cost Mechanical	% Cost Electrica
35,400	163.18	10.2	6.8
70,000	73.79	24.3	13.1
78,900	140.42	13.3	14.3
80,100	113.03	16.9	10.5
100,000	113.45	21.1	9.8
166,000	76.67	16.1	11.2
298,400	72.48	11.3	11.9

COMPLETE COLLEGE FACILITIES

Project Size Gross S.F.	Project Cost $/S.F.	% Cost Mechanical	% Cost Electrical
95,300	116.19	19.4	11.7
450,000	140.00	18.6	10.8

ELEMENTARY SCHOOL

Project Size Gross S.F.	Project Cost $/S.F.	% Cost Mechanical	% Cost Electrical
18,000	83.85	19.1	13.8
30,800	77.74	16.0	10.0
31,600	65.86	12.1	11.3
35,700	96.11	17.3	8.7
40,000	82.80	18.5	13.4
40,500	69.68	22.1	11.3
57,000	63.33	22.1	7.5
69,700	97.97	20.6	9.3
91,400	79.24	18.2	7.6

HIGH SCHOOL

Project Size Gross S.F.	Project Cost $/S.F.	% Cost Mechanical	% Cost Electrical
116,400	87.03	18.1	12.9
133,000	75.30	17.8	10.7
184,000	129.12	23.0	10.0
217,200 R	70.37	26.4	10.6
254,000 R	55.65	17.2	14.6
431,700	83.85	13.1	9.6

A = Addition R = Remodel

For more information subscribe to **Design Cost & D**

SQUARE FOOT TABLES

EDUCATIONAL (Cont.)

JUNIOR HIGH SCHOOL

Project Size Gross S.F.	Project Cost $/S.F.	% Cost Mechanical	% Cost Electrical
26,000	108.71	9.5	9.3
28,100	67.31	11.9	9.5
52,800	86.62	18.3	8.6
91,600	107.63	21.2	11.0
123,700	87.86	29.1	9.1

LABORATORY/RESEARCH

Project Size Gross S.F.	Project Cost $/S.F.	% Cost Mechanical	% Cost Electrical
9,200	187.83	25.4	4.5
80,300	150.22	20.2	15.5

LIBRARY

Project Size Gross S.F.	Project Cost $/S.F.	% Cost Mechanical	% Cost Electrical
6,900	101.11	17.9	10.3
8,200	92.75	18.8	10.6
12,000	120.62	19.4	15.9
15,000	105.62	15.7	18.7
16,300	87.59	11.3	7.4
28,600	79.43	15.7	9.0
30,100	107.70	13.0	11.0
37,700	84.74	14.6	8.0
43,500	71.70	16.6	6.7
47,900	125.88	29.7	9.0
51,400	127.67	13.1	12.4
63,400 A	96.07	13.5	8.3
64,000	89.44	10.9	11.4
74,000	102.18	17.2	8.0
75,600	136.94	12.5	16.7
176,000	75.43	11.2	9.8

SPECIAL NEEDS FUNCTION

Project Size Gross S.F.	Project Cost $/S.F.	% Cost Mechanical	% Cost Electrical
15,200	85.38	16.6	8.4
27,900	102.96	19.4	11.2

For more detailed information see **Building News Square Foot 1993 Costbook.**

STUDENT CENTER/MULTIPURPOSE

Project Size Gross S.F.	Project Cost $/S.F.	% Cost Mechanical	% Cost Electrical
90,000	129.99	16.2	10.0
49,600	96.07	18.2	9.2
187,700	144.68	15.3	8.8
194,800	76.16	17.2	8.5

HOTEL/MOTEL

CONVENTION/CONFERENCE CENTER

Project Size Gross S.F.	Project Cost $/S.F.	% Cost Mechanical	% Cost Electrical
8,600 A	113.03	20.7	16.3
71,900	125.40	12.5	13.8
433,800	65.47	22.1	8.3

HOTEL

Project Size Gross S.F.	Project Cost $/S.F.	% Cost Mechanical	% Cost Electrical
19,900 A	64.47	16.8	5.8
25,875 R	58.78	11.8	10.5
48,400 A	116.15	23.6	8.2
64,300 R	155.72	20.3	10.1
104,200 A	72.59	15.0	8.8
108,040	61.94	13.0	7.0
110,100	79.64	18.4	10.5
132,000	150.01	16.4	5.4
135,900 A	92.17	15.2	7.7
144,100 A	105.61	19.3	11.5
231,000	116.25	15.2	8.4
449,800 A	64.89	13.1	7.0

HOTEL/INN

Project Size Gross S.F.	Project Cost $/S.F.	% Cost Mechanical	% Cost Electrical
57,400	74.69	13.0	10.0
73,000	47.93	24.4	18.1
75,900	64.42	16.5	7.6
162,000	77.37	17.5	8.0
197,000	74.53	15.7	7.7
277,900	56.20	18.8	9.4

A = Addition R = Remodel

SQUARE FOOT TABLES

INDUSTRIAL

MANUFACTURING

Project Size Gross S.F.	Project Cost $/S.F.	% Cost Mechanical	% Cost Electrical
14,300	65.84	13.0	7.0
18,500	98.09	20.5	13.5
26,600	29.56	6.1	12.1
31,400	77.32	18.9	17.7
33,400	44.77	23.1	15.8
37,300	22.15	3.0	24.0
43,400	42.98	14.7	11.7
45,400	72.45	15.9	15.0
79,800	76.16	16.0	14.5
81,100	80.29	8.2	8.4
137,400	50.25	41.6	13.6
179,600	57.99	26.3	13.6
186,000	54.15	19.5	11.7

RESEARCH AND DEVELOPMENT

Project Size Gross S.F.	Project Cost $/S.F.	% Cost Mechanical	% Cost Electrical
89,140	102.65	20.8	9.0
100,400	135.94	36.1	25.1
114,200	131.89	21.4	9.8
125,000	90.70	20.8	8.4
140,000	125.73	20.1	11.2

WAREHOUSE W/OFFICE

Project Size Gross S.F.	Project Cost $/S.F.	% Cost Mechanical	% Cost Electrical
14,000	27.25	6.5	9.2
19,000	17.07	2.3	1.8
19,700	35.18	9.5	7.0
31,200	27.55	7.0	7.0
40,500	34.71	6.6	10.5
47,500	12.96	2.5	4.8
62,000	45.93	10.8	10.0
96,200	21.07	2.2	6.3
105,000	20.07	5.3	11.1
149,800	21.44	14.5	8.6

For more detailed information see **Building News Square Foot 1993 Costbook.**

WAREHOUSE W/OFFICE (Cont.)

Project Size Gross S.F.	Project Cost $/S.F.	% Cost Mechanical	% Cost Electrica
168,600	27.76	11.4	6.3
209,600	25.33	4.9	10.3
402.400	34.82	14.9	8.3

MEDICAL

EDUCATION CENTER

Project Size Gross S.F.	Project Cost $/S.F.	% Cost Mechanical	% Cost Electrica
35,400	173.60	10.2	6.8

HOSPITALS

Project Size Gross S.F.	Project Cost $/S.F.	% Cost Mechanical	% Cost Electrica
9,300 R	174.60	29.5	12.0
15,900 A	126.86	27.4	15.3
16,600 R	43.51	14.7	9.6
22,000	293.27	31.6	17.5
39,100	145.00	23.3	7.9
63,800	115.77	23.4	7.8
98,000 A	167.07	20.5	17.0
100,200 A	267.36	22.2	9.6
103,900 A	147.90	31.1	11.7
109,300	146.43	20.7	15.7
148,700	115.18	25.2	11.8
154,700	187.00	19.3	10.5
165,484	162.50	21.5	14.8
165,700 A	158.14	26.2	17.7
179,400 A	96.55	28.9	20.3
182,800	143.74	19.1	14.0
265,000	171.44	31.2	14.1
281,100	164.23	24.1	14.6
435,000	170.29	21.5	13.7
694,300	94.39	28.9	9.7
772,300	172.29	31.5	13.4

A = Addition R = Remodel

SQUARE FOOT TABLES

MEDICAL (Cont.)

MEDICAL OFFICES/CENTERS

Project Size Gross S.F.	Project Cost $/S.F.	% Cost Mechanical	% Cost Electrical
3,000	72.79	13.7	11.5
5,500	75.75	8.5	18.7
10,000	95.82	13.9	9.4
10,600	117.52	13.2	8.9
16,300	97.08	21.8	13.2
18,300	52.37	7.1	13.4
20,600	75.32	14.3	6.4
24,900	133.83	21.0	15.1
27,000	133.71	19.7	10.1
28,400	91.62	13.6	9.3
30,500	102.32	17.2	12.7
32,000	82.36	18.1	10.5
44,300	82.07	24.4	16.2
50,200	46.30	9.8	4.9
51,200	118.40	21.0	12.0
64,600	56.25	13.2	6.4
66,000	52.46	9.8	8.8
80,000	33.50	8.2	8.3
137,175	75.64	12.8	6.6

NURSING HOMES

Project Size Gross S.F.	Project Cost $/S.F.	% Cost Mechanical	% Cost Electrical
11,600 A	221.22	53.2	7.9
16,800	143.62	33.3	7.8
31,900 A	108.77	22.0	11.0
64,100	89.96	20.6	11.1
290,000	127.99	16.1	13.6

RESEARCH

Project Size Gross S.F.	Project Cost $/S.F.	% Cost Mechanical	% Cost Electrical
34,600	121.88	18.1	3.0

For more detailed information see **Building News Square Foot 1993 Costbook.**

PUBLIC FACILITIES

ANIMAL CENTER

Project Size Gross S.F.	Project Cost $/S.F.	% Cost Mechanical	% Cost Electrical
20,000	155.59	20.7	4.6
39,100	116.82	22.9	6.8
44,300	110.71	8.1	5.8

AUTO DEALERSHIP

Project Size Gross S.F.	Project Cost $/S.F.	% Cost Mechanical	% Cost Electrical
7,700	73.53	16.7	9.4

BROADCASTING

Project Size Gross S.F.	Project Cost $/S.F.	% Cost Mechanical	% Cost Electrical
20,000	214.37	20.0	13.0
29,500	164.97	16.6	15.0
45,000 R	119.88	15.0	13.0

CIVIC CENTER

Project Size Gross S.F.	Project Cost $/S.F.	% Cost Mechanical	% Cost Electrical
6,000	118.51	9.2	2.8
23,900	148.11	12.3	10.9
34,400	67.42	3.5	17.8
69,800 A	179.19	17.7	8.8
206,500	95.76	15.0	11.0

CORRECTIONAL FACILITIES

Project Size Gross S.F.	Project Cost $/S.F.	% Cost Mechanical	% Cost Electrical
44,600	138.95	20.9	13.1
66,000	85.96	15.0	23.0
257,800 A	150.43	20.9	10.7
360,000	99.65	32.7	13.2

FIRE STATION

Project Size Gross S.F.	Project Cost $/S.F.	% Cost Mechanical	% Cost Electrical
6,900	125.88	12.4	9.7
7,600	95.91	16.0	9.5
8,430	130.84	12.8	9.6
9,600	115.93	13.5	11.8

A = Addition R = Remodel

SQUARE FOOT TABLES

PUBLIC FACILITIES (Cont.)

GOVERNMENT BUILDINGS

Project Size Gross S.F.	Project Cost $/S.F.	% Cost Mechanical	% Cost Electrical
12,300	89.86	13.5	10.7
23,500	149.90	11.1	15.6
27,300	116.19	19.5	9.3
31,600	130.78	27.1	11.3
46,600	123.15	17.4	13.8
72,100	133.85	24.4	10.7
78,200	110.93	19.7	15.2
332,900	108.19	16.2	14.2
364,100	140.95	14.6	12.9
771,000	155.49	17.6	11.3

MUSEUM

Project Size Gross S.F.	Project Cost $/S.F.	% Cost Mechanical	% Cost Electrical
27,600	109.56	17.8	14.1
30,100	125.04	18.6	7.9
43,264	116.40	11.1	8.3
63,000	126.09	8.8	18.1

PARKING GARAGE

Project Size Gross S.F.	Project Cost $/S.F.	% Cost Mechanical	% Cost Electrical
66,000	31.60	2.8	3.1
169,000	19.70	10.3	3.7
562,700	17.59	2.4	6.2

TRANSPORTATION

Project Size Gross S.F.	Project Cost $/S.F.	% Cost Mechanical	% Cost Electrical
7,300	201.10	15.1	3.2
14,300	164.23	13.3	16.6
23,000	112.87	9.6	13.7
35,500	94.97	1.0	19.0
49,100	139.58	35.5	11.6

For more detailed information see **Building News Square Foot 1993 Costbook.**

TRANSPORTATION (Cont.)

Project Size Gross S.F.	Project Cost $/S.F.	% Cost Mechanical	% Cost Electrical
288,100	80.69	8.3	13.3
2,160,000	133.05	23.3	11.5

OFFICES

BANKS

Project Size Gross S.F.	Project Cost $/S.F.	% Cost Mechanical	% Cost Electrical
2,900	188.98	5.3	4.3
3,100	83.85	5.9	6.7
3,300	118.83	10.3	16.4
3,600	91.54	10.3	12.2
4,000	104.55	8.0	9.0
4,100	103.24	21.4	13.0
4,200	110.50	8.6	14.21
4,400	127.67	12.2	12.7
4,500	80.69	12.4	13.5
4,900	127.58	11.7	11.2
5,900	94.39	9.3	13.8
6,000	106.71	11.6	7.3
6,100	142.53	11.0	8.0
7,000	182.87	6.0	9.0
7,300	117.56	11.9	11.3
7,700	129.62	8.0	7.5
7,800	144.53	11.0	11.7
8,000	73.21	10.0	14.0
9,200	116.19	9.9	12.1
9,400	84.17	11.7	11.8
10,200	175.61	12.6	12.6
12,600	62.68	7.0	18.0

A = Addition R = Removal

For more information subscribe to **Design Cost & D**

SQUARE FOOT TABLES

OFFICES (Cont.)

BANKS (Cont.)

Project Size Gross S.F.	Project Cost $/S.F.	% Cost Mechanical	% Cost Electrical
13,300	110.19	9.5	8.3
13,800	98.71	10.0	9.7
15,000	84.80	15.4	12.4
15,200	67.52	9.2	12.9
15,500	86.28	9.8	10.3
16,000	56.67	13.4	23.1
20,100	53.20	13.0	11.0
21,700	112.82	8.8	11.3
44,800	96.39	13.0	8.2
53,200	174.08	14.9	7.2
62,100	99.97	10.1	7.9
95,100	131.41	13.5	4.3

OFFICE BUILDINGS

Project Size Gross S.F.	Project Cost $/S.F.	% Cost Mechanical	% Cost Electrical
2,600	119.35	17.2	9.4
3,400	97.70	10.5	11.3
3,800	96.18	16.4	11.8
4,400	94.70	12.8	8.5
4,500	81.43	13.0	7.0
5,100	59.10	16.6	10.2
5,200	77.11	8.0	5.7
6,700	114.30	16.8	10.4
7,500	124.09	10.1	8.0
7,900	96.18	17.4	9.0
8,100	149.16	10.6	11.0
10,600	85.85	9.9	9.7
10,900	48.04	13.8	10.8

For more detailed information see **Building News Square Foot 1993 Costbook.**

OFFICE BUILDINGS (Cont.)

Project Size Gross S.F.	Project Cost $/S.F.	% Cost Mechanical	% Cost Electrical
11,300	98.28	17.0	6.0
13,000	67.48	15.0	9.0
14,400	88.59	19.8	12.9
14,500	66.48	17.3	12.5
17,000	96.11	14.7	7.9
17,800 A	52.25	10.4	11.0
18,100	101.23	22.5	11.6
19,300	51.51	11.1	8.1
24,600	46.46	18.5	14.1
27,700	59.31	19.6	5.5
27,800	110.50	12.7	5.1
27,800	59.52	17.8	10.3
32,500 R	101.34	13.9	6.4
35,400	58.36	15.0	12.0
36,500	47.61	10.2	10.2
42,300	63.10	10.3	7.7
44,400	87.12	23.5	14.7
44,400	44.88	11.0	5.0
44,500	62.94	11.5	3.1
45,400	53.72	19.1	13.2
47,300	54.46	18.5	8.0
49,700	93.02	22.1	7.2
50,000	88.38	19.4	15.6
50,400	95.86	23.2	7.8
52,200	62.57	18.3	7.8
52,900	73.32	4.4	3.9
53,700	112.61	15.4	9.0

A = Additional R = Removal

SQUARE FOOT TABLES

OFFICES (Cont.)

OFFICE BUILDINGS (Cont.)

Project Size Gross S.F.	Project Cost $/S.F.	% Cost Mechanical	% Cost Electrical
54,000	44.88	14.4	2.6
56,000	51.30	10.6	6.2
56,500	69.68	19.1	11.0
72,000 R	23.49	12.9	20.2
74,000	47.25	13.5	6.8
80,800	49.41	12.0	6.0
81,800	67.95	21.1	9.5
81,900	58.15	19.9	9.0
82,000	79.64	14.0	4.2
83,100	94.02	16.0	8.6
85,400	82.43	18.7	8.6
86,200	64.50	14.7	11.0
99,900	74.58	16.6	6.6
100,000	83.38	10.6	12.1
100,000	49.72	16.3	10.4
116,400	79.64	12.6	10.0
134,500	161.38	14.1	12.4
140,000	125.73	20.1	11.2
155,700	156.75	11.1	6.1
171,000	73.32	24.0	7.7
174,300	89.44	12.3	9.1
203,300	122.25	19.0	13.2
265,800	169.29	11.0	10.0
287,300	43.14	10.7	4.1
319,800	73.42	13.4	5.6
350,000	108.50	20.0	13.0
360,900	65.79	14.2	11.1
394,000	51.83	22.5	8.7

For more detailed information see **Building News Square Foot 1993 Costbook.**

OFFICE BUILDINGS (Cont.)

Project Size Gross S.F.	Project Cost $/S.F.	% Cost Mechanical	% Cost Electrical
430,000	78.11	21.8	9.3
490,000	83.22	11.2	7.5
588,400	170.92	15.0	9.0
606,000	68.26	9.4	8.8
620,000	191.09	12.6	12.8
733,500	47.83	10.8	4.2

RECREATIONAL

ARENA

Project Size Gross S.F.	Project Cost $/S.F.	% Cost Mechanical	% Cost Electrical
315,200	195.73	10.4	8.00
385,800	109.77	13.2	8.80
727,000	98.24	14.9	7.40

HEALTH CLUB

Project Size Gross S.F.	Project Cost $/S.F.	% Cost Mechanical	% Cost Electrical
15,900	44.03	8.7	9.7
21,800	75.32	10.3	10.3
30,100	116.25	11.4	19.4
66,400	56.09	10.7	8.5

RECREATIONAL CENTER

Project Size Gross S.F.	Project Cost $/S.F.	% Cost Mechanical	% Cost Electrical
9,900	104.82	6.6	9.8
14,000	68.79	11.2	5.0
14,000	83.64	11.3	16.7
15,700	59.04	17.8	14.3
20,000	144.37	19.1	8.9
21,200	80.85	16.0	9.6

A = Addition R = Removal

SQUARE FOOT TABLES

RECREATIONAL (Cont.)

RECREATIONAL CENTER (Cont.)

Project Size Gross S.F.	Project Cost $/S.F.	% Cost Mechanical	% Cost Electrical
26,000	92.39	9.3	7.0
53,400 A	110.50	11.7	6.4
69,800	179.19	17.7	8.8

RELIGIOUS

CHURCH

Project Size Gross S.F.	Project Cost $/S.F.	% Cost Mechanical	% Cost Electrical
4,100	121.78	8.8	13.5
10,400 R	171.18	12.8	8.1
11,100	73.42	10.9	12.7
13,400	111.24	5.5	6.0
14,500	76.00	12.8	7.4
15,200	98.71	18.3	8.
15,700	85.38	14.4	8.4

For more detailed information see **Building News Square Foot 1993 Costbook.**

CHURCH (Cont.)

Project Size Gross S.F.	Project Cost $/S.F.	% Cost Mechanical	% Cost Electrical
16,000	138.84	14.1	9.6
20,900	86.49	16.0	14.0
21,500	95.07	7.3	8.0
22,900	85.70	12.1	9.5
30,600	64.79	15.5	8.0
42,700	74.19	18.6	7.7

MULTI-PURPOSE

Project Size Gross S.F.	Project Cost $/S.F.	% Cost Mechanical	% Cost Electrical
4,400	77.27	11.5	17.5
5,800 A	98.50	8.9	7.5
6,400	134.47	15.6	15.8
9,000	65.84	7.7	5.9
9,000	93.97	16.0	6.6
10,100	62.78	11.1	12.0
12,000	123.99	13.1	10.7
18,400	67.95	10.8	10.1
19,500	119.67	16.0	17.3

A = Addition R = Removal

BNi Building News™

INDEX

- D -

- T -

- U -

The Complete Estimating Reference Book

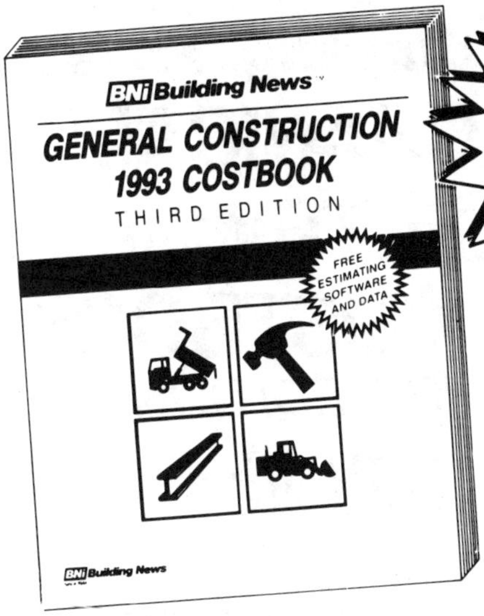

FREE Estimating Software And Data See Order Form

With Your BNi Building News™ 1993 Costbook Purchase You Can Receive FREE . . . CONSTRUCTION ESTIMATOR With Electronic Data

For Expanded Software Offerings See Pages 12, 13 & 14 Of This Catalog

Over 550 Pages

BNi Building News™

GENERAL CONSTRUCTION 1993 COSTBOOK

THIRD EDITION

$39.95

(1557010706)

Edited by A. M. Fogarty & Associates
Construction Estimating Consultants
Serving The Industry For Over 35 Years.

•Accurate •Authoritative • Up-To-Date

All New For 1993!

Available For Delivery In December

- *Every Price Looked At And Changed As Needed*
- *Over 10,000 Construction Items*
- *Material and Labor Costs Broken Out*
- *CSI MASTERFORMAT — All 16 Divisions*
- *All Trades Covered In Detail*
- *Man-Hour Tables — For Scheduling And Productivity*
- *Geographic Cost Modifiers*
- *Affordable — About-A-Dime-A-Day!*

This extensive resource provides current, accurate construction costs and man-hour data for all aspects of building construction. Costs are based on "contractors' prices" for materials and prevailing rates for labor. Overhead and profit are included in all costs — there's no confusion!

All data is organized in the 16 division CSI MASTERFORMAT — the industry standard. This simple, easy-to-use format along with the detailed index means you can instantly find the costs you need for fast, reliable estimates.

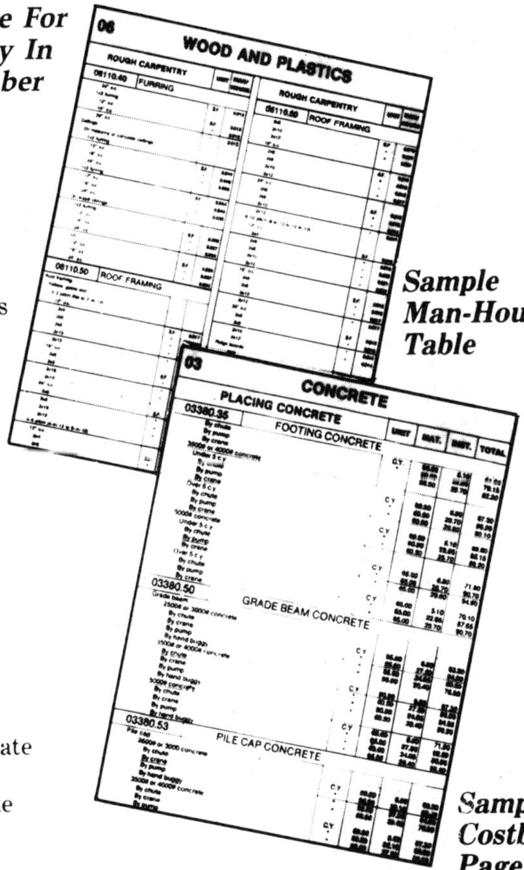

Sample Man-Hour Table

Sample Costbook Page

Helpful 'Extras' To Add To The Costbook's Usefulness

- **Easy-To-Follow Instructions** in no nonsense language.
- **A Detailed Index** for quick reference. You can find cost categories fast.
- **Geographic Cost Modifiers** mean you can customize your estimate by location — by multiplying by an adjustment factor.
- **Square Foot Tables,** page after page of them, allow you to make quick budget estimates based on actual projects.
- **Construction Fact Tables For Fast Reference** Allowance and measurement charts, tips, standard sizes, do's and don'ts . . . and more!

Data For Your Most Technical Estimating

Over 500 Pages

FREE Estimating Software And Data See Order Form

With Your BNi Building News™ 1993 Costbook Purchase You Can Receive FREE . . . CONSTRUCTION ESTIMATOR With Electronic Data

For Expanded Software Offerings See Pages 12, 13 & 14 Of This Catalog

Co-Sponsored by

Consulting-Specifying Engineer

BNi Building News™

MECHANICAL/ ELECTRICAL 1993 COSTBOOK

THIRD EDITION

$39.95

(1557010730)

Edited by A. M. Fogarty & Associates Construction Estimating Consultants Serving The Industry For Over 35 Years.

All New For 1993!

Available For Delivery In December

Every Price Looked At And Changed As Needed

- *Plumbing*
- *Fire Protection*
- *HVAC*
- *Electrical*
- *Material and Labor Costs*
- *Man-Hour/Labor Unit Tables*
- *Geographic Cost Modifiers*
- *Detailed Index*

Accurate, Authoritative Cost Data

From pipe to duct, conduit to receptacle, this detailed reference book provides extensive coverage of the most technical aspects of building construction. With thousands of current, reliable mechanical and electrical costs at your fingertips, you can estimate quickly and accurately. All data is organized according to the industry standard — the CSI MASTERFORMAT — so you can find the costs you need . . . fast! Geographic Cost Modifiers allow you to tailor your estimates to specific areas of the country.

Typical Installation Labor For Thousands Of Construction Items

These valuable Man-Hour Tables take all activities and all conditions into account.

HOURS:
Called out by estimating units in an easy-to-read format.

DESCRIPTION:
Construction terms to identify and quantify each task.

CSI FORMAT:
The industry standard, so you know the "language".

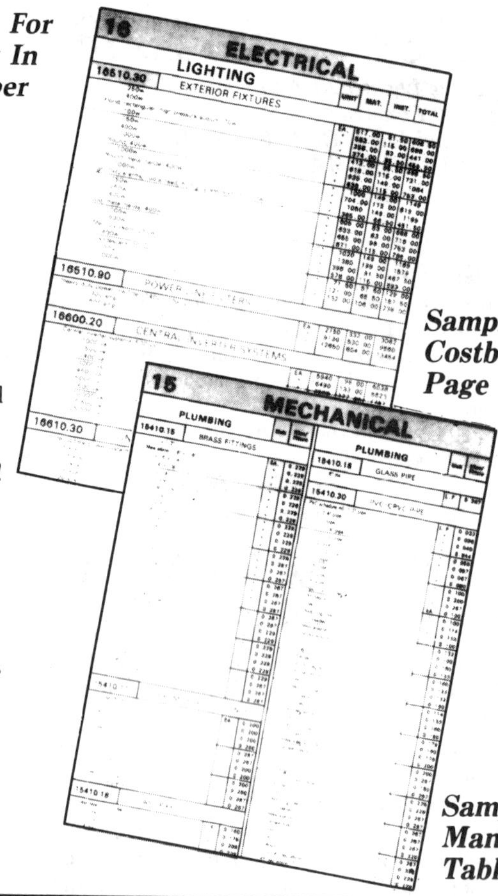

Sample Costbook Page

Sample Man-Hour Table

Finally . . . An Estimating Costbook Designed For The Unique Requirements Of The Home Builder

FREE Estimating Software And Data See Order Form

With Your BNi Building News™ 1993 Costbook Purchase You Can Receive FREE . . . CONSTRUCTION ESTIMATOR With Electronic Data

For Expanded Software Offerings See Pages 12, 13 & 14 Of This Catalog

Over 350 Pages

BNi Building News™

HOME BUILDER'S 1993 COSTBOOK

FIRST EDITION

$29.95

(1557010749)

Co-Published with the HOME BUILDER PRESS

Publishing Arm of the National Association of Home Builders

Edited by A.M. Fogarty & Associates Construction Estimating Consultants Serving the Industry for Over 35 Years.

Available For Delivery In December

- ***Man-Hour Data***
- ***New And Remodel***
- ***Foundations to Finish***
- ***Material And Labor Costs***

Published in conjunction with the National Association of Home Builders, the largest construction organization in the U.S., this new costbook is the first in years devoted entirely to the home building industry.

This complete reference is designed for the small to mid-size home builder. It contains material and labor costs and installation times for all components of house construction: concrete, framing, insulation, painting, flooring, plumbing, wiring, etc.

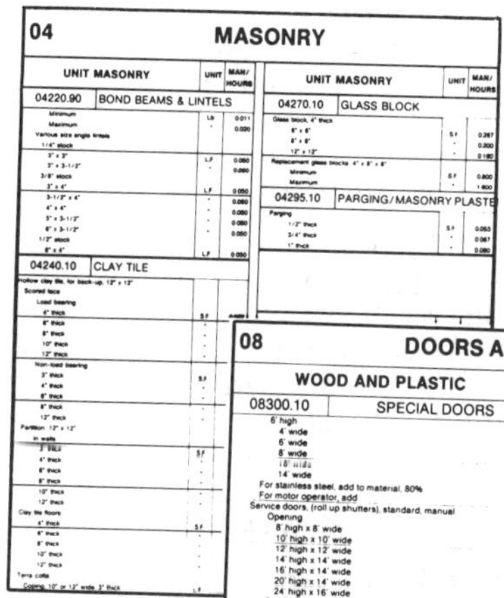

Sample Man-Hour Table

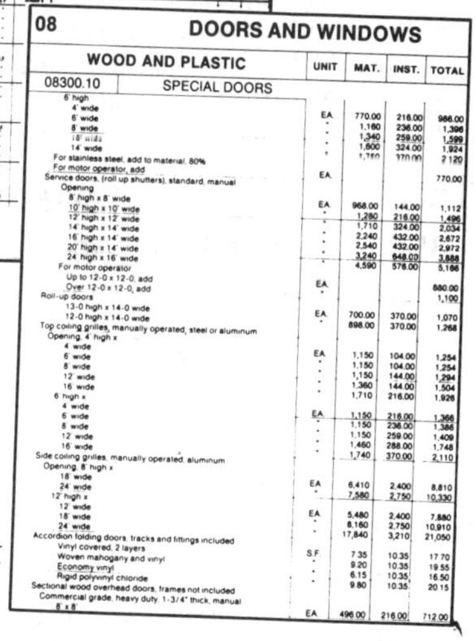

Sample Costbook Page

Helpful "Extras" To Add To The Costbook's Usefulness

- **Easy-to-Follow Instructions** in no nonsense language.
- **A Superb, Detailed Index.** You can find cost categories fast.
- **Geographic Cost Modifiers** means you can customize your estimate by location — by multiplying by an adjustment factor.
- **Square Foot Tables,** page after page of them, allow you to make quick budget estmates based on actual projects.
- **Construction Fact Tables for Fast Reference.** Allowamce and measurement charts, tips, standard sizes, do's and don'ts . . . and more!

Help For Your Toughest Job . . . Estimating

Over 500 Pages

FREE Estimating Software And Data See Order Form

With Your BNi Building News™ 1993 Costbook Purchase You Can Receive FREE . . . CONSTRUCTION ESTIMATOR With Electronic Data

For Expanded Software Offerings See Pages 16, 17 & 18 Of This Catalog

Co-Sponsored by

Professional Builder & Remodeler

BNi Building News™

REMODELING 1993 COSTBOOK

THIRD EDITION

$39.95

(1557010714)

Edited by A. M. Fogarty & Associates Construction Estimating Consultants Serving The Industry For Over 35 Years.

Includes The Most Difficult Items To Estimate

- *Demolition — Cutting and Patching*
- *Minimum Quantities and Costs*
- *Working Around Existing Conditions*
- *CSI MASTERFORMAT — All Divisions*
- *Man-Hour Data*
- *Geographic Cost Modifiers*

Available For Delivery In December

Remodeling is the toughest type of construction to estimate. You need all the help you can get. **Now** you have a resource you can turn to for fast, accurate cost information.
Hidden costs can be easily identified with the help of the CSI MASTERFORMAT, numbering systems and the clear, concise presentation format . . . and you can have access to this information every day of the year — for about A-Dime-A-Day!

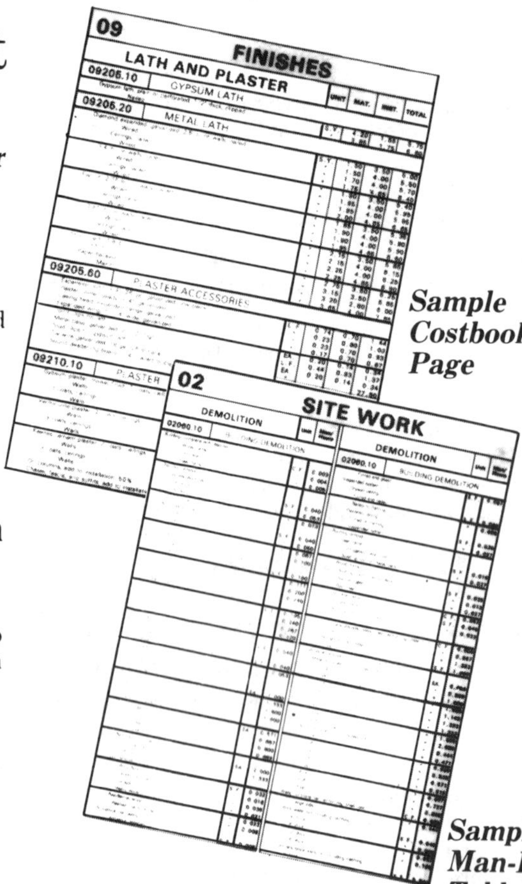

Sample Costbook Page

Sample Man-Hour Table

Thousands Of Costs Broken Out By Easy-To-Work-With Categories For Fast, Accurate Figuring

✔ **Material Costs:**

Represent national averages for prices a contractor would expect to pay, plus allowances for handling. Researched as close to publication time as possible for maximum accuracy and timeliness.

✔ **Labor Costs:**

Prevailing rates — appropriate for each trade and type of work. Eliminates costly "guesstimates."

✔ **Overhead and Profit:**

Commonly applicable taxes, insurance and other factors are included. No confusion about what a number includes and what it doesn't.

Square Foot Cost Data Compiled From Actual Projects

FREE Estimating Software And Data See Order Form

With Your BNi Building News 1993 Costbook Purchase You Can Receive FREE . . . CONSTRUCTION ESTIMATOR With Electronic Data

For Expanded Software Offerings See Pages 16, 17 & 18 Of This Catalog

Fully Illustrated

BNi Building News™
SQUARE FOOT 1993 COSTBOOK
THIRD EDITION

$44.95

(1557010722)

Over 475 Pages

Co-Sponsored by

BUILDING DESIGN & CONSTRUCTION

Edited by A. M. Fogarty & Associates
Construction Estimating Consultants
Serving The Industry For Over 35 Years.

Available For Delivery In December

Complete Project Costs For 10 Building Types

- *Commercial*
- *Residential*
- *Educational*
- *Hotel/Motel*
- *Industrial/Warehouse*
- *Medical*
- *Public Facilities*
- *Offices*
- *Recreational*
- *Religious*

Hundreds of actual projects. Thousands of actual costs. All data in this one-of-a-kind book is based on actual building projects thoroughout the country.
Each project is broken down into the 16 CSI MASTERFORMAT divisions. Costs are presented as Cost by Division, Percent of Total and Cost per Square Foot.
You can use this data with confidence for figuring of budgets, conceptual estimates, cost comparisons, design development . . . any time you need fast, accurate project costs.

Sample Pages

Use With Confidence For Fast, Accurate Figuring

- *Budgets*
- *Conceptual Estimates*
- *Cost Comparisons*
- *Appraisals*
- *Replacement Costs*
- *Design Development*

Compiled By Estimating Experts

Brought to you by "Building Design & Construction" magazine and Building News, serving your industry with construction/building information since 1946.

The Ninth Triennial Edition of the "Green Book"

Standard Specifications For Public Works Construction

$29.95
(1557010609)

630 Pages

1992 Supplement Available

Easy To Use Construction Product Tables

Charts To Help You Better Understand Product Make-Up And Composition

Clear And Easy To Read Text, Written By Professionals "Who Do The Work"

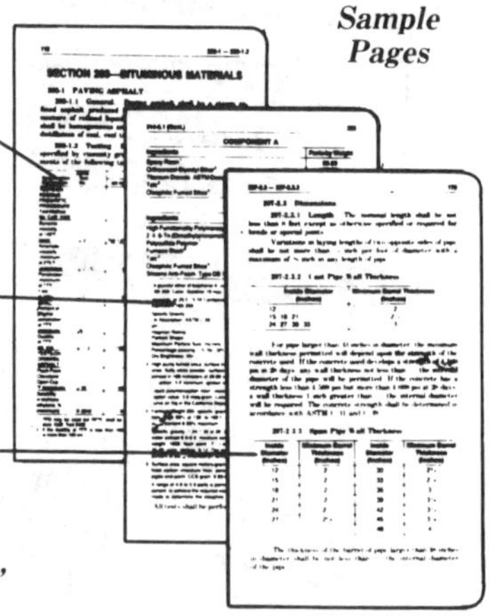

Sample Pages

Updated And Expanded To Include The Latest Construction Techniques And Technology

The Standard Specifications For Public Works Construction, a document updated and published every three years, is the official Public Works Specifications contract document for the majority of cities, counties and other public agencies in California, as well as a number of major jurisdictions and regions of the United States.

This document has been written and promulgated by a Joint Cooperative Committee of the Southern California Chapter of America Public Works Association and the Southern California Districts of Associated General Contractors of California. The original research and development of these standards was done by a task force of over 400 experts on the Joint Cooperative Committee.

1992 Supplement (05001692) **$2.95**

The Indispensable Resource For Public Works

Public Works Inspectors' Manual

Revised 3rd Edition

$29.95
(0070052980)

670 Pages

Industry Grading Charts To Assist You In Your Profession

Forms And "How To Fill Them Out" Take The Problems Out Of Everyday Happenings

Specifications And Graphics Combine To Help You Understand The Components Of Construction

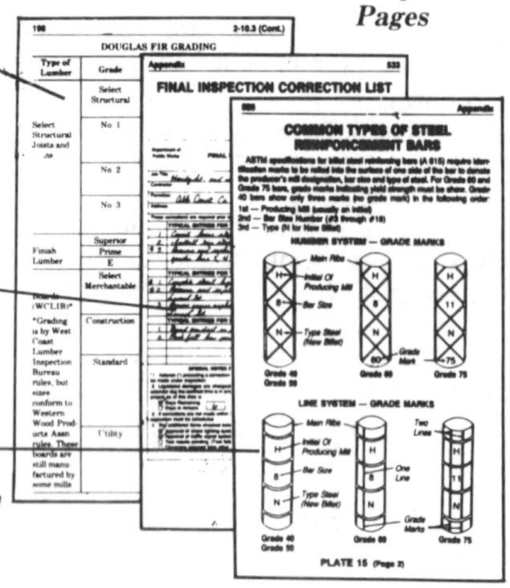

Sample Pages

A Valuable Reference For The Public Works Official, Engineer And Contractor

A complete guidebook for professionals inspecting all types of public works construction for city, county, state and federal agencies. The most comprehensive and authentic text of its kind ever written and published. Of considerable value not only to inspection personnel, but als to contractors, engineers, architects and students.

Not only is this book a complete guidebook for inspection personnel, but it is one of the best treatises on the whole general subject of construction, both public and private. As such, it's a must for every professional in Public Works Construction.

Graphic Companion To The "Green Book"

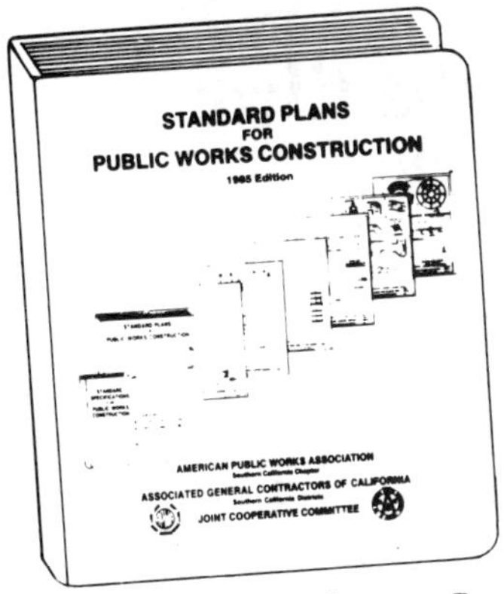

Over 300 Pages Loose Leaf

2nd Edition

Standard Plans For Public Works Construction

$39.95

(1557010552)

Includes 1992 Supplement

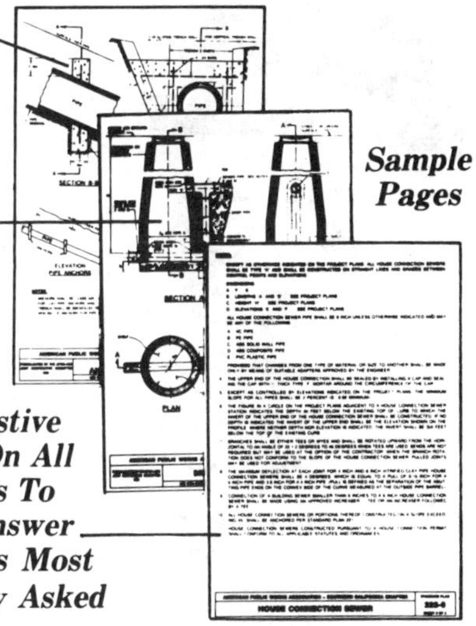

Details And Dimensions To Assist You In Your Job

Cross Sections To Better Visualize The Specified Item

Sample Pages

Exhaustive Notes On All Details To Help Answer Questions Most Commonly Asked

Hundreds Of Standardized Drawings And Details

The culmination of seven years of intensive work by a multi-governmental agency sub-committee of the Joint Cooperative Committee, this publication fulfills a vital need as a companion document to the Standard Specifications for Public Works Construction by providing standardized drawings and design criteria for the primary structures, facilities and devices referenced in the Standard Specifications.

The Standard Plans are a important adjunct to any user of the Standard Specifications, including public agencies, design professionals, contractors and materials and equipment suppliers.

1992 Supplement only (05002292) **$4.95**

State of California

Updated 1990 Edition Illustrated

CAL/TRANS Manual of Traffic Controls

For Construction And Maintenance Work Zones

$6.95

(050012)

This Manual of Traffic Controls for Construction and Maintenance Work Zones is authored by the State of California, Department of Transportation (Cal/Trans), and is issued to provide the basic standards for uniform types of warning signs, lights and devices to be placed upon any public highway or street by any person engaged in performing work which interferes with or endangers the safe movement of traffic upon such highway or street, in accordance with Section 21400 of the California Vehicle Code.

WATCH Booklet

Newly Updated Fully Illustrated

Work Area Traffic Control Handbook

7th Edition

$4.95

(050025-7)

The purpose of this handbook is to set basic principles and to serve as a standard for control of traffic work areas for those who perform work in public streets. These standards attempt to provide safe and effective work areas and to warn, control, protect and expedite vehicle and pedestrian traffic.

Large Orders Can Be Customized With Your Logo

Design Guide To The 1991 Uniform Building Code

By Alfred Goldberg

$55.00

(0961480890)

The Guide provides analysis and discussion of specific code provisions essential to the design of buildiong. With illustrations and technical appendix. Incorporates all code changes from the last three years.

1991 Uniform Building Code Standards

Over 1500 Pages

$93.30

(90009)

This publication contains all of the pertinent standards and specifications referred to in the 1991 Uniform Building Code.

1991 Uniform Fire Code

Loose Leaf

$55.10

(91008)

Soft Cover (900015) **$47.80**

1991 Standard Building Code

$60.00

(900026)

The Standard Building Code, developed by the Southern Building Code Congress International, is the model code for the majority of the southern states in the U.S. Adopts by reference nationally recognized standards.

1991 Uniform Building Code

$76.85

(90008)

Loose Leaf

The Uniform Building Code is the most widely adopted model building code in the world. Published triennially, it provides complete regulations covering all major aspects of building design and construction relating to life and fire safety and structural safety. Construction requirements reflect the latest technological advances available in the building industry. It is the official building code of the majority of cities, counties and states in the West.

Soft Cover (90007) **$67.00**

1991 Uniform Mechanical Code

Loose Leaf

$48.55

(900036)

This code provides complete provisions for heating, ventilating and air-conditioning. Complete standards for this code are within its appendix. Jointly published by the International Association of Plumbing and Mechanical Officials and the International Conference of Building Officials.

Soft Cover (900035) **$42.10**

1991 Uniform Plumbing Code

Loose Leaf

$55.90

(91006A)

The Uniform Plumbing Code is promulgated and published by the International Association of Plumbing and Mechanical Officials and covers all aspects of plumbing, including requirements for plumbing materials, and IAPMO installation Standards.

Soft Cover (90020A) **$47.80**

Produce More Accurate Estimates In Less Time With . . .

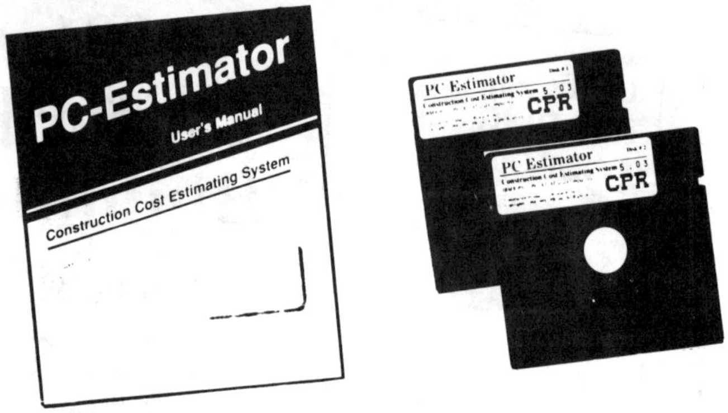

PC-ESTIMATOR

Only $99.00 (CPR101)

Affordable, Basic Construction Estimating Software

Features Include:

- *Complete Stand-Alone System.*
- *Standard Cost Database (CSI).*
- *dBase III Compatible.*
- *Built-In Takeoff Calculator.*
- *Automatic File Update Feature.*
- *Summary and Detail Reporting.*
- *Pull Down Menus, Lookup Windows.*
- *For IBM PC/XT/AT, PS-2 or Compatibles.*

As a construction estimator, you want a system that is easy to setup and use. Yet, it must be fast, powerful, comprehensive and contain all the functions and features that you need. You want a built-in costcode library containing material unit prices and unit labor rates.

With the purchase of PC-Estimator, or any of the software products on these facing pages, you not only get a full-feature and comprehensive cost estimating system, but you also get our unlimited support of the system including free telephone support, program and data update notices. Designed for General Construction Cost Estimating.

An Estimating Wonder For Smaller Projects

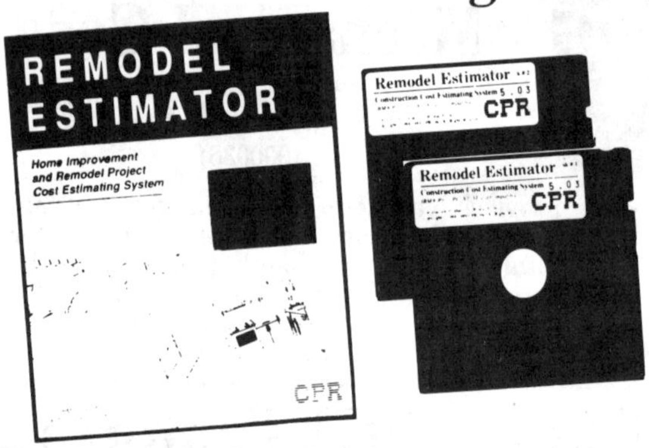

REMODEL/ESTIMATOR

Only $89.00 (CPR102)

The Lowest Price For A Complete Estimating Package

Features Include:

- *Specially Designed for Home Building And Home Improvement Work.*
- *Pre-packaged "Basic Pak" Database with Typical Home Improvement Projects.*
- *Built-In Takeoff Calculator.*
- *Remodel Cost Database.*
- *For IBM PC/XT/AT, PS-2 or Compatibles.*

Finally there is a cost estimating software that is developed specifically for remodelers and home builders. Designed for estimating residential and commercial construction and developed in collaboration with professional estimators, if now brings tailor-made productivity to your IMB or compatible computer.

Remodel Estimator provides in an easy-to-understand, easy-to-use and yet comprehensive package.

Civil Engineering Handbook

1184 Pages

4th Edition

$97.50

(0070661480)

The complete coverage of civil engineering given here, ranging from routine technical data to the most advanced construction methods, makes this book the standard reference in the field. *It is the one engineering tool that you can't afford to be without.*

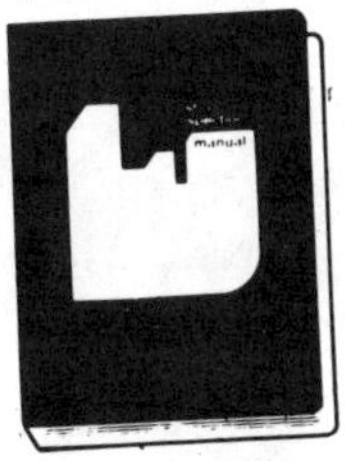

Construction Inspection Manual

5th Edition

320 Pages
Loose Leaf

$24.95

(050003)

The accepted national guide used by the construction industry for improvement of specification writing and inspection procedures. Features a series of technical items checklists, covering basic elements in a construction project, formatted to correspond to the 16 divisions of the MASTERFORMAT of the Construction Specifications Institute.

Underwriters Laboratories

New

1992

UL Fire Resistance Directory
2 Volumes
(7501692) **$43.95**

UL Building Materials Directory
(54008092) **$26.50**

UL Roofing Materials & Systems
(54008192) **$25.00**

Building Design & Construction Handbook

4th Edition
1472 Pages
662 Illustrations

$106.50

(0070415218)

Completely updated and revised to reflect the latest recommended procedures and materials, this is the only book that covers all aspects of building construction and design. Important topics include systems design, value engineering, energy savings, specification preparation, hazard protection, and much, much more.

AISC Manual of Steel Construction

1100 Pages

9th Edition

$63.00

(1564240002)

The essential steel reference for engineers, architects, detailers, contractors, building officials and fabricators. Sections include Dimensions and Properties, Beam, Girder and Column Design, Connections, Specs and Codes, and Data and Tables.

Standard Handbook for Electrical Engineers

12th Edition

$99.75

(0070209758)

With 2248 pages and 1863 illustrations, the newly updated reference includes the latest advancements and changes in electrical engineering. All topics covered including generation, high and low voltage, conservation and more.

1991 NFPA 101 Life Safety Code

$48.75

(20026)

This code, adopted by many states and municipalities, is the national guideline that establishes minimum requirements and standards for fire safety in buildings and structures.

1991 Life Safety Code Handbook — Contain the complete 1991 NFPA 101 (0877653798) **$72.00**

New

1992 ASTM Standards In Building Codes

29th Edition

Four Volumes — 7080 Pages

$395.00

(0803117515)

This massive reference contains over 1300 ASTM specs as adopted by all model codes including BOCA, UBC, CABO and Standard Codes. All standards are 60% new or revised. All construction materials and practices are covered.

BNi Building News™ ORDER FORM

All prices are subject to change. Normal delivery will be approximately 10-14 business days from receipt of order.

QUANTITY	TITLE	ITEM #	PRICE	TOTAL CHARGE

Charge To Credit Card
Or Attach Check And
Mail To:

BNi Building News™
Los Angeles, CA 90051
(310) 202-7775
FAX: (310) 202-1129

SEND PREPAID ORDER TO:

☐ VISA ☐ MASTERCARD Exp. Date ____________

NO. ____________________________

____
(Cardholder's Signature)

MINIMUM CREDIT CARD ORDER $10.00
For credit card orders call toll free:

1-800-873-6397

462M

SUBTOTAL....$__________

7.75% SALES TAX TO CALIF. DESTINATIONS....$__________

8.25% to L.A. (COUNTY DESTINATIONS)....$__________

ADD SHIPPING/HANDLING CHARGES OF $5.75 FOR THE FIRST ITEM AND $1.00 FOR EVERY ITEM OVER ONE....$__________

GRAND TOTAL....$__________

NAME ____________________ COMPANY ____________________

STREET ____________________ SUITE/APT. #__________ PHONE ____________________
(Please use street address only to expedite delivery; do not use Post Office Box)

CITY ____________________ STATE ____________________ ZIP__________
(Please type or print)